About this edition:
"Social Neuroscience has vigorously established itself as one of the newest and most exciting sub-disciplines of psychology. Ward's pioneering *Student's Guide* is now updated covering new insights in the biological basis of social behaviour and their relevance to everyday life. Down-to-earth and imaginatively linked with web based materials, it can't fail to inspire the next generation of students."

Chris and Uta Frith, University College London

About the previous edition:
"I stopped using textbooks more than a decade ago, but that's about to change. Given that Ward's is the very first textbook focusing on social neuroscience, I am extremely impressed. It will be the best around for years to come. It is current, broad, and precise. The writing style will be accessible to undergraduates, graduates, and even professors. It is the perfect introduction to this exciting new field."

Matthew D. Lieberman, University of California, Los Angeles

The Student's Guide to Social Neuroscience

Social neuroscience is an expanding field that, by investigating the neural mechanisms that inform our behavior, explains our ability to recognize, understand, and interact with others. Concepts such as trust, revenge, empathy, prejudice, and love are now being explored and unraveled by neuroscientists. This engaging and cutting-edge text is an accessible introduction to the complex methods and concepts of social neuroscience, with examples from contemporary research and a blend of different pedagogical features helping students to engage with the material, including essay questions, summary and key points, and further reading suggestions.

The second edition of this groundbreaking text is thoroughly revised and expanded to reflect the growing volume of evidence and theories in the field. Notable additions include a greater emphasis on genetics and hormones, and the expansion of topics such as cultural neuroscience, emotion regulation, biological markers of autism, power and status, social categorization of faces and people, and new accounts of mirror neuron functioning. The book is supported by a new and updated companion website, including useful features such as lecture recordings, multiple choice questions and web links, as well as PowerPoint slides for lecturers.

Richly illustrated in attractive full-color, with figures, boxes, and 'real-world' implications of research, this text is the ideal introduction to the subject for undergraduate and postgraduate students in fields such as psychology and neuroscience.

Jamie Ward is Professor of Cognitive Neuroscience at the University of Sussex, UK, and Co-Director of Sussex Neuroscience. He has published over 100 scientific papers and several books including the *Student's Guide to Cognitive Neuroscience* (now in its third edition) and *The Frog who Croaked Blue: Synesthesia and the Mixing of the Senses* (now translated into 3 languages), and was the Founding Editor of the journal *Cognitive Neuroscience*.

Routledge
Companion Websites

Enhancing online learning
and teaching.

www.routledge.com/cw/ward

The Student's Guide to Social Neuroscience

Second Edition

Jamie Ward

Routledge
Taylor & Francis Group

LONDON AND NEW YORK

Second edition published 2017
by Routledge
2 Park Square, Milton Park, Abingdon, Oxon OX14 4RN

Simultaneously published in the USA and Canada
by Routledge
711 Third Avenue, New York, NY 10017

Routledge is an imprint of the Taylor & Francis Group, an informa business

First edition published 2012 by Psychology Press

British Library Cataloguing in Publication Data
A catalogue record for this book is available from the British Library

Library of Congress Cataloging-in-Publication Data
Names: Ward, Jamie, 1972– author.
Title: The student's guide to social neuroscience / Jamie Ward.
Description: 2nd Edition. | New York : Psychology Press, 2016. |
   Revised edition of the author's The student's guide to social
   neuroscience, 2012.
Identifiers: LCCN 2016020634 (print) | LCCN 2016021734 (ebook) |
   ISBN 9781138908611 (hardback) | ISBN 9781138908628 (pbk.) |
   ISBN 9781315694306 (ebk)
Subjects: LCSH: Perception. | Social interaction.
Classification: LCC BF311 .W26947 2016 (print) | LCC BF311 (ebook) |
   DDC 153—dc23
LC record available at https://lccn.loc.gov/2016020634

ISBN: 978-1-138-90861-1 (hbk)
ISBN: 978-1-138-90862-8 (pbk)
ISBN: 978-1-315-69430-6 (ebk)

Typeset in Times
by Apex CoVantage, LLC

Contents

About the author

Jamie Ward is Professor of Cognitive Neuroscience at the University of Sussex, UK, and the co-director of Sussex Neuroscience. He has published more than 100 scientific papers and several books including the *Student's Guide to Cognitive Neuroscience* (now in its third edition) and *The Frog who Croaked Blue: Synesthesia and the Mixing of the Senses* (now translated into three languages). He was the founding editor of *Cognitive Neuroscience*, a journal from Psychology Press.

Preface

This textbook came about through a desire to create an accompanying text to *The Student's Guide to Cognitive Neuroscience* specifically in the area of social neuroscience. Cognitive neuroscience may be the parent discipline of social neuroscience, but it was becoming increasingly clear over the last few years that social neuroscience had now grown up and was trying to establish a home of its own. For example, there are now several excellent journals dedicated to it and many universities have introduced social neuroscience onto the undergraduate curriculum as a separate module distinct from cognitive neuroscience. This textbook aims to reflect the new maturity of this discipline and attempts to convey the excitement of this field to undergraduate and early stage postgraduate students.

My own interest in the field stemmed from the claims surrounding mirror systems, empathy, and theory of mind. At the start of this project, I imagined that this would form the core of the textbook. However, the more that I delved into the literature the more I was taken aback by the volume and quality of research in other areas such as prejudice, morality, culture, and neuro-economics. The resulting book is, I hope, a more balanced view of the field than I initially anticipated. As with my previous textbook, it isn't an exhaustive summary of the field. It isn't my aim to teach students everything about social neuroscience but it is my aim to provide the intellectual foundations to acquire that knowledge, should they wish to become researchers themselves. My ethos is to try to present the key findings in the field, to develop critical thinking skills, and to instill enthusiasm for the subject.

In the absence of previous textbooks on social neuroscience, it was an interesting exercise deciding how to carve the field into chapters and how to order the chapters. For example, the chapter on relationships appeared and disappeared several times (with these sections being divided amongst the 'Interactions' and 'Development' chapters). The first two chapters begin with an overview of the topic and a summary of the methods used in social neuroscience. The 'methods' chapter is a condensed, but updated, version of the more extensive chapters in *The Student's Guide to Cognitive Neuroscience*. The chapter uses examples from the social neuroscience literature to illustrate the various methods. The third chapter covers the evolution of social intelligence and culture. It introduces mirror neurons in the context of imitation, social learning, and tool use. The fourth and fifth chapters deal with the 'primitive' building blocks of social processes, namely emotions and motivation (Chapter 4), and recognizing others (Chapter 5). Chapter 6 is concerned with empathy, theory of mind, and autism. The next two chapters consider social interactions and then relationships, dealing with issues such as altruism, game theory, attachment, and social exclusion. Chapter 9 is concerned with groups and identity, covering the notion of 'the self', prejudice, and religion. Chapter 10 covers antisocial behavior, aggression, and morality. The final chapter considers social development from infancy through to adolescence.

The second edition is thoroughly revised and expanded to reflect the growing volume of evidence and theories. In doing so, I have been careful to ensure that the 'big picture' is not lost in the details. Notable additions to the second edition include a greater emphasis on genetics and hormones, and the expansion of topics such as cultural neuroscience, emotion regulation, biological markers of autism, power and status, social categorization of faces and people, and new accounts of mirror neuron functioning.

Finally, I'd like to thank the many reviewers who provided constructive feedback on drafts of chapters and for Routledge for being so accommodating.

Jamie Ward
Brighton, UK, May 2016

For Katie

CHAPTER 1

CONTENTS

Introduction to social neuroscience

Imagine two participants lying in different rooms, each with their heads placed in a very large magnetic field. Crucially, the two participants are interacting with each other in order to win money and this interaction requires trust. By trusting money to the other person they stand a greater chance of getting more money returned to them in the future, but they also run the risk of exploitation. As their brains engage in the decision to trust or not to trust, there are subtle changes in blood flow corresponding to these different decisions that can be detected. The fact that different patterns of thought should result in different patterns of brain activity is perhaps not surprising. The fact that we now have methods that can attempt to measure this is certainly note-worthy. What is most interesting about studies such as these is the fact that activity in regions of one person's brain can reliably elicit activity in other regions of another person's brain during this social interaction. For instance, in a trusting relationship, when one person makes a decision the other person's brain 'lights up' their reward pathways, even before any reward is actually obtained – as illustrated in Figure 1.1 (King-Casas et al., 2005). Cognition in an individual brain is characterized by a network of flowing signals between different regions of the brain. However, social

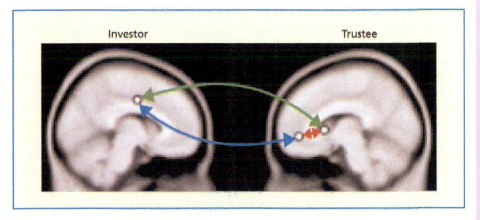

Figure 1.1 The technique of *hyperscanning* records from two or more different brains simultaneously (such as MRI scanners): for example, whilst participants in the scanners engage in a social activity (Montague et al., 2002). The details of this particular study, involving a game of trust, are not important here (they are covered in Chapter 7) and hyperscanning is a relatively rare methodology. What is of interest is that neural activity in different regions correlates not only within the same brain (due to physical connections; depicted in red) but also across brains (due to mutual understanding; depicted in blue and green). From King-Casas et al. (2005). Copyright © 2005 American Association for the Advancement of Science. Reproduced with permission.

KEY TERMS

Social psychology
An attempt to understand and explain how the thoughts, feelings, and behaviors of individuals are influenced by the actual, imagined, or implied presence of others.

Cognitive psychology
The study of mental processes such as thinking, perceiving, speaking, acting, and planning.

interactions between different individuals can be characterized by the same principle: a kind of 'mega-brain' in which different regions in different brains can have mutual influence over each other. This is not caused by a physical flow of activity between brains (as happens between different regions in the same brain) but by our ability to perceive, interpret, and act on the social behavior of others.

This introductory chapter will begin by providing a brief overview of the (brief) history of social neuroscience. It will then go on to consider what kind of mechanisms could constitute the 'social brain' and how they might relate to nonsocial brain processes. Finally, it will consider how different levels of explanation are needed to derive a complete understanding of social behavior, and it will discuss how neuroscience can be combined with other approaches.

THE EMERGENCE OF SOCIAL NEUROSCIENCE

Allport (1968) defined **social psychology** as 'an attempt to understand and explain how the thoughts, feelings, and behaviors of individuals are influenced by the actual, imagined, or implied presence of others'. By extension, a reasonable working definition of social neuroscience would be:

> *an attempt to understand and explain, using neural mechanisms, how the thoughts, feelings, and behaviors of individuals are influenced by the actual, imagined, or implied presence of others.*

Based on this definition, one could regard social neuroscience as being a subdiscipline within social psychology that is distinguished only by its adherence to neuroscientific methods and/or theories. Whilst this may be perfectly true, most researchers working within the field of social neuroscience do not have backgrounds within social psychology but tend to be drawn from the fields of **cognitive psychology** and neuroscience. Indeed social neuroscience has also gone by the name 'social cognitive neuroscience' (the term is less commonly used now). Cognitive psychology is the study of mental processes such as thinking, perceiving, speaking, acting, and planning. It tends to dissect these processes into different sub-mechanisms and explain complex behavior in terms of their interaction. Cognitive psychology has an important role to play in social neuroscience because it aims to decompose complex social behaviors into simpler mechanisms (operating in individual minds) that are amenable for exploration using neuroscientific methodologies. Social neuroscience links together all these disciplines: linking cognitive and social psychology, and linking 'mind' (psychology) with brain (biology, neuroscience). Of course, these divisions themselves are arbitrary. They serve as convenient ways of categorizing research programs, and they become embedded in the way they are taught (lecture courses, textbooks, etc.).

The term *social neuroscience* can be traced to an article by Cacioppo and Berntson (1992) entitled 'Social psychological contributions to the decade of the brain: Doctrine of multi-level analysis'. The term appears twice: once in a footnote, and once in a heading accompanied by a question mark – i.e. 'Social Neuroscience?'. Their particular interest in the topic stemmed from research showing that psychological processes such as perceived social support can affect immune functioning. However, many other areas of study that now fall under the social neuroscience umbrella were already active areas of study prior to 1992. In cognitive psychology, there was a mature literature on face perception. However, this literature was primarily concerned with understanding faces as a type of visual object rather than treating faces as cues to social interactions. There were also detailed accounts of how social behavior breaks down as a result of acquired brain damage (Damasio, Tranel, & Damasio, 1990; Eslinger & Damasio, 1985) or in developmental conditions such as autism (e.g. Frith, 1989). In behavioral neuroscience, there was a longstanding interest in emotional processes such as fear (e.g. LeDoux, Iwata, Cicchetti, & Reis, 1988), aggression (e.g. Siegel, Roeling, Gregg, & Kruk, 1999), and separation distress (e.g. Panksepp, Herman, Vilberg, Bishop, & DeEskinazi, 1980). In social psychology, the field of 'social cognition' applied the approach and methods of cognitive psychology (e.g. response time) to social psychology questions. Finally, the 1990s saw the refinement of the newly established methods of cognitive neuroscience, such as functional magnetic resonance imaging (fMRI) and transcranial magnetic stimulation (TMS), and these methods were directed to social processes as well as to the more traditional areas within cognitive psychology.

By the year 2000, social neuroscience could be recognized as a relatively coherent entity with a core set of research issues and methods and as reflected in prominent reviews of the time (e.g. Adolphs, 1999; Frith & Frith, 1999; Ochsner & Lieberman, 2001). The first journals dedicated to this field, *Social Neuroscience* and *Social, Cognitive and Affective Neuroscience (SCAN)*, both appeared in 2006. The Society for Social and Affective Neuroscience (SANS; www.socialaffectiveneuro.org) and Society for Social Neuroscience (S4SN; www.s4sn.org) were established in 2008 and 2010 respectively. Both societies welcome student members. The first edition of this textbook, published in 2012, was the first single-authored study guide aimed at undergraduate students.

Stanley and Adolphs (2013, p. 822) provide a compelling summary of the current state of the field and point to its future evolution. They summarize it as follows:

Although we have moved from regions to networks, the next key step is to identify the flow of information through these networks to follow social information processing from stimulus through to response. This requires an understanding

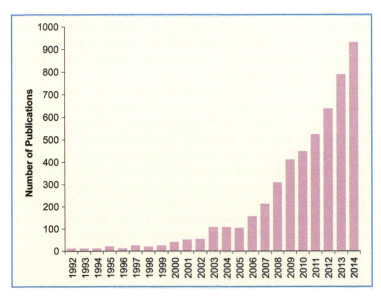

Figure 1.2 The number of publications incorporating the term 'social' and 'neuroscience' has increased dramatically since 2000. This data is based on a search of the Web-of-Knowledge database searching for a conjunction of 'social' and 'neuroscience' in the topic field.

of the detailed computations implemented by the different nodes in the networks as well the dynamic interplay between them. One could make the analogy of moving from words (brain areas) to sentences (networks) to propositions (arrangements of network dynamics) to conversations (brains interacting). We are still solidly in the age of sentences and are only beginning to enter the age of propositions and conversations.

Many of these ideas are explored in more detail throughout the chapter and, indeed, the book. Can we assign unique functions to brain regions or use activity in a given brain region to infer the nature of information processing (e.g. emotional versus rational)? Are there brain regions or networks that can be understood specifically in terms of their contribution to social functioning or do these regions/networks also participate in similar ways in non-social cognition? How can social neuroscience study realistic social interactions? On the latter point, Schilbach et al. (2013) have pointed out that most previous research in social neuroscience has tended to involve observing and interpreting other people and that the neural mechanisms underlying social interactions per se can (metaphorically) be considered the 'dark matter' of social neuroscience. Complementary to this, Willingham and Dunn (2003, p. 669) cautioned social psychologists against changing their research agenda just to make them amenable to a neuroimaging approach:

Some of the topics of interest to social psychologists are not amenable to brain localization techniques because of the complexity of the processes; they have embedded in them subprocesses that interact, and such complex processes are difficult to localize. It would be a pity if, in their justifiable enthusiasm for this powerful tool [i.e. neuroimaging], social psychologists subtly shifted their research programs to problems that are amenable to brain localization or shifted their theoretical language to constructs that are localizable.

What should we make of criticism such as this? Willingham and Dunn (2003) are correct to point out that it is important not to shift the whole social psychology research agenda to fit with trendy neuroscience methods. To a large extent, the shift has to come from the development of neuroscientific techniques that can tackle the questions that matter. However, their characterization of social neuroscience in terms of localization of functions is inaccurate (or, at least, outdated). Social neuroscience should be concerned primarily with the underlying mechanisms, and these are unlikely to be localized to discrete brain regions.

Stanley and Adolphs (2013) also report a number of surveys conducted on Social Neuroscience researchers attending international conferences. Figure 1.3 shows a summary of a survey asking researchers what they presently work on, what is presently lacking in the field, and what the future of social neuroscience will consist of. Current researchers tend to work on topics such as emotion, self-regulation, and decision-making but feel that the discipline as a whole needs more statistical and methodological rigor, needs to be more *ecologically valid*, and needs more interdisciplinary integration. The future of social neuroscience, in their eyes, lies both in terms of real-world applications and also in terms of an additional level of sophistication afforded by computational approaches to brain networks.

KEY TERM

Ecological validity
An approach or measure that is meaningful outside of the laboratory context.

What Do Social Neuroscientists Say?		
Current Research Interests	**Social Neuroscience Is Currently Lacking**	**Future of Social Neuroscience**
Emotion	Statistical/Methodological Rigor	Applied Science
Clinical Disorders	Ecological Validity	Computational Approaches
Self-Regulation	Interdisciplinary Integration	Networks in the Brain
Development	Computational Approaches	Real-World Behaviors
Decision-Making	Theory	Social Interaction

Figure 1.3 What do social neuroscientists say about their discipline? Stanley and Adolphs (2013) asked researchers in the field about their current interests (left), what is lacking in the field (middle), and what the future should hold (right). The degree of shading represents the rank ordering of responses, with some responses being tied.

THE SOCIAL BRAIN?

One overarching issue within social neuroscience is the extent to which the so-called 'social brain' can be considered distinct from all the other functions that the brain carries out – talking, walking, planning, etc. In other words, is the 'social brain' special in any way? This will be a recurring theme in the book, although Chapter 3 considers it in detail from an evolutionary perspective.

One possibility is that there are particular neural substrates in the brain that are involved in social cognition but not in other types of cognitive processing. This relates to the notions of **modularity** and **domain specificity**. A module is the term given to a computational routine that responds to particular inputs and performs a particular computation on them, that is, a routine that is highly specialized in terms of what it does to what (Fodor, 1983). One core property that has been attributed to modules is domain specificity, namely that the module processes only one kind of input (e.g. only faces, only emotions). One contemporary claim is that there is a module that responds to the sight of faces, but not the sight of bodies or the sounds of people's voices or indeed to any non-face stimuli (e.g. Kanwisher, 2000). Another claim is that there is a module for reasoning about mental states (e.g. people's desires, beliefs, knowledge) but not other kinds of reasoning (e.g. Saxe, 2006). Yet another claim is that there is a module for detecting cheating (Cosmides, 1989). In this modular view, the social brain is special by virtue of brain mechanisms that are specifically dedicated to social processes. Moreover, it is claimed that these mechanisms evolved to tackle specific challenges within the social environment (e.g. the need to recognize others, the need to detect when you are being exploited). To some critics, this view of the mind and brain resembles phrenology (Uttal, 2001) – see Figure 1.4. However, such criticisms are not entirely fair given that modern approaches to addressing the issue of domain-specific social processes are subjected to experimental rigor that was never applied to Nineteenth century ideas of the brain localization of function.

The alternative, diametrically opposite, approach is to argue that the 'social brain' is not, in fact, specialized uniquely for social behavior but is also involved in non-social aspects of cognition (e.g. reasoning, visual perception, threat detection). The evolution of general neural and cognitive mechanisms that increase intellect, such as having bigger brains, may make us socially smarter too (e.g. Gould, 1991). Of course, it is also possible that the reverse is true – namely that the evolutionary

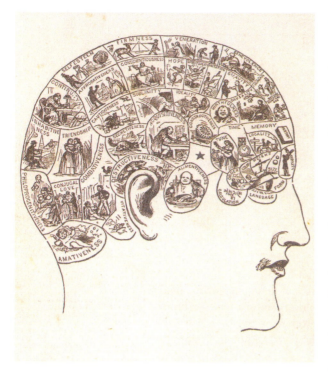

Figure 1.4 The phrenologist's head was used to represent highly localized functions of the brain in the early nineteenth century. It is an extreme form of a modularity view, albeit based on outdated notions of what the core functions are likely to be (e.g. love of animals, conscientiousness) and how individual differences in these functions represent themselves biologically (larger brain regions giving larger bumps on the skull). To what extent is the 'social brain' a set of specialized modules?

need to be socially smarter leads to general cognitive advances in other domains (e.g. Humphrey, 1976). Under these accounts social cognition and non-social cognition evolved hand-in-hand (albeit with one factor driving the other) but, crucially, they did not necessarily lead to highly specialized routines in the brain for dealing with social problems.

Needless to say, there are other positions that lie in-between these two extremes. Mitchell (2009) notes that there are certain regions of the brain (e.g. the medial prefrontal cortex) that are activated in fMRI studies by a wide range of social phenomena such as evaluating attitudes, interpreting other's behavior, and emotional experience. Rather than arguing for a narrowly defined module in this region, he suggests that social psychology is a 'natural kind' that distinguishes itself from other aspects of cognition because it relates to concepts that are less stable and less definite than those involved in, say, perception and action. In this account, the 'social brain' is special because of the nature of the information that is processed (more fuzzy) rather than because it is social (i.e. interpersonal) per se.

Another possibility is that it is not particular regions of the social brain that are 'special' but rather that there are particular kinds of neural mechanisms especially suited to social processes. For example, Frith (2007) claims: 'I have speculated about the role of various components of the social brain, but in most cases, I believe that these processes are not specifically social. The exception is the brain's mirror system.' Similarly, Ramachandran (2000) predicts that 'mirror neurons will do for psychology what DNA did for biology: they will provide a unifying framework and help explain a host of mental abilities that have hitherto remained mysterious and inaccessible to experiments.' Mirror neurons respond both when an animal sees an action performed by someone else and when they perform the same action themselves (e.g. Rizzolatti & Craighero, 2004). The key insight, with regard to social neuroscience, is that there may be a simple mechanism – implemented at the level of single neurons – that enables a correspondence between self and other. Mirror neurons have been implicated in imitation (see Chapter 3), empathy, and 'mind reading' (see Chapter 6). Although they were originally discovered for actions, it is possible that mirroring is a general property of many neurons (e.g. those processing pain, emotion, etc.) and they may not be tightly localized to one region (Mukamel, Ekstrom, Kaplan, Iacoboni, & Fried, 2010). Whereas some researchers have argued that mirror neurons serve a specifically social function (Frith, 2007) others have suggested that they arise primarily out of associative learning between action and perception in both social and non-social contexts such as observing one's own actions (Heyes, 2010).

Barrett and Satpute (2013) offer a useful overview of this general debate concerning the nature of the social and emotional brain that is illustrated in Figure 1.5. They consider three broad ways in which the 'social brain' may be implemented. The first

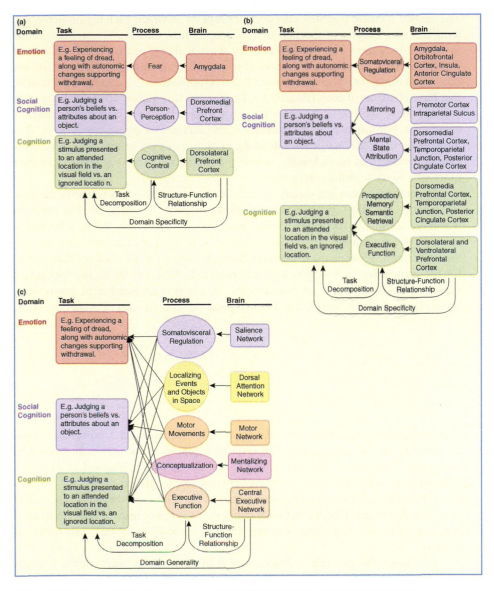

Figure 1.5 Three different ways in which different brain structures might be mapped to different functions (tasks and processes). In (a) there is a one-to-one association between brain structure and function whereas in both (b) and (c) a network of regions may make differing contributions to a given function. In (b) the network consists of specialized units that interact, but in (c) the network consists of interactions between non-specialized units. From Barrett and Satpute (2013).

scenario is a simple domain-specific view consisting of brain regions that are specialized for processing particular kinds of social information (e.g. person perception) and non-social information (e.g. cognitive control). Few, if any, contemporary researchers would endorse such a view. The second and third scenarios involve the idea of brain networks and are more compatible with contemporary ideas in the literature.

The second scenario postulates networks of regions in which each region in the network has a high degree of specialization (e.g. specific to social information), whereas in the third scenario (the one endorsed by these authors) neither brain regions nor individual brain networks are functionally specialized or segregated into social and non-social functions. Of course, it would also be possible to imagine hybrid scenarios that have elements of each (Fedorenko, Duncan, & Kanwisher, 2013). Adjudicating between these different scenarios will require linking evidence from a wide range of techniques. For instance, whilst evidence from functional imaging, reviewed by Barrett and Satpute (2013), often points to the existence of very generic networks, other evidence (e.g. from brain lesions, or using brain stimulation) has tended to reveal more specificity within the networks.

In summary, there is a variety of views concerning the broad nature of the neural mechanisms that support human social behavior. At one end, there is the view that there are highly specialized neural mechanisms. These may be very limited in the type of information they process (e.g. faces, beliefs). At the other end, there is the view that the mechanisms that support social behavior are used for many other functions (possibly including non-social cognition). Whereas the highly specialized viewpoint tends to have been linked to the idea of a small number of contributing brain regions (localizability), it is not incompatible with the idea of brain networks.

IS NEUROSCIENCE AN APPROPRIATE LEVEL OF EXPLANATION FOR STUDYING SOCIAL BEHAVIOR?

Perhaps the most general criticism that could be leveled at social neuroscience is that the brain is not the most appropriate level of explanation for understanding social processes. Surely social processes need to be studied and understood at the social level – that is, at the level of interactions between people, groups of people, and societies. There have been some fascinating studies on neural responses to Black faces by White American students, but what could we ever really learn about racism from brain-based measures without situating them in a social, economic, and historical context?

Of course, this presents a distorted view of what social neuroscience is really all about. Most researchers in the field do not take a strongly reductionist approach. **Reductionism** implies that one type of explanation will become replaced with another, more basic, type of explanation over time. In a reductionist framework the language of social psychology (e.g. attitudes, relationships, conformity) will be replaced by the concepts of neuroscience (e.g. oxytocin, plasticity, medial prefrontal cortex). However, most researchers in social neuroscience are attempting to create bridges between different levels of explanation rather than replace one kind of explanation with another – see Figure 1.6. For example, social neuroscience studies may combine questionnaire measures (the bread-and-butter of social psychology research) with neuroscience data.

Another common way in which neuroscience data are used to bridge levels of explanation has been termed the **reverse inference** approach (Poldrack, 2006). The reverse inference approach is an attempt to infer the nature of cognitive processes from neuroscience (notably neuroimaging) data. Examples of this abound in the social neuroscience literature. For example, activity within the amygdala may be taken to imply the involvement of a fear-related (or more broadly emotion-related)

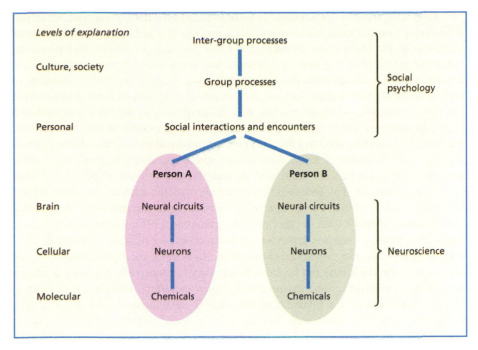

Figure 1.6 Social psychology and neuroscience employ different levels of explanation. Social neuroscience aims to create bridges between these different levels of explanation.

mechanism in studies of race processing (Phelps et al., 2000). The nature of various moral dilemmas has been inferred on the basis of whether the dilemmas activate regions of the brain implicated in emotion or in higher order reasoning (Greene, Sommerville, Nystrom, Darley, & Cohen, 2001). If the hippocampus is activated, then long-term memory is involved; if the right temporo-parietal junction is activated, then 'theory of mind' is involved; etc. Is reverse inference necessarily good practice? It goes without saying that the reliability of this inference depends on what is known about the functions of given regions. If these regions turned out to have very different functions then the inference would be flawed. Also the function of regions is not resolutely fixed but depends on the context in which they are employed. Poldrack (2006) argues that reverse inference may be improved by examining networks of regions or examining more precise regions (e.g. not just the 'frontal lobes'). Another more general methodological point is the importance of not being over-reliant on neuroimaging data, but to look at other sources of evidence such as TMS in which behavior itself is normally measured (and hence does not suffer from the problems of reverse inference in the same way). Reverse inference is a legitimate approach, but it is not problem free.

An example of a forward and reverse inference in social neuroscience. Even if the forward inference is correct, then it doesn't necessarily imply that the reverse inference will be.

Forward inference: If someone is frightened their amygdala is activated.

Reverse inference: If the amygdala is activated then someone is frightened.

Logically, there is one scenario in which brain-based data could have no significant impact on our understanding of social processes – and that is the **blank slate** scenario. In the blank slate scenario, the brain just accepts, stores, and processes whatever information is given to it without any pre-existing biases, limitations, or knowledge. According to the blank slate, the brain is not completely redundant (it still implements social behavior) but the nature of social interactions themselves is entirely attributable to culture, society, and the environment. According to the blank slate, the structure of our social environment is created entirely within the environment itself, reflecting arbitrary but perpetuated historical precedents. Thus, culture, society, and the nature of social interactions invent and shape themselves. A more realistic scenario is that the brain, and its underlying processes, creates constraints on social processes. For example, it is claimed that the number of close friends that we have is predicted by the size of the human brain, extrapolating from known group sizes and brain sizes in other primates (Dunbar, 1992). The tendency to form monogamous attachments is dependent on brain chemistry (Carter, DeVries, & Getz, 1995). Apparently arbitrary social conventions, such as the rules governing right and wrong (e.g. the law), may not be entirely arbitrary but may reflect a basic tendency to empathize with others and reason about causes and effects (Haidt, 2012). Even in the first few hours of life, infants appear to treat social and non-social stimuli differently. They enter the world with a preference for social stimuli and even appear to have rudimentary knowledge about how faces should be structured (Macchi Cassia, Turati, & Simion, 2004). Social processes are *all* in the brain, but some of them are created by environmental constraints and historical accidents (and learned by the brain) whereas others may be caused by the inherent organization, biases, and limitations of the brain itself.

Aggression as an example of interacting levels of explanation

To give a feel for this debate, consider the topic of aggression. Many current social psychology textbooks (e.g. Hogg & Vaughan, 2011) are dismissive of the role of biological factors in aggression, noting, for instance, the huge variability in levels of aggressive acts such as murder across cultures. However, we can consider this in terms of two questions: What causes aggression and what causes variability in levels of aggression? These questions may generate quite different answers. To give a non-social analogy, the typical number of fingers that we have on our hands (i.e. ten) is almost entirely down to our biology, whereas the *variability* in the number of fingers we have on our hands is almost entirely down to environment, such as industrial accidents (this example is from Ridley, 2003). Figure 1.7 shows that there is indeed huge cross-cultural variability in murder rates of developed countries. This is likely to reflect differences across cultures: USA and Finland have some of the highest rates of gun ownership whereas Singapore has the lowest. However, the pattern is not random and is linked to income inequality (the magnitude of difference between the highest and lowest earners). Whilst income inequality is itself cultural, and not biological, the fact that aggression is linked to resource control and perceived injustice is likely to be independent of culture. Cultural differences may act as an 'accelerator' or 'brake' on biological tendencies. To give a specific example of the interaction between cultural and biological factors, consider the effects of the hormone testosterone (a biological factor) and socio-economic status (SES, a cultural factor). Levels

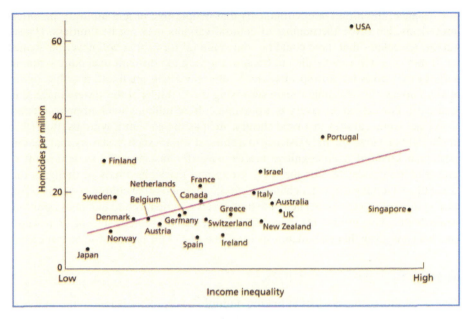

Figure 1.7 Murder rates vary considerably from country to country, but does this imply that we should understand murder from a purely social perspective? There may be underlying factors, perhaps reducible to particular kinds of cognitive mechanisms (e.g. perceived inequality), that explain this variability rather than reflecting random variability. From Wilkinson and Pickett (2009). Copyright © 2009 Penguin Books Ltd. Reproduced with permission.

of testosterone in males are correlated with levels of aggression in people of low SES individuals but not high SES (Dabbs & Morris, 1990). So high SES acts as a brake on biological influences (these people are already at the top of the social ladder), but low SES may bring to the fore biologically bound instincts for achieving status and securing resources.

A biological basis for culture?

The same logic can be applied to other domains, including culture itself. The answer to the question 'what causes culture?' might be something like 'a set of mechanisms that enables people to transfer skills, beliefs, and knowledge from each other and retain these as a relatively stable pattern across individuals' (this being a cognitive mechanistic explanation). A more neuroscientific answer could be 'neural mechanisms that respond to the repeated patterns of behavior in others, whom we affiliate positively with, and increase the likelihood that our own neural mechanisms will generate those behaviors'. This is not intended as a truly accurate answer, but merely conveys what a reductive neuroscience concept of conformity (a central aspect of culture) *might* look like. But note that it would be an entirely circular argument to say that culture creates itself. To take that argument to a logical absurdity, culture cannot create itself in the absence of appropriate biological entities! As to the question of what creates *variability* in culture, the answer could be quite different. It may reflect, for instance, the different environments that people live in and arbitrary historical precedents. However, the number of cultural variants may not be limitless. Hauser (2009) speculates that there could be some cultural forms that will never be created or, if they are, will rapidly die out because they are too difficult to acquire – that is, biology may go as far as to specify which cultural variants are likely, possible, or virtually impossible. This might seem surprising if one thinks of the variety of cultures that exist. For example, slavery is a possible culture (although abhorrent to modern eyes) and some cultures that used slavery, such as the ancient Egyptians, flourished for millennia. However, the existence of a cultural variant such as slavery may require particular kinds of neurocognitive mechanisms: for instance, the switching off of empathic processes towards the slave group and particular kinds of thoughts that drive this switching off (e.g. dehumanization). An impossible culture could therefore be a system of slavery associated with high levels of empathy and humane cognitions towards the slave group. The impossibility is created by the nature of brain-based mechanisms, even though it manifests itself in terms of the nature of social processes.

Gene-culture co-evolution

One good illustrative attempt at linking multiple levels of explanation in social neuroscience comes from a newly coined sub-discipline termed **cultural neuroscience** (e.g. Han et al., 2013; Kim & Sasaki, 2014). Cultural neuroscience is an interdisciplinary field bridging cultural psychology, neurosciences, and neurogenetics that explains how neurobiological processes give rise to cultural values, practices, and beliefs as well as how culture shapes neurobiological processes (Chiao, 2010). That is, it explicitly assumes that not only will cultural differences influence the brain (the top-down in Figure 1.6) but also that the brain will impact on culture itself (the less intuitive bottom-up approach in Figure 1.6). The scope of cultural neuroscience encompasses such things as examining how immersion in different cultural systems (e.g. collectivism, individualism; large-scale or small-scale communities) affects the functioning of different brain networks (e.g. those involved in trust or compassion), and also how differences in biology (e.g. genetic differences) might be linked to cultural practice. Perhaps not surprisingly, there has been some healthy skepticism to this approach. Denkhaus and Bos (2012) have argued that: 'Putting subjects from the United States and China in an MR tomograph and scanning their brains while they are performing a set of specialized tasks is not exactly what most people would regard as a promising way of unpacking the complexities of culture.' Of course it is an open question as to whether any genuinely novel insights will emerge from this approach, but already there are intriguing results. In this section, I shall consider the evidence for **gene-culture co-evolution**, and Chapter 9 considers cross-cultural and group differences more generally. Those less familiar with the basic principles of genetics are referred to the box in Chapter 2.

According to the principle of gene-culture co-evolution (e.g. Boyd & Richerson, 1985) certain genotypes may predispose people to create particular features in their environment (thus influencing cultural selection) and – at the same time – aspects of a given culture may tend to favor individuals of a given genotype (thus influencing genetic selection). The outcome of this iterative process is that there is a good fit between a particular genotype and a particular cultural practice. A commonly cited example in the literature is the genetic disposition to lactose tolerance (which has occurred relatively recently in human history and is not culturally universal) and the cultural practice of cattle domestication and dairy farming.

More recently, researchers have investigated the prevalence of various genetic sub-types linked to social sensitivity in cultures that vary in their degree of individualism and collectivism (see Way & Lieberman, 2010). In a **collectivist culture** the goals of the social group are emphasized over individual goals (e.g. in East Asian countries such as China). In an **individualist culture** the goals of the individual are emphasized over the social group (e.g. in Western countries such as USA). There is evidence that genes linked to increased social sensitivity are more prevalent in collectivist cultures, whereas genes linked to reduced social sensitivity are more prevalent in individualist cultures (see Figure 1.8). One possible conclusion is that genes and culture have co-evolved. The serotonin transporter gene occurs in two variants, or alleles, termed short and long. Carriers of the short allele show more mental health problems (e.g. depression) following a negative life event (such as divorce, Caspi et al., 2003), but also show more responsiveness to positive life events particularly in the social realm (Way & Taylor, 2010). That is, the short gene confers social sensitivity rather than being, say, a 'gene for depression'. The short is more prevalent in collectivist cultures (Chiao & Blizinsky, 2010). The mu-opioid receptor exists in two allelic variants (termed G and A), and the G version is linked to greater sensitivity to social rejection as measured by fMRI and questionnaires (Way, Taylor, & Eisenberger, 2009). The G version is more prevalent in collectivist cultures (Way & Lieberman, 2010). Finally, monoamine oxidase A (MAO-A) is an enzyme that breaks

KEY TERMS

Collectivist culture
The goals of the social group are emphasized over individual goals.

Individualist culture
The goals of the individual are emphasized over the social group.

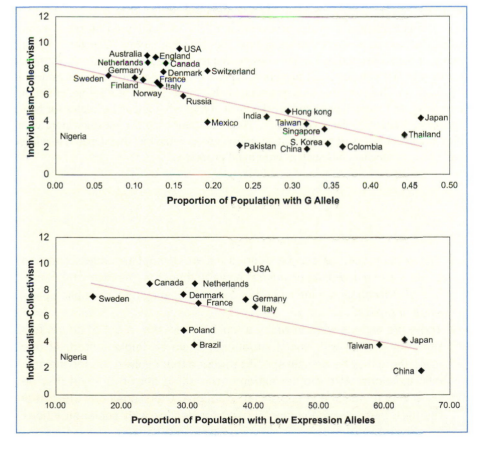

Figure 1.8 The G allele of the mu-opioid receptor gene and the low expression allele of the MAO-A gene are both linked to increased social sensitivity. Their prevalence in a given country is correlated with the degree of collectivism in that culture. Is this evidence for gene-culture co-evolution? From Way and Lieberman (2010).

down serotonin and dopamine and exists in different variants. The low expressing variant has been linked to anti-social behaviors following negative life events (such as neglect/abuse; Caspi et al., 2002) but this genotype often reports the lowest psychological problems following positive life experiences (Belsky et al., 2009). That is, it conveys social sensitivity. It is also more common in collectivist cultures (Way & Lieberman, 2010).

There is an obvious, and hard to disprove, criticism of these findings: namely that the evidence is all correlational in nature. However, the fact that the pattern occurs across multiple genes that convey social sensitivity makes it unlikely to be a chance occurrence. If true, it suggests that differences at the lowest level of analysis in neuroscience (e.g. a single change in the genetic code) interact with the highest level of analysis in social psychology (whole cultures) – albeit interacting over multi-generational timescales.

OVERVIEW OF SUBSEQUENT CHAPTERS

This chapter has offered a taster to some of the big issues that dominate research in social neuroscience, as well as outlining its conceptual and historical underpinnings. The chapter can perhaps be best appreciated by reading it again after tackling some of the detailed topics covered later on. From an assessment perspective, most essays on social neuroscience could be enhanced by insightfully discussing concepts such as reverse inference, localization, appropriate levels of explanation (and so on), together with the relevant evidence-base for the specific topic of the essay.

The next chapter provides an overview of the methods used by social neuroscientists, and also offers further critique of their limitations. Chapter 3 revisits many of the issues outlined in the current chapter specifically from an evolutionary (cross-species) perspective. Chapters 4, 5, and 6 discuss what can be regarded as the 'core' mechanisms that underpin social behavior (e.g. face processing, emotions, mirroring, and mentalizing) and are the most commonly touted candidates for being specialized and modular mechanisms. The remaining chapters consider more complex social behaviors (e.g. altruism, morality, group behavior) that depend, to a large part, on core mechanisms such as emotional processes.

SUMMARY AND KEY POINTS OF THE CHAPTER

- Social neuroscience can be defined as an attempt to understand and explain, *using neural mechanisms*, how the thoughts, feelings, and behaviors of individuals are influenced by the actual, imagined, or implied presence of others.
- There are various ways in which a 'social brain' (i.e. a set of neural routines for dealing with social situations) could be implemented. At one level, there may be domain-specific routines that evolved for serving specific functions. At the other extreme, the same set of routines may be used in both social and non-social situations. Other positions are neural routines that are predominantly used for dealing with social situations

but serve a more generic function, or non-modular solutions implemented throughout the brain (e.g. mirror systems).

- Social neuroscience aims to create bridges between different levels of explanation of social behavior. The brain, and its workings, is likely to create causal constraints on the way that social interactions are organized rather than merely soaking up the social world (as a blank slate).

- Gene-culture co-evolution is one example of interactions between different levels of explanation that determines how culture selects genes and, more controversially, for how genes select cultures

EXAMPLE ESSAY QUESTIONS

- Is the 'social brain' highly modular?
- How can neuroscience and social psychology inform each other?
- Is cultural neuroscience likely to be a promising way of unpacking the complexities of culture?

RECOMMENDED FURTHER READING

- Frith, C. D. (2007). The social brain? *Philosophical Transactions of the Royal Society, 362,* 671–678.

- Mitchell, J. P. (2009). Social psychology as a natural kind. *Trends in Cognitive Sciences, 13,* 246–251.

- Stanley, D. A., & Adolphs, R. (2013). Toward a neural basis of social behavior. *Neuron, 80,* 816–826. An interesting review of the current state of social neuroscience and how it is likely to develop in the future.

- Willingham, D. T., & Dunn, E. W. (2003). What neuroimaging and brain localization can do, cannot do, and should not do for social psychology. *Journal of Personality and Social Psychology, 85,* 662–671.

ONLINE RESOURCES

- Become a student member of an academic society: the Society for Social and Affective Neuroscience (SANS; www.socialaffectiveneuro.org) or Society for Social Neuroscience (S4SN; www.s4sn.org).
- References to key papers and readings
- Talks and lectures by Joan Chiao, Lisa Feldman Barrett, Ralph Adolphs, V. S. Ramachandran, and others
- Recorded lecture given by textbook author, Jamie Ward
- Multiple choice questions and interactive flashcards to test your knowledge
- Downloadable glossary

CHAPTER 2

CONTENTS

The methods of social neuroscience

Social neuroscience is too recent a field to have developed a distinct methodology of its own. As such its methods are borrowed from disciplines such as psychology (both cognitive and social psychology) and neuroscience (particularly cognitive neuroscience). The chapter will begin by considering various psychological methods such as performance measures (e.g. response times), observational studies, and questionnaires. It then goes on to consider methods linked to cognitive neuroscience – psychophysiological responses (e.g. skin conductance response) and electrophysiological responses – before turning to functional imaging, effects of brain lesions, and brain stimulation. Most of these methods are covered in more detail in Ward (2015). However, specific examples from the field of social neuroscience are used to illustrate the different methods and to explain the complementary nature of the different methods.

The main methods of cognitive neuroscience can be placed on a number of dimensions as illustrated in Table 2.1 below and Figure 2.1:

- The **temporal resolution** refers to the accuracy with which one can measure when an event is occurring. The effects of brain damage are permanent and so this has no temporal resolution as such. Methods such as electroencephalography/ event-related potential (EEG/ERP), magnetoencephalography (MEG), TMS, and single-cell recording have millisecond resolution. Positron emission tomography (PET) and functional magnetic resonance imaging (fMRI) have temporal resolutions of minutes and seconds, respectively, that reflect the slower hemodynamic response.
- The **spatial resolution** refers to the accuracy with which one can measure where an event is occurring. Lesion and functional imaging methods have comparable resolution at the millimeter level, whereas single-cell recordings have spatial resolution at the level of the neuron.

TABLE 2.1 THE DIFFERENT METHODS USED IN COGNITIVE NEUROSCIENCE

Method	Method type	Invasiveness	Brain property used
EEG/ERP	Recording	Non-invasive	Electrical
Single-cell (and multi-unit) recordings	Recording	Invasive	Electrical
TMS	Stimulation	Non-invasive	Electromagnetic
tDCS	Stimulation	Non-invasive	Electrical
MEG	Recording	Non-invasive	Magnetic
PET	Recording	Invasive	Hemodynamic
fMRI	Recording	Non-invasive	Hemodynamic

Figure 2.1 The methods of cognitive neuroscience can be categorized according to their spatial and temporal resolution. Adapted from Churchland and Sejnowski (1988).

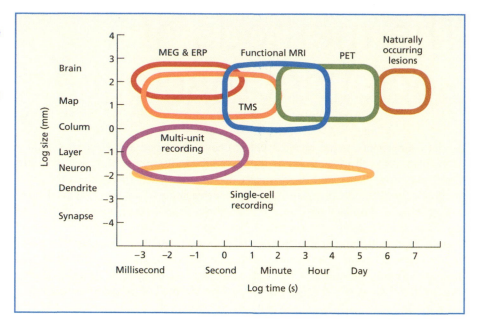

- The **invasiveness** of a method refers to whether or not the equipment is located internally or externally. PET is invasive because it requires an injection of a radiolabeled isotope. Single-cell recordings are performed on the brain itself and are normally only carried out in non-human animals. Methods such as TMS are not strictly invasive (because the coil is located entirely outside the body) even though it leads to stimulation of the brain.

MEASURING BEHAVIOR AND COGNITION: PSYCHOLOGICAL METHODS

Almost all experiments in social neuroscience measure behavior in some way, given that it is social behavior that they are trying to explain. In functional imaging experiments, the participant is given a set of instructions on how to respond even if the main dependent measure is brain activity rather than behavior per se. In social neuroscience, it is also common to correlate neurophysiological responses (e.g. during functional imaging) when performing a task with individual differences on a psychological measure such as empathy or personality (assessed outside the scanner using a questionnaire).

In this section, an overview will be provided of three different ways of measuring behavior and cognition: performance-based measures (where the dependent measures are typically response times or error rates); observation-based measures (where the dependent measure is often a frequency count of how often something occurs); and first-person-based measures (where the dependent measure may be scores on a questionnaire).

Performance-based measures: response times and accuracy rates

Mental chronometry can be defined as the study of the time-course of information processing in the human nervous system (Posner, 1978). The basic idea is that changes in the nature or efficiency of information processing will manifest themselves in the time it takes to complete a task. For example, participants are faster at verifying that 4 + 2 = 6 than they are in verifying that 4 + 3 = 7, and this is faster than verifying that 4 + 5 = 9 (Parkman & Groen, 1971). What can be concluded from this? First of all, it suggests that mathematical sums such as these are not just stored as a set of facts. If this were so, then all the reaction times would be expected to be the same because all statements are equally true. It suggests, instead, that the task involves a stage in processing that encodes numerical size together with the further assumption that larger sums place more limits on the efficiency of information processing (manifested as a slower verification time). To give an example more relevant to social neuroscience, it has been found that the response time to identify a face (e.g. by naming it) depends on whether or not the face displays an emotional expression – decisions are faster if the face is smiling (Gallegos & Tranel, 2005). This is shown in Figure 2.2. What can we conclude from this? First of all, many models of face processing assume that recognizing who a person is and recognizing their expression are separate (Bruce & Young, 1986; Haxby, Hoffman, & Gobbini, 2000). These results speak against this, to some extent. However, there are various possibilities. One is that known faces tend to be stored in the brain in an expressive pose (e.g. smiling) rather than a neutral pose as generally assumed. This would make them more efficient to process. Similarly facial identity can be computed, in part, from idiosyncratic facial movements, and smiles (even static smiles) may provide this additional information – that is, it might be motor/movement cues rather than emotion itself that drive the effect. An alternative is that familiar face recognition and expression recognition really are separate but can interact, such that the latter can provide a boost to the former in certain situations. These competing ideas could be explored with further response time studies (e.g. comparing smiling vs. angry expressions) or with other methods such as EEG, which can be used to determine whether the effect is early or late in time (i.e. consistent with an interaction at either the perceptual or decision-making stage).

Aside from response times, the other main performance measure is accuracy. This can be measured in terms of error rates, percentage correct, or percentile performance in which individual scores are recalculated relative to the population mean (e.g. IQ scores). Accuracy is obviously crucially related to whether certain knowledge is present/absent rather than to processing efficiency (which is more closely related to response time). However, accuracy and

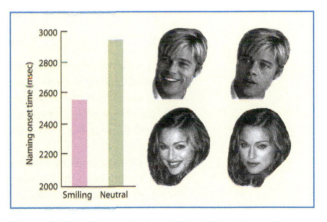

Figure 2.2 People are faster at identifying faces when they are smiling relative to a neutral pose. How can evidence from response times, such as this, be used to guide theories in social neuroscience? What might such a result mean and how could we explore these hypotheses using other studies? Graph based on data from and images taken directly from Gallegos and Tranel (2005). Copyright © 2005 Elsevier. Reproduced with permission.

Figure 2.3 Many cognitive psychology studies instruct participants to be 'as fast and accurate as possible'. In practice, these two factors tend to be in opposition – faster responses tend to be less accurate, and very accurate responses tend to be slower. This is termed speed–accuracy trade-off.

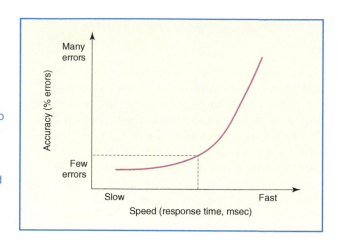

efficiency are related in certain circumstances. For example, if people are forced to respond faster they will tend to be less accurate, a so-called **speed–accuracy trade-off**. This is shown in Figure 2.3.

Summary of performance measures

- *Advantages*: They reflect actual behavior; they are simple to analyze and interpret.
- *Disadvantages*: They are hard to link directly to neural substrates (unless combined with other measures). There is not always a clear relationship between laboratory tasks and real-world behavior.

Observational measures

If performance-based measures measure 'how well' or 'how fast' something is done, observational measures tend to code 'what' is being done or 'how often' something is done through one person observing the behavior of others. There are certain situations in which observational measures are used in place of the more common performance-based measures:

- Observational measures are the norm in infancy research because the infant cannot be trained or instructed to perform a task.
- Observational methods may often be used for understanding non-human species for similar reasons to those used for human infants. Although training is possible here, there is still a need to know how (untrained) animals behave in the wild. For example, researchers have documented how often different primate species engage in deception in the wild and have correlated this with brain size (Byrne & Corp, 2004).
- Observational measures might be appropriate when the experimenter does not want the participant to know the true nature of a task. For instance, one study in human adults scored their behavior in terms of how often they imitate a given action (e.g. nose rubbing) whilst performing a cooperative task (e.g. Chartrand & Bargh, 1999).

Two specific observational methods in the infant literature are preferential looking and habituation. In **preferential looking** paradigms the infant is presented with a number of stimuli (normally two) and the amount of time that the infant spends looking at each of them is scored. A deviation from chance (e.g. 50/50 for two stimuli) implies that the infant is able to discriminate between the two stimuli (i.e. can tell they are not the same) and has a preference for one (although the reason for the preference is harder to infer). In **habituation** paradigms, the same stimulus (or the same kind of stimulus) is presented repeatedly and the infant's attention towards the stimulus (measured in terms of looking time) diminishes. The critical phase of a habituation experiment occurs when a new stimulus is presented. If the infant's attention is increased it implies that he/she can recognize that it is different, whereas if it is not it implies that he/she treats it as the same. Results using preferential looking and habituation reveal that infants have a preference for social stimuli (e.g. faces) over nonsocial stimuli (e.g. Johnson, Dziurawiec, Ellis, & Morton, 1991). Coding the imitative behavior of infants (e.g. whether they produce tongue protrusions or lip rounding) is another example of the use of observational methods in this group (Meltzoff & Borton, 1979; Meltzoff & Moore, 1977).

There are several methodological problems that need to be borne in mind when using observational measures, primarily because the scoring system is open to human error. First, there is the issue of **inter-rater (or inter-observer) reliability**, that is, the extent to which two independent observers would generate the same answers. This is typically dealt with by recording the experiment and having two people independently scoring a randomly selected subset of the behaviors. The second issue is whether the observer knows the hypothesis and might be biased to report what they expect to see. In such instances, the observer should perform **blind scoring** of behavior. For example, in preferential looking paradigms the observer typically would not know which stimulus the infant is being presented with on a given trial. To some extent, this problem

www.CartoonStock.com

Figure 2.4 ". . . I can therefore conclude that the primates are indeed social animals."

KEY TERM

Masking
The presentation of junk visual material after a stimulus (to eliminate persistence of a visual image).

can be overcome by having computerized scoring methods, but this only applies in some domains (e.g. infant eye movements) and not others (e.g. primatology field research).

Summary of observational measures

- *Advantages*: They can be used when it is impossible or inappropriate to give instructions to a participant; they can be used in naturalistic settings.
- *Disadvantages*: There are difficulties associated with scoring and observer biases (although methods such as eye tracking can limit this).

HOW TO MEASURE THE UNCONSCIOUS

Much of what we know is computed prior to us becoming aware of knowing it. Moreover, there is some information that we never become aware of but that can still guide behavior. Here, we will consider how we can measure the effects of stimuli that have been processed unconsciously.

The standard way of presenting a visual stimulus unconsciously is to present it for a brief duration (e.g. less than 50 ms) and follow it with junk visual material (termed **masking**). This prevents an afterimage of the briefly presented stimulus from persisting. This method is an example of subliminal perception and research in this area has shown that people can detect whether a stimulus was present/absent above chance even when they claim to be guessing (Cheesman & Merikle, 1984) and that subliminally presented stimuli are subsequently judged as more pleasant than non-presented stimuli (Zajonc, 1980). Both studies imply that the stimulus was seen, because it influences behavior, but in the absence of conscious report. An alternative methodology in this literature is to present stimuli for longer durations, but such that they remain outside of the locus of attention (e.g. Simons & Chabris, 1999). This normally requires that the attended task is demanding (e.g. Lavie, 1995). This method can apply to hearing as well as vision.

How can we know whether something was conscious or unconscious? One strategy is to rely on verbal reports: for example, analyzing only those trials in which participants claim to be guessing whether something was seen or not. Another method that has recently been used is wagering in which participants are asked to bet on their performance on a given trial (Persaud, McLeod, & Cowey, 2007). A rather different approach is to use measures of which the participant has no (or very little) volitional control, such as the skin conductance response, electromyography, eyeblink startle responses, and so on. In this case, the participant may (or may not) be conscious of the stimulus, but they are unlikely to be able to consciously influence this response.

Survey measures: questionnaires and interviews

Survey methods involve questioning participants using questions and a set of responses that are fixed in advance (e.g. most questionnaires) or questions and a range of responses that are open-ended (e.g. interviews). These are first-person methods in that the participant is expressing his/her own thoughts that cannot be objectively labeled as right or wrong (in contrast to performance measures described above). For example, contemporary assessments of individual differences in personality (e.g. Costa & McCrae, 1985) or empathy (e.g. Davis, 1980) involve presenting participants with a list of statements (e.g. 'I get easily distressed by the sight of someone else crying') and participants are asked the extent to which they agree or disagree with them.

The **reliability** of questionnaire measures can be assessed by asking participants to repeat the same questionnaire at another time point, and/or by including items in the questionnaire that tap the same knowledge but may require a different response (e.g. 'I like caring for others' / 'I do not like caring for others'). The latter is important because there is a tendency for people to opt for 'agree' more than 'disagree' during surveys. This is termed an **acquiescence bias**. Survey methods can also be used to explore whether lay concepts such as empathy can be fractionated into several underlying variables. For example, if one devises a questionnaire with 40 items on it, one may find 20 questions that are reliably answered in the same way and another 20 questions that also are reliably answered in the same way but differ from the first set. In this example, this would imply an influence of two different underlying variables (statistically, this is assessed using a method called **factor analysis**).

Whereas observational methods measure how people actually behave, survey methods ask people how they think they might behave. As such, one could argue that survey methods have lower **external validity** than observational methods. However, survey methods do have some advantages. Many researchers are interested in what people think and feel, rather than simply how they behave. The fact that our thoughts and behavior may sometimes appear to contradict each other is of interest in its own right, rather than necessarily reflecting a methodological flaw. Pragmatically, questionnaires are easier to carry out, especially using the internet. The external validity may be improved by administering the surveys anonymously and confidentially (the latter being the normal ethical standard). This is because participants may be more inclined to give an honest answer in these situations, rather than presenting themselves in a positive light.

Although survey methods play a central part in social psychology, they have a more supporting role in social neuroscience. In social neuroscience, questionnaire results tend to be correlated with other measures (e.g. from fMRI, EEG) in order to identify the neural correlates of attitudes, feelings, and traits. One particular challenge for this approach stems from the fact that in methods such as fMRI the brain is divided into tens of thousands of regions (termed voxels) and statistical tests may be performed on each and every voxel. As such, the chance of getting a significant, but meaningless, result somewhere in the brain becomes high – called a **Type I error** (contrast with a **Type II error** in which a null result is obtained even though there is a real effect). Type I and Type II errors are a problem for *all* inferential statistics no matter the discipline. However, one specific problem linked to social neuroscience has been extensively debated and has been termed 'voodoo correlations'.

Vul, Harris, Winkielman, and Pashler (2009) noted that the reliability of many questionnaire measures is no more than about .8 (i.e. if the same questionnaire is repeated twice, the correlation between answers on the two occasions is about .8). The reliability of fMRI is of a similar magnitude or less, at around .7 (i.e. if the same experiment is done twice then the correlation between activity levels on the different occasions is about .7). However, many studies in social neuroscience report correlations between brain activity and questionnaire measures greatly in excess of what is

KEY TERMS

Reliability
The extent to which the same measure would yield the same results if repeated.

Acquiescence bias
A tendency to respond affirmatively in surveys, irrespective of the content of the question

Factor analysis
A statistical method for reducing a data set (e.g. in questionnaires, 20 questions may be grouped into a smaller number of factors)

External validity
The extent to which a measure relates to something useful in 'real life'

Type I error
Getting a significant result in a statistical test when, in fact, there is no real effect

Type II error
Getting a nonsignificant result when in fact there is a real effect

considered theoretically possible ($\sqrt{.8 \times .7} = .74$). This suggests that some key findings are Type I errors or, at least, an inflation of the true size of the effect. This has informally become known in the social neuroscience literature as a 'voodoo correlation'. There are various steps that one can take to minimize this when correlating questionnaires with brain imaging data relating, for instance, to whether the correlation is performed on a voxel that has already been selected for its statistical significance (Lieberman & Cunningham, 2009; Vul et al., 2009). However, as a general point it is worth noting that social neuroscience methods – despite technological sophistication – are not invulnerable to flawed designs or analyses.

Summary of survey measures

- *Advantages*: They can be used in situations where an experimental manipulation is not possible or unethical (e.g. exposure to repeated violence); they measure thoughts and attitudes rather than behavior.
- *Disadvantages*: Participants' self-reports may not reflect their true behavior; much social cognition may occur unconsciously.

STRUCTURE AND FUNCTION OF THE NEURON

All **neurons** have basically the same structure. They consist of three components: a cell body (or soma), **dendrites**, and an **axon**. All neurons have the same basic structure and function (see Figure 2.5). The cell body contains the nucleus and other organelles. The nucleus contains the genetic code, and this is involved in protein synthesis (e.g. of certain neurotransmitters). Neurons receive information from other neurons and they make a 'decision' about this information (by changing their own activity) that can then be passed on to other neurons. From the cell body, a number of branching structures called dendrites enable communication with other neurons. Dendrites receive information from other neurons in close proximity. The number and structure of the dendritic branches can vary significantly depending on the type of neuron (i.e. where it is to be found in the brain). The axon, by contrast, sends information to other neurons. Each neuron consists of many dendrites but only a single axon (although the axon may be divided into several branches called collaterals).

The terminal of an axon flattens out into a disc-shaped structure. It is here that chemical signals enable communication between neurons via a small gap termed a **synapse**. The two neurons forming the synapse are referred to as pre-synaptic (before the synapse) and post-synaptic (after the synapse),

KEY TERMS

Neurons
A type of cell that makes up the nervous system.

Dendrites
Branching structures that receive information from other neurons.

Axon
A branching structure that carries information away from the cell body towards other neurons and transmits action potentials.

Synapse
The small gap between neurons in which neurotransmitters are released, permitting signaling between neurons.

reflecting the direction of information flow (from axon to dendrite). When a pre-synaptic neuron is active, an electrical current (termed an **action potential**) is propagated down the length of the axon. When the action potential reaches the axon terminal, chemicals are released into the synaptic cleft. These chemicals are termed **neurotransmitters**. (Note that a small proportion of synapses, such as retinal gap junctions, signal

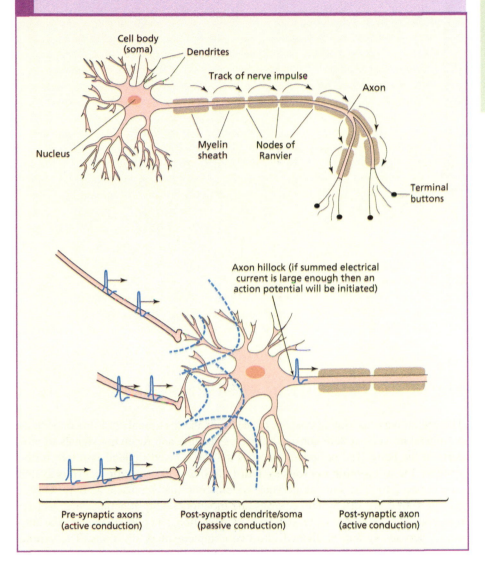

Figure 2.5 The structure and electrical functioning of neurons (top and bottom, respectively). Neurons consist of three basic features: a cell body, dendrites that receive information and an axon that sends information. In this diagram the axon is myelinated to speed up the conduction time. Electrical currents are actively transmitted through pre-synaptic axons by an action potential. Electrical currents flow passively through dendrites and soma of neurons but will initiate an action potential if their summed potential is strong enough at the start of the axon of the post-synaptic neuron (called the hillock).

electrically and not chemically.) Neurotransmitters bind to receptors on the dendrites or cell body of the post-synaptic neuron and create a synaptic potential. Depending on the nature of the chemical reaction, the potential can either be excitatory (i.e. promote further firing) or inhibitory (i.e. reduce the likelihood of further firing). The synaptic potential is conducted passively (i.e. without creating an action potential) through the dendrites and soma of the post-synaptic neuron. If these passive currents are sufficiently strong when they reach the beginning of the axon in the post-synaptic neuron, then an action potential (an active electrical current) will be triggered in this neuron. It is important to note that each post-synaptic neuron sums together many synaptic potentials, which are generated at many different and distant dendritic sites (as opposed to a simple chain reaction between one neuron and the next). Passive conduction tends to be short range because the electrical signal is impeded by the resistance of the surrounding matter. Active conduction enables long-range signaling between neurons by the propagation of action potentials.

The amplitude of an action potential does not vary, but the number of action potentials propagated per second varies along a continuum. This rate of responding (also called the 'spiking rate') relates to the informational 'code' carried by that neuron. For example, some neurons may have a high spiking rate in some situations (e.g. during speech) but not others (e.g. during vision), whereas other neurons would have a complementary profile. Neurons responding to similar types of information tend to be grouped together. This gives rise to the functional specialization of brain regions.

KEY TERMS

Autonomic Nervous System (ANS)
A set of nerves located in the body that controls the activity of the internal organs.

Somatic nervous system
Part of the peripheral nervous system that coordinates muscle activity.

Sympathetic system
A division of the ANS that increases arousal (increased heart rate, breathing, pupil size) but decreases functions such as digestion.

Parasympathetic system
A division of the ANS that has a resting effect (decreased heart rate, breathing, pupil size) but increases functions such as digestion.

MEASURING BODILY RESPONSES

The central nervous system consists of the brain and the spinal cord. In contrast, the peripheral nervous system consists of nerves sending and receiving signals to other parts of the body. The peripheral nervous system is itself divided into two further systems: the **autonomic nervous system (ANS)** and the **somatic nervous system** (see Figure 2.6). The somatic nervous system coordinates muscle activity whereas the autonomic nervous system controls and monitors bodily functions such as heart rate, digestion, respiration rate, salivation, perspiration, and pupil diameter. The autonomic nervous system is divided into two complementary divisions. The **sympathetic system** increases arousal (e.g. increased heart rate, breathing, pupil size) and decreases functions such as digestion. The **parasympathetic system** has a resting effect (decreased heart rate, breathing, pupil size) and increases functions such as digestion (see Figure 2.7).

Several methods in social neuroscience rely on measurements related to the peripheral nervous system. Electromyography (EMG) is an electrical measure of muscle contraction, implemented by the somatic nerves. Measures of autonomic system functioning include the skin conductance response (SCR), measures of heart rate

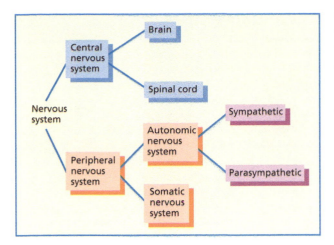

Figure 2.6 Various methods, including skin conductance response, electromyography (EMG), and pupil size, rely on measures of activity of the peripheral nervous system.

and breathing (e.g. the traditional lie detector, or polygraph, measures various autonomic functions including these), and also pupilometry (measuring changes in pupil dilation). The use of EMG and SCR is considered in more detail here.

The skin conductance response (SCR)

A common way of measuring increased activity of the sympathetic system is to monitor small changes in conductivity as a result of mild sweating (Berry Mendes, 2009). Heightened arousal can lead to more sweat even without overt sweating taking place, and this sweating response (from eccrine glands) is separate from the thermo-regulatory sweating response. The SCR is measured by applying a weak electrical current to the skin. During a sweating response (e.g. elicited by an emotional stimulus) there is decreased conductivity of the skin and the electrical signals flow more easily. This is termed the **skin conductance response (SCR)** or galvanic skin response (GSR). The electrodes are normally placed on two adjacent fingertips with gel in-between the fingers and electrodes to improve contact. A peak SCR occurs between one and five seconds after stimulus presentation and this is normally recalculated relative to some baseline (e.g. pre-stimulus) activity (see Figure 2.8). Boucsein et al. (2012) report a consensually agreed set of experimental and publishing guidelines for researchers using this method.

Tranel and Damasio (1995) report how the SCR is affected by a number of brain lesions. Lesions to the ventromedial frontal lobes abolish SCR to psychological stimuli (e.g. risk) but not physical stimuli (e.g. bangs), whereas lesions to the anterior cingulate cortex abolish both. Functional imaging also points to a key role for the anterior cingulate in the production of the SCR (Critchley, Elliott, Mathias, & Dolan, 2000).

Electromyography (EMG)

Facial **electromyography (EMG)** has been used in social neuroscience research to measure muscle activity associated with emotional expressions in response to seeing expressions in others (e.g. Dimberg, Thunberg, & Elmehed, 2000;

KEY TERMS

Skin Conductance Response (SCR)
Small changes in conductivity as a result of mild sweating

Electromyography (EMG)
A method for assessing electrical activity associated with muscle movement

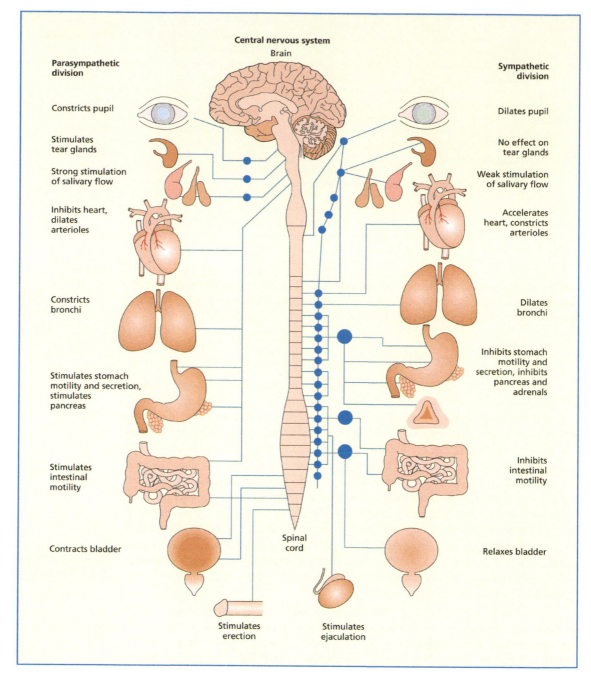

Figure 2.7 The autonomic nervous system (ANS) is divided into two complementary divisions. The sympathetic system increases arousal and decreases functions such as digestion. The parasympathetic system has a resting effect and increases functions such as digestion.

Hess & Blairy, 2001) or as a potentially implicit measure of prejudice (Vanman, Paul, Ito, & Miller, 1997). It is also used to measure the **eyeblink startle response**, which is elicited by a startling sound but is further modulated by the participant's present emotional state (Lang, Bradley, & Cuthbert, 1990).

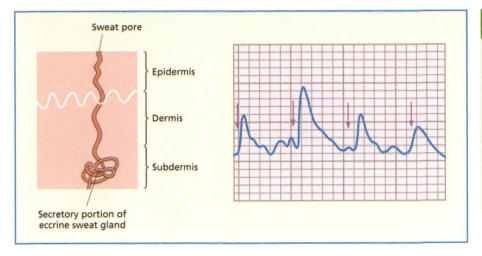

Figure 2.8 The skin conductance response (SCR) method involves recording changes in electrical conductivity on a person's skin on the hand. A person's SCR can be plotted as a continuous trace throughout the experiment. A peak SCR occurs between 1 and 5 seconds after stimulus presentation.

Electromyography (EMG) is a measure of electrical activity associated with muscle contraction (Fridlund & Cacioppo, 1986; Hess, 2009). These changes come about because of an increase in the number of action potentials in muscle fibers during muscle contraction. The greater the force produced by the muscles, the greater the electrical activity (Lawrence & DeLuca, 1983). However, individual action potentials are not measured. Instead the EMG signal is the sum of many such potentials, including those that cancel out because the muscle fibers are not completely aligned. The EMG signal is recorded by placing two small electrodes close to each other and measuring the potential difference (in microvolts) between them. The frequency and amplitude range of the EMG signal are comparable with those of electrical signals generated by the brain (in EEG) and heart (ECG or electrocardiogram). In order to reduce the influence of the latter two sources, the two measurement electrodes are compared to a third electrode (the ground) placed elsewhere (e.g. the center of the forehead is often used in facial EMG). Fridlund and Cacioppo (1986) offer guidance on the placement of EMG electrodes associated with various facial muscles (see Figure 2.9). In addition to electrical activity from other bodily sources it is also common practice to reference the EMG signal to a baseline measure, such as a rest phase or the period before a stimulus is presented. This is because muscle activity is rarely zero and may fluctuate over time (e.g. due to tension in the participants).

Summary of measures of bodily responses

- *Advantages*: Bodily responses are often present in the absence of awareness of a stimulus and may occur in the absence of a specific task; they are relatively easy to record and analyze.
- *Disadvantages*: It is not straightforward to link bodily responses to brain and cognition.

Figure 2.9
Recommended placement of pairs of electrodes for recording facial EMG. Adapted from Fridlund and Cacioppo (1986).

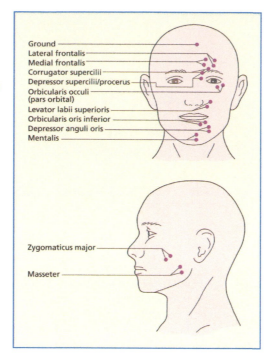

ELECTROPHYSIOLOGICAL METHODS

By measuring changes in the responsiveness of a neuron to changes in a stimulus or changes in a task, it is possible to make inferences about the building blocks of cognitive processing. The action potential is directly measured in the method of single-cell recording, whereas EEG is particularly sensitive to post-synaptic dendritic electrical activity.

Single-cell recording

Single-cell recordings can be obtained by implanting a very small electrode either into the axon itself (intracellular recording) or outside the membrane (extracellular recording) and counting the number of times that an action potential is produced (spikes per second) in response to a given stimulus (e.g. a face). This is an invasive method. As such, the procedure is normally conducted on experimental animals only (see Figure 2.10). It is occasionally conducted on humans undergoing brain surgery (Engel, Moll, Fried, & Ojemann, 2005). It is impossible to measure action potentials from a single neuron non-invasively (i.e. from the scalp) because the signal is too weak and the noise from other neurons is too high. Technology has now advanced such that it is possible to simultaneously record from 100 neurons in multi-electrode arrays. This is termed multi-cell recording.

To give one example from the literature, Quiroga, Reddy, Kreiman, Koch, and Fried (2005) recorded the firing rates of neurons in humans undergoing brain surgery.

The participants were shown images of famous people and buildings whilst recordings were taken from cells in the medial temporal lobe (a region of the brain implicated in memory). Many neurons showed a high degree of specificity in their response pattern – that is, they tended to respond to some stimuli (measured in spikes per second) more than others. For instance, one neuron responded to images of Jennifer Aniston in a variety of different poses, although not when she appeared next to her ex-husband, Brad Pitt (see Figure 2.11). Another neuron responded to various images of Halle Berry, but not to Jennifer Aniston. It responded to Halle Berry dressed as catwoman, but not to another actress dressed as catwoman, and it even responded to the printed name 'Halle Berry'. Another neuron responded to images of the Sydney Opera House but not the Eiffel Tower or Golden Gate Bridge. Studies such as these show how stimuli in the 'outside' world can be represented in a neural code, using a simple biological parameter

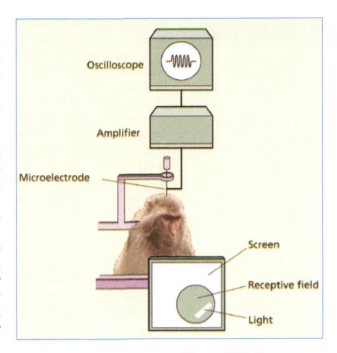

Figure 2.10 An illustration of a typical experimental set-up for single-cell recording

Figure 2.11 Neurons learn to 'tune in' to (i.e. become specialized in) familiar things in the environment. This neuron, recorded from the human medial temporal lobes, responds to different images of Jennifer Aniston but not to images of other actresses, buildings, or even Jennifer Aniston if she is accompanied by Brad Pitt. The red bars show firing rate (i.e. number of action potentials) across time (on the x-axis). The blue bars above it represent individual trials from which the neuron was presented with that image. From Quiroga et al. (2005). Copyright © 2005 Nature Publishing Group. Reproduced with permission.

(rate of action potentials). However, it need not mean that Jennifer Aniston is represented by a single neuron in our heads (perhaps there are many neurons that respond in this way) and nor does it necessarily mean that these neurons only respond to this particular person (given that the researchers only presented a relatively small number of stimuli).

The results of this study can be classified as **rate coding** of information by neurons, in that a given stimulus/event is associated with an increase in the rate of neural firing. An alternative way for neurons to represent information about stimuli/events is in terms of **temporal coding**, in that a given stimulus/event is associated with greater synchronization of firing across different neurons (Engel, Konig, & Singer, 1991). In temporal coding, the information is contained in the phase of firing (i.e. it depends on whether neurons are firing at the same time).

Summary of single-cell recording

- *Advantages*: This measure is directly related to neural activity (compare fMRI); it has excellent spatial and temporal resolution.
- *Disadvantages*: It is invasive so is very rarely performed on humans; information is limited to the regions probed (not a whole-brain technique).

Electroencephalography (EEG)

The physiological basis of the EEG signal originates in the post-synaptic dendritic currents rather than the axonal currents associated with the action potential (Nunez, 1981). **Electroencephalography (EEG)** records electrical signals generated by the brain, through electrodes placed at different points on the scalp. A typical EEG cap is shown in Figure 2.12. As the procedure is non-invasive and involves recording (not stimulation), it is completely harmless as a method. For an electrical signal to be detectable at the scalp a number of basic requirements need to be met in terms of underlying neural firing. First, a whole population of neurons must be active in synchrony to generate a large enough electrical field. Second, this population of neurons must be aligned in a parallel orientation so that they summate rather than cancel out. Fortunately, neurons are arranged in this way in the cerebral cortex. However, the same cannot necessarily be said about all regions of the brain. For example, the orientation of neurons in the thalamus may render its activity invisible to this recording method.

To gain an EEG measure one needs to compare the voltage between two or more different sites. A reference site is often chosen that is likely to be relatively uninfluenced by the variable under investigation. One common reference point is the mastoid bone behind the ears or a nasal reference; another alternative is to reference to the average of all electrodes. It is important to stress that the activity recorded at each location cannot necessarily be attributed to neural activity near to that region. Electrical activity in one location can be detected at distant locations. In general, EEG is not best equipped for detecting the location of neural activity.

Rhythmic oscillations in the EEG signal

The EEG signal, when observed over a sufficiently long timescale, has a wave-like structure. The EEG signal tends to oscillate at different rates (also called frequency bands) that are named after letters of the Greek alphabet: thus alpha waves reflect

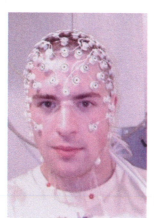

Figure 2.12 A participant in an EEG experiment

oscillations in the 7–14 Hz range, beta in the 15–30 range, and gamma in the 30 Hz and above range (and so on). These oscillations arise because large groups of neurons tend to be in temporal synchrony with each other in terms of their firing (action potentials) and in terms of their slower dendritic potentials (which forms the basis of the EEG signal). It has long been established that different rates of oscillation characterize different phases of the sleep–wake cycle. In recent decades, attempts have been made to link the relative amount of oscillations (the 'power') in different bands to different kinds of cognitive function during normal wakefulness. To give one example, it is found that oscillations in the alpha frequency range that originate from the sensory-motor cortex (these are also referred to by the Greek letter, mu, to distinguish them from other kinds of alpha waves) are greatest when the participant is at rest but decrease in power when performing a movement. What is particularly interesting from a social neuroscience perspective is that this decrease in power (or mu suppression) also occurs when observing someone else perform an action – so it is a potential signature of the mirror neuron system (Pineda, 2005).

KEY TERM

Event-related potentials (ERPs)
An averaged set of EEG recordings that are time-locked to a particular event

Event-Related Potentials (ERPs)

The most common usage of EEG in cognitive neuroscience is in the context of electrophysiological changes elicited by particular stimuli and cognitive tasks. These are referred to as **event-related potentials** (**ERPs**). The EEG waveform reflects neural activity from all parts of the brain. Some of this activity may specifically relate to the current task (e.g. reading, listening, calculating) but most of it will relate to spontaneous activity of other neurons that do not directly contribute to the task. As such, the signal-to-noise ratio in a single trial of EEG is very low (the signal being the electrical response to the event and the noise being the background level of electrical activity). The ratio can be increased by averaging the EEG signal over many presentations of the stimulus (e.g. 50–100 trials), relative to the onset of a stimulus (Figure 2.13). This eliminates the wave-like rhythmic oscillations (discussed previously) that occur spontaneously throughout the brain, and leaves an electrophysiological signature that is specific to the stimulus and task. The results are represented graphically by plotting time

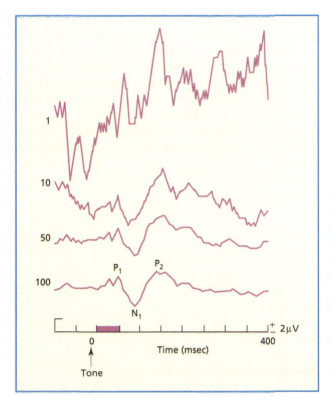

Figure 2.13 When different EEG waves are averaged relative to presentation of a stimulus (e.g. a tone) then the signal-to-noise ratio is enhanced and an ERP is observed. From Kolb & Whishaw (2002). Copyright © 2002 by Worth Publishers. Reproduced with permission.

KEY TERM

N170

An event-related
potential (negative
peak at 170 ms) that
has been linked to
face perception

(milliseconds) on the x-axis and electrode potential (microvolts) on the y-axis. The graph consists of a series of positive and negative peaks, with an asymptote at 0 microvolts. This is done for each electrode, and each will have a slightly different profile. The positive and negative peaks are labeled with 'P' or 'N' and their corresponding number. Thus, P1, P2, and P3 refer to the first, second, and third positive peaks, respectively. Alternatively, they can be labeled with 'P' or 'N' and the approximate timing of the peak. Thus, P300 and N400 refer to a positive peak at 300 ms and a negative peak at 400 ms (not the 300th positive and 400th negative peak!). It is to be noted that the polarity of the peaks (i.e. whether positive or negative) is of no real significance either cognitively or neurophysiologically (e.g. Otten & Rugg, 2005). It depends, for instance, on the baseline electrical activity and position of the reference electrode. What is of interest in ERP data is the timing and also the amplitude of the peaks.

How can ERP recordings be used to inform theory? Consider one example from the face processing literature.

There is evidence for an ERP component that is relatively selective for the processing of faces compared with other classes of visual objects. This has been termed the **N170** (a negative peak at 170 ms) and is strongest over right posterior temporal electrode sites (e.g. Bentin, Allison, Puce, Perez, & McCarthy, 1996). This, in itself, is interesting for several reasons. First, it suggests that there is a mechanism in the brain that is relatively specialized for faces more than objects. In this respect it is on a par with other methods such as fMRI or patient-based neuropsychology. However, in other respects it reveals something new: it suggests *when*, in time, faces become treated as special. This cannot be revealed by fMRI or neuropsychology. Nor can it be revealed by response-time methods. Unlike response-time measures, ERP measures can tell *when* something is happening before any response to it has been made. They are not, however, the best method for revealing *where* in the brain something is happening because the electrical activity conducts itself through the brain to distant sites.

One can then ask additional questions about the N170 to understand its nature and – by inference – to understand the nature of early mechanisms of face processing. For example, is it just found for human faces? No it is not, as shown in Figure 2.14 (Rousselet, Mace, & Thorpe, 2004). Is it found for schematic 'smiley' faces? Yes, it is (Sagiv & Bentin, 2001). Does it depend on whether a face is famous or not? No, it does not (Bentin & Deouell, 2000). The details of these studies are not of relevance here. The main point is that ERPs do not just tell us *when* things are happening in

Figure 2.14 The N170 is observed for both human faces (purple) and animal faces (blue) but not other objects (green). From Rousselet et al. (2004). Reproduced with permission of ARVO.

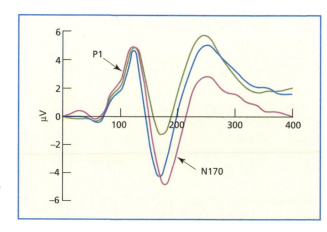

the brain; they can be taken a step further in order to understand *how* things happen in the brain.

Summary of ERPs

- *Advantages*: Excellent temporal resolution; direct measure of neural activity
- *Disadvantages*: Poor spatial resolution; some subcortical brain regions are impossible to investigate

Magnetoencephalography (MEG)

All electric currents, including those generated by the brain, have an associated magnetic field that is potentially measurable. As such, magnetoencephalography (MEG) can be regarded as a parallel method to EEG that is similar in many regards. For instance, one can examine either rhythmic neural oscillations or stimulus-evoked changes. There are some key differences too. The biggest potential advantage of MEG over EEG is that it permits a much better spatial resolution in addition to the excellent temporal resolution (e.g. Hari, Parkkonen, & Nangini, 2010). To give an example from the social neuroscience literature, it is possible to detect an MEG equivalent to the N170 linked, in ERP research, to structural encoding of faces (MEG components tend to be prefaced by an 'm/M'). One MEG study showed that the M170 is sensitive both to facial expressions (angry, happy, neutral) and to facial orientation that may indicate status (aloof or downcast) (Arviv et al., 2015). However, the authors were able to demonstrate that different brain regions were influencing this early process rather than reflecting a single localized mechanism and this conclusion rested on the enhanced spatial resolution of MEG relative to EEG.

In terms of practicalities, MEG is a more challenging and costly enterprise than EEG. The size of magnetic field generated by the brain is very small relative to the ambient magnetic field of the earth. As such, the development of magnetoencephalography (MEG) had to wait for suitable technological advances to become a viable enterprise. This technological advance came in the form of superconducting devices termed SQUIDs (an acronym of Superconducting Quantum Interference Device). A whole-head MEG contains 200–300 of these devices. The apparatus used requires extreme cooling, using liquid helium, and isolation of the system in a magnetically shielded room.

Summary of MEG

- *Advantages*: Both excellent temporal and spatial resolution
- *Disadvantages*: Not equally sensitive across the whole brain (differences between gyri and sulci, and deep and shallow dipoles); expensive and limited availability

THE ORGANIZATION AND STRUCTURE OF THE BRAIN

Neurons are organized within the brain to form **white matter** and **gray matter**. Gray matter consists of neuronal cell bodies. White matter consists of axons and support cells (**glia**). The brain consists of a highly

KEY TERMS

Gyri
The raised folds of the cortex

Sulci
The buried grooves of the cortex

convoluted folded sheet of gray matter (the cerebral cortex), beneath which lies the white matter. In the center of the brain, beneath the bulk of the white matter fibers, lies another collection of gray matter structures (the subcortex), which includes the basal ganglia, the limbic system, and the diencephalon. White matter tracts project between different cortical regions within the same hemisphere, between different cortical regions in different hemispheres (the most important being the corpus callosum), and between cortical and subcortical structures.

Anterior and posterior refer to directions towards the front and the back of the brain, respectively. These are also called rostral and caudal, respectively, particularly in other species that have a tail (caudal refers to the tail end). Directions towards the top and the bottom are referred to as superior and inferior, respectively; they are also known as dorsal and ventral, respectively. The terms anterior, posterior, superior, and inferior (or rostral, caudal, dorsal, and ventral) enable navigation in two dimensions: front–back and top–bottom. This is shown in Figure 2.15. Needless to say, the brain is three-dimensional, and so a further dimension is required. The terms lateral and medial are used to refer to directions towards the outer surface and the center of the brain, respectively.

The cerebral cortex consists of two folded sheets of gray matter organized into two hemispheres (left and right). The raised surfaces of the cortex are termed **gyri** (or gyrus in the singular). The dips or folds are called **sulci** (or sulcus in the singular). The lateral surface of the cortex of each hemisphere is divided into four lobes: the frontal, parietal, temporal, and occipital lobes. Other regions of the cortex are observable only in a medial section, for example the cingulate cortex. Finally, an island of cortex lies buried underneath the temporal lobe; this is called the insula (which literally means 'island' in Latin).

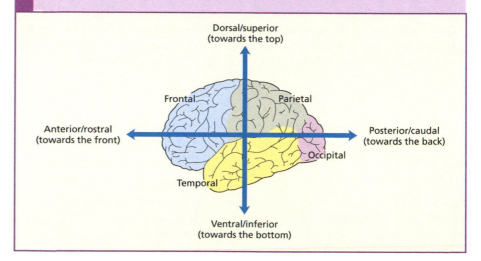

Figure 2.15 Terms of reference in the brain and the four lobes of the lateral surface. Note also the terms lateral (referring to the outer surface of the brain) and medial (referring to the central regions).

Beneath the cortical surface and the intervening white matter lies another collection of gray matter nuclei termed the subcortex. The subcortex is typically divided into a number of different systems with different evolutionary and functional histories (see Figures 2.16 and 2.17). These include the basal ganglia, the limbic system, and the diencephalon:

- The **basal ganglia** are large rounded masses that lie in each hemisphere. The main structures comprising the basal ganglia are the caudate nucleus (an elongated tail-like structure), the putamen (lying more laterally), and the globus pallidus (lying more medially). Different circuits passing through different regions in the basal

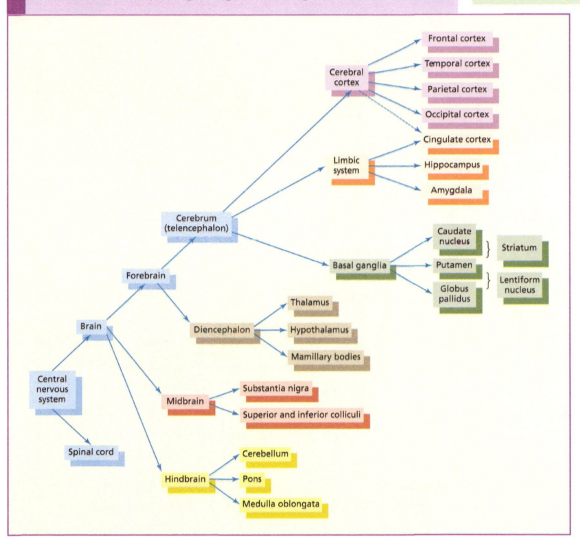

Figure 2.16 The central nervous system (CNS) is organized hierarchically. The upper levels of the hierarchy, corresponding to the upper branches of this diagram, are the newest structures from an evolutionary perspective.

ganglia connect to the thalamus and prefrontal cortex and midbrain structures. These serve somewhat different functions. For example, one circuit is involved in regulating motor activity and the learning of actions (e.g. skills and habits) and another is involved in processing rewards.

- The **limbic system** is important for relating the organism to its environment based on current needs and the present situation, and based on previous experience. It is involved in the detection and expression of emotional responses. For example, the amygdala has been implicated in the detection of fearful or threatening stimuli and parts of the cingulate gyrus have been implicated in the detection of emotional

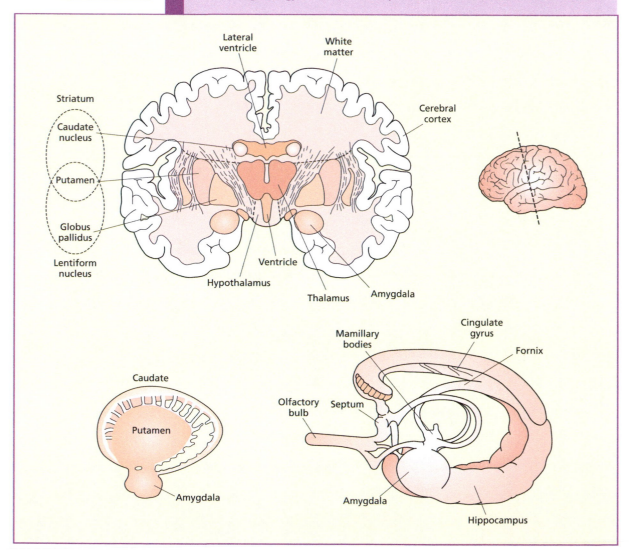

Figure 2.17 A coronal cross-section of the brain showing various subcortical regions (top). The bottom panels show the structure of the basal ganglia from a single hemisphere in a lateral view (left) and the structure of the limbic system across both hemispheres in a lateral view (right).

and cognitive conflicts. The hippocampus is particularly important for learning and memory.

- The two main structures that make up the **diencephalon** are the **thalamus** and the **hypothalamus**. The thalamus consists of two interconnected egg-shaped masses that lie in the center of the brain and appear prominent in a medial section. The thalamus is the main sensory relay for all senses (except smell) between the sense organs (eyes, ears, etc.) and the cortex. It also contains projections to almost all parts of the cortex and the basal ganglia. At the posterior end of the thalamus lie the lateral geniculate body and the medial geniculate body. These are the main sensory relays to the primary visual and primary auditory cortices, respectively. The hypothalamus lies beneath the thalamus and consists of a variety of nuclei that are specialized for different functions primarily concerned with the body. These include body temperature, hunger and thirst, sexual activity, and regulation of endocrine functions.

FUNCTIONAL IMAGING: HEMODYNAMIC MEASURES

Functional imaging methods such as PET and fMRI measure the dynamic physiological changes in the brain that are associated with different patterns of thought and behavior. This can be contrasted with structural imaging, which measures the stable properties of the brain (e.g. the distribution of white and gray matter) and includes CT scanning and conventional MRI. It is important to emphasize at the outset that PET and fMRI are not measuring the activity of neurons directly but, rather, are measuring a downstream consequence of neural activity: namely, changes in blood

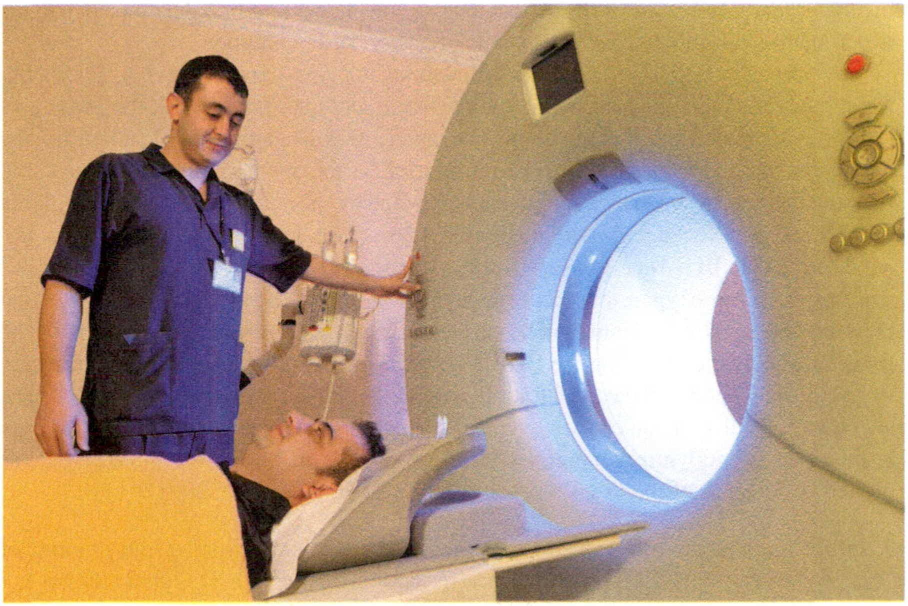

KEY TERMS

Diencephalon
Subcortical gray matter including the thalamus and hypothalamus

Thalamus
A major subcortical relay center; for instance, it is a processing station between all sensory organs (except smell) and the cortex

Hypothalamus
Consists of a variety of nuclei that are specialized for different functions that are primarily concerned with the body and its regulation

Figure 2.18 In fMRI experiments, a strong magnetic field is applied constantly during the scanning process. This is harmless, although it is important to follow safety procedures (e.g. removing metal from pockets).

flow/blood oxygen to meet metabolic needs of neurons. It is for this reason that they are termed **hemodynamic methods**, in contrast to methods such as EEG that measure the electrical fields generated by the activity of neurons themselves.

The main advantage of these methods is their spatial resolution. The whole brain is divided into tens of thousands of regions, termed **voxels**, each of the same size (e.g. 3 × 3 × 3 mm) and the 'activity' of each of these voxels in various tasks can be assessed in order to draw inferences about the functioning of different brain regions. In fMRI the larger the magnet, the smaller the voxel size that can be obtained. The strength of the magnetic field is measured in units called tesla (T). Typical scanners have field strengths between 1.5 T and 3 T; the Earth's magnetic field is of the order of 0.0001 T.

Basic physiology underpinning functional imaging

The brain consumes 20% of the body's oxygen uptake; it does not store oxygen and it stores little glucose. Most of the brain's oxygen and energy needs are supplied from the local blood supply. When the metabolic activity of neurons increases, the blood supply to that region increases to meet the demand (Attwell & Iadecola, 2002; Raichle, 1987). Techniques such as PET measure the change in blood flow to a region directly, whereas fMRI is sensitive to the concentration of oxygen in the blood.

The brain is always physiologically active. Neurons would die if they were starved of oxygen for more than a few minutes. This has important consequences for using physiological markers as the basis of neural 'activity' in functional imaging experiments. It would be meaningless to place someone in a scanner, with a view to understanding cognition, and simply observe which regions were receiving blood and using oxygen, because this is a basic requirement of all neurons, all of the time. As such, when functional imaging researchers refer to a region being 'active', what they mean is that the physiological response in one condition is greater relative to another. There is a basic requirement in all functional imaging studies that the physiological response must be compared to one or more control responses. A good understanding of the hypothesized mechanisms underlying the behavior is needed to ensure that the baseline task is appropriately matched to the experimental task, otherwise the results will be very hard to interpret.

What is the 'signal' measured in functional imaging experiments? In PET studies, the participant is injected with a radioactive tracer and the signal is the amount of radioactivity in each voxel of the brain. The major breakthrough in fMRI came from the realization that no tracer needed to be introduced for this method, but rather one could measure a magnetic resonance signal that is affected by the amount of deoxyhemoglobin in the blood in different regions of the brain (Ogawa, Lee, Kay, & Tank, 1990). This signal is termed the **BOLD response**, standing for Blood Oxygen-Level Dependent. When neurons consume oxygen they convert oxyhemoglobin to deoxyhemoglobin. Deoxyhemoglobin has strong paramagnetic properties and this introduces distortions in the local magnetic field, reducing the BOLD signal. This distortion can itself be measured to give an indication of the concentration of deoxyhemoglobin present in the blood. The way that the BOLD signal evolves over time in response to an increase in neural activity is called the **hemodynamic response function (HRF)**. The HRF has three phases, as plotted in Figure 2.19 and discussed below (see also Hoge & Pike, 2001):

1 *Initial dip*. As neurons consume oxygen there is a small rise in the amount of deoxyhemoglobin, which results in a reduction of the BOLD signal (this is not always observed in smaller 1.5 T magnets).

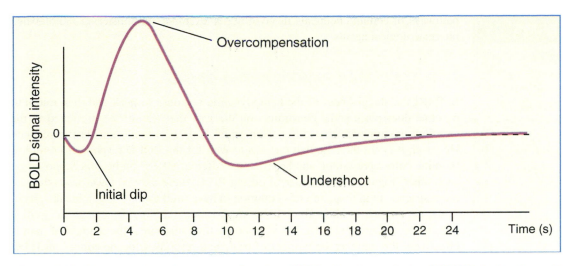

Figure 2.19 The hemodynamic response function (HRF) is the change in the fMRI BOLD signal over time as a result of an increase in neural activity in that region. It has a number of distinct phases that reflect the changing levels of deoxyhemoglobin in response to oxygen consumption and increased blood supply to that region.

2 *Overcompensation.* In response to the increased consumption of oxygen, the blood flow to the region increases. The increase in blood flow is initially greater than the increased consumption, which means that the BOLD signal increases substantially. This is the component that is normally measured in fMRI.

3 *Undershoot.* Finally, the blood flow and oxygen consumption dip before returning to their original levels. This may reflect a relaxation of the venous system, causing a temporary increase in deoxyhemoglobin again.

The hemodynamic signal changes are small – approximately 1–3% of the total signal with moderately sized magnets (1.5 T). The HRF is relatively stable across sessions with the same participant in the same region, but is more variable across different regions within the same individual and more variable between individuals (Aguirre, Zarahn, & D'Esposito, 1998). However, the shape of the HRF in different regions and different people can be estimated during the fMRI analysis, and this can be entered into the analysis.

The temporal resolution of fMRI is several seconds and related to the rather sluggish hemodynamic response. The temporal resolution of PET is related to the time it takes for radioactivity levels in the bloodstream to peak (around 30 s for a 'heavy water' isotope). Although the temporal resolution is better in fMRI than PET it is still slow compared to the speed at which cognitive processes take place. However, at this temporal resolution fMRI offers far more freedom in the choice of experimental designs (enabling event-related designs, described below). The spatial resolution of MRI is also better than PET, at around 1 mm depending on the size of the voxel.

Over the last 10 years, fMRI has largely taken over from the use of PET in functional imaging experiments. The key advantages of fMRI over PET are the better temporal and spatial resolution, the fact that event-related designs are possible, and the fact that it does not use radioactivity. PET, however, can be used to trace the

Event-related design
All trials are randomly (or semi-randomly) interspersed during stimulus presentation but are then regrouped at the analysis stage.

Block design
Trials that belong together are grouped together during stimulus presentation.

pathways of certain chemicals in vivo by, for example, administering radiolabeled pharmacological agents.

Constraints on experimental design

In fMRI the sluggishness of the hemodynamic response to peak and then return to baseline does place some constraints on the way that stimuli are presented in the scanning environment that differ from equivalent tasks done outside the scanner. However, it is not the case that one has to wait for the BOLD response to return to baseline before presenting another trial, as different HRFs can be superimposed on each other (Figure 2.20). In general during fMRI, there may be fewer trials that are more spaced out in time, and it is common to have 'null events' (e.g. a blank screen) in the experiment to allow the HRF to dip towards baseline. In standard cognitive psychology experiments (e.g. using response time measures) the amount of data is effectively the same as the number of trials and responses. In the equivalent fMRI experiment the amount of data is related to the number of *brain volumes* acquired rather than the number of trials or responses.

The way that different kinds of trials are grouped together can be broadly classified into either **event-related designs** or **block designs**. In a block design, trials that belong together are grouped together during stimulus presentation. In an event-related design, all trials are randomly (or semi-randomly) interspersed during stimulus presentation but are then treated separately at the analysis stage. Block designs are the only option in PET studies, but fMRI studies may use either. This is illustrated in Figure 2.21.

To give an example, imagine that one wanted to investigate the neural basis of gender judgments on names and faces (i.e. deciding whether a name or face was male or female). A block design could involve the presentation of a block of 20 faces, followed by a separate block of 20 names. An event-related design might involve presenting all 40 stimuli (20 faces, 20 names) randomly. There is no objectively 'right' or 'wrong' design to choose from but there are important differences. In this example, the event-related design would mean that the participant could not predict

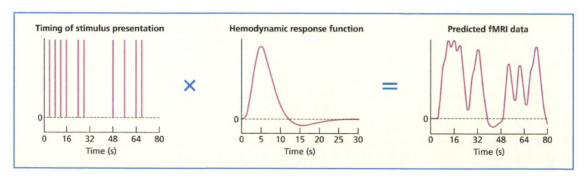

Figure 2.20 Unless the stimuli are presented far apart in time (e.g. every 16 s) the predicted change in BOLD response will not resemble a single HRF but will resemble many superimposed HRFs. Statistically, we are trying to find out which voxels in the brain show the predicted changes in the BOLD response over time, given the known design of the experiment and the estimated shape of the HRF. To achieve this there has to be sufficient variability in the predicted BOLD response (big peaks and troughs). Figure from http://imaging.mrc-cbu.cam. ac.uk/imaging/DesignEfficiency. Reproduced with permission from the author.

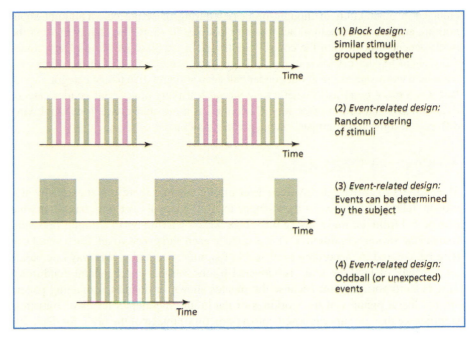

(1) *Block design:* Similar stimuli grouped together

Time

(2) *Event-related design:* Random ordering of stimuli

Time

(3) *Event-related design:* Events can be determined by the subject

Time

(4) *Event-related design:* Oddball (or unexpected) events

Time

Figure 2.21 Different conditions (either stimuli or tasks) can be presented in two ways in fMRI experiments: either in a block design or an event-related design.

whether he/she would see a face or name, whereas in the block design it would be predictable (and this predictability, or lack of, will have its own neural substrate). Block designs have more statistical power (i.e. are better able to detect small but significant effects) than truly random designs (e.g. Josephs & Henson, 1999). However, in many situations event-related designs are favored or are the only option. Sometimes there is no way of knowing in advance how trials should be blocked. For example, one could present a participant with a list of 40 words to remember during scanning. Maybe 10 words will be subsequently forgotten and 30 remembered, and one could then go back to the scanning data and see if there were any differences in brain activity at the initial stage that are predictive of later remembering versus forgetting (e.g. Wagner et al., 1998).

Finally, perhaps the most important aspect of experimental design in functional imaging research is in the selection of control conditions. For example, the simple experiment concerning gender categorization is inadequate in this regard. If one were to compare activity when performing gender judgments to faces relative to gender judgments to names (i.e. names is the control condition) then it would only reveal activity in regions implicated in face perception (and perception of images) and not in regions that may be involved in gender judgments. To test for the neural substrates of gender judgments to faces, better control conditions could be either presenting faces and requiring no judgment at all or presenting faces and requiring some other judgment (e.g. age). The 'no judgment at all' condition could be problematic if it is compared with a condition in which an actual response is made (e.g. a button press) because the comparison would reveal lots of regions involved in motor production, which probably has little to do with gender judgment. Passive viewing of stimuli, as a control condition, also has its own problems because participants may become engaged in other kinds of thought processes during these periods. Thinking about nothing at all is not a natural act for most people. In short, the claims that can be made depend crucially on the control condition. In one case we would learn something about gender categorization relative to face perception, and in the other case we would learn something about gender categorization relative to age categorization. But it is a moot point as to whether we would learn anything about gender categorization itself in a context-free way. Of course, as more studies are done employing different designs and contrasts, such a picture may indeed emerge.

Analysis of fMRI data

The analysis of fMRI data involves determining whether there is a statistically significant relationship between the changes in BOLD signal over time based on what is expected from the study design (i.e. given the known timings of stimulus presentations in the various conditions). This is done at each and every voxel. Each voxel can then be 'colored in' according to its level of significance and in this way one builds up a picture of how the brain is activated by the various experimental conditions. This is an important point because the images shown in functional imaging papers are not literal pictures of the workings of the brain: they depict levels of statistical significance that are superimposed onto a (structural) image of the brain for depictive purposes.

To get from the raw data (which consists of a set of two-dimensional images from one volume of the brain) to being able to perform a statistical analysis of the data involves a number of stages that are termed **pre-processing**. The following stages of pre-processing normally occur:

1 One needs to link the brain images (in each brain volume) to the timing of the stimulus presentation.
2 Correction for head movement. If a person moves their head, even a millimeter or so, then the regions of activity will also shift around, making them harder to detect. Fortunately, one can use the raw images themselves in order to infer how the person has moved over time and then use these so-called movement parameters to partial out these effects from the data.
3 **Stereotactic normalization** maps the voxels on an individual's brain onto the equivalent regions in a standard brain. This is needed in any analysis in which several different brains are entered, and is needed because of anatomical variability in brain size and shape. For example, the position of the sulci can vary by a centimeter or more in different people (Thompson, Schwartz, Lin, Khan, & Toga, 1996). The standard brain that is often used is based on an average of 305 brains provided by the Montreal Neurological Institute – the **MNI template** (Collins, Neelin, Peters, & Evans, 1994). Thus, a voxel in one person's brain can be compared to an equivalent voxel in another person's brain if they share the same coordinates in this standard space. Another template that is used, particularly for reporting data in journals, is based on the atlas of Talairach and Tournoux (1988). These are referred to as **Talairach coordinates** and are based on detailed data from a single postmortem brain.
4 **Smoothing** involves increasing the spatial extent of activity in voxels (see Figure 2.22). Specifically, a normal distribution is superimposed on each voxel such that most activity remains in the voxel itself but some spreads to neighbors. One reason why this is done is to cope with individual differences in the site of brain

KEY TERMS

Pre-processing
In functional imaging, the stages between data collection and data analysis

Stereotactic normalization
The procedure of mapping the voxels on an individual's brain onto the equivalent regions in a standard brain

MNI template
A 'standard' brain based on an average of 305 brains provided by the Montreal Neurological Institute

Talairach coordinates
The coordinate system of a 'standard' brain based on the atlas of Talairach and Tournoux (1988)

Smoothing
Increasing the spatial extent of activity in voxels

activity. Common regions of activity between participants are easier to find when the activity itself is more spatially diffuse. A second reason why smoothing is done is to improve the signal-to-noise ratio. If there are collections of voxels nearby with high activity then that activity will be enhanced by smoothing, whereas if there is a single active voxel with no active neighbors then its activity will be diminished. Smoothing is almost always done when comparing several brains in an analysis, but it need not be done if the analysis is done on one individual's brain. For example, there are various techniques that explore how the *pattern* of activity in a region relates to cognition (rather than the *amount* of activity in a region) in which smoothing is not used (e.g. Haynes & Rees, 2006).

After pre-processing each person, it is then possible to conduct statistical tests. The fact that the brain is divided into tens of thousands of voxels creates a statistical challenge in itself. If one were to use the standard threshold of statistical significance used in psychology ($P < .05$, meaning that there is a 5% probability of getting a significant result by chance) then there could be thousands of brain regions active by chance. However, the activities in different voxels are not strictly independent from each other (voxels tend to have similar activity to their neighbors). This has led to the development of mathematical models that choose a level of significance based on assumptions of spatial smoothness using so-called random field theory. This generates a statistical threshold called the **familywise error (FWE)**. This threshold is related to the number of statistical tests that are run. An alternative method is based on considering the actual number of positive results obtained, and is termed the **false discovery rate (FDR)**. A comparison that produces lots of positive results (i.e. lots of activity) would have a different proportion of expected false positives than a study with only a few positive results. The FDR method takes this into account, whereas the FWE method only takes into account the number of tests performed.

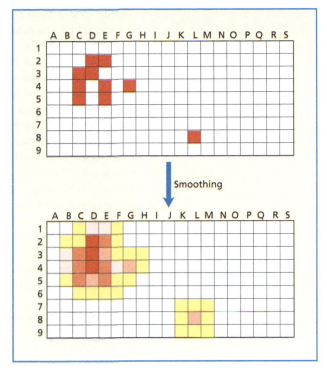

Figure 2.22 Smoothing spreads the activity across voxels – some voxels (e.g. D4) may be enhanced whereas others (e.g. L8) may be reduced.

Network analyses in fMRI

Most of the functional imaging studies described in this book could be labelled as studies of functional specialization. Functional specialization implies that a region responds to a limited range of stimuli/conditions and that this distinguishes it from the responsiveness of other regions. It is not strictly the same as localization, in that it is not necessary to assume that the region is solely responsible for performance on a given task or to assume that other regions may not also respond to the same stimuli/conditions (Phillips, Zeki, & Barlow, 1984). Functional integration, on the other hand, refers to the way in which different regions communicate with each other – i.e. *networks* of regions working together (e.g. Friston, 2002). It is to be noted that these kinds of connectivity analyses are based solely on the BOLD signal. As such, its

KEY TERMS

Familywise Error (FWE)
A statistical threshold used in functional imaging based on assumptions of spatial smoothness

False Discovery Rate (FDR)
A statistical threshold used in functional imaging based on random permutations of the data

KEY TERMS

Resting state paradigm
A technique for measuring functional connectivity in which correlations between several regions (networks) are assessed while the participant is not performing any task

Default mode network
A set of brain regions that is more hemodynamically active during rest than during tasks

relationship to structural connectivity (i.e. synapses, white matter tracts) within the network cannot be assumed.

The basic pre-processing stages used in network analyses are the same as those described previously, but the kinds of statistical techniques applied to the data are different. The other main difference is that whereas standard fMRI (i.e. based on functional specialization) requires the participant to be engaged in some kind of task, in network analyses a task is not always required during the data acquisition. These task-free scanning protocols are known as **resting state paradigms**.

Network analyses based on fMRI data acquired whilst the participant performs a task is based on correlations between the BOLD signal in different regions of the brain. So it might be the case that, say, when activity in the amygdala goes up during a task then activity in the visual cortex also goes up. This type of correlation is termed *functional connectivity*: one can't infer the causes of the correlation (i.e. whether the amygdala is driving the visual cortex or vice versa) or whether the connectivity is direct or indirect via some third region. Functional connectivity is testing for the presence of a correlation against a null hypothesis (i.e. assuming no correlation). An alternative approach involves setting up different models, such that the fit of one model is contrasted directly with another (rather than with the null hypothesis). So one could compare a model in which the amygdala directly connected with visual cortex against a model in which it did so via a parietal region. This is termed *effective connectivity* and it uses statistical methods such as Structural Equation Modelling (SEM).

One commonly used approach for network analyses does not use any task at all when the BOLD data is acquired. These are known as resting state paradigms. Participants are merely asked to lie back and rest. In the absence of a task, the fluctuations in brain activity are little more than noise. However, in brain regions that are functionally connected the noise levels tend to correlate together. This has enabled researchers to identify sets of networks in the brain, consisting of spatially separated regions, for which fluctuations in activity tend to be shared (Damoiseaux et al., 2006). For instance, one commonly studied network is called the **default mode network** of the brain and is implicated in internalized thoughts: for instance, it tends to be more active when *not* engaged in an experimental task (Raichle et al., 2001). Differences in the way that these networks operate and are constructed are found in various conditions such as schizophrenia and autism (Buckner, Andrews-Hanna, & Schacter, 2008).

Summary of fMRI

- *Advantages:* Very good spatial resolution; generally safe and non-invasive
- *Disadvantages:* Poor temporal resolution; not a direct measure of neural activity

SAFETY ISSUES IN FMRI RESEARCH

Before entering the scanner, all participants should be given a checklist that asks them about their current and past health. People with metal body parts, cochlear implants, embedded shrapnel, or pacemakers will not be allowed to take part in fMRI experiments. In larger magnets, eye make-up should not be worn (it can heat up, causing symptoms similar to sunburn) and women wearing contraceptive coils are not normally tested. Before going into the scanner both the researcher and participant should put to one side all metal objects such as keys, jewelry, and coins, as well as credit cards, which would be wiped by the magnet. Zips and metal buttons are generally okay, but metal spectacle frames should be avoided. It is important to check that participants do not suffer from claustrophobia because they will be in a confined space for some time. Participants wear ear protectors, given that the scanner noise is very loud. Larger magnets (> 3 T) can be associated with dizziness and nausea and participants need to enter the field gradually to prevent this. Participants have a rubber ball that can be squeezed to signal an alarm to the experimenter, who can terminate the experiment if necessary. The standard safety reference is by Shellock (2014) and updates can be found at www.magneticresonancesafetytesting.com.

LESION METHODS AND BRAIN STIMULATION

Whilst electrophysiological and hemodynamic methods *correlate* some measure of brain activity with some aspect of cognition or behavior, there are a set of methods that directly *manipulate* the functioning of the brain in order to measure a change in performance. These methods are therefore more appropriate for exploring the causal relationship between brain and behavior. Lesion methods impair the functioning of the brain and typically impair rather than enhance performance. Brain stimulation, on the other hand, may either impair or enhance performance. The brain may be stimulated chemically (through the administration of psychoactive substances) or electromagnetically. Two methods relating to the latter that are considered in this section are **transcranial magnetic stimulation (TMS)** and **transcranial direct current stimulation (tDCS)**.

Neuropsychology: the effects of naturally occurring lesions in humans

The basic premise behind the approach is that, by studying the abnormal, it is possible to gain insights into normal function. This is a form of 'reverse engineering', in which one attempts to infer the function of a region by observing what the rest of the cognitive system can and cannot do when that region is removed. The effects of brain damage on cognition are explored in the field of cognitive neuroscience known as **neuropsychology**. The experimenter obviously has no experimental control over the nature and location of brain damage, but can exert control in terms of the selection criteria for participation in studies. There are various causes of brain damage, including:

1 Strokes or CVAs (cerebrovascular accidents). These are disruptions to the blood supply to the brain due to blockages or ruptures, leading to neuronal death. The probability of suffering a stroke is age related (increasing sharply in the elderly).

2 Traumatic head injuries may be of either an 'open' nature (the skull is fractured) or 'closed' nature (as commonly found in road traffic accidents). They tend to be most prevalent in younger males. Brain damage typically results from compression and bruising.

3 Tumors are caused when new cells are formed in a poorly regulated manner and in the brain they are formed from the supporting cells, such as the meninges and glia (termed meningioma and glioma, respectively). The extra material puts pressure on neurons, resulting in possible cell death.

4 Neurosurgery tends not to be favored over pharmacological treatments but is sometimes used (e.g. to remove the focus of an epileptic seizure) and results in localized brain damage.

5 Viral infections may target specific cells in the brain, possibly leading to cell death.

6 Neurodegenerative disorders such as dementia of Alzheimer's type (DAT) are becoming increasingly more prevalent as the population ages.

The logic of patient-based neuropsychology rests, to some degree, on the notion of functional specialization (i.e. the notion that different regions of the brain perform different kinds of computation or process different kinds of information). In this respect, it is similar to other methods such as functional imaging. In other respects it differs in important ways from the logic of functional imaging. The data in neuropsychology are behavioral (typically error rates, or degree of impairment) and the independent variable (i.e. conditions manipulated) is the lesion itself. In functional imaging, the data are localized differences in brain activity and the independent variable is behavior itself (i.e. the instructions given to the participant). This is illustrated in Figure 2.23. These differences are important because it suggests that one method cannot simply substitute for the other in a simple way. It has also been argued that neuropsychology, but not functional imaging, is able to make claims that a region is *necessary* for a task to be performed successfully (Kosslyn, 1999). This is because there is an assumed causal connection between the lesion and the impaired behavior. In functional imaging, activity in a particular brain region can depend on many factors, such as how well the control condition is matched (e.g. for difficulty) and the particular strategy that a person uses (the strategy itself would have its own neural activity but may be unrelated to how well the task is performed).

The notion of functional specialization leads to the prediction that localized brain damage will impair some but not all cognitive functions (unless, of course, the region is crucial to all aspects of cognition). These are termed **dissociations**. For example,

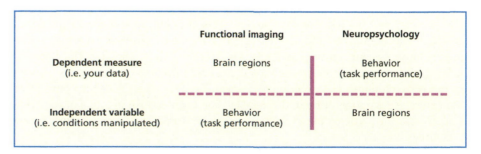

Figure 2.23 Functional brain imaging and patient-based neuropsychology (or TMS) are logically different types of methodology. It is unlikely that one will supplant the other.

patients have been reported who – after sustaining damage to the amygdala – have difficulties in recognizing facial expressions of fear (Adolphs, Tranel, Damasio, & Damasio, 1994; Calder et al., 1996). This could be taken to imply that there is a specialized region for processing fear. However, it is not the only conclusion that could be reached. It could be the case that facial expressions of fear are more difficult to recognize than other facial expressions. Evidence against this comes from **double dissociation**, in which recognition of fear expressions is preserved but other emotions (such as disgust) are hard to recognize (Calder, Keane, Manes, Antoun, & Young, 2000). This is shown in Figure 2.24. However, even with a double dissociation one cannot necessarily conclude that there is a specialized region for recognizing fear expressions. It could be, for example, that recognizing fear in faces requires attention to particular regions of a face and that disrupting this face-attention mechanism will disrupt fear recognition (Adolphs et al., 2005). Nor can one conclude that this brain region is not involved in any other cognitive function (one would need to test in other domains to establish this). It would be wrong to take away from this argument that nothing can be concluded from double dissociations from patients. We can draw some conclusions using these methods (e.g. that fear recognition and disgust recognition use partially distinct neural resources). However, the conclusions we draw do depend on the theories we are aiming to test (i.e. what those neural resources are believed to be computing).

One interesting aspect of patient-based neuropsychology is that it can still be possible to draw inferences in the absence of knowing where the lesion is. From the double dissociation mentioned previously we can conclude that fear recognition and disgust recognition are partially separable without knowing *where* in the brain they may be. This does not mean that neuropsychology cannot be used to address the 'where' question, but it is not the only question that can be asked. However, to find out where the crucial region (or regions) is typically involves either patients with very focal brain lesions or large groups of patients (with non-focal lesions) who can be assessed on a particular task and their group performance linked (e.g. on a voxel-by-voxel basis) to their known lesions (e.g. Rorden & Karnath, 2004). In this way, patient-based neuropsychology benefits from the advances in imaging science in the same way as fMRI.

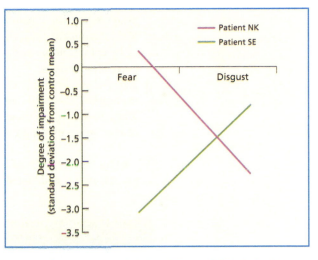

Finally, the logic of patient-based neuropsychology may not hold for brain damage sustained in childhood relative to adulthood. Lesions in the same region may have different consequences in childhood because other regions may be better able to take on lost functions (e.g. Ballantyne, Spilkin, Hesselink, & Trauner, 2008). However, it remains unclear whether moving a function to a different location in the brain (e.g. the intact hemisphere) necessarily results in a change in *how* the function is performed. Is language qualitatively the same if it shifts into the right hemisphere as a result of a childhood left lesion?

Figure 2.24 A double dissociation: Patient NK is impaired at recognizing disgust (but not fear) whereas patient SE is impaired at recognizing fear (but not disgust). The y-axis shows the degree of impairment relative to controls measured in terms of standard deviations from the control mean (0 = no impairment, negative = impaired). Data from Calder et al. (1996, 2000).

Summary of human neuropsychology

- *Advantages:* Lesion methods enable inferences of brain–behavior causality.
- *Disadvantages:* There is no experimental control over lesions.

Experimentally induced lesions in animal models

Although lesion methods in humans rely on naturally occurring lesions, it is possible – surgically – to carry out far more selective lesions on other animals. Unlike human lesions, each animal can serve as its own control by comparing performance before and after the lesion. It is also common to have control groups of animals that have undergone surgery but received no lesion, or a control group with a lesion in an unrelated area. There are various methods for producing experimental lesions in animals (Murray & Baxter, 2006):

1 *Aspiration.* The earliest methods of lesioning involved aspirating brain regions using a suction device and applying a strong current at the end of an electrode tip to seal the wound. These methods could potentially damage both gray matter and the underlying white matter that carries information to distant regions.
2 *Transection.* This involves the cutting of discrete white matter bundles, such as the corpus callosum (separating the hemispheres) or the fornix (carrying information from the hippocampus).
3 *Neurochemical lesions.* Certain toxins are taken up by selective neurotransmitter systems (e.g. for dopamine or serotonin) and once inside the cell they create chemical reactions that kill the cell. A more recent approach involves toxins that bind to receptors on the surface of cells, allowing for even more specific targeting of particular neurons.
4 *Reversible 'lesions'.* Pharmacological manipulations can sometimes produce reversible functional 'lesions'. For example, scopolamine produces a temporary amnesia during the time in which the drug is active. Cooling of parts of the brain also temporarily suppresses neural activity.

Whilst the vast majority of behavioral neuroscience research is conducted on rodents, some research is conducted on non-human primates. In many countries, including in

Figure 2.25 A family of macaque monkeys

the EU, neuropsychological studies of great apes (e.g. chimpanzees) are not permitted. More distant human relatives used in research include three species of macaque monkeys (rhesus monkey, cynomolgus monkey, and Japanese macaque) and one species of New World primate (the common marmoset).

Summary of animal neuropsychology

- *Advantages:* Lesion methods enable inferences of brain–behavior causality; precision is possible through anatomically or chemically selected lesions.
- *Disadvantages:* The range of behaviors studied is limited relative to humans; there are concerns over animal welfare.

Transcranial magnetic stimulation (TMS)

Attempts to stimulate the brain electrically and magnetically have a long history. Electrical currents are strongly reduced by the scalp and skull and are therefore more suitable as an invasive technique on people undergoing surgery. In contrast, magnetic fields do not show this attenuation by the skull. However, the limiting factor in developing this method has been the technical challenge of producing large magnetic fields, associated with rapidly changing currents, using a reasonably small stimulator (for a historical overview, see Walsh & Cowey, 1998). It was not until 1985 that adequate technology was developed to magnetically stimulate focal regions of the brain (Barker, Jalinous, & Freeston, 1985). Since then, the number of publications using this methodology has increased rapidly.

Typically, the effects of TMS are small, such that they alter reaction time profiles rather than elicit an overt behavior. But there are instances of the latter. If the coil is

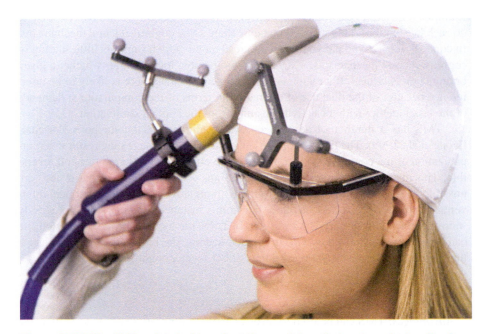

Figure 2.26 The TMS coil is held against the participant's head, and a localized magnetic field is generated during performance of the task. University of Durham/Simon Fraser/Science Photo Library.

placed over the right visual cortex, then the subject may report visual sensations or 'phosphenes' on the left side (given that the right visual cortex represents the left side of space). Even more specific examples have been documented. Stewart, Battelli, Walsh, and Cowey (1999) stimulated a part of the visual cortex dedicated to motion perception (area V5) and reported that these particular phosphenes tended to move. Stimulation in other parts of the visual cortex produces static phosphenes.

TMS works by virtue of the principle of electromagnetic induction that was first discovered by Michael Faraday. A change in electrical current in a wire (the stimulating coil) generates a magnetic field. The greater the rate of change in electrical current, the greater the magnetic field. The magnetic field can then induce a secondary electrical current to flow in another wire placed nearby. In the case of TMS the secondary electrical current is induced, not in a metal wire, but in the neurons below the stimulation site. The induced electrical current in the neurons is caused by making them 'fire' (i.e. generate action potentials) in the same way as they would when responding to stimuli in the environment. The use of the term 'magnetic' is something of a misnomer because the magnetic field acts as a bridge between an electrical current in the stimulating coil and the current induced in the brain. Pascual-Leone, Bartres-Faz, and Keenan (1999) suggest that 'electrodeless, non-invasive electric stimulation' may be more accurate.

TMS causes neurons underneath the stimulation site to be activated. If these neurons are involved in performing a critical cognitive function, then stimulating them artificially will disrupt that function. Although the TMS pulse itself is very brief (less than 1 ms), the effects on the cortex may last for several tens of milliseconds. As such, the effects of a single TMS pulse are quickly reversed. Although this process is described as a 'virtual lesion' or a 'reversible lesion', a more accurate description would be in terms of interference. In the cognitive psychology literature, dual-task interference paradigms are used to determine whether two tasks share cognitive resources. For example, it is hard to pat your head whilst rubbing your tummy (although it is easy to do each in isolation). This suggests that they share some cognitive/neural mechanisms. In contrast, it is easy to pat your head and read aloud, which suggests little sharing of cognitive/neural mechanisms. TMS uses a comparable logic to infer whether a given brain region is critical. If a region is critical for a task, then there is likely to be interference because of the dual use of the region in terms of the computational demands of the task together with the activity ensuing from the applied stimulation.

TMS has a number of advantages over traditional lesion methods (Pascual-Leone et al., 1999). The first advantage is that real brain damage may result in a reorganization of the cognitive system, whereas the effects of TMS are brief and reversible. This also means that within-subject designs (i.e. with and without lesion) are possible in TMS that are very rarely found with organic lesions (neurosurgical interventions are an interesting exception, but in this instance the brains are not strictly 'normal' given that surgery is warranted). In TMS, the location of the stimulated site can be removed or moved at will. In organic lesions, the brain injury may be larger than the area under investigation and may affect several cognitive processes.

Summary of TMS

- *Advantages*: TMS can be used to investigate the timing of cognition as well as the location of cognition; a 'virtual lesion' can be moved within the same participant.
- *Disadvantages*: TMS can only stimulate certain regions; it is hard to predict how TMS affects the functioning of distant sites.

SAFETY ISSUES IN TMS RESEARCH

Whereas single-pulse TMS is generally considered to be safe, repetitive-pulse TMS carries a very small risk of inducing a seizure (Wassermann, Cohen, Flitman, Chen, & Hallett, 1996). Given this risk, participants with epilepsy or a familial history of epilepsy are normally excluded. Participants with pacemakers and medical implants should also be excluded. Credit cards, computer discs, and computers should be kept at least one meter away from the coil. The number of pulses that can be delivered to a participant in a given testing session has been established, by consensus (Rossi, Hallett, Rossini, Pascual-Leone, & TMS Safety Consensus Group, 2009). The intensity of the pulses that can be delivered is normally specified with respect to the 'motor threshold' – the intensity of the pulse, delivered over the motor cortex, that produces a just noticeable motor response (for a discussion of problems with this, see Robertson, Theoret, & Pascual-Leone, 2003).

During the experiment, some participants might experience minor discomfort due to the sound of the pulses and facial twitches. Although each TMS pulse is loud (~100 dB), the duration of each pulse is brief (1 ms). Nonetheless, the ears should be protected with earplugs or headphones. When the coil is in certain positions, the facial nerves (as well as the brain) may be stimulated, resulting in involuntary twitches (e.g. blinking, jaw clamping). Participants should be warned of this and told they can exercise their right to withdraw from the study if it causes too much discomfort.

Transcranial direct current stimulation (tDCS)

The method of transcranial direct current stimulation (tDCS) uses a very weak electric current applied between two electrodes. Direct current involves the flow of electric charge from a positive site (an anode) to a negative site (a cathode) – see Figure 2.27. In tDCS, a stimulating pad (either anodal or cathodal) is placed over the region of interest and the other pad is placed in a site of no interest (sometimes on the front of the forehead, or sometimes on a distant site such as the shoulders). After a period of stimulation (e.g. 10 minutes) a cognitive task is performed and this can be compared with sham stimulation, or anodal and cathodal stimulation can be directly contrasted. **Cathodal tDCS** stimulation tends to disrupt performance (i.e. it is conceptually equivalent to a virtual lesion approach) whereas **anodal tDCS** stimulation tends to enhance performance (Nitsche et al., 2008).

To give an example from the social neuroscience realm, Santiesteban, Banissy, Catmur and Bird (2012) explored the automatic tendency to imitate. Participants saw a hand moving in which either the index finger or middle finger moved. At the same time they were instructed to move their own index or middle finger when the number '1' or '2' was displayed. When their own finger movement is incongruent with the observed finger movement then they are slower to initiate the movement, and this slowing is assumed to reflect an automatic tendency to imitate. However, anodal tDCS over the right temporo-parietal junction, rTPJ (a region implicated in social

KEY TERMS

Cathodal tDCS
Decreases cortical excitability and decreases performance

Anodal tDCS
Increases cortical excitability and increases performance

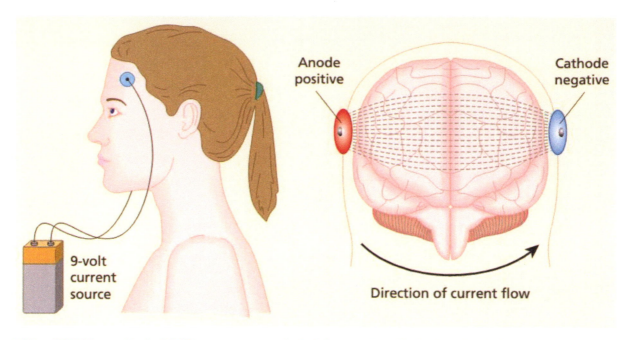

Figure 2.27 The method of tDCS uses a very weak electric current applied using stimulating pads attached to the scalp. Direct current involves the flow of electric charge from a positive site (an anode) to a negative site (a cathode). Adapted from George and Aston Jones.

perception and understanding others) reduces this interference: they perform better by being less susceptible to automatic imitation.

Stagg and Nitsche (2011) provide a summary of the likely neurophysiological mechanisms. It is important to consider the immediate effects of direct current stimulation and the aftereffects separately. Animal models of direct current stimulation followed by single-cell recordings have shown that anodal stimulation increases the spontaneous firing rate of neurons whereas cathodal stimulation reduces the firing rate. The *immediate* effects of stimulation are believed to occur on the resting membrane potential rather than modulation at the synapse. However, the *aftereffects* of stimulation are likely to occur due to changes in synaptic plasticity influencing learning and perhaps affecting different neurotransmitter systems. Anodal stimulation affects the GABA system (this neurotransmitter has inhibitory effects) whereas cathodal stimulation affects the glutamate system (this neurotransmitter has excitatory effects).

The current safety guidelines recommend upper limits on the size of the current and the surface area of the stimulating electrodes (Nitsche et al., 2003). If the current is concentrated on a small electrode, then it can cause skin irritation. However, unlike TMS, participants often cannot tell whether the machine is switched on or used as sham (there is no sound or twitching). As such there is very little discomfort.

Summary of tDCS

- *Advantages*: can enhance as well as impair performance
- *Disadvantages*: low spatial resolution and poor temporal resolution

A GENETICS PRIMER FOR THE UNINITIATED

DNA is present in every cell of the body and consists of two twisted strands (the infamous 'double helix') made up of four different kinds of chemical bases (referred to as A, T, G, and C) that are arranged in very long sequences. Parts of this sequence act as a 'code' for making proteins (via another chemical called mRNA). Each of these sections of code is referred to as a **gene** – see Figure 2.28. Proteins serve a vast number of functions around the body, but within the nervous system they can act – amongst other things – as neurotransmitters, as hormones, as enzymes (i.e. proteins that act on other chemicals, such as breaking down neurotransmitters), as ion channels in electrical signaling, and as receptors in chemical signaling. Proteins also act as transcription factors: i.e. they bind to parts of the genome in order to regulate the expression of genes. Most of the genome, however, contains non-gene segments. The spacing of these segments differs from person to person although it is unknown whether these spacing differences contribute to observable individual differences.

KEY TERM

Gene
A basic physical unit of heredity consisting of a distinct sequence of DNA that can be translated into proteins

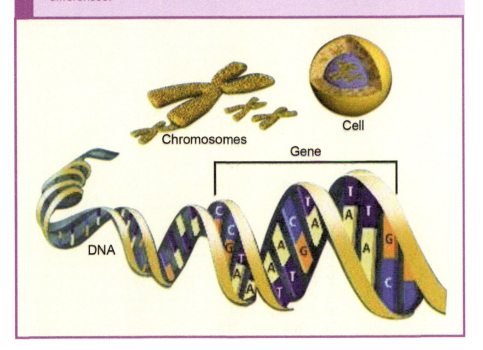

Figure 2.28 The human genome is made up of the molecule DNA and is packaged in to 23 pairs of chromosomes. Some sections of the genome can be translated into other molecules (ultimately creating proteins) and these sections are referred to as genes. Although the genetic sequence of an individual is fixed at conception, molecules may bind to the DNA strand (epigenetic markers) that affect the expression of the gene as a consequence of the organism's interaction with its environment.

KEY TERMS

Chromosome
A packaged and organized structure containing most of the DNA of an organism

Allele
A naturally occurring variant of a gene

Genotype
The particular set of alleles, relating to a given trait, that an organism possesses

Phenotype
The particular trait coded by a given genotype (e.g. blue eyes)

The human genetic code is organized on to 23 pairs of **chromosomes**, making a total of 46 chromosomes. One of the chromosomes of each pair comes from the mother and one from the father. In each individual there are two copies of each gene normally present, one on each chromosome. However, genes may exist in different forms, termed **alleles**. The different alleles represent changes (or mutations) in the sequence of the gene that is propagated over many generations unless natural selection intervenes (e.g. a mutation is lethal). Many different allelic forms are common and benign but they account for the individual differences that are found between humans as well as differences between species. For instance, some genes confer greater sensitivity to social events (e.g. Way & Lieberman, 2010). A different allele may mean that the end product encoded by the gene (such as enzymes) works less efficiently, more efficiently, or not at all. Alternatively, it may mean that the gene works in an entirely novel way. The particular set of alleles that a person possesses is referred to as a **genotype** and the trait that these genes confer (whether psychological or physical) is referred to as the **phenotype**. Most behavioral traits will be an outcome of the concerted action of many genes. Even though a given gene may exist in only a small number of discrete allelic types, when many such genetic variants are combined together they may produce an outcome that is continuously distributed – such as the normal distribution found for height or IQ. Disorders such as autism and schizophrenia also appear to be polygenic in nature (Tager-Flusberg, 2003).

Although the *structure* of the genetic code for each person is fixed at conception, the *functioning* (or 'expression') of the genetic code is highly

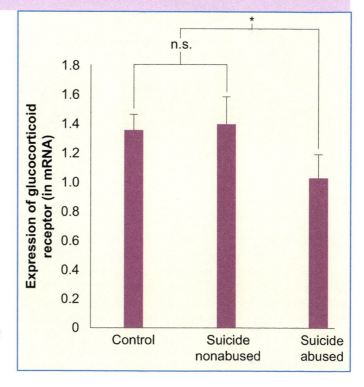

Figure 2.29 Abuse in early childhood, rather than suicide-related mental illness, is linked to lower expression of the stress-related glucocorticoid receptor gene in the human hippocampus. This is evidence of an influence of the social environment on the functioning of the genome, i.e. epigenetics. From McGowan et al. (2009).

dynamic. Different genes become active or inactive at different times of life (e.g. causing one to go through puberty, or our hair to turn gray). The expression of the genetic code is also influenced by the environment – a phenomenon termed *epigenetics*. In epigenetic marking the genes are not changed but get tagged with a chemical marker that dampens (e.g. a methyl group) or accentuates (e.g. an acetyl group) their expression. Perhaps the best-known example of this in the social neuroscience domain is the epigenetic effect of neglect in early development. In rats, mothers vary in the amount of care (licking and grooming) given to their pups. Low care is linked to an increased, and lasting, stress response in the pups to both neutral and stressful events (e.g. Meaney, 2001). The effect has been related to an epigenetic reduced expression of a gene coding for a glucocorticoid receptor, leading to an increased stress response (Weaver et al., 2004). In humans, McGowan et al. (2009) examined the post-mortem brains of people who had committed suicide versus control brains. They found epigenetic influences on the gene coding for the glucocorticoid receptor in suicide victims who had experienced early neglect/abuse, but this was not found on the other suicide victims or the controls – see Figure 2.29. This confirms that the epigenetic effects were linked to early abuse rather than other factors contributing to suicide.

MEASURING BIOLOGICAL INDIVIDUAL DIFFERENCES

Individual differences in social functioning will reflect a complex interaction between biological differences (e.g. genetic differences, differences in brain structure, hormonal differences) and differences in the environment (e.g. exposure to trauma, different cultures). As discussed in Chapter 1, the brain is the system in which all these differences are played out and such influences are likely to be bi-directional. The environment influences the operation of genes within a lifetime (epigenetics) as well as the selection of certain genes across generations, and the presence of certain genetic variants may predispose certain environments to predominate. The present chapter has already considered how individual differences in the social realms can be measured (performance measures, questionnaires, observational techniques). The final section of this chapter will consider how various biological individual differences can be captured experimentally.

Genetic differences

The genetic differences that are measured in humans tend to be structural differences in the genetic code itself (e.g. variations in the DNA sequence) rather than functional differences in the way genes are expressed (epigenetics). At present, epigenetic markers have to be explored in vitro. A lack of availability of human brain tissue, and the difficulty in linking it to cognitive or behavioral measures, has resulted in limited progress in human epigenetics (Miller, 2010). It is also unclear how epigenetic

KEY TERM

Genotype-first
An analysis
approach in which
different genotypes
(e.g. different alleles)
are used to explore for
phenotypic variation

differences that can easily be measured in the DNA of other tissues (e.g. blood cells) would relate to those of neurons that are not so easily accessed. Thus, epigenetic differences tend to be studied in animal models (e.g. Meaney, 2001).

Taking cheek swabs remains the most common and most simple way of extracting cells for human genetic analysis. Cells on the inside of the cheek are loose and can be removed by light abrasion with a swab. Recall that the genetic code is the same across all the cells in the body so it makes no difference whether it is extracted from a cheek cell or a neuron itself. Sterile kits are cheaply available and swabs can be obtained remotely with participants returning their DNA sample via post. Of course the types of analyses that can then be conducted on the DNA sample are rarely simple. The two approaches that one could adopt for analysis of genetic differences are called genotype-first and phenotype-first.

An example of a **genotype-first** approach would be to take a single gene that is known to exist in multiple variants (polymorphisms) and that may be relevant to your given research question (i.e. based on previous research). In the social neuroscience literature, the study of Way, Taylor and Eisenberger (2009) examined a variation in the gene encoding the mu-opioid receptor that has been implicated, by previous research, in individual differences in the susceptibility to physical pain. These researchers then examined whether the same polymorphism conveyed susceptibility to 'social pain' (being rejected by others) and found that it did. The advantage of this genotype-first approach is that the genetic analysis is limited to one specific gene and so is relatively straightforward to conduct (commercial companies can provide this service to the neuroscience community). It also avoids the problem of multiple comparisons when testing multiple genes (the problem of Type I errors discussed previously). In conducting this kind of research it is important to obtain accurate information about the

Figure 2.30 DNA can be obtained from cheek cells using commercially available swab kits. The sequence of DNA in an individual is the same in all the cells of their body – so cheek cells are as useful as neurons for this purpose.

ethnicity of your genetic sample (or have a homogeneous sample) as the prevalence of many polymorphisms can vary considerably depending on race.

An example of a *phenotype-first* approach would be to take a given trait that is known to vary in the population (e.g. empathy) or to take a clinically defined condition (e.g. autism spectrum disorder) and determine which portions of the genome contribute most to variations in that trait (as a continuous measure) or the presence/absence of a condition (as a binary measure). One method that uses this approach is termed a *genome-wide association study (GWAS)* and tends to involve many thousands of participants. This method is based on the fact that there are many small variations in the genome across individuals termed single nucleotide polymorphisms (SNPs, pronounced 'snips'). These SNPs are not of interest in their own right but provide useful clues as to which parts of the genome contain a 'hot spot' (i.e. regions where individuals with the same phenotype have genetic similarities that deviate from chance). To give an example of this approach, Ma et al. (2009) studied 487 Caucasian families (1537 individuals) with autism and examined their genomes using over one million SNPs. Figure 2.31 shows the way that this data can be visualized: the genome is plotted on the x-axis and (log) probability values on the y-axis. They found 96 genetic locations (SNPs) have $p < 10^{-4}$ (.0001) and 6 that have $p < 10^{-5}$ (.00001). This suggests that genes close to these locations are commonly implicated in the development of autism.

KEY TERMS

Phenotype-first
An analysis approach in which different phenotypes are used to explore genetic differences

Genome-Wide Association Study (GWAS)
A phenotype-first approach in which the presence/absence, or continuous variation, in a trait is linked to variations at many different sites in the genetic code

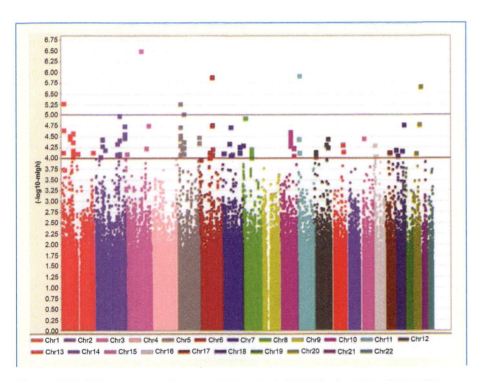

Figure 2.31 This genome-wide association study (GWAS) of autism divides the genome into different regions (Chr 1 to 22 refer to the different chromosomes) and measures the statistical probability (plotted on the y-axis) that variability in the presence of the trait is linked to variability in that region of the genome. The values of 4.00 and 5.00 refer to $p < 10^{-4}$ (.0001) and $p < 10^{-5}$ (.00001) respectively.

Summary of genetic measures

- *Advantages*: can be both genotype-first and phenotype-first; once genes have been discovered it would enable social and cognitive levels of explanation to be integrated with cellular and molecular levels
- *Disadvantages*: unlikely to be simple one-to-one relationship between genes and behavior; epigenetic differences not straightforward to study in humans

Neuroendocrine differences

Hormones are messenger molecules that are released into the bloodstream by specialized neurons in the brain or by certain glands (for a good coverage see Neave, 2008). This is illustrated in Figure 2.32. The message that they convey depends on the receptors they bind to. This ultimately means that the same hormone may carry multiple messages: a one-to-many relationship. Thus, the same hormone that affects cognition when it binds to receptors in the brain may also affect water retention when it binds to receptors in the kidneys.

Hormones can have two kinds of effects on brain and behavior: long-lasting organizational effects and shorter-lasting activational effects. Organizational effects include the effects of sex hormones at certain key stages of development (during puberty, and during prenatal development). These affect both the development of the body and of the brain circuitry. Neuroendocrine differences relating to organizational effects are hard to measure experimentally, although there are some interesting proxy markers of this process outside of the nervous system. For instance, the ratio of the lengths of the second (index) finger to the fourth (ring) finger, sometimes termed 2D:4D, is taken as a proxy for levels of fetal testosterone (Lutchmaya, Baron-Cohen, Raggatt, Knickmeyer, & Manning, 2004) and has been linked to adult traits such as

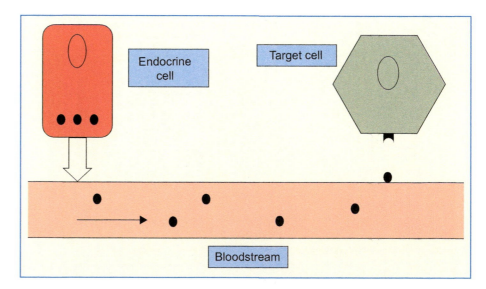

Figure 2.32 Hormones are chemical messengers that are carried in the bloodstream. In terms of its interface with the nervous system, some neurons function as endocrine cells (i.e. manufacturing and releasing hormones) and some neurons function as target cells (hormones may bind to receptors on the surface of the neuron or enter the neuron). From Neave (2008)

physical aggression (Bailey & Hurd, 2005). Similarly, the facial width in males has been associated with testosterone levels (Lefevre, Lewis, Perrett, & Penke, 2013) and to dominance-related social behaviors (Carre & McCormick, 2008).

Social neuroscientists tend to be primarily concerned with the shorter-term activational effects of hormones that are easier to study experimentally (e.g. through saliva or blood samples). Common hormones that are studied by social neuroscientists include:

- *Cortisol*. This hormone is released by the adrenal glands (next to the kidneys) but is initiated by neural circuits in the hypothalamus, which is responsible for the release of a hormone termed CRH (corticotrophin releasing hormone). Cortisol is linked to the stress response, and can be measured from a salivary sample.
- *Testosterone*. This hormone is synthesized in the testes (of males) and the adrenal glands (of males and females). Testosterone levels in blood plasma are 10–15 times higher in adult males. The activational effects of testosterone are linked to competitive behavior. Testosterone can be measured from a salivary sample.
- *Oxytocin and vasopressin*. These are peptide hormones with a similar chemical structure linked to affiliative behavior and reproduction. They are manufactured in certain nuclei in the hypothalamus and bind to specialized receptors on cell membranes (of both neurons and other kinds of cell in the body). It cannot be extracted from the saliva (the molecule is too large to pass through the cell membrane) and is normally extracted via a blood sample.

Commercially manufactured assays can be purchased for determining the concentration of a given hormone in a sample. These either contain radioactive chemicals or (safer to use) fluorescent/luminescent markers (see Schultheiss & Stanton, 2009, for a detailed overview of the methods). Note that the saliva samples used for hormone assays are not related to the check swabs used for DNA assays, despite the superficial similarity.

One experimental consideration for researchers working in this area is that the level of hormones such as cortisol, testosterone, and estrogen follow a circadian rhythm (peaking in the morning and declining through the day). Hence it is important to control the time of day of testing or, at the very least, to document it so that such effects can be reported or statistically controlled for. The menstrual cycle and the oral contraceptive pill are also significant influences on certain hormones (e.g. progesterone and estradiol).

It is also possible to measure the effect of social environment on neuroendocrine differences as well as vice versa. Thus, one can measure the *change* in hormone after a task/event relative to before it (as opposed to simply correlating with the resting level of hormone). For example, losing a competition can lower testosterone levels (Elias, 1981) and watching a romantic movie can increase progesterone levels (Schultheiss, Wirth, & Stanton, 2004). Similarly, one can administer hormones to participants (e.g. nasally or intravenously), relative to a placebo, and examine their effects on social behavior or on brain circuits using other measures (e.g. fMRI; Hermans et al., 2010). As such, one can use neuroendocrine differences to examine causality and not just as a correlational approach.

Summary of neuroendocrine measures

- *Advantages*: correlational approach that can also be adapted to examine causation (via administration of hormones); effects of hormones relatively well established from animal models

- *Disadvantages*: hormone differences in blood or saliva may not reflect those in the brain; hormone levels are naturally very variable (circadian, menstrual cycles)

Differences in brain structure measured with MRI

Small-scale individual differences (at the millimeter level) in the organization and concentration of white matter and gray matter can now be analyzed noninvasively using MRI. This is providing important clues about how individual differences in brain structure are linked to individual differences in social cognition. Two important methods are **voxel-based morphometry, or VBM**, and **diffusion tensor imaging, or DTI**.

Voxel-based morphometry (VBM) capitalizes on the ability of structural MRI to detect differences between gray matter and white matter (Ashburner & Friston, 2000). VBM divides the brain into tens of thousands of small regions, several cubic millimeters in size (called voxels) and the concentration of white/gray matter in each voxel is estimated (based on the lightness of the voxel – white matter is whiter). It is then possible to use this measure to compare across individuals by asking questions such as these: if a new skill is learned, such as a second language, will gray matter density increase in some brain regions? Will it decrease in other regions? How does a particular genetic variant affect brain development? Which brain regions are larger, or smaller, in people with good social skills versus those who are less socially competent? Kanai and Rees (2011) provide a review of this method in relation to cognitive differences. In the domain of social neuroscience, one may take measures of social functioning, such as empathy questionnaires (Banissy, Kanai, Walsh, & Rees, 2012) or number of friendships (Kanai, Bahrami, Roylance, & Rees, 2012), and correlate these differences with localized variation in gray matter. As with imaging in general, this is a correlational approach and it can be hard to infer direction of causation. Thus, the positive correlation between the size of the amygdala and the number of Facebook friends one has (Kanai et al., 2012) could either be due to variations in brain structure influencing ones social behavior, or variations in social behavior leading to different brain development.

Diffusion tensor imaging (DTI) is different from VBM in that it measures the white matter connectivity between regions (Le Bihan et al., 2001). (Note: VBM measures the *amount* of white matter without any consideration of how it is connected.) It is able to do this because water molecules trapped in axons tend to diffuse in some directions but not others. Specifically, a water molecule is free to travel down the length of the axon but is prevented from traveling out of the axon by the fatty membrane. When many such axons are arranged together it is possible to quantify this effect with MRI (using a measure called **fractional anisotropy**). As an example of a social neuroscience study using DTI, Bjornebekk et al. (2012) examined the relationship between a questionnaire measure of social reward processing and white matter integrity. People scoring high on this measure tend to be eager to help and please others, persistent, sympathetic, sentimental, sensitive to praise, and social cues; whereas individuals scoring low are described as socially detached, practical, tough minded, emotionally cool, and independent. The authors reported a negative relationship between social reward processing and fractional anisotropy (degree of white matter coherence) in frontal regions and around the striatum (i.e. high social reward processing linked to less white matter coherence). The negative direction of the correlation is, perhaps, surprising. However, note that DTI is a measure of *organization* (rather than volume) so fibers that project in many directions (e.g. projecting to diffuse targets) would have lower measures.

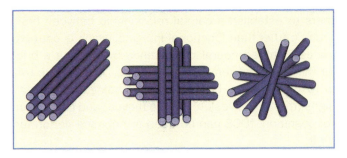

Figure 2.33 Diffusion tensor imaging (DTI) measures the degree of organization of white matter tracts using a measure called fractional anisotropy (FA). The image on the left has an FA close to 1, the image on the right has an FA close to 0 and the image in the middle is intermediate in FA.

Summary of VBM and DTI

- *Advantages*: enables assessment of whole brain; MRI widely available
- *Disadvantages*: primarily correlational in nature (direction of cause and effect hard to ascertain)

SUMMARY AND KEY POINTS OF THE CHAPTER

- The neuroscientific methods used by social neuroscience (e.g. fMRI) do not preclude the use of more traditional psychological methods (e.g. response time, questionnaires) and a combination of the two approaches is commonplace (e.g. linking questionnaire data to neural function).
- Methods relying on activity of the peripheral nervous system include the skin conductance response (SCR), electromyography (EMG), and pupilometry. These measures are often used as 'unconscious' measures of behavior or affective processes.
- Electrophysiological methods offer excellent temporal resolution. Single-cell recordings additionally offer excellent spatial resolution and are important for understanding how neurons code information, but they have the disadvantage of being usually limited to non-human animals. EEG methods measure electrical activity at the scalp and have an advantage over response time measures in that changes in neural activity can be detected without waiting for (or requiring) a response.
- The functional imaging methods of PET and fMRI are not direct measures of neural activity but are hemodynamic methods, dependent on changes in blood flow and blood oxygenation resulting from neural activity. They offer good spatial resolution but poor temporal resolution.
- Neuropsychological methods in humans rely on naturally occurring lesions to the brain that may selectively impair certain aspects of cognition; in non-human animals, more selective lesions are possible. They enable

researchers to establish a causal relationship between brain structure and function (rather than functional imaging, which is correlational).

- TMS creates localized neural interference by creating a brief magnetic field over the skull that temporarily disrupts performance (measured by errors or response time). Unlike neuropsychology, it enables within-subject designs (the 'virtual lesion' can be moved) and can explore the time-course of cognition. tDCS can be used to enhance performance as well as impair it.
- Genetic differences can be measured either by taking a candidate gene that is known to vary (several alleles) and exploring how it produces different traits ('genotype-first') or by taking a phenotype that varies and exploring which parts of the genome co-vary with it ('phenotype-first').

EXAMPLE ESSAY QUESTIONS

- Do questionnaires and other methods relying on subjective report have any place in social neuroscience?
- Describe how three different types of methodology in social neuroscience can provide complementary insights into our understanding of face processing.
- What can functional imaging reveal about the social brain that other methods cannot?
- Do the methods of social neuroscience make it impossible to study naturalistic social behavior?
- How can individual differences at the biological level be linked to individual differences in social behavior?

RECOMMENDED FURTHER READING

- Harmon-Jones, E., & Beer, J. S. (2009). *Methods in Social Neuroscience*. New York: Guilford Press. A good collection of chapters that include some methods not covered here (e.g. salivary hormones, genomic imaging, neuroendocrine manipulations).

- Senior, C., Russell, T., & Gazzaniga, M. S. (2009). *Methods in Mind*. Cambridge, MA: MIT Press. A good collection of chapters that include some methods not covered here (e.g. magnetoencephalography, MEG).

- Ward, J. (2015). *The Student's Guide to Cognitive Neuroscience* (3rd edition). New York: Psychology Press. The methods chapters in this book have been condensed to form the basis of this chapter.

ONLINE RESOURCES

- Links to online neuroanatomy and basic neuroscience resources
- References to key papers and readings
- Demonstrations and tutorials on various methodologies including fMRI, EEG, MEG, and brain stimulation
- Interviews and lectures given by leading figures including Geoffrey Aguire, Steve Luck, Vincent Walsh, Read Montague, Elizabeth Warrington, and others
- Recorded lecture given by textbook author, Jamie Ward
- Multiple choice questions and interactive flashcards to test your knowledge
- Downloadable glossary

CHAPTER 3

CONTENTS

Evolutionary origins of social intelligence and culture

Modern humans, Homo sapiens, emerged as a distinct species only 200,000 years ago. Over time, this new species developed a variety of tools, produced elaborate art, and began to bury their dead in ornate rituals. In the last few hundred years, they invented computers, visited the moon, and discovered the basic physical laws that govern the universe. We are separated from our nearest living ancestor, the chimpanzee, by only 1.6% of DNA (King & Wilson, 1975), and we shared a common ancestor with the chimpanzee around six or seven million years ago (see Figure 3.1). What is it in this 1.6% of DNA that has enabled humans to achieve such a level of

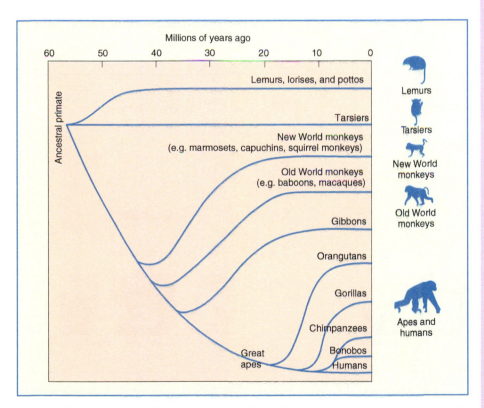

Figure 3.1 A history of primate evolution. The extinction of the dinosaurs occurred around 65 million years ago and, at this time, the primates' ancestral relative was like a tree-dwelling shrew. Changes in the genetic code between species act as a 'molecular clock' that enables more accurate timing of branching.

technological and cultural complexity? According to one idea, the main evolutionary pressure for human intellectual development is not the ability to be smarter per se but rather the ability to understand and predict complex social interactions and to outwit our peers – so-called **social intelligence** (Humphrey, 1976). According to this view, evolutionary pressures to be socially smarter would lead to more general changes (e.g. larger brain size) that would lead to increased intellect in other, non-social, domains. For example, the study of Herrmann, Call, Hernandez-Lloreda, Hare, and Tomasello (2007) – illustrated in Figure 3.2 – compared the physical and social intelligence of 2- to 3-year-old human children with juvenile chimpanzees and orangutans. Human children, as opposed to adults, were used to limit the effects of culturally based formal training. They found that humans excelled in the social domain on tasks such as social learning (solving a problem after a solution is demonstrated), communication (pointing to receive a hidden reward), and gaze following. They did not, however, excel in the physical domain on tasks such as spatial memory (to locate a reward), quantity discrimination, and tool use (using a stick to get a hidden reward). This was taken to support the view that the evolution of human intelligence has primarily occurred in the social domain.

If social intelligence is the thing that sets humans apart from other primates at the level of *individual* minds and brains, then it is complex culture itself that appears to set humans apart in terms of *group* behavior. **Culture** can be defined as a shared set of values, skills, artifacts, and beliefs amongst a group of individuals. There is an inherent social dimension to culture. Culture is *shared* amongst members of a group

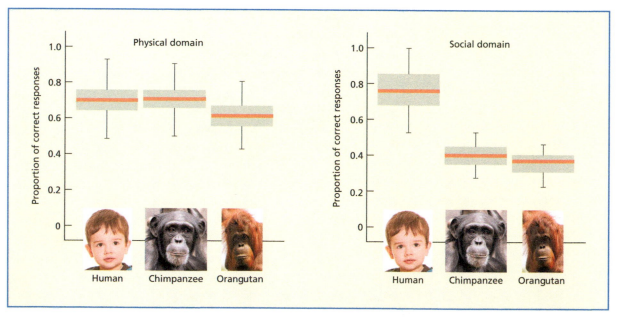

Figure 3.2 Herrmann, Call, Hernàndez-Lloreda, Hare, and Tomasello (2007) compared the physical and social intelligence of 2–3-year-old human children with that of somewhat older chimpanzees and orangutans. They found that humans excelled in the social domain. They did not, however, excel in the physical domain. This was taken to support the view that the evolution of human intelligence has primarily occurred in the social domain. Figure adapted from Herrmann et al. (2007). Copyright © 2007 American Association for the Advancement of Science. Reproduced with permission.

via a process of **social learning** from person to person, both within and across generations. Differences in culture also form one way of distinguishing between social groups. It is to be noted that culture in this context does not mean 'high culture' (opera, art, etc.), although this would be a component of it. Culture encompasses far more than this, including skills (e.g. literacy), technology (e.g. tool-making from spears to computers), and beliefs (including but not limited to religious beliefs). One tends to think of culture as an environmental rather than a genetic effect. We are not born predisposed to speak a particular language or believe a particular religion. However, whilst the *differences* between cultures are largely attributable to environmental factors (our time and place of birth), the *similarities* between cultures (including the fact that we are all immersed in one) are almost certainly down to biology and evolution. Our brains have developed in such a way that allows us to both create and absorb shared knowledge, skills, and beliefs.

This chapter begins by examining the hypothesis that primate intelligence evolved to deal with increasing social complexity. It will then go on to consider how this could lead to non-genetic social transmission of knowledge via culture, and it will consider evidence for culture in other species. Finally, the chapter will discuss how certain aspects of culture are represented in the brain through a consideration of material symbols (written words and numbers, symbolic art) and tool use. Other aspects of culture, such as culturally different perceptions of self, are dealt with in Chapter 9.

> **KEY TERMS**
>
> **Social learning**
> The transmission of skills and knowledge from person to person
>
> **Social intelligence hypothesis**
> Evolutionary pressures to be socially smarter lead to more general changes (e.g. increased brain size) resulting in increased intellect in non-social domains.

THE SOCIAL INTELLIGENCE HYPOTHESIS

As already noted, the **social intelligence hypothesis** argues that evolutionary pressures to be socially smarter lead to more general changes (e.g. larger brain size) resulting in increased intellect in non-social domains (Humphrey, 1976). This has also been termed the *social brain hypothesis* (Dunbar, 1998) and the *Machiavellian intelligence hypothesis* (Whiten & Byrne, 1988). (Machiavelli was a Renaissance politician renowned for cunning and deceit.) However, when one comes to consider this hypothesis in detail it can be shown to have several different meanings. Whiten and van Schaik (2007) considered three different interpretations of this basic idea:

1 'Intelligence is manifested in social life.' This is the weakest interpretation of the hypothesis. Historically, animal intelligence was studied by looking at the ability of individual animals in a laboratory to solve problems. This weaker interpretation merely states that intelligence should be more broadly construed to include problem solving in one's social life. This idea is now widely accepted (Whiten & van Schaik, 2007).

2 'Complex society selects for enhanced intelligence.' This stronger interpretation argues that there is something particularly demanding about problem solving in the social realm that leads to a need for greater intelligence. The more complex the society, the greater the need. In this interpretation, 'intelligence' is regarded as a more general capacity rather than a specialized set of functions that deal with social life.

3 'Complex society selects the specific characteristics of intelligence.' This interpretation is a stronger form of the one above. It suggests that social pressures select not only for the *amount* of intelligence but also the *type* of intelligence. For

example, one might imagine relatively specialized mechanisms in the brain for dealing with social problems (e.g. imitation, theory of mind) that are not necessarily reducible to general intellect.

To some extent, the debate between the second and third interpretations will be a recurring theme of this book. This chapter will consider the issue with particular reference to evidence from non-human species.

Social intelligence and brain size in primates

Several studies have tested the social intelligence hypothesis by examining the relationship between social intelligence and brain size. The prediction is that the more complex the species' social world, the larger the brain will need to be to cope with such complexity. As Seyfarth and Cheney (2002) put it, primates 'live in large groups where an individual's survival and reproductive success depends on its ability to manipulate others within a complex web of kinship and dominance relations' (p. 4141).

There are several inherent problems with examining a link between brain size and social intelligence. First, there are different ways of defining 'social intelligence' as noted above. Second, it is not easy to measure it in the natural social settings of, say, an orangutan or capuchin monkey. Finally, this approach assumes that brain size is a useful index of general intellect. Typically, the size of the whole brain is not measured, but a ratio is taken between the amount of neocortex (most of the gray matter surface of the brain) and the rest of the brain. The reliance on these measures of brain size has been criticized by some (e.g. Healy & Rowe, 2007). Despite these differences and difficulties, several studies using somewhat different measures have converged on a common answer – that there is a strong correlation between social intelligence and brain size.

In one of the earliest attempts to explore this issue, Dunbar (1992) used social group size of various primates as an approximate measure of social complexity and found a significant correlation with neocortex ratio. This is shown in Figure 3.3. The implication is that the larger the brain, the greater the number of social relationships that can be sustained. Extrapolating from other species to humans, Dunbar (1992) estimated that humans are adapted to an optimal group size of about 150 people. Although most of us know many more people than this, the claim is that our brains can only support *active* relationships (based on regular exchanges) with around 150 others. There is some evidence that supports this number, from clan sizes in traditional societies (see Dunbar, 1998) to the exchange of Christmas cards in Western societies (Hill & Dunbar, 2003). The claim is that populations above 150, which is what most of the human race lives in, can only be maintained by creating special roles that enforce social cohesion (police, lawyers, bureaucrats, etc.) rather than via the knowledge and maintenance of multiple one-to-one social bonds. Knowing 150 people intimately might sound trivially easy, but knowing the full set of social relations between 150 people is a huge challenge (the number of relations is given by $150 \times 149 \times 148 \ldots \times 1$). Cooperation between people can be maintained, within small groups, based on first-hand experience of each other's behavior (e.g. their tendency to cheat or be generous), direct and third-party retaliation (e.g. attacking the attacker of my friend), and alliance formation (e.g. the enemy of my enemy is my friend). In non-human primates, this may be the sole mechanism for keeping the group together, and if the group gets too big to maintain, it will split into two. In

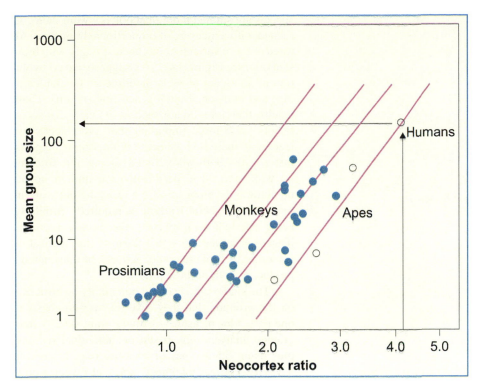

Figure 3.3 There is a relationship between the size of social groups (plotted here on a logarithmic scale) and the size of the neocortex (relative to the rest of the brain). Each point represents a different primate species with open circles depicting the great apes (note the general rightward shift in neocortex size for these species). The group size of humans is estimated at around 150 based on the known neocortex ratio. From Dunbar (1992).

humans, larger-scale groups exist because of the development of cultural rules of cooperation that individuals collectively agree on (e.g. legal, moral, and religious norms) in addition to those based on direct experience. These facilitate interactions between strangers who have never met before and will never meet again. Other species that live in large social groups, such as insects, are governed by biologically specified rules that dictate their role and behavior in the group (Queen, worker, etc.). Maintaining an insect social group does not require the same cognitive mechanisms found in primates (social learning of dominance and alliance relationships between individuals) and humans (learning of culturally specified rules of behavior), and nor do they require bigger brains (Chittka & Niven, 2009).

Other research has used a more direct measure of social intelligence than group size. Byrne and Corp (2004) found a correlation between frequency of tactical deception in different primate species and neocortex ratio. Deception is a complex social skill involving an appreciation of another's knowledge and the ability to manipulate it. In another study, Reader and Laland (2002) measured the number of times that researchers had documented, in natural habitats, examples of social learning, innovation (coming up with novel solutions), and tool use in 106 species of primates. These figures, scaled according to number of opportunities for observation (i.e. some

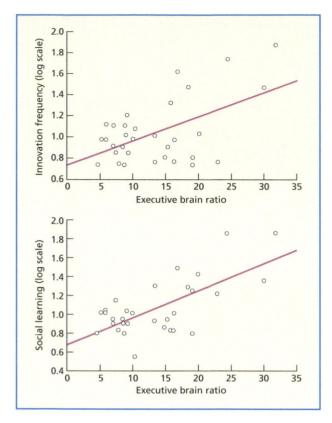

Figure 3.4 The frequency of social learning and frequency of innovation (both on logarithmic scales) correlate with executive brain ratio (defined here as ratio of neocortex and striatum volume relative to brainstem volume). Each data point represents a different primate species, and the frequency counts are corrected according to 'research effort' (i.e. taking into account the fact that some species are studied more than others giving more opportunity for positive observations). From Reader and Laland (2002). Copyright © 2002 National Academy of Sciences, USA. Reproduced with permission.

species are studied more than others), were then correlated with a measure of proportional brain size. All three of these variables correlated strongly with brain size (e.g. see Figure 3.4). This suggests the co-evolution of an aspect of social intelligence (social learning) and non-social intelligence (innovation). These results, therefore, do not support the view that social factors were more important than other factors in leading to increased intellect / brain size. It suggests, instead, that both were crucial. Being able to come up with innovative ideas will have limited impact on cultural development if the ideas die out with the inventor (i.e. social learning is required). Similarly, being able to learn from each other is only important if there is something worth learning (i.e. innovation is required). Tool use could, perhaps, be considered a product of both of these processes.

The studies above could potentially be criticized on the grounds that they give an over-emphasis to one particular measure – namely brain size. A finer grained analysis (e.g. based on individual regions) plus using other types of measure (e.g. brain connectivity, cortical thickness) may yet reveal a quite different picture. One important factor that goes hand-in-hand with evolutionary increases in brain size is the length of immaturity (Joffe, 1997). Humans take an unusually long time to reach adulthood and this comes at a significant cost in terms of provision of food and protection of offspring. One likely benefit to this is that it provides an extended window for learning and adapting to one's environment and culture. Greater dependency in early life provides rich opportunities for learning from one another. The emphasis on social learning puts a new twist on the nature–nurture debate. If intelligence is related (at least in part) to our ability to learn from each other, then it is equally a product of our nature (a genetic disposition to learn from each other) and nurture (our accumulating knowledge of the world) – without one, it is not possible to have the other.

Social intelligence and brain size in non-primates

Examining the relationship between social intelligence and brain evolution in various non-primate species such as cetaceans (including dolphins and porpoises) and birds is important for assessing the generality of this relationship. If there is a relationship between social intelligence and various markers of brain development in species very distantly related to primates, then this provides potential evidence for convergent evolution. **Convergent evolution** occurs when the same evolutionary pressures (e.g. to be socially smarter) create the same outcome independently in different species. This can be contrasted with divergent evolution in which an association can be traced back to a common ancestor possessing both characteristics.

The first problem one faces when considering non-primate species is that of translating markers of social intelligence across different species. For instance, social group size can be harder to measure in non-primate species or may be a poor marker of social intelligence if the groups are not stable or if they are not organized via cooperative allegiances (e.g. does flocking in birds require social intelligence?). Bottlenose dolphins, for instance, do not form strongly stable groups over time (the groups exhibit fission and fusion behavior). Nevertheless, at a given point in time they may have 60–70 associates which is comparable to the largest non-human primate groups (Connor, 2007). Dolphins also have unusually large brains relative to other cetaceans (Connor, 2007) and display social intelligence on other measures, such as recognizing themselves in a mirror to investigate marked parts of their body (Reiss & Marino, 2001) – a trait found in apes but not monkeys (Schilhab, 2004).

Whilst other species of mammal, such as the elephant, fit the profile of having larger brains and high social intelligence (McComb, Moss, Durant, Baker, & Sayialel, 2001), other mammals such as the spotted hyena (Holekamp, Sakai, & Lundrigan, 2007) and bats (Kerth, Perony, & Schweitzer, 2011) display much better social intelligence than might otherwise be predicted from measures of brain size. Van Schaik, Isler, and Burkart (2012) discuss possible reasons for this mismatch. Having larger brains is energetically very expensive, and requires other kinds of adaptations in the organism to support the evolution of bigger brains: such as a shift in diet (as happened in human evolution from predominantly vegetarian to omnivorous), or a slowing down of metabolism and reduced energy demands elsewhere (e.g. as manifest in a longer lifespan). However, other factors may pose limitations. For instance, the high energy costs of flight in bats may pose limits on brain growth, or strong seasonal variations in food supply may pose limits on the ability of some species to grow larger brains (despite the advantages that this may ultimately afford).

With regards to birds, Emery, Seed, von Bayern, and Clayton (2007) failed to find a linear relationship between group size and relative brain size across many species of birds, suggesting that the pattern observed in primates does not necessarily hold elsewhere. Of course, it may also be the case that measures of group size are a poor reflection of the nature of the social interactions, in the same way as most humans living in large cities only have reciprocal social alliances with a small subset of that group. However, they did find an association between bird brain-size and the type of mating relationship that they engage in – see Figure 3.6. Bigger brains were linked to cooperative mating systems and longer-term monogamy (over several years), relative to having multiple partners at once or serial monogamy (changing partner every breeding season). They suggest that maintaining

Figure 3.5 Dolphins may affiliate to each other via gentle rubbing with their pectoral fins (Connor, Wells, Mann, & Read, 2000) and through synchronous displays such as joint surfacing (Herman, 2002). Such behavior may be analogous to primate grooming and imitation, respectively.

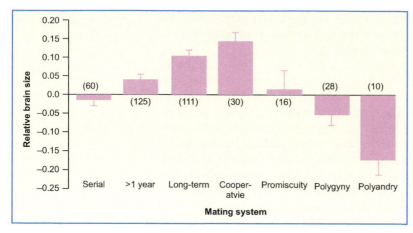

Figure 3.6 Across a large range of bird species (N in brackets), cooperative mating and long-term monogamy (over 2+ years) are associated with the largest relative brain size. Other mating strategies include finding a new mating partner each year (serial), or for a second season (> 1 year), non-selective mating (promiscuity), multiple female partners (polygyny), and multiple male partners (polyandry). From Emery et al. (2007).

these relationships requires social intelligence (e.g. recognizing and responding to others' needs) that resembles those seen in primate alliances.

The family of crows (corvids), including ravens and rooks, are one well-studied example of big-brained birds. Ravens offer consolation to a distressed partner (Fraser & Bugnyar, 2010), show reconciliation to a partner after an aggressive encounter (Fraser & Bugnyar, 2011), and are sensitive to changes in dominance status of other birds created experimentally (Fraser & Bugnyar, 2011). Given that primates and corvids have not shared a common ancestor for hundreds of millions of years, the similarities in social intelligence are taken as evidence for convergent evolution (Seed, Emery, & Clayton, 2009).

In sum, evidence from non-primates is not necessarily inconsistent with the social intelligence hypothesis but, rather, throws a spotlight on to the potential mechanisms both in terms of specific evolutionary selection pressures and in terms of evolved cognitive processes. With regards to the latter, there is a current interest in the kinds of cognitive mechanism that supports *cooperation* and bonding between individuals (Dunbar & Shultz, 2010; Tomasello, 2009). This can perhaps be contrasted with the earlier evolutionary accounts that emphasized Machiavellian strategies such as deception and exploitation.

Language evolution and the social intelligence hypothesis

There are a number of interesting parallels to be made between various theories of language evolution and the social intelligence hypothesis. For instance, there is a debate concerning whether language arose from non-specific evolutionary changes such as increased brain size. This view has been most famously championed by the evolutionary biologist Stephen Jay Gould (e.g. Gould, 1991) and the linguist Noam Chomsky (e.g. Chomsky, 1980). According to this view, language arose out of general selection pressures to be smarter – either socially, cognitively, or probably both. This pressure then led to general changes (e.g. in brain size) from which language emerged. One specific idea along these lines is Dunbar's (2004) proposal that language evolved to facilitate the bonding of large social groups. He argues that language evolved due to social pressures to live in large groups and that language enables greater cohesion of groups. Dunbar draws an explicit comparison between grooming behavior in primates and social use of language in humans (Figure 3.7). Human language, according to Dunbar, evolved to enable humans to keep 'in touch'

Figure 3.7 Grooming serves a social function in primate societies. It is related to social complexity (larger groups engage in more grooming) and alliance formation (grooming releases endorphins, the brain's natural opiates). Why do humans not engage in this activity? According to Dunbar (2004), human language serves the social function of keeping in touch (without literally 'keeping in touch') and sharing information about who is doing what to whom.

without literally being in touch! The contrary view to Chomsky, Gould, and Dunbar, espoused for example by Pinker and Bloom (1990), is that language did arise from selection pressures relating specifically to communicative needs rather than as a by-product of other, more general changes such as brain size or social group size.

Recent papers take the more realistic position that language should be considered not as a single entity but as being multi-faceted, consisting of speech production, syntax, semantic concepts, and so on (Fitch, Hauser, & Chomsky, 2005; Pinker & Jackendoff, 2005). Each of these different aspects of language can then be considered from either a generalist or a language-specific perspective on the merits of its own case. To give a flavor of the debate, consider two key aspects of language: the descent of the human **larynx**, and **syntax**. Humans have evolved a descended larynx that is not present in other primates and comes at a significant survival cost. Unlike most other animals, adult humans cannot breathe and swallow at the same time and thus have elevated risk of choking. However, a descended larynx is also crucial for human speech. This is acknowledged by both camps. Where the disagreement lies is in whether larynx descent occurred because of the need to increase the repertoire of speech (i.e. it evolved for the function of communication) or whether it occurred for some other reason not related to communication. For example, a descended larynx makes an animal sound bigger, thus making it more attractive to a mate and better able to achieve social dominance (e.g. Reby et al., 2005). This would set up an 'arms race' in which the larynx descends further and further until it ceases to have benefit. It is found in at least one other species that lacks language – red deer (Fitch & Reby, 2001). Thus, whilst a descended larynx enables human speech, it is impossible to

KEY TERMS

Larynx
An organ in the neck of mammals involved in sound production

Syntax
The rules by which words are combined to make meaningful sentences

say for certain whether it evolved for that specific reason. Similarly for syntax (i.e. the grammatical rules that specify how words are combined), one suggestion is that the hierarchical nature of syntactic representation (in which words are clustered into phrases, and phrases into sentences, etc.) may have been driven by the need to mentally represent complex social groups and hierarchies rather than for communicative purposes (Hauser, Chomsky, & Fitch, 2002).

Evaluation

There is good evidence that evolutionary increases in relative brain size have been accompanied by increased complexity in the social domain – larger social groups, more deception, and more social learning. This provides support for the social intelligence hypothesis. What remains unclear, and keenly debated, is the extent to which specific processes (such as language or theory of mind) arose out of these more general changes or were specifically shaped during the course of evolution. Another important avenue for future research will be to identify genes linked to brain growth that have changed during primate and human evolution (e.g. Dorus et al., 2004). This offers a way of linking change at the behavioral/cognitive level to changes in the genetic control of brain development.

HUMAN FRIENDSHIPS IN THE CYBER AGE

Online social media enable us to keep in touch with people like never before. According to a 2014 survey, well over half of the American population use Facebook with the average number of friends of users being 338. How are our brains, which evolved to cope with small-scale hunter–gatherer societies, able to manage the volume of social information in the cyber age? Will our brains change as a result of this profound cultural shift?

Of course friends differ in their degree of emotional closeness to us, and anthropologists such as Dunbar (2010) conceptualize social groups in terms of nested layers varying in emotional closeness: in humans, it may correspond to groups of size 5, 15, 50, 150, and beyond in similar ratios (500, 1500, 5000. . .). The figure of ~150 individuals has been given as an estimate of the number of people with whom there is a mutual understanding of reciprocity: i.e. they feel they could call on you in a time of need and vice versa. It is entirely possible to know and keep in contact with many more people ('friends'), but whether there is an equal emotional connectedness to them all is doubtful. Instead the number of Facebook friends may reflect a desire to manage ones reputation (i.e. perceived popularity).

One factor that strongly determines how people rank their closest friends is how high they think that these friends would rank them (DeScioli & Kurzban, 2009): thus friendships are based on the perception of an alliance, even after taking into account factors such as similarity and degree of liking. In this respect, the use of the term 'friendship' is not anthropomorphic and can reasonably be applied to other animals (Seyfarth & Cheney, 2012), and can be understood in terms of the functioning of the social brain (Brent, Chang, Gariepy, & Platt, 2014).

Several studies have examined individual differences in the size of human friendship networks and correlated them with gray matter structural differences in the brain using VBM. Using measures of number of friends in the real world (e.g. how many people you could ask a favor of and expect it to be granted), it was found that the size of amygdala – linked to emotional processing – was positively correlated with the number of friends (Bickart, Wright, Dautoff, Dickerson, & Barrett, 2011; Kanai, Bahrami, Roylance, & Rees, 2012). The study of Kanai et al. (2012) also looked at the number of Facebook friends and found other regions to be correlated with this measure, including the superior temporal sulcus (involved in person perception) and the entorhinal cortex (involved in memory, e.g. for face-name associations) – see Figure 3.8. As the study is correlational, it is impossible to know whether variations in brain size are driving the number of friendships formed or vice versa.

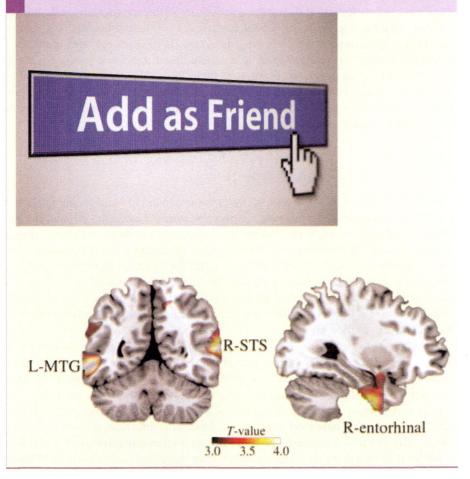

Figure 3.8 The number of friends one has on Facebook correlates with the number of friends one has in face-to-face interactions, and also correlates with gray matter density in several regions of the brain: the entorhinal cortex, right superior temporal sulcus, and left medial temporal gyrus. Lower image from Kanai et al. (2012).

EVOLUTIONARY ORIGINS OF CULTURE

It is strange, but almost certainly true, that if you were to take (with the aid of a time machine) a newborn baby born tens of thousands of years ago and raise it in the modern age it would have no difficulties in learning our language, performing algebra, driving a car, or surfing the net. What has primarily evolved in this period of time is not our biology but our culture. However, we should not fall into the trap of thinking that some human societies have 'more' culture than others; rather they have different systems of culture. Hunter–gatherers are cultured – they have certain skills, tools, and beliefs that enable them to survive in *their* environment. A Westerner transplanted into their environment may quickly die of poisoning, predation, or starvation. However, one also cannot gloss over the fact that some cultures have developed far more advanced forms of technology than others. It has been claimed (Blackmore, 1999; Diamond, 1997) that certain cultural trends will tend to dominate over others when they come into contact with each other in a kind of cultural 'survival of the fittest' – think guns versus spears, or the ease with which one can perform sums with modern number notation (e.g. $54 \times 10 = 540$) compared to Roman numerals (e.g. LIV $\times$ X = DXL). Blackmore (1999; following Dawkins, 1976) uses the term **meme** as a deliberate analogy to the term gene, to denote cultural ideas that pass themselves, socially, from person to person according to their 'fitness' level (i.e. the benefits they convey or are believed to convey). Certain skills or ideas may be more valued by particular members of a group than others and these skills or ideas may be more likely to be passed on until surpassed by something more appropriate.

Culture in non-human species

Other species do not have the complexity of culture found in humans, but important similarities can nonetheless be found. To give one concrete example, Whiten, Horner, and de Waal (2005) taught two individual chimpanzees from two different groups one of two ways of obtaining food from the same apparatus – either by poking with a stick to remove a block, or using the stick to lift the block, as shown in Figure 3.9. When these individuals were released back into the group, almost all group members adopted the particular skill of their group rather than the other group. Even when subsequently a new way of getting the food was taught, individuals tended to conform to the social norm of their group.

Whiten and van Schaik (2007) described the evolution of culture in terms of a 'culture pyramid' consisting of four tiers (Figure 3.10). The use of a pyramid shape conveys the fact that those aspects of culture on the broader, lower tiers are more pervasive in the natural world than those higher up. At the lowest tier there is 'social information transfer' in which animals may, for example, learn from each other by watching where they hide food or forage for food. This information tends to be used temporarily and then discarded. The next level up is termed **traditions**. A tradition is considered to be a distinctive pattern of behavior shared by two or more individuals in a social group. For example, the same species of bird in one location may have a different song structure from the same birds living in another location. The use of two ways of achieving the same reward, perpetuated within a group, is another example of a tradition as in the study of chimpanzees described above. The tier above traditions in the 'culture pyramid' consists of culture itself, which is construed here as a collection of traditions. Cumulative cultures are those in which traditions are

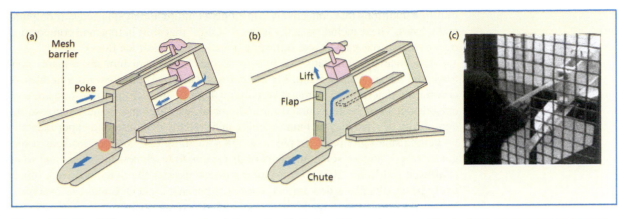

Figure 3.9 Two chimpanzees were taught how to obtain food from the same device either using a poke action (a, c) or a lift action (b). When introduced to two groups, other individuals learned using the conventional method and were resistant to change even when shown the alternative. This study demonstrates social learning of traditions and conformity to social norms once learned. From Whiten et al. (2005). Copyright © 2005 Nature Publishing Group. Reproduced with permission.

gradually enhanced or modified over time, such as moving from stone-based to metal-based tools or from Roman to Arabic numbers.

The most basic level of the pyramid, social information transfer, is found in many mammals, birds, fish, and even invertebrates such as bees (Whiten & van Schaik, 2007). Evidence for traditions in a wide variety of species has not been collected. Whilst there are many examples of social learning and traditions in birds (e.g. Lefebvre & Bouchard, 2003), there is little evidence of multiple patterns of behavior being transmitted together. According to the criteria above, birds would have traditions but not culture. One potential exception is the New Caledonian crow (*Corvus moneduloides*), which exhibit two types of tool use involving both twigs and strips cut from leaves (e.g. Hunt & Gray, 2003). However, others have suggested that this may be an innate skill rather than socially learned (Kenward, Weir, Rutz, & Kacelnik, 2005).

The most convincing examples of multiple traditions come from apes, but there is evidence from

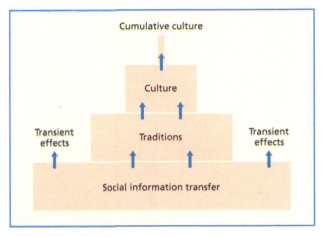

Figure 3.10 Whiten and van Schaik (2007) categorize socially learned behavior amongst different species into a 'culture pyramid'. Lower levels are more prevalent in nature, and higher levels develop from the lower levels. From Whiten and van Schaik (2007). Copyright © 2007 The Royal Society. Reproduced with permission.

some species of monkeys too. Perry et al. (2003) studied social traditions in capuchin monkeys (*Cebus capucinus*) involving a variety of social games, such as the 'toy game' (putting non-food objects in each other's mouths and removing them, taking-turns), the 'hand sniff game' (place another's hand or foot over own face and, with eyes closed, inhale deeply and repeatedly for more than one minute), and so on. These multiple traditions had unique distributions amongst different capuchin communities. This suggests that the games are culturally learned rather than part of their innate repertoire of behaviors. Whitehead and Rendell (2014) also report multiple traditions that collectively constitute a culture amongst species of dolphins and whales. These include socially learned 'songs' that shift in tone and composition over time, various games, and skills (e.g. knocking seals off ice floes).

Evidence for cumulative culture in species other than humans remains controversial. Dean, Vale, Laland, Flynn, and Kendal (2014) put forward a number of criteria for determining cumulative culture: it must involve multiple transmission episodes, it must be transmitted through social learning, and it must increase the complexity or efficiency. In humans, this can be readily demonstrated experimentally by getting small groups of people to design something (e.g. a paper airplane, a spaghetti tower), and for some members of that group to teach new members, and so on (Caldwell & Millen, 2008). Over multiple transmissions the outcomes were objectively better (the plane flew further, the tower became higher). Candidate examples, from field observations in chimpanzees, include cracking a nut using one stone to hit it and using another stone to steady it. Here the putative transmission would have involved directly hitting a nut against a stone, modified by using another stone as an implement, modified by using another stone to steady it. However, it is equally conceivable that this entire skill could have been learned by single individual rather than via multiple modifications by different animals over different episodes. Attempts to teach cumulative culture in chimpanzees and capuchin monkeys using multi-stage problem-solving tasks have also failed, despite the same task being completed by human children who are unable to solve the task individually but succeed cooperatively (Dean, Kendal, Schapiro, Thierry, & Laland, 2012). There are several possible cognitive mechanisms that might enable cumulative culture: the degree of innovation in a species; different social learning mechanisms (e.g. language, imitation); and the desire to be cooperative and prosocial. These mechanisms are discussed in more detail in the next section.

KEY TERM

Biological anthropology
Study of the behavioral and anatomical evolution of the human species

PREHISTORIC ORIGINS OF HUMAN CULTURE

Although comparisons between humans and our closest living relatives offer a window into social and cognitive evolution, another crucial line of evidence comes from comparisons with our now *extinct* ancestral relatives, in the field of **biological anthropology** (Leakey, 1994, offers an accessible review, and the information here comes from this source unless otherwise stated). The evolutionary branch that separates the different species of humans that once existed from other great apes is characterized by bipedalism (walking upright). It was initially speculated by Darwin (1871) and others that bipedalism may have arisen due to a selection pressure for using the hands to manipulate tools – that is, those early apes that were better at using their hands for tools would have been more likely to survive, and so this trait would be gradually enhanced over time. This view is no longer accepted. The first bipedal hominids emerged around 6–7 million years ago, whereas the first evidence of stone tools emerged only 2–3 million years ago. The emergence of stone tool use was associated with an evolutionary expansion in brain size, not bipedalism (e.g. de Sousa & Cunha, 2012). The earliest bipedal hominids had small brains and probably had similar cognitive and social intellect to modern-day great apes.

With the arrival of the big-brained Homo erectus around two million years ago, there was a major advance in stone tool manufacture and evidence of dietary change towards meat eating (from the tooth fossil record). Both the increase in brain size and the need to cooperate when hunting for meat are likely to have been associated with a shift in social complexity. There is evidence of right-handedness in tool manufacture in this period, suggesting laterality changes in brain organization and possibly protolanguage (Toth, 1985). The other great cultural skill in this period, aside from stone tools, was the use of fire around 700,000 years ago. This would have enabled cooking of meat and boiling of water, as well as offering warmth and protection. It is not unreasonable to imagine that those who were better at cooperating during hunting and those better at learning new skills (tools, fire) would have increased their survival chances, leading to further enhancement of these abilities in future generations.

Modern humans (Homo sapiens) did not emerge until around 200,000 years ago in Africa, and this was accompanied by another major advance in tool manufacture. The most significant radiation out of Africa, from which all other present-day humans descend, was

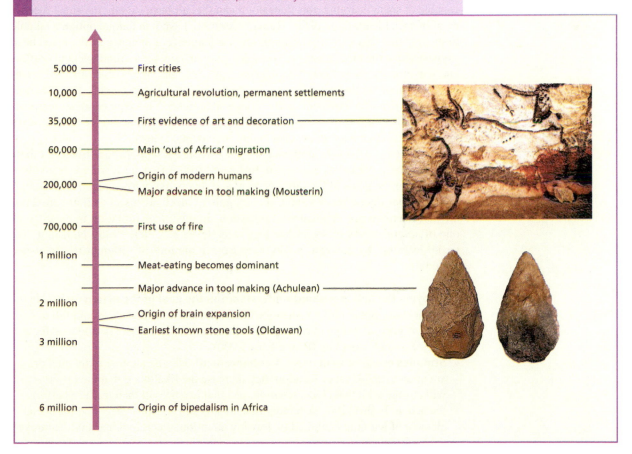

Figure 3.11 Some significant milestones in human evolution and human culture

as recent as 60,000 years ago (Forster, 2004). For many prominent anthropologists, the most significant human cultural change cannot be linked to a genetic modification but rather to a change in lifestyle that occurred independently in several places in the world from 10,000 years ago – namely, the shift from hunting–gathering to permanent settlements (Mithen, 2007; Renfrew, 2007). This would have led to a more diverse division of labor, requiring greater organization and institutional control. It would have created opportunities for trade and also a greater reliance on material culture than that found in portable communities in which the material world had to be transported or created anew in each location.

KEY TERMS

Imitation
Social learning based on an understanding of the goals, intentions, and mental states of other individuals

Stimulus enhancement
Having another individual draw attention to an object may increase the likelihood that the observer will engage with that object.

Local enhancement
Having another individual draw attention to a location may increase the likelihood that the observer will engage with that location.

Contagion
Repetition of behaviors that are innate rather than learned (e.g. yawning)

Social learning versus imitation

Just because birds, dolphins, monkeys, apes, and humans (to name a few) are all capable of social learning, this does not mean that all have exactly the same mechanism of social learning in place. Humans certainly have one unique option available to them in that they can acquire traditions via language. Language aside, it has been suggested that the mechanisms of social learning in humans are fundamentally different than other species. According to some, only humans are capable of social learning based upon an understanding of the goals, intentions, and mental states of other individuals (Penn & Povinelli, 2007; Tomasello, 1999). This type of social learning, termed **imitation**, is not straightforward to spot. The challenge lies in finding ways to observe, via behavior, the unobservable (i.e. mental states).

Imitation involves the understanding and reproduction of the actions of others and could be regarded as a more sophisticated form of social learning. It involves reproducing the goals of the other person, and this is likely to entail an understanding of his/her intentions. As such, sociocognitive mechanisms as well as sensorimotor mechanisms are implicated in imitation. In this view, imitation is regarded as one of several forms of social learning (see Heyes & Galef, 1996). Non-imitative social learning, by contrast, could arise from a number of different mechanisms, including:

- **Copying the action without understanding the goal of the action** (also called mimicking). Some bird vocalizations that mimic human speech would fall under this category, although in certain well-documented cases there is evidence for a degree of understanding (Pepperberg, 2000).
- **Stimulus enhancement** or **local enhancement**. Having another individual draw attention to an object or location may increase the likelihood that the observer will engage with that object/location, and that he/she will then learn to perform the action. In this view, each individual engages in self-discovery and the social element of learning is limited to drawing attention to certain important features in the environment.
- **Contagion**. This refers to the repetition of behaviors that are innate rather than learned, such as yawning and laughing. If you see someone else yawning you

are also likely to yawn, and canned laughter on comedy shows can lead to more smiling and laughter (Provine, 1996).

In humans there is evidence for 'true' imitation. For instance, if asked to reproduce a complex action sequence, participants often reproduce the end-state but not the means to the end (Wohlschlager, Gattis, & Bekkering, 2003). Even human infants show evidence of goal-based imitation (Gergely, Bekkering, & Kiraly, 2002). In this study, the infants watched an adult press a button on a table by using their forehead. In one condition, the adult's hands and arms are bound up under a blanket and in the other condition the adult's hands are free. This is shown in Figure 3.12. When the adult's hands are free, the infants copy the action directly – they use their foreheads too. But when the adult's hands are not free the infants imitate the goal but not the action (i.e. the infants use their hands rather than their head). The implication is that the infants understand that the goal of the action is to press the button, and they assume that the adult would have used his/her hands if they had been free. This is often called 'taking the **intentional stance**', in that it involves attributing intentions to another person to account for their actions (Dennett, 1983).

What about other primates? According to Tomasello (1999) cultural traditions such as washing sand off potatoes in a nearby stream by Japanese macaque monkeys (Kawai, 1965; Kawamura, 1959) may arise via self-discovery facilitated by stimulus enhancement and location enhancement. When using tools, Tomasello (1999) argued that chimpanzees use trial-and-error learning to achieve a goal, rather than imitation. For example, when watching another chimp being rewarded with food (e.g. after poking a stick in a hole), he/she will attempt to get a reward too, but without necessarily inferring that the other chimp *knew* there was food in the hole or that the other chimp *intended* to get it. One problem that arises from linking imitation closely with inferring of mental states (or 'theory of mind') is that whereas imitation emerges in humans in the first year of life, accurate performance on most tests of reasoning about mental states emerges between 3 to 4 years. The problem could be resolved by arguing, as Tomasello (1999) did, for a distinction between conscious attribution of mental states (e.g. as assessed in many theory-of-mind tasks) and unconscious goal attribution required for imitation. This idea is discussed more extensively in Chapter 11.

Other evidence suggests that chimpanzees and other apes are capable of 'true' imitation, and some former skeptics have now changed their position (Call & Tomasello, 2008). Chimpanzees (Custance, Whiten, & Bard, 1995) but not macaque monkeys (Mitchell & Anderson, 1993) are capable of learning a 'do-as-I-do' game

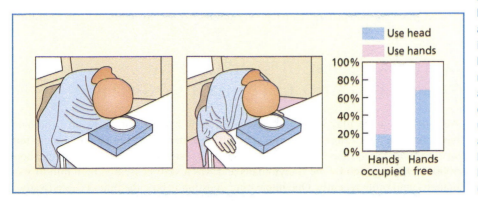

Figure 3.12 Infants imitate the goal of actions, rather than the motor aspects of actions. If the experimenter presses a button with his/her head because their arms are occupied, the infants 'copy' the action by using their hands rather than heads – they appear to infer that the experimenter would have used his/her hands to achieve the goal if they had been free. Drawing based on Gergely et al. (2002).

to produce complex actions (e.g. grab thumb of other hand). This does require considerable training (humans do not need to be trained to imitate), but this research represents a proof of principle that arbitrary acts can be imitated by non-human apes. Buttelmann, Carpenter, Call, and Tomasello (2007) adapted the human infant study of Gergely et al. (2002) in which actions were performed with an unusual body part. When the hands are occupied by the demonstrator then the chimpanzees did use their hands, implying that not only did they understand the goal but they also understood why the other person did not use their hands. Similarly, Horner and Whiten (2005) studied imitation in young chimpanzees. They observed a familiar person ram a stick several times into a hole in the top of a box, and then insert the stick into a front hole in order to extract a food reward. In one condition, the top of the box was transparent and it could be seen that the first stage was meaningless (i.e. the top hole was not connected to the reward). In another condition, the top of the box was covered except for the hole. Young chimpanzees in the transparent condition omitted the first step and went straight for the reward by putting the stick in the front. Young chimpanzees in the covered-box condition performed both steps. It suggests that the chimpanzees in this task are imitating, based on an understanding of goals and perhaps intentions.

If human imitation and ape imitation are cognitively equivalent, as many researchers now agree, one then needs to ask why human imitation (and culture) is far more prolific than that found in apes. One possibility is human ability in other domains. If humans are more creative and innovative (and the evidence suggests so) then there could simply be more things that are worth imitating. Another possibility is that there are different rewards to imitation in humans versus apes. Apes may imitate in order to obtain a material reward, such as food. Humans may imitate each other because imitating, and being imitated, is a reward in itself. As such, imitation may serve to bind human social groups together in ways that are less apparent in other species (Dijksterhuis, 2005). As the proverb goes, 'imitation is the sincerest form of flattery' and there may be truth in that.

Evaluation

Natural selection, brought about by variations in the gene pool, enables species to adapt slowly to their environments. However, humans and other species are also able to adapt to their environments via a much faster mechanism – social learning. When coupled with innovation and other cognitive skills, it enables complex systems of culture to evolve a 'life of their own' insofar as they are modified over time and come and go according to how useful they are. These cultural traditions expand our cognitive capacities (as described in the next section) and physical capacities (through tools and technology), and provide a means for establishing group and individual identities.

MATERIAL SYMBOLS: NEURONAL RECYCLING AND EXTENDED COGNITION

Although language may have been selected for by evolutionary pressures, humans have created a wide range of material symbols that are products of culture – including writing, number systems, and art. For example, writing was first invented about 5000–6000 years ago by the Babylonians and was a skill possessed by a minority of humans until quite recently. One interesting question is how the brain is able to adapt to incorporate such information. For example, to what extent do different cultural manifestations of these symbols (e.g. in different writing systems) lead to different brain-based solutions? One might imagine that a purely cultural invention, such as literacy, may end up using different brain circuits in different people if, for example, we imagine that our brains are highly plastic, such that any new information can be slotted into any under-used region. However, this does not appear to be the case and the neural circuits for writing and calculation appear to be quite conserved across individuals and across cultures (Dehaene & Cohen, 2007). Dehaene and Cohen (2007) refer to this as 'neuronal recycling'. Their assumption is that neural resources, set aside for other functions in our evolutionary past, may be recruited by cultural knowledge.

Reading involves a number of cognitive capacities: for recognizing written words, for translating these words into speech, and for understanding the meaning of the words. In their review, Dehaene and Cohen (2007) concentrate particularly on the system for recognizing written words and on a region in the left ventral visual stream termed the **visual word form area** (**VWFA**). This region responds to visual presentation of letter strings more than other objects, including made-up letters (so-called false fonts) and letters from unfamiliar writing systems (Cohen et al., 2002). A number of cross-cultural studies now show that the same region is activated by Roman script, Chinese characters, and Japanese Kana and Kanji (Bolger, Perfetti, & Schneider, 2005). Dehaene and Cohen (2007) speculate that this region may have evolved for certain types of object recognition. They suggest, for example, that the common developmental confusion between b/d stems from the fact that object recognition systems tend to treat mirror-images as the same (e.g. a cup is a cup irrespective of where the handle is) whereas this is not true for letters.

Similarly, for numerical cognition there is a region in the parietal lobes (intraparietal sulcus) that responds during arithmetic tasks, and when viewing different types of numerical symbols (digits, dot patterns, number words) both within (e.g. Piazza, Izard, Pinel, Le Bihan, & Dehaene, 2004) and across cultures (Tang et al., 2006). This may represent a core semantic representation of number (i.e. an approximate code for 'how many?'). This basic system may not only be cross-cultural but may also exist in other species. For example, monkeys contain neurons in this region that respond to different numbers of objects such as dots in an array (Nieder, 2005). However, humans can augment this basic ability through the additional use of numerical symbols (represented in other regions of the brain) such as written digits, number names, tallies, and so on, which extends their numerical abilities beyond other species (Dehaene, Dehaene-Lambertz, & Cohen, 1998) – that is, cognition itself is transformed by the availability of certain culturally learned symbols. For example, cultures that lack number words for numbers above four (using a term corresponding to 'many' for all quantities greater than four) appear to have some difficulties in understanding large exact quantities but can understand large approximate quantities (Pica, Lemer, Izard, & Dehaene, 2004). For example, if asked to add together 5 stones with 7 stones they may choose an answer that is approximately 12 (e.g. 11, 12, or 13) but are less likely to choose a more distant number (e.g. 8 or 20) – see Figure 3.13. Similarly, even in highly numerate cultures, certain forms of higher math (e.g. algebra, multi-digit calculations) can be performed with ease using pen and paper (or calculator and computer) because these systems effectively function as externalized working memories, enabling humans to escape the confounds of our own limited capacity and error-prone memory systems. When viewed in this way, symbols and tools are quite literally 'mind expanding' (Clark, 2008). By offloading certain cognitive capacities (e.g. for remembering, calculating, reaching) onto

KEY TERM

Visual Word Form Area (VWFA)
A brain region in the ventral visual stream that responds to visual presentation of letter strings more than it responds to other objects

Figure 3.13 Although certain cognitive abilities are shaped by biological evolution (e.g. the ability to judge the approximate number of items in an array), the cultural 'evolution' of ideas and symbols can extend this capacity. In this example, the numerical abilities of an Amazonian tribe (the Mundurukú) with no names for large numbers (but a generic name meaning 'many') is assessed on tasks involving putting different numbers of counters into a tin (for addition) and/or taking them out (for subtraction). It is still possible to perform approximate arithmetic without any words for large numbers, but not exact arithmetic with large numbers. Adapted from Pica et al. (2004).

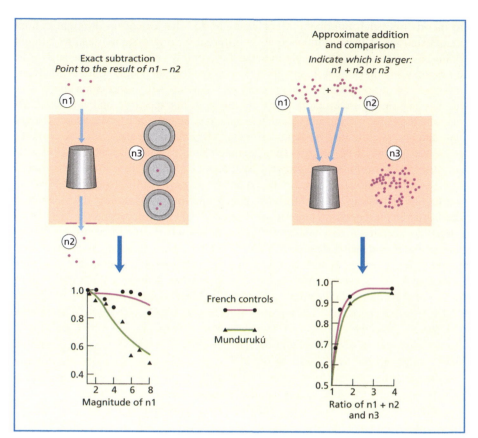

external technology, it is claimed that we create an **extended cognition** that bridges the brain-based and material-based worlds (Clark, 2008).

Although systems of writing and number representation can be considered social in the narrow sense of having been invented and passed on by the collective action of many minds, they have had more direct influences on the nature of social interactions. The most obvious example is money. Indeed most of the earliest written records were for trade transactions rather than, say, poetry or stories. According to some contemporary thinkers (e.g. Lea & Webley, 2006), the function of money is essentially social. It may serve two broad social functions: as a means of social exchange (related to the notion of reciprocal altruism: 'if you scratch my back, I'll scratch yours') and also as a way of displaying or achieving a higher social standing through conspicuous consumption or benevolence. This is an interesting example of how our culture mirrors our biology.

CULTURAL SKILLS: TOOLS AND TECHNOLOGY

If culturally based symbols enable us to escape the constraints of our own minds (e.g. escaping the limits of our working memory), then cultural tools could be said

KEY TERM

Extended cognition
The use of external technologies (e.g. writing systems, tools, computers) to increase cognitive capacities (e.g. for remembering, calculating)

to free us from the constraints of our own bodies. They enable us to fly (e.g. in an airplane or spacecraft) and they enable us to perform extraordinary feats of strength (e.g. chopping trees, killing larger animals). The human body is not adapted for flying or chopping trees, but our brains are adapted to create useful objects (**tools**) and transmit this information, socially, from person to person. As noted above, some have argued that cultural use of symbols and tools enables new kinds of thought. For instance, Clark (2003) dismisses the notion, popular in evolutionary psychology, that modern-day humans are stuck with the Stone Age minds that were selected for in our earliest ancestors. For Clark (2003) we are 'natural-born cyborgs' capable of soaking up and creating complex technologies. The technology and the ideas behind them are themselves passed from person to person (and modified over time), not in the genes but by social and cultural transmission. But, crucially for his argument, in taking on such technology our minds and brains are themselves transformed.

Modifying the brain by using tools and technology

There are certain neurons in the brain that respond both when a particular body part is touched and when a visual stimulus is moved near to the same body part. These neurons are found in both frontal (Graziano, 1999) and parietal (Graziano, Cooke, & Taylor, 2000) regions and they can be said to be multi-sensory insofar as they receive input from more than one sensory system. They are normally studied in monkeys via the method of single-cell electrophysiology, which records how often a neuron 'fires' (i.e. produces an action potential) in response to a particular stimulus. If the neuron produces a large response (relative to some baseline, such as spontaneous activity), then it is concluded that the neuron codes information related to that stimulus. Some of these multi-sensory neurons might fire when the monkey's hand is touched, even if the hand cannot be seen, and even when the hand is moved around in space. These same neurons also fire when a visual stimulus is placed on or near the hand, again irrespective of where the hand is. The region of space that elicits a neuronal response is termed the neuron's **receptive field**, and in this example we can say that the receptive field is centered on the hand rather than being at some fixed coordinate relative to the eyes.

Iriki, Tanaka, and Iwamura (1996) noted that the visual receptive fields of these neurons changed as a result of the monkey using a tool (a rake for getting peanuts out of reach). As a result of using the tool, the receptive field was no longer centered on the arm but was elongated down the length of the tool itself. It was as if the monkey's neural representation of its body had been stretched to incorporate the tool, as illustrated in Figure 3.14. There was also an important control condition in which the monkey passively held the tool but did not use it. In this condition, the receptive field was not extended. This control condition physically resembles the tool-use condition but is cognitively equivalent to the no-tool-use condition.

In humans there is evidence that multi-sensory processing of space is extended by tool use. When sighted people are trained to use a blind-person's cane there is evidence that multi-sensory space becomes expanded along the length of the cane (Serino, Bassolino, Farne, & Ladavas, 2007). A sound emanating near the end of the cane (1.25 m away) can facilitate detection of a weak tactile stimulus applied to the hand. However, when the sighted person passively holds the cane (without using it as a tool) this does not occur. Thus, their brains temporarily adapt to cane use. Blind

KEY TERMS

Tools
Objects, normally hand-held, used to manipulate secondary objects

Receptive field
The region of space that elicits a neuronal response

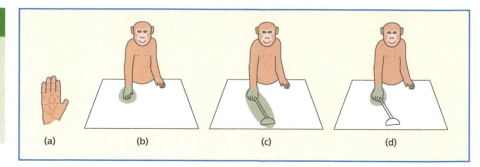

Figure 3.14 Some neurons respond to touch to the hand (a) and also to the sight of a visual stimulus near the hand, shown in green (b). After the monkey has been trained to use a tool the neuron may also respond to visual stimuli along the length of the tool (c). It is as if the tool is an extension of the body. Merely holding a tool does not have this effect (d). Note that (c) and (d) are physically equivalent but have very different neural responses. From Iriki & Sakura (2008). Reproduced with permission.

people who have extensive experience of cane use show evidence of extended body space even from passive holding, suggesting that their brains have undergone more permanent adaptation. In the visual, rather than auditory, domain flashes of light both near the hand and at the end of a tool can facilitate detection of a tactile stimulus on the hand after tool use (Holmes, Calvert, & Spence, 2007). Prior to tool use, only a flash of light on or near the hand (but not the end of the tool) has this effect.

Having considered one example of how an individual's brain may be modified via tool use, I shall go on to consider how this process may spread, at the neural level, via imitation.

Mirror neurons, action understanding, and imitation

One of the most fascinating discoveries in cognitive neuroscience over the last decade has been of the **mirror neuron** system. Rizzolatti and colleagues found a group of neurons in the monkey premotor cortex (area F5) that respond both during the performance and during observation of the same action (e.g. di Pellegrino, Fadiga, Fogassi, Gallese, & Rizzoloatti, 1992; Rizzolatti, Fadiga, Fogassi, & Gallese, 1996). Thus, the response properties of mirror neurons disregard the distinction between self and other, and this may provide a crucial basis for imitation. They respond to actions performed by the experimenter or another monkey as well as to actions performed themselves – see Figure 3.15. The response properties of these neurons are quite specific. They are often tuned to precise actions (e.g. tearing, twisting, grasping) that are goal-directed. They do not respond to mimicked action in the absence of an object, or if the object moves robotically without an external agent. This suggests that it is the purposeful nature of the action rather than the visual/motoric elements that is critical. As such, mirror neurons have been likened to 'intention detectors' (Iacoboni et al., 2005).

Mirror neurons respond if an appropriate action is implied as well as directly observed. Umilta et al. (2001) compared viewing of a whole action versus viewing of

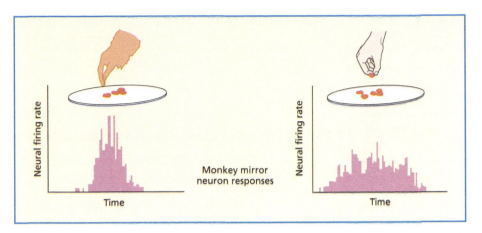

Figure 3.15 Mirror neurons respond both when the animal performs an action (left) and when the animal sees someone else perform the action (right). This idea of self–other similarity, operating at the neural level, has been very influential in social neuroscience, including for theories of social learning and imitation. From Rizzolatti et al. (2006). Reproduced with permission from Lucy Reading-Ikkanda for *Scientific American Magazine*.

the same action in which a critical part (the hand–object interaction) was obscured by a screen – see Figure 3.16. These findings suggest that the premotor cortex contains abstract representations of action intentions that are used both for planning one's own actions and interpreting the actions of others.

Umilta et al. (2008) have shown that neurons that respond to grasping with the hand will also respond when a tool (pliers) are used to grasp. This occurs both with normal pliers, in which the action is to squeeze the tool, and reverse pliers in which the action is to relax the grip – see Figure 3.17. In this example, the action is different, but the goal is the same, and the neural response is determined by the goal. Studies such as these have been used to argue that mirror neurons enable understanding of at least one mental state: intentions.

Subsequent research has found mirror neurons in other parts of the macaque brain, such as the parietal lobes (Fogassi et al., 2005), but they do not necessarily have the same functional properties as those described in the premotor cortex. Mirror neurons in the parietal lobe tend to be more sensitive to the wider context in which an action is situated, for instance responding to a grasping action differently depending on whether the subsequent goal is to eat it or put it in a container (Bonini et al., 2010). The primary motor cortex itself contains neurons with motor and visual properties, but they respond to the mechanics of particular movements rather than more abstract features such as goals (Dushanova & Donoghue, 2010). By contrast, other regions such as the superior temporal sulcus also respond to specific movements of body parts but have a purely visual component (Perrett et al., 1989) that may act as input to the mirror neuron system. Figure 3.18 summarizes the main regions linked to mirror neurons.

The evidence above is derived from non-human primates. What is the evidence that humans possess such a system? The human analogue of area F5 is believed to be in Broca's area (specifically in Brodmann's area 44) extending into the premotor area (Rizzolatti, Fogassi, & Gallese, 2002). This region is activated by the observation

Figure 3.16 Mirror neurons respond to *inferred* goal-directed actions as well as those observed. In this example: (1) the monkey *sees* a goal-directed action to an object; (2) the monkey *sees* the same action but without an object; (3) the monkey *knows* that the object is there, because it has previously seen it, even though it cannot see it now; and (4) the monkey knows the object is not there. Note that conditions (3) and (4) are visually identical but only in condition (3) does the mirror neuron respond. The hand and object in (3) and (4) cannot be seen, but are drawn here for illustrative purposes. From Rizzolatti et al. (2006). Reproduced with permission from Lucy Reading-Ikkanda for Scientific American Magazine.

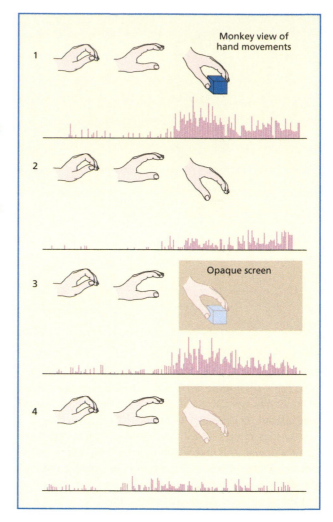

Figure 3.17 Mirror neurons respond to the same goal rather than the same action. Mirror neurons in monkeys responded similarly after training with both normal pliers and reverse pliers (maximum responding at point of grasping the food), even though both required different actions. From Umilta et al. (2008). Copyright © 2008 Proceedings of the National Academy of Science, USA. Reproduced with permission.

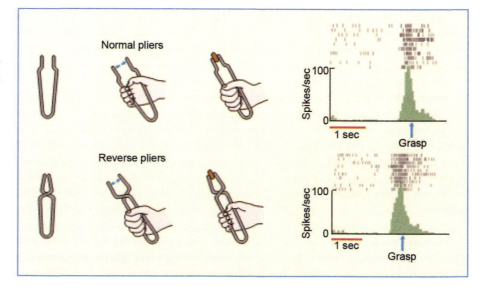

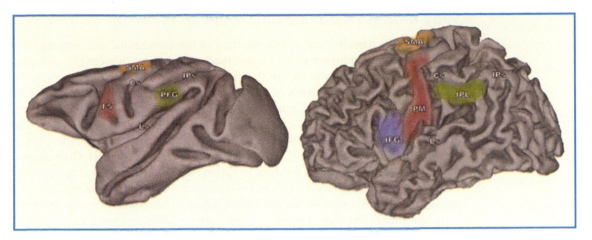

Figure 3.18 The main regions linked to mirror neuron activity in macaque and human brain (left and right respectively). Colors show homologous regions in different species. PM= premotor; SMA = supplementary motor area; IPL = inferior parietal lobule; IFG = inferior frontal gyrus. From Vanderwert, Fox, and Ferrari (2013).

of hand movements, particularly when imitation is required (Iacoboni et al., 1999), and also the observation of lip movements within the human repertoire (e.g. biting and speaking but not barking) (Buccino et al., 2004). TMS applied over the primary motor cortex increases the amplitude of motor-evoked potentials elicited in the hands/arms when participants also observed a similar action (Strafella & Paus, 2000). This suggests that action observation biases activity in the primary motor area itself. Direct evidence of mirror neurons in humans, in terms of the firing response properties of individual neurons, was lacking until much later (Mukamel et al., 2010). These were patients undergoing surgery, and this limited the regions that could be explored. Although Broca's area and the premotor area were not studied, mirror neurons were found in other regions of the brain (e.g. medial temporal region).

Mirror neurons have been subjected to intense critical scrutiny in recent times, with one book referring to *The Myth of Mirror Neurons* (Hickok, 2014). The controversy surrounds their functionality (what they actually do) rather than their existence (which isn't disputed). For instance, some researchers argue that mirror neurons can arise via associative learning (Cook, Bird, Catmur, Press, & Heyes, 2014). When we move our own bodies then we see the visual consequences of our actions and learn to associate action observation and action execution together. In this view, mirror neurons are not genetically pre-programmed for imitation or any other function. Similarly, one can debate whether mirror neurons give rise to action understanding, in its entirety, or whether there is a separate system involved in action understanding with mirror neurons acting downstream rather centrally. One theory argues that mirror neurons function to predict what the consequences of actions are, such as how the limbs will move through space in order to achieve a goal, but that other regions represent the intended goals (Kilner, 2011). The role of mirror neurons in social behavior will be considered at multiple points throughout the book (notably Chapter 6). The next section will focus specifically on tool use and imitation in non-human primates.

Why monkeys do not use tools

There is one potentially fatal flaw in the story of mirror neurons, imitation, and tool use – namely, that mirror neurons are assumed to be present in monkeys, chimpanzees, and humans, but evidence for tool use in the wild is virtually nonexistent for monkeys, common in chimpanzees, and extensive in humans. Macaque monkeys are able to perform some kinds of imitation-like behavior. For instance, neonate monkeys reproduce basic facial gestures such as tongue protrusion and mouth opening (Ferrari et al., 2006). Within the taxonomy presented above, this could perhaps reflect contagion (of innate motor programs) rather than imitation based on goals. Chimpanzees are capable of more complex forms of imitation (see above), but it perhaps does not serve the same social functions as imitation in humans. So what is missing between a macaque with mirror neurons but minimal imitation, a chimp with mirror neurons and some evidence of imitation, and humans with mirror neurons and boundless imitation?

Iriki and colleagues have suggested a potential solution to this problem (Iriki, 2006; Iriki & Sakura, 2008). Under their account, mirror neurons may be a necessary precursor to imitative tool use but are not sufficient (see also Rizzolatti, 2005). First of all, macaque monkeys *can* use tools in the laboratory but only after extensive training. It takes a minimum of 10–14 days of intensive training using a specially developed training regime (e.g. Hihara et al., 2006). This training regime involves systems of reward and gradual modification of behavior rather than imitation or social learning of tool use. This long length of time plus the nature of the training may be sufficient to prevent all but very minimal tool use in the wild by these monkeys (e.g. bending a branch to get a fruit at the end), but as a proof of principle, they can use tools. Moreover, when they have achieved it their performance is swift and effortless and shows some degree of flexibility. Without hesitation, the monkeys can use a short rake to pull a long rake to get a more distant food (i.e. a chaining process of successive tools) (Hihara, Obayashi, Tanaka, & Iriki, 2003). The key question is what are the differences in the brains of macaque monkeys who have acquired tool use versus those that have not? The answer, according to Iriki and Sakura (2008), lies in the way that two particular regions are connected. In monkeys who are proficient tool-users there are extra connections between the intraparietal sulcus and the temporo-parietal junction that are absent in monkeys who cannot use tools (Hihara et al., 2006). The intraparietal region contains neurons whose visual receptive fields are extended via tool use and also mirror neurons. The temporo-parietal junction, in humans, has been implicated both in theory of mind (Frith & Frith, 2003) and in feelings of embodiment, for instance when contrasting physical perspectives between self and other (Blanke et al., 2005). Changes in gene expression in the intraparietal region accompany learning of tool use (Ishibashi et al., 2002) and presumably trigger the connectivity changes. The implication of this finding, in evolutionary terms, is that the human brain may have evolved (via genetic modification) stable connections between these two regions that are normally absent in many other primates. This may enable humans to link neural mechanisms related to tool use (e.g. multi-sensory visuo-tactile neurons) with mechanisms related more closely to social cognition (perspective taking, theory of mind). The question of whether this adaptation is specific to tool use or a response to more general evolutionary pressures is unknown.

Evaluation

Tool use is a particularly interesting example of culture for a number of reasons. First, there are obvious parallels between human tool use and that found in other species.

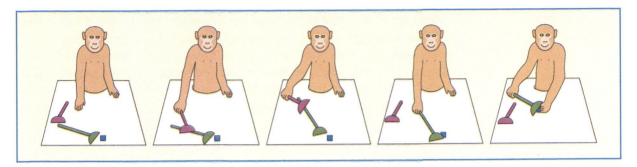

Figure 3.19 This monkey has been trained to use a single tool to reach a food reward. However, when the tool is too short (purple) but can be used to reach a longer tool (green) the monkey – without further training – is able to use the short tool to get the long tool to get the food reward. From Iriki & Sakura (2008). Reproduced with permission.

However, there are important differences too. Second, tool use is now beginning to be understood in terms of basic neuroscience, offering the possibility of linking together different levels of explanation from genes through to the behavior of individuals and groups. Mirror neurons may be an important starting point for imitation and the spread of tools and technology, but recent evidence suggests that other aspects of brain function are crucial too.

SUMMARY AND KEY POINTS OF THE CHAPTER

- The social intelligence hypothesis argues that evolutionary pressures to be socially smarter lead to more general changes (e.g. larger brain size) resulting in increased intellect in non-social domains. Evidence for this comes from the correlation between relative brain size in different primates and factors such as the size of social groups, the degree of deception, and the amount of social learning.
- It remains keenly debated whether the need to deal with social complexity led to general changes linked to intellect (e.g. bigger brains) or shaped intelligence in a more precise way (e.g. by creating specialized neural circuits to represent the social world).
- Culture can be defined as a shared set of values, skills, artifacts, and beliefs amongst a group of individuals. Many species have elements of culture, including social learning (e.g. of food caches) and traditions (e.g. different tribes of chimpanzee use stick tools in different ways).
- There are likely to be different mechanisms for social learning. Humans, and possibly some other primates, may learn from each other by inferring intentions (i.e. imitation).
- Mirror neurons respond both when an animal performs an action and when it sees someone else performing the action. Their response

depends on the goal rather than the movement per se and it has therefore been suggested that it provides the neural basis of imitation.

- Monkeys, who are known to have mirror neurons, do not necessarily use tools in the wild and they show limited evidence for spontaneous imitation. This suggests that mirror neurons may be necessary but not sufficient for imitative tool use by animals.

EXAMPLE ESSAY QUESTIONS

- What is the 'social intelligence hypothesis' and what is the evidence for and against it?
- Do non-human animals have culture?
- Modern human culture is too recent to have been shaped by genetic evolution, so what kind of primitive mental capacities have made it possible?
- What kind of neural and cognitive mechanisms enable imitation by humans and other animals?

RECOMMENDED FURTHER READING

- Emery, N., Clayton, N., & Frith, C. (2008). *Social Intelligence: From Brain to Culture.* Oxford: Oxford University Press. This extensive collection of papers was originally published in *Philosophical Transactions of the Royal Society B* (2007, vol. 362, pp. 485–754).

- Hurley, S., & Chater, N. (2005). *Perspectives on Imitation: From Neuroscience to Social Science.* Cambridge, MA: MIT Press. A collection of essays in two volumes dealing with humans and other animals.

- Rizzolatti, G., Cattaneo, L., Fabbri-Destro, M., & Rozzi, S. (2014). Cortical mechanisms underlying the organization of goal-directed actions and mirror neuron–based action understanding. *Physiological Reviews, 94*(2), 655–706. A thorough review of the neurobiology of mirror neurons from the group who discovered them. For the contrary view, an accessible read is Hickok, G. (2014). *The Myth of Mirror Neurons.* New York: W. W. Norton and Company.

ONLINE RESOURCES

- References to key papers and readings
- Interviews and lectures given by leading figures including Michael Tomasello, Giacomo Rizzolatti, Robin Dunbar, and others

- Recorded debate on 'Do Mirror Neurons Explain Anything?' with Vittorio Gallese and Gregory Hickok
- Recorded lecture given by textbook author, Jamie Ward
- Multiple choice questions and interactive flashcards to test your knowledge
- Downloadable glossary

CHAPTER 4

CONTENTS

Emotion and motivation

The classic science fiction depiction of androids such as C3PO in *Star Wars* and Data in *Star Trek: Next Generation* is of superhuman intelligent beings able to speak many languages and store vast amounts of information. Nevertheless, such intelligence does not enable them to fully understand the eccentric behaviors of their human colleagues who constantly place themselves in danger, fall in love, and tell jokes. These androids lack **emotions**. Reading between the lines of these popular depictions we might conclude various things. We might conclude that emotions are 'what makes us human, force us to make illogical decisions, and that we could do without them if redesigned from scratch. Needless to say, these conclusions are incorrect. Emotional processes have a long evolutionary history and are by no means unique to humans. What may 'make us human' is our ability to consciously reflect on our emotions and share them socially via our language and culture, but not our emotions per se. If a new organism were redesigned from scratch it would still be helpful to have early warning routines for danger and fast-acting mechanisms that prepare it to fight or flee. It would still be helpful to devote greater attention to stimuli that are necessary for survival. These are all considered functions of emotions. Finally, emotions do sometimes lead to decisions that may not have occurred via more deliberative reasoning but this, by itself, does not make them illogical. For instance, in cooperative games with another person we often make decisions based on social values of fairness rather than maximizing individual financial gain (e.g. Sanfey, Rilling, Aaronson, Nystron, & Cohen, 2003). This is not necessarily illogical – there may be good survival reasons, honed by evolution, that promote such cooperation. For humans, many social stimuli and situations are rewarding (e.g. imitation, cooperation) or punishing (e.g. social exclusion). As such, both social stimuli and non-social stimuli are likely to have been selected as having survival value in our evolutionary past.

KEY TERMS

Emotions
States associated with stimuli that are rewarding or punishing

Reward
An outcome that one is willing to work to obtain

Punishment
An outcome that one is willing to work to avoid

Mood
An emotional state that is extended over time (e.g. anxiety is a mood and fear is an emotion)

Hedonic value
The subjective liking or disliking of a stimulus/event

SOME CHARACTERISTICS OF EMOTIONS

- An emotion is a state associated with stimuli that are *rewarding* (i.e. that one works to obtain) or *punishing* (i.e. that one works to avoid). These stimuli often have inherent survival value.
- Emotions are transient in nature (unlike a **mood**, which is where an emotional state becomes extended over time), although the emotional status of stimuli is stored in long-term memory.
- An emotional stimulus directs attention to itself, to enable more detailed evaluation or to prompt a response.
- Emotions have a **hedonic value**, that is, they are subjectively liked or disliked.

- Emotions have a particular 'feeling state' in terms of an *internal* bodily response (e.g. sweating, heart rate, hormone secretion).
- Emotions elicit particular *external* motor outcomes in the face and body, which include emotional **expressions**. These may prepare the organism (e.g. for fighting) and send signals to others (e.g. that one intends to fight).

An emotion can be regarded as a state that can have various facets – conscious and unconscious; internal and external; automatic or controlled. The precise nature of the state may vary according to the stimulus, learned history, and current context. An emotional stimulus may also affect processing in other more basic cognitive mechanisms – for example, by making a memory more memorable (e.g. Cahil, Prins, Weber, & McGaugh, 1994) and by directing attention to certain objects or locations (e.g. Vuilleumier, 2005). As such, a theory of the 'neuroscience of emotions' is likely to entail a range of different interacting brain processes, in the same way as contemporary theories of vision divide processing amongst routines specialized for shape, color and motion, visually guided action, and so on.

Why are some stimuli associated with emotions and others are not? The standard answer to this question is that some stimuli are more important than others (e.g. because they enhance or threaten survival chances). Emotions are one way of tagging these stimuli to ensure that they receive priority treatment and are responded to appropriately. Broadly speaking, they can be tagged in one of two ways: either as something that is to be sought (i.e. a rewarding stimulus) or avoided (i.e. a punishing stimulus). As such, many theories closely tie emotions with the concept of **motivation**. For example, Rolls (2005) defines emotions as states elicited by rewards and punishers, whereas motivation is defined as states in which rewards are sought and punishers are avoided. Importantly, emotions are not just tied to stimuli but also to predicted stimuli. Thus, the omission of an expected reward can lead to emotions (e.g. anger), as can omissions of expected punishment (e.g. relief). For humans, we can make the further claim that we *like* rewards and we *dislike* punishers, and that we are motivated to seek the things we like and avoid the things we dislike. (For animals, we tend to avoid the terms like/dislike and adopt more neutral terminology such as seek/avoid because we cannot know their subjective feelings.) Although we may be born with a core set of basic likes and dislikes (e.g. we like sweet things and dislike pain), it is possible to arbitrarily learn new emotional associations by pairing neutral stimuli with emotive responses. We may come to be afraid of flying in airplanes, or we may come to like certain painful stimuli (e.g. eating chilies, fetishes). As such, emotional learning is a highly flexible system that is not limited to stimuli in our evolutionary past and extends beyond stimuli with obvious survival value.

This chapter will first consider different historical accounts of emotion. It will then consider whether or not there are discrete 'basic emotions' in the brain. Finally, it will go on to consider the role of emotions in motivation and goal-directed behavior. Throughout, I will present examples of emotional processes in social and nonsocial contexts and consider how they may be related.

HISTORICAL PERSPECTIVES ON THE EMOTIONS

Darwin's evolutionary theory of emotion

In 1872, Charles Darwin published *The Expression of the Emotions in Man and Animals* (Darwin, 1872/1965). For much of this work Darwin was concerned with documenting the outward manifestations of emotions – expressions – in which animals produce facial and bodily gestures that characterize a particular emotion such as fear, anger, or happiness. Darwin noted how many expressions are conserved across species: anger involves a direct gaze with mouth opened and teeth visible, and so on. This is shown in Figure 4.1. He claimed that such expressions are innate 'that is, have not been learnt by the individual'. Moreover, such expressions enable one animal to interpret the emotional state of another animal – for example, whether an animal is likely to attack, or is likely to welcome a sexual advance. For Darwin, an emotional expression was a true reflection of an inner state: 'They reveal the thoughts and intentions of others more clearly than do words, which may be falsified.'

Darwin's contribution was to provide preliminary evidence as to how emotions may be conserved across species. His reliance on expressions resonates with some contemporary approaches, such as Ekman's attempts to define 'basic' emotions from cross-cultural comparisons of facial expressions (e.g. Ekman, Friesen, & Ellsworth, 1972). This is covered in detail later in the chapter. More recent research has elucidated the functional origins of some of these expressions; for instance; a posed fear expression increases the visual field and nasal volume and leads to faster eye movements (adaptive for detecting danger), whereas a disgust expression has an opposite effective (adaptive for avoiding contaminants) (e.g. Susskind et al., 2008).

Freud and unconscious emotional motivations

For Freud, our minds could be divided into three different kinds of mechanisms: the id, the ego, and the super-ego. The **id** was concerned with representing our 'primitive'

Figure 4.1 Darwin argued that many emotional expressions have been conserved by evolution.

EMOTION AND THE 'RIGHT BRAIN'

This notion of the right hemisphere being more emotional than the left has had an enduring influence on popular scientific views of the brain. Following Broca's discovery that language is a predominantly left-hemisphere faculty, nineteenth-century neurologists speculated on possible complementary specializations for the right hemisphere. Emotions were considered a good candidate for right hemispheric specialization given the long-held, but misguided, view that emotions are the opposite of logic. However, this 'right brain hypothesis' of emotions is essentially incorrect.

Contemporary theories concerning the neural basis of emotions assume that both hemispheres are crucially involved in emotional processing. However, there is evidence of subtle laterality differences between the cerebral hemispheres. When *producing* emotional expressions, there is evidence from muscle recordings that the left side of the face (controlled by the right hemisphere) is more expressive than the right side of the face (Dimberg & Petterson, 2000). This was found for both a positive expression (smiling) and a negative one (anger). When *recognizing* emotional expressions, there is evidence that the valence of the emotion is important. The left side of a face is judged to be sadder and the right side happier (Nicholls, Ellis, Clement, & Yoshino, 2004). This implies a right-hemisphere bias for sadness recognition and a left-hemisphere bias for happiness recognition. Note that in these examples, both hemispheres are implicated in producing and recognizing emotions, but one hemisphere may have a small relative advantage over the other.

KEY TERMS

Id
Unconscious motivations that represent 'primitive' urges from our nonhuman ancestry (in Freudian theory)

Ego
The conscious self operating according to reason rather than passion (in Freudian theory)

Super-ego
The ideal self such as our cultural norms and our aspirations (in Freudian theory)

urges that connect us to non-human ancestry. It includes motivations to meet our basic emotional needs for sex, food, warmth, and so on. The id was concerned with unconscious motivations, but these ideas would sometimes be consciously accessible via the **ego**, which operates according to reason rather than passion. The **super-ego**, by contrast, represents the ideal self, such as our cultural norms and our aspirations. For Freud (and many of his clients), there was a perceived conflict between the super-ego, for which sexual behavior was tightly regulated by cultural norms, and the id, for which more unbridled sexual impulses were considered as desirable.

The specific details of Freud's theory no longer have contemporary currency. Many of his ideas (e.g. relating to childhood sexual fantasies) were derived from anecdotes and speculation rather than scientific testing. However, the basic idea that emotions are an unconscious bias in our behavior is very much relevant. For example, simple emotional reactions can be elicited from stimuli that are presented too briefly to be consciously seen (Tamietto & De Gelder, 2010). Most cognitive models of emotions assume that the majority of emotional processing occurs unconsciously. Freud's other enduring influence is the notion that many psychiatric disorders can be understood as emotional disturbances. Freud was particularly interested in neuroses, or what would now be called anxiety disorders, and today many of these are understood as emotional disturbances (e.g. LeDoux, 1996).

The James–Lange theory

One of the founding fathers of psychology, William James, proposed a theory of emotion that placed the somatic (i.e. bodily) response of the perceiver at its center (James, 1884). This theory later became known as the **James–Lange theory** of emotion. According to this theory, it is the self-perception of bodily changes that produces emotional experience. Thus, changes in bodily state precede the emotional experience rather than the other way around. We feel sad because we cry, rather than we cry because we feel sad: see Figure 4.2. This perspective seems somewhat radical compared to the contemporary point of view elaborated thus far. For instance, it raises the question of what type of processing leads to the change in bodily states and whether or not this early process could itself be construed as a part of the emotion. Changes in the body are mediated by the autonomic nervous system (ANS), a set of nerves located in the body that controls activity of the internal organs.

There is good empirical evidence to suggest that changes in somatic state, in themselves, are not sufficient to produce an emotion. Schacter and Singer (1962) injected participants with epinephrine (also termed adrenaline), a drug that induces autonomic and visceral changes. They found that the presence of the drug by itself did not lead to self-reported experiences of emotion, contrary to the James–Lange theory. However, in the presence of an appropriate cognitive setting (e.g. an angry or happy man enters the room), the participants did self-report an emotion. A cognitive setting, but without epinephrine, produced less intense emotional ratings. This study suggests that bodily experiences do not create emotions (contrary to the James–Lange theory) but they can enhance conscious emotional experiences.

> **KEY TERM**
>
> **James–Lange theory**
> The self-perception of bodily changes produces emotional experience (e.g. one is sad because one cries).

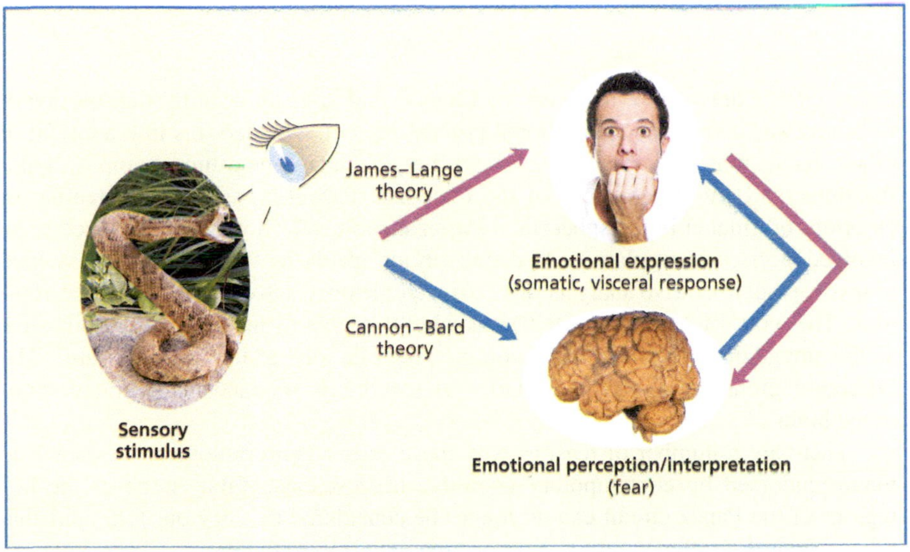

Figure 4.2 According to the James–Lange theory, bodily reactions occur first and emotional processing occurs after (as the perception/interpretation of those reactions). According to the Cannon–Bard theory, the emotional perception/ interpretation occurs first and the bodily reaction occurs after.

There are several contemporary theories that bear similarity to the James–Lange theory, most notably Damasio's (1994) suggestion that bodily responses linked to emotions guide decision-making. Although the James–Lange theory states that these bodily responses must be consciously perceived, Damasio (1994) takes the different view that they are unconscious modifiers of behavior.

The Cannon–Bard theory

The **Cannon–Bard theory** of emotions that emerged in the 1920s argued that bodily feedback could not account for the differences between the emotions (Cannon, 1927). According to this view, the emotions could be accounted for solely within the brain, and bodily responses occur after the emotion itself. The Cannon–Bard theory was inspired by neurobiology. Earlier research had noted that animals still exhibit emotional expressions (e.g. of rage) after removal of the cortex. This was considered surprising given that it was known that cortical motor regions are needed to initiate most other movements (Fritsch & Hitzig, 1870). In a series of lesion studies, Cannon and Bard concluded that the hypothalamus is the centerpiece of emotions. They believed that the hypothalamus received and evaluated sensory inputs in terms of emotional content, and then sent signals to the autonomic system (to induce the bodily feelings discussed by James) and to the cortex (giving rise to conscious experiences of emotion).

Although it has not stood the test of time (e.g. the hypothalamus is not a central nexus of emotions, although it does regulate bodily homeostasis), the theory was important historically in providing an alternative to the James–Lange theory and also for the development of another important theory: namely the Papez circuit and the limbic brain hypothesis.

The Papez circuit and the limbic brain

Papez (1937) drew upon the work of Cannon and Bard in arguing that the hypothalamus was a key part of emotional processing, but extended this into a circuit of other regions that included the regions of the cingulate cortex, hippocampus, hypothalamus, and anterior nucleus of the thalamus. Papez argued that the feeling of emotions originated in the subcortical **Papez circuit**, which was hypothesized to be involved in visceral regulation. A second circuit, involving the cortex, was assumed to involve a deliberative analysis that retrieved memory associations about the stimulus. The work of MacLean (1949) extended this idea to incorporate regions such as the amygdala and orbitofrontal cortex, which he termed the 'limbic brain'. The different regions were hypothesized to work together to produce an integrated 'emotional brain'.

There are a number of reasons why these earlier neurobiological views are no longer endorsed by contemporary cognitive neuroscience. First, some of the key regions of the Papez circuit can no longer be considered to carry out functions that relate primarily to the emotions. For example, the role of the hippocampus in memory was not appreciated until the 1950s (e.g. Scoville & Milner, 1957). Second, contemporary research places greater emphasis on different types of emotion (e.g. fear versus disgust). Each basic emotion may form part of its own circuit, and different parts of the circuit may make different cognitive contributions.

A QUICK TOUR OF THE EMOTIONAL BRAIN

In this chapter, five regions of the brain are considered in detail, and their basic architecture and functions are summarized below for reference.

The amygdala

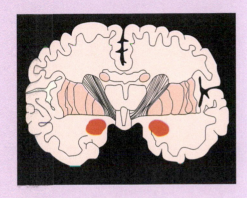

- A collection of nuclei buried bilaterally in the anterior temporal poles
- It receives connections mainly from the overlying temporal lobes (involved in higher stages of sensory processing and conceptual knowledge). There may be some inputs from early sensory areas (e.g. auditory inputs via medial geniculate nucleus; LeDoux, 1996) and subcortical inputs including parts of the hippocampus, hypothalamus, and olfactory structures. Its outputs include the hypothalamus, ventral striatum, and temporal, orbitofrontal and insula cortex (for a review see Amaral, Price, Pitkanen, & Carmichael, 1992).
- Involved in learning the emotional value of stimuli (e.g. via classical conditioning) and coding emotional salience

The insula

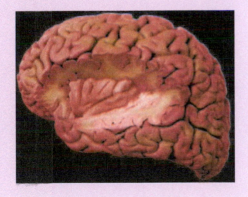

- A region of cortex lying beneath the temporal lobes
- Connects anteriorly to the orbitofrontal cortex, limbic structures, and basal ganglia; the posterior region receives connections from sensory thalamus and parietal and temporal association cortex.
- Anterior portion of the insula is considered to be involved in interoceptive awareness (e.g. detection of heartbeat) and bodily feelings in general.
- Some evidence of a preferential involvement in disgust perception

The anterior cingulate cortex

- Located around the anterior corpus callosum on the medial surface of the brain and often divided into dorsal regions and ventral regions.

- It has connections from medial thalamic areas (concerned with pain perception), orbitofrontal cortex, amygdala, and insula. It has output connections to the periaqueductal gray area (linked to pain), dorsal motor nucleus of the vagus (elicits autonomic effects), and ventral striatum (for a review see Van Hoesen, Morecraft, & Vogt, 1993).

- Involved in the production of certain bodily responses elicited by an emotional stimulus, such as skin conductance response (Tranel & Damasio, 1995) and changes in heart rate and blood pressure (Critchley et al., 2003)

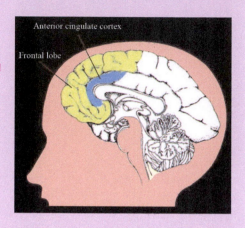

- Dorsal region involved in monitoring of responses, for instance in terms of whether a response is incorrect and in terms of whether responses are rewarded or punished.

- Ventral region of cingulate is adjacent to medial prefrontal cortex region implicated in 'mentalizing', but the specific role of this cingulate region is not agreed upon.

The orbitofrontal cortex

- Located on the ventral surface of the frontal lobes, above the eye sockets (orbits)
- Receives connections from cortical sensory areas, and has reciprocal connections with areas such as the amygdala, hippocampus, insula, and cingulate cortex (Cavada, Company, Tejedor, Cruz-Rizzolo, & Reinoso-Suarez, 2000)
- Computes the motivational value of rewards (e.g. whether I would like chocolate now rather than whether I like chocolate per se), and changes the value of rewards according to context (e.g. reversal learning)

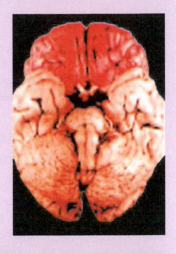

The ventral striatum

- Part of the basal ganglia and includes the nucleus accumbens
- Involved in a 'limbic circuit' connecting the orbitofrontal cortex, basal ganglia, and thalamus
- Important for operant conditioning, for example learning to press a lever when a certain tone is heard in order to obtain a reward
- Responds to rewards and the anticipation of rewards. The latter has been used to argue that it computes a reward prediction error (i.e. the discrepancy between actual reward and expected reward).

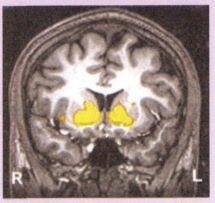

DIFFERENT CATEGORIES OF EMOTION IN THE BRAIN

One of the first challenges faced by the empirical study of emotions is how to go about categorizing them or, indeed, whether it makes more sense to treat all emotions as a single entity. Are some types of emotion (e.g. happy, sad) more basic or primary than other types (e.g. love, jealousy)? Are emotional categories independent of language and culture?

Emotions as basic kinds versus constructions/appraisals

One of the most influential ethnographic studies of the emotions concluded that there are six **basic emotions** that are independent of culture (Ekman & Friesen, 1976; Ekman et al., 1972): happy, sad, disgust, anger, fear, and surprise (shown in Figure 4.3). This study was based on comparisons of the way that facial expressions are categorized and posed across diverse cultures. Ekman (1992) considers other characteristics for classifying an emotion as 'basic' aside from universal facial expressions, such as: each emotion having its own specific neural basis; each emotion having

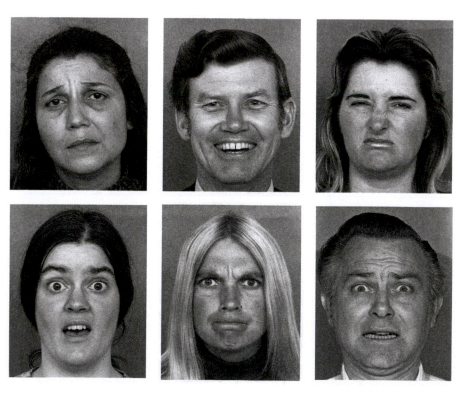

Figure 4.3 Ekman (e.g. 1992) has argued that there are six basic emotions that manifest themselves as universal (i.e. cross-cultural) facial expressions: happy, sad, fear, anger, disgust, and surprise. Can you match the expression with the emotion? Copyright © Paul Ekman. Reproduced with permission.

evolved to deal with different survival problems; and occurring automatically. The list of emotions is not considered closed. For instance, Ekman (1992) considers adding embarrassment, awe, and excitement as basic emotions and dropping surprise. Johnson-Laird and Oatley (1992) examined the words that we have for emotions and came up with a list of five basic emotions that overlap closely with Ekman's but does not contain surprise.

What of other candidate emotions? One possibility is that different candidate emotions are different shades of the same basic emotion. For example, happiness might include amusement, relief, pride, satisfaction, and excitement (Ekman, 1992). Another possibility is to consider some emotions as being comprised of two or more basic emotions. Plutchik (1980) offers a detailed account along these lines. He proposes eight basic emotions (surprise, sadness, disgust, anger, anticipation, joy, acceptance, fear) that may be combined in various ways: for example,

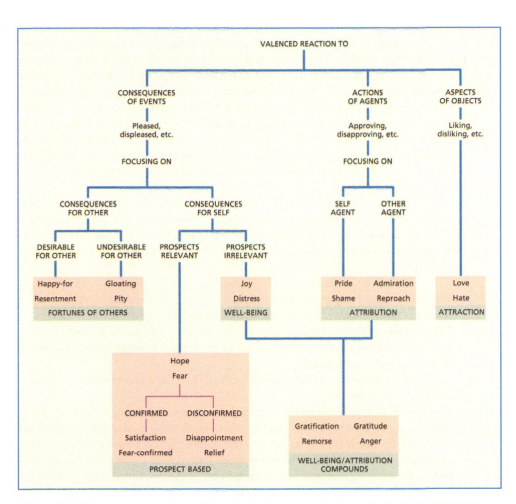

Figure 4.4 Researchers such as Ortony categorize emotions in terms of qualitatively different kinds of cognitive appraisal that may occur either consciously or unconsciously. From Clore and Ortony (2000). Copyright © 2000 Oxford University Press. Reproduced with permission.

joy + fear = guilt and fear + surprise = alarm. A third possibility is that some emotions should be construed in terms of a basic emotion(s) plus a non-emotional cognitive appraisal. These cognitive + emotional blends might be needed to account for complex emotions such as jealousy, pride, embarrassment, and guilt. Such emotions might involve attribution of mental states that imply awareness of another person's attitude to oneself, or awareness of oneself in relation to other people. As such, they have been referred to as **moral emotions** (e.g. Haidt, 2003). Along these lines, Smith and Lazarus (1990) argue that pride, shame, and gratitude might be uniquely human emotions. Darwin (1872/1965) also believed that blushing (linked to shame or embarrassment) might be a uniquely human expression.

Not all models of emotion assume that some emotions are more basic. Three different accounts along these lines are considered here: that of Ortony and colleagues (Ortony, Clore, & Collins, 1988; Ortony & Turner, 1990); that of Rolls (2005); and that of Barrett (2006; Lindquist & Barrett, 2012). Ortony and colleagues argue that *all* emotions are appraisals based on a valenced reaction (i.e. positive vs. negative) to a given stimulus and event. The range of emotions is limited by the range of appraisals that one can deploy, rather than consisting of some pre-determined number. These appraisals can occur unconsciously as well as consciously. For example, an emotion such as 'shame' would be an outcome of various appraisals such as: it has a negative valence; it refers to the action of people; and it is self-focused. This is illustrated in Figure 4.4. Although this theory is not couched in terms of neuroscience, it is not hard to imagine how such a model would translate. It may, for instance, involve a small set of regions involved in generating the 'valenced reaction' that interact with a more distributed set of regions involved in different kinds of appraisal. Indeed, we see examples of this kind of model in the theories presented below.

Rolls (2005) offers an account of different emotions that arise out of different aspects of reward and punishment, but he does not assume a core set of basic emotions in the same way as many, indeed most, other theories do. Instead, he argues that different types of emotion emerge from a consideration of a small set of principles, including:

- Whether a reward or punishment is applied (e.g. pleasure vs. fear); whether a reward is taken away (e.g. anger) or a punishment is taken away (e.g. relief).
- The intensity of the above (e.g. rage, anger, sadness, or frustration) could be different emotional outcomes arising out of having a rewarding stimulus removed or unexpectedly not appearing.
- Different combinations of the above (e.g. guilt) may be a combination of reward and punishment learning.
- The context in which an emotional stimulus appears. For example, whether the stimulus is social or not (i.e. related to other people) may determine whether the emotion feels like love, anger, jealousy (emotions implying another agent) versus enjoyment, frustration, or sadness (emotions that need not imply another agent). Indeed the eliciting stimulus is considered part of the emotional state, so love for one person may be different to love for another person just because the individual is different.

Finally, the theory of Barrett and colleagues (Barrett & Wager, 2006; Barrett, 2006; Lindquist & Barrett, 2012) assumes that all emotions tap into a system termed *core affect* that is organized along two dimensions: pleasant–unpleasant and high/low arousal. The latter is also termed activation. This is illustrated in Figure 4.5. Evidence that emotional experience can be classified along these two dimensions comes from studies employing factor analysis of current mood ratings (Yik, Russell, & Barrett, 1999). This study found that all subjective moods fall somewhere within this two-dimensional space. In biological terms, this is linked to bodily feelings of emotion and linked to limbic structures such as medial temporal lobes, cingulate and orbitofrontal cortex (Lindquist & Barrett, 2012). This echoes the older ideas of Papez and Maclean. The novel aspect of the model is the idea that categories of emotion are constructed (and can be differentiated from each other) because they tap the core affect system in somewhat different ways and because they are linked to certain kinds of information processed outside of the core affect system, including executive control

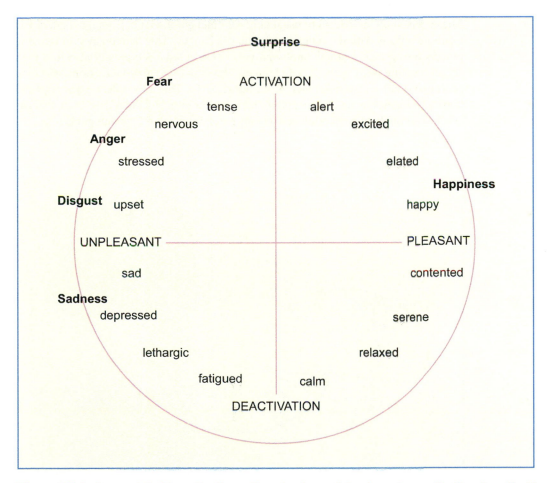

Figure 4.5 In the model of Barrett, all emotions (and mood) involve a 'core affect' system that is organized along two dimensions corresponding to pleasantness and activation (or arousal). Different categories of emotion are points in that space (and linked to associated cognitions – language, memory, perception, theory of mind) but are not afforded a special status. From Russell and Barrett (1999).

KEY TERMS

Amygdala
Collection of nuclei
buried bilaterally in
the anterior temporal
poles

**Kluver–Bucy
syndrome**
Behaviors associated
with lesions in the
amygdala region of
primates, including an
unusual tameness,
emotional blunting, a
tendency to examine
objects with the
mouth, and dietary
changes

(for regulating and appraising emotions), language (for categorizing and labeling), theory of mind (for conceptualizing emotions in terms of other agents), and so on.

The following sections will present evidence for and against the 'basic emotions' hypothesis, by considering whether different emotional categories have their own neural substrate (favoring the 'basic emotions' position) or not (favoring more distributed models of the emotions).

The amygdala and fear

The **amygdala** (from the Latin word for almond) is a small mass of gray matter that lies buried in the tip of the left and right temporal lobes as shown in Figure 4.6. It lies to the front of the hippocampus and, like the hippocampus, is believed to be important for memory – particularly for the emotional content of memories (Richardson, Strange, & Dolan, 2004) and for learning whether a particular stimulus/response is rewarded or punished (Gaffan, 1992). In monkeys, bilateral lesions of the amygdala have been observed to produce a complex array of behaviors that have been termed the **Kluver–Bucy syndrome** (Kluver & Bucy, 1939; Weiskrantz, 1956). These behaviors include an unusual tameness and emotional blunting, a tendency to examine objects with the mouth, and dietary changes. This is explained in terms of objects losing their learned emotional value. The monkeys typically also lose their social standing (see Figure 4.7). In humans, the effects of amygdala lesions are not as profound (Paul, Corsello, Tranel, & Adolphs, 2010). This may reflect either a greater cortical influence on emotional and social behavior or the fact that the earlier monkey studies are likely to have produced lesions extending beyond the amygdala.

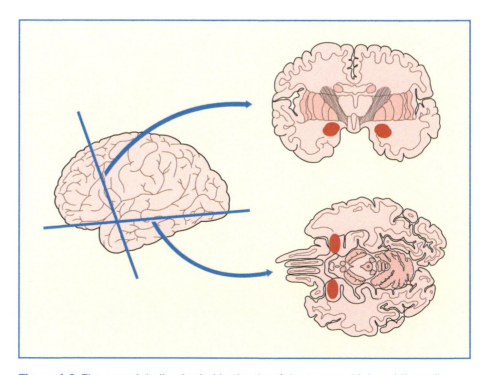

Figure 4.6 The amygdala lies buried in the tip of the temporal lobes, bilaterally.

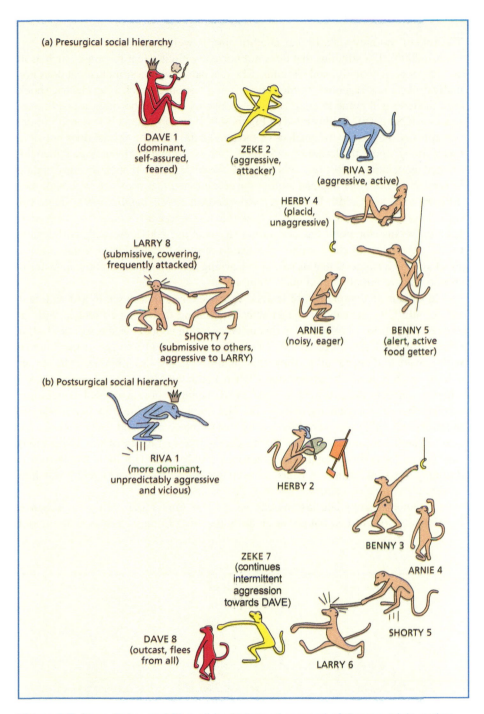

Figure 4.7 Dave, Zeke, and Riva all had bilateral removal of the amygdala region (each animal was operated on in that order, two months apart) resulting in changes in the social hierarchy. The amygdala is involved in fear processing but also modulates the fight-or-flight response linked to aggressive behavior. From Rosvold, Mirsky, and Pribram (1954). Copyright © 1954 American Psychological Association. Reproduced with permission.

Evidence for the role of the amygdala in fear

The role of the amygdala in fear conditioning is well established (LeDoux, 1996; Phelps, 2006). If a stimulus that does not normally elicit a fear response, such as an auditory tone (unconditioned stimulus, CS−), is paired with a stimulus that does normally evoke a fear response (termed conditioned response), such as an electric shock, then the tone will come to elicit a fear response by itself (it becomes a conditioned stimulus, CS+). This is illustrated in Figure 4.8. If the amygdala is lesioned in mice (specifically the basolateral nucleus of the amygdala) then the animal does not show this learning, and if the lesion is performed after the animal has been trained then this learned association is lost (e.g. Phillips & LeDoux, 1992) – that is, the amygdala is important for both learning and storing the conditioned fear response (although for a different view see Cahill, Weinberger, Roozendaal, & McGaugh, 1999). Single-cell recordings suggest that different cells within the amygdala could be involved in learning versus storage of the association (Repa et al., 2001). Animals with lesions to the amygdala still show a fear response to normal fear-evoking stimuli (such as shocks), which suggests that its role is in learning and storing the emotional status of stimuli that are initially emotionally neutral.

In humans, a comparison of learned fear responses to a shock (CS+) with neutral stimuli (CS−) reveals amygdala activation during fMRI that correlated with the degree of conditioned response, in this instance a skin conductance response (LaBar, Gatenby, Gore, LeDoux, & Phelps, 1998). Bechara et al. (1995) report that humans with amygdala damage fail to show this conditioned response, but nevertheless are able to verbally learn the association ('when I saw the blue square I got a shock'), whereas amnesic patients with hippocampal damage show a normal conditioned response but cannot recall the association. This suggests that the association is stored in several places: in the amygdala (giving rise to the conditioned fear response) and also in the hippocampus (giving rise to declarative memories of the association). fMRI studies also show that the amygdala may also be important for fear-related conditioning in social settings in which participants learn fear associations by watching someone else receive a shock (Olsson & Phelps, 2004).

In humans, amygdala lesions can selectively impair the ability to recognize facial expressions of fear but not necessarily the other Ekman categories of emotion

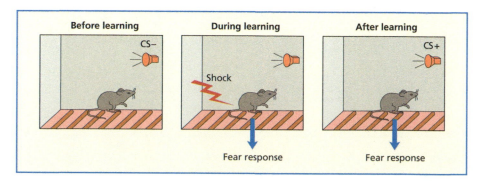

Figure 4.8 The basic procedure in fear conditioning involves presenting an initially neutral stimulus (the CS−, e.g. a tone) with a shock. After sufficient pairings, the stimulus will elicit a fear response without an accompanying shock (it has become a CS+).

(e.g. Adolphs et al., 1994; Calder et al., 1996). For example, patient DR suffered bilateral amygdala damage and subsequently displayed a particular difficulty with recognizing fear (Calder et al., 1996). She was also impaired to a lesser degree in recognizing facial anger and disgust. She could imagine the facial features of famous people, but not of emotional expressions. She could recognize famous faces and match different views of unfamiliar people, but could not match pictures of the same person when the expression differed (Young, Hellawell, Van de Wal, & Johnson, 1996). DR also shows comparable deficits in recognizing vocal emotional expressions, suggesting that the deficit is related to emotion processing rather than modality-specific perceptual processes (Scott et al., 1997).

Functional imaging studies generally support, and extend, these conclusions. Morris et al. (1996) presented participants with morphed faces on a happy–neutral–fearful continuum. Participants were required to make male–female classifications (i.e. the processing of emotion was incidental). Left amygdala activation was found only in the fear condition; the happy condition activated a different neural circuit. Winston, O'Doherty, and Dolan (2003) report that amygdala activation was independent of whether or not subjects engaged in incidental viewing or explicit emotion judgments. However, other regions, including the ventromedial/orbitofrontal cortex, were activated only when making explicit judgments about the emotion. This was interpreted as reinstatement of the 'feeling' of the emotion.

Some researchers have argued that the ability to detect threat is so important, evolutionarily, that it may occur rapidly and without conscious awareness (LeDoux, 1996). Ohman, Flykt, and Esteves (2001) report that people are faster at detecting fear-related stimuli such as snakes and spiders amongst flowers and mushrooms than the other way around. When spiders or snakes are presented subliminally to people with spider or snake **phobias**, then participants do not report seeing the stimulus but show a skin conductance response indicative of emotional processing (Ohman & Soares, 1994). In these experiments, arachnophobics show the response to spiders, not snakes; and ophidiophobics show a response to snakes but not spiders. In terms of neural pathways, it is generally believed that there is a fast subcortical route from the thalamus to the amygdala and a slow route to the amygdala via the visual cortical pathways (Adolphs, 2002; Morris, Ohmann, & Dolan, 1999). This is illustrated in Figure 4.9. Functional imaging studies suggest that the amygdala is indeed activated by unconscious fearful expressions, presented too briefly to be consciously seen (Morris et al., 1999). This is consistent with a subcortical/fast route to the amygdala, although it is to be noted that the temporal resolution of fMRI does not enable any direct conclusions to be drawn about processing speed.

Activation of a fear response by the amygdala may trigger changes elsewhere in the brain that enable the threat to be evaluated and responded to, if necessary. There are connections from the amygdala to the autonomic system (LeDoux et al., 1988). These may help prepare the body for fight and flight by increasing the heart and breathing rates. The anterior cingulate is believed to be involved in this process (Critchley et al., 2003), and it too is selectively activated by fear relative to happiness (Morris et al., 1996). In addition, there is a strong relationship between the level of fear and increases in activity in regions of the visual cortex (Morris et al., 1998). Thus, the detection of potential threat by the amygdala may trigger more detailed perceptual processing of the threatening stimulus, enabling further evaluation. Other, more frontal, regions may also be important for deciding whether to act on this information. In conclusion, although the amygdala may be essential for the evaluation of

KEY TERM

Phobia
Long-term fear and avoidance of particular stimuli or situations

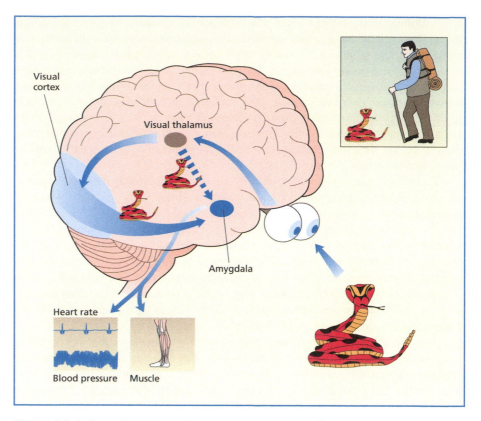

Figure 4.9 LeDoux has argued that the amygdala has a fast response to the presence of threatening stimuli such as snakes.

potential danger, its role should be construed in terms of its influence upon a wider circuit of emotional processing.

Evidence for the role of the amygdala in other emotions

The most convincing evidence for a specialized role of the amygdala in fear comes from functional imaging studies that compare fear expressions with other emotional expressions, and studies of human patients with damage to the amygdala who show relatively selective deficits in recognizing fear. However, the conclusion that the amygdala is specialized for fear may still be premature. First, the interpretation of these findings hinges on the assumption that the stimuli were appropriately matched. If fear-related stimuli are simply more difficult or more arousing (e.g. because a fearful face has more survival value than happy or sad faces) then the data could be explained without assuming specialization for fear. However, evidence that damage to other regions does *not* selectively affect fear (e.g. insula lesions and disgust) speaks against this more general account. Second, the amygdala might be specialized for some other process that just happens to be more relevant for fear. For instance, it has been suggested that selective impairments in fear may arise because of a failure to attend closely to the eyes (Adolphs et al., 2005). However, evidence that the amygdala is involved in fear in other domains (music, speech) speaks against this account (Gosselin, Peretz, Johnsen, & Adolphs, 2007; Scott et al., 1997).

With regard to learning of stimulus–emotion associations there is evidence that the amygdala is involved in learning positive associations, based on food rewards, as well as fear conditioning (Baxter & Murray, 2002). However, the amygdala system for positive associations operates somewhat differently to fear conditioning (and, hence, could be argued to be independent of the fear-based system). For example, selective lesions of the amygdala in animals do *not* affect learning of classically conditioned light–food associations (i.e. the animal learns to approach the food cup when the light comes on) (Hatfield, Han, Conley, Gallagher, & Holland, 1996), although such lesions are known to affect learning that a light predicts a shock. However, amygdala lesions do affect other aspects of reward-based learning, such as **second-order conditioning** (using one conditioned stimulus to learn about another). For instance, an animal may initially learn that a light predicts food (first-order) and then subsequently learn (second-order) that tone + light predicts an absence of food (Hatfield et al., 1996). Different nuclei within the amygdala also have rather different roles in fear learning relative to reward learning (Baxter & Murray, 2002).

Recent functional imaging studies that compare stimuli with learned positive and negative associations relative to emotionally neutral ones but do not rely on facial expressions have revealed amygdala activation to negative and positive affective stimuli. This includes comparing positive, negative, and neutral tastes (Small et al., 2003), smells – see Figure 4.10 — (Winston, Gottfried, Kilner, & Dolan, 2005), pictures, and sounds (Anders, Eippert, Weiskopf, & Veit, 2008), and comparing the emotional intensity of personally held attitudes to concepts such as 'welfare' and 'abortion' (Cunningham, Raye, & Johnson, 2004).

Summary

There is good evidence that the amygdala is crucial for the perception of fear. This includes facial expressions of fear but is not limited to faces. In addition, it appears to have a more general role in learning and storing the emotional value of stimuli. This includes both positive associations to stimuli (e.g. pleasant smells and tastes) as well as negative associations (e.g. stimuli paired with pain). These emotional associations extend to the emotional content of episodic memories.

The insula and disgust

The **insula** is a small region of cortex buried beneath the temporal lobes (it literally means 'island'), as shown in Figure 4.11. It is involved in various aspects of bodily perception, including important roles in pain perception and taste perception. The word **disgust** literally means 'bad taste' and this category of emotion may be evolutionarily related to contamination and disease through ingestion.

Patients with Huntington's disease can show selective impairments in recognizing facial expressions of disgust (Sprengelmeyer et al., 1997) and

KEY TERMS

Second-order conditioning
A form of learning in which a stimulus is first made meaningful through an initial step of learning, and then that stimulus is used as a basis for learning about some new stimulus

Insula
A region of cortex lying beneath the temporal lobes

Disgust
A category of emotion that may be evolutionarily related to contamination and disease through ingestion

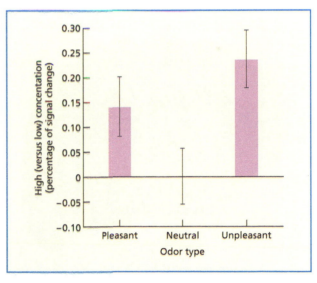

Figure 4.10 There is evidence that the amygdala responds to pleasant and unpleasant smells (but not neutral smells). This suggests a wider role of the amygdala in emotion processing, in contrast to the commonly held assumption that it is specific to fear. From Dolan (2007). Reproduced with permission.

relative impairments in vocal expressions of disgust (Sprengelmeyer et al., 1996). Huntington's disease is a genetic disorder with symptoms arising in mid-adulthood and including excessive movements, cognitive decline, and structural atrophy in the brain, particularly in regions such as the basal ganglia. However, the degree of the disgust-related impairments in this group correlates with the amount of damage in the insula (Kipps, Duggins, McCusker, & Calder, 2007). Selective lesions resulting from brain injury to the insula can affect disgust perception more than recognition of other facial expressions (Calder et al., 2000). In healthy participants undergoing fMRI, facial expressions of disgust activate this region but not the amygdala (Phillips et al., 1997). Feeling disgust oneself and seeing someone else disgusted activates the same region of the insula (Wicker et al., 2003).

According to a 'basic emotion' viewpoint, a separate neural substrate for disgust may have evolved to deal with one particular situation – contamination. This may also explain why disgust has its particular anatomical location, close to the primary gustatory cortex involved in early cortical processing of taste. However, we use the word 'disgust' in at least one other context, namely to refer to social behavior that violates moral conventions. Disgusting behavior is said, metaphorically, to 'leave a bad taste in the mouth'. But is there more to this than metaphor? Some have argued that moral disgust has evolved out of non-social, contamination-related disgust (e.g. Tybur, Lieberman, & Griskevicius, 2009). Moral disgust also results in activity in the insula (Moll, Zahn, de Oliveira-Souza, Krueger, & Grafman, 2005) and moral disgust is associated with subtle oral facial expressions characteristic of disgust more generally (Chapman, Kim, Susskind, & Anderson, 2009).

The insula is generally considered to have a wider role in emotional processing, in addition to a more specific involvement in disgust. Specifically, it is regarded as monitoring (probably both consciously and unconsciously) for bodily reactions that are characteristic of emotional states (Craig, 2009; Singer et al., 2009). When the bodily reactions are consciously perceived they may constitute the 'feeling' of an emotion. Damasio et al. (2000) report insula activity in response to recalling emotional memories from various categories (sadness, happiness, anger, fear) relative to emotionally neutral memories (note that disgust was not studied). This monitoring of bodily states does not occur in isolation but rather attempts to link actual bodily states with those predicted from the current context and sensory inputs (Critchley, Wiens, Rotshtein, Ohman, & Dolan, 2004). For example, it shows greater activity in risky decisions in which outcomes are less certain (Paulus, Rogalsky, Simmons, Feinstein, & Stein, 2003).

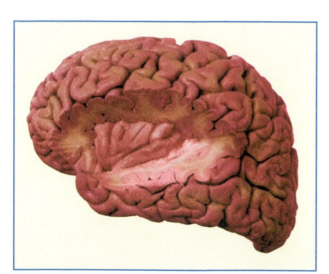

Figure 4.11 The insula is an island of cortex lying, bilaterally, underneath the temporal lobes. It is implicated in the creation of bodily feelings associated with emotions, and in the perception of disgust in particular. From Singer, Critchley, and Preuschoff (2009). Copyright © 2009 Elsevier. Reproduced with permission.

Anger

A selective deficit in recognizing anger has been reported following damage to the ventral striatal region of the basal ganglia (Calder, Keane, Lawrence, & Manes, 2004). The dopamine system in this region has been linked to the production of aggressive displays in rats (van Erp & Miczek, 2000), as well as in reward-based motivation more generally. In this latter context anger/ aggression could be construed as a motivated behavior to obtain or defend rewards. Functional imaging studies of anger also show activity in several regions linked to emotional processing, and are not suggestive of a unique neural signature. For example, one study insulted participants and then asked them to ruminate on the insult in the scanner (Denson, Pedersen, Ronquillo, & Nandy, 2009). Activity in the anterior cingulate correlated with self-reported anger, and initial activity in this region, the insula, and hippocampus predicted the degree of rumination. Anger and aggression are considered in detail in Chapter 10.

Evaluation

The concept of a 'basic emotion' rests, at least in part, on the assumption that there are separate neural foundations for different emotions. The clearest examples concern the role of the amygdala in fear, and the role of the insula in disgust. In both instances, the evidence suggests that the particular brain region is critically involved in that emotion. However, other evidence suggests that both the amygdala and insula are involved in the processing of other emotions too. How can these seemingly contradictory findings be reconciled? One possibility is that the processing of some emotions is more distributed across the brain than other emotions. Thus, damage to one part of the circuit for a more distributed emotion could be partly compensated for elsewhere. Another (related) possibility is that the same brain region is involved in processing of different emotions but performs different computations for each emotion. For example, there is some evidence that the amygdala performs rather different roles in fear conditioning relative to reward conditioning, even though it is relevant to them both.

The idea of basic emotions is hard to prove definitively correct or incorrect. In general, there are a number of key difficulties with the 'basic emotion' approach. These have been discussed by Barrett (2006) and Panksepp (2007). Some of the salient points are listed here:

- Emotions can be 'basic' in one respect but not another (e.g. love does not have a facial expression but may have evolved to meet specific needs).
- Some 'basic' emotions appear to be more basic than other emotions – for example, fear has more specialized neural substrates than happiness even though both are considered equally basic.
- The tendency for researchers in social neuroscience to focus on basic emotions has left many other (arguably more social) aspects of emotion under-researched, such as pride, guilt, and jealousy.

MOTIVATION: REWARDS AND PUNISHMENT, PLEASURE AND PAIN

Motivation makes one work to obtain a reward, or work to avoid a punishment (Rolls, 2005). A motivational state is one in which a goal is *desired*, whereas an emotional state is elicited when a goal is *obtained* (or not). For example, hunger is a motivational state (related to the goal of eating) and happiness and disgust are emotional states that may be an outcome of eating. Unlike eating, many of our goals have a social dimension to them, such as a need for love and group affiliation. One influential attempt to categorize different aspects of motivation is Maslow's (1943) **hierarchy of needs** (see Figure 4.12). At the lowest level are physiological needs such as food and sex. At the top of the hierarchy is our need to realize the full potential of our abilities (self-actualization). In between, Maslow lists safety needs (e.g. financial security, good health), social needs (e.g. for friendship, family), and self-esteem (our need to be liked and valued by others). This is reminiscent of Freud's hierarchy of id, ego, and super-ego. There is little evidence that such needs are organized hierarchically (Wahba & Bridgewell, 1976) but it provides a useful description of the various motivators of behavior.

Innate versus conditioned likes and dislikes

As noted earlier, the emotional characteristics of stimuli, in terms of whether they are rewarding or punishing, can either be innate or learned. A reinforcer is a stimulus that increases or decreases a particular pattern of behavior. **Primary reinforcers** act as rewards or punishers without any learning. **Secondary reinforcers** act as rewards or punishers as a result of learning. For example, in learning that a particular tone predicts a shock (as in fear conditioning) the secondary reinforcer is the tone, which was previously paired with a primary reinforcer (a painful shock). As already discussed, the amygdala is considered important for learning the secondary reinforcement properties of aversive stimuli (e.g. fear conditioning; LeDoux, 2000) and in some aspects of learning about rewarding stimuli (e.g. the amygdala is involved in learning that a pleasant secondary reinforcer has been devalued but not in the initial Pavlovian conditioning of neutral stimulus with reward; Baxter & Murray, 2002).

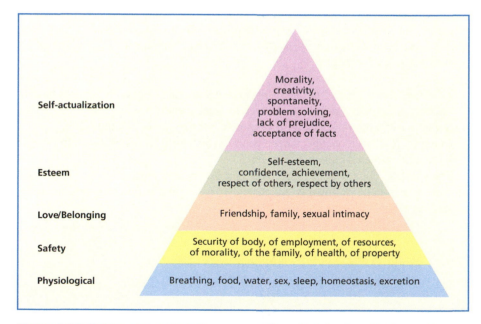

Figure 4.12 Maslow argued that there were different kinds of motivation and he also suggested that they could be arranged hierarchically. Although it is reasonable to suggest different forms of motivation, there is little evidence for this organization.

Primary reinforcers consist of certain tastes (e.g. sweet is rewarding, bitter is punishing), smells (e.g. putrefying is a punisher; pheromones are rewarding), touch (e.g. pain is a punisher, stroking is rewarding), and so on. Within the social domain, Rolls (2005) provides a list of possible reward-related primary reinforcers in humans, including attachment (e.g. to one's parent, partner, and child), cooperation (or reciprocal altruism), group acceptance, and 'mind reading'. To this list, one could add social exclusion as a possible punishment-related primary reinforcer (see Chapter 8) and imitation as a possible reward-related primary reinforcer (see Chapter 3).

Facial expressions also act as primary reinforcers. If human infants are given a novel object, their behavior will be influenced by the response of their primary caregiver – a phenomenon termed **social referencing** (e.g. Klinnert, Campos, & Source, 1983). If the caregiver displays disgust, then the object will be avoided, but if the caregiver smiles, then the child will interact with the object (Figure 4.13). The object itself has acquired an emotional value and is now classed as a secondary reinforcer. In adults, facial expressions of fear (Mineka & Cook, 1993), sadness (Blair, 1995), and disgust (Rozin, Haidt, & McCauley, 1993) can all be used as negative reinforcers. Happy expressions can be used as positive reinforcers (Matthews & Wells, 1999). Disgust expressions are often employed in the context of food. Sickness itself can also be a very powerful reinforcer for food. Novel food eaten during a period of illness, for example during chemotherapy (Fredrikson et al., 1993), may elicit a highly durable subsequent avoidance of that food – called conditioned taste aversion. In this instance, the sickness is the primary reinforcer and the food paired with the sickness becomes a learned secondary reinforcer. Anger, by contrast, tends not to act as an unconditioned stimulus but, rather, is used to curtail ongoing behavior by implying violation of social norms (Blair & Cipolotti, 2000).

KEY TERM

Social referencing
Use of emotional cues in others (e.g. expressions) to learn the rewarding/punishing nature of initially neutral stimuli

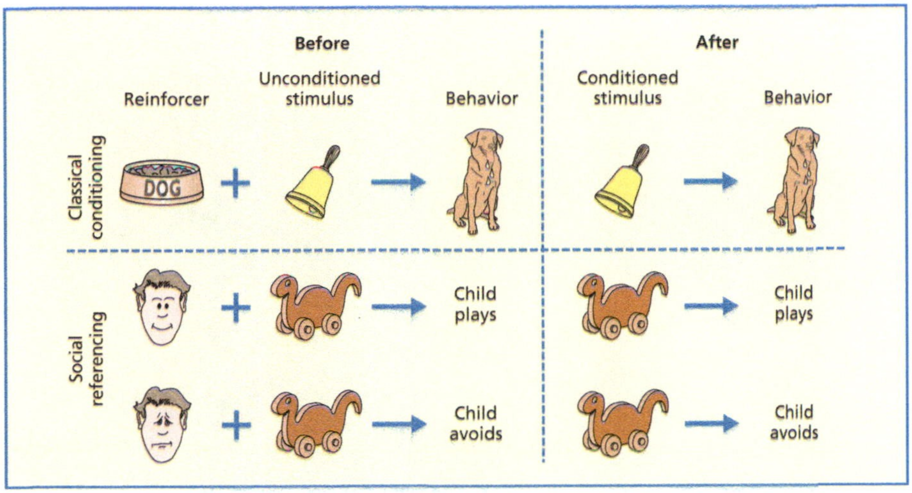

Figure 4.13 Facial expressions can act as primary reinforcers in that they modulate behavior towards an (initially) affectively neutral stimulus (unconditioned stimulus), which may then act as a secondary reinforcer (conditioned stimulus).

The orbitofrontal cortex computes the motivational value of rewards

The most basic anatomical division within the prefrontal cortex is that between the three different cortical surfaces: lateral, medial, and orbital. The lateral prefrontal cortex is more closely associated with sensory inputs than the orbitofrontal cortex. It receives visual, somatosensory, and auditory information, as well as receiving inputs from multi-modal regions that integrate across senses. In contrast, the medial and orbital prefrontal cortex is more closely connected with medial temporal lobe structures critical for long-term memory and processing of emotion.

The orbitofrontal cortex (OFC) consists of the ventral surface of the frontal lobes above the eye sockets (the orbits). It is shown in Figure 4.14. It consists of Brodmann's areas 10, 11, 12, 13, and 14 (Walker, 1940), with area 12 being co-extensive with area 47 and areas 11 and 14 extending into the medial surface. The ventral part of the medial prefrontal cortex (VMPFC), including areas 25 and 32, has similar connections to the orbital surface and is often similarly affected by strokes. These two regions (orbital and ventromedial) tend to act as a functional network (Öngür & Price, 2000).

One general function of the OFC is in computing the *current* value of a stimulus (i.e. how rewarding the stimulus is within the current context). For example, chocolate may be a rewarding stimulus, but it may not be *currently rewarding* if one is full-up or if eating it may incur the anger of someone else. Small, Zatorre, Dagher, Evans, and Jones-Gotman (2001) asked participants to eat chocolate during several blocks of PET scanning. Initially, the chocolate was rated as pleasant and participants were motivated to eat it, but the more they ate the less pleasant it became and they were less motivated to eat it. This change in behavior was linked to changes in activity in orbitofrontal regions. Specifically, there was a shift in activity from medial regions (pleasant/wanting) to lateral regions (unpleasant/not-wanting). Other studies

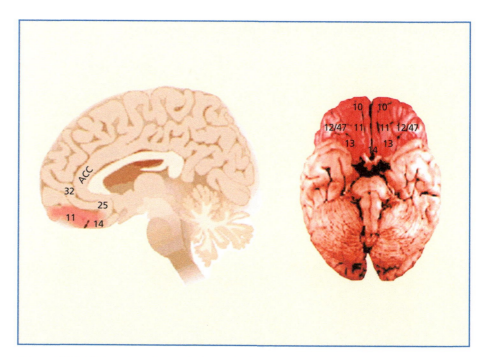

Figure 4.14 Different regions of the orbitofrontal cortex (and neighboring ventromedial frontal cortex), displayed in terms of Brodmann's areas. The left image is a medial view, and the right image is viewed from the underside of the brain. ACC = anterior cingulate cortex.

are consistent with different regions of OFC coding rewards and punishments (e.g. for a review see Kringelbach, 2005). For instance, activation of lateral OFC is found when a rewarding smile is expected but an angry face is presented instead (Kringelbach & Rolls, 2003) and is correlated with amount of monetary loss on a trial (O'Doherty, Kringelbach, Rolls, Hornak, & Andrews, 2001).

The OFC may enable flexible changes in behavior to stimuli that are normally rewarding (or recently rewarding) but suddenly cease to be. This can account for its role in **reversal learning** (in which rewarded and non-rewarded stimuli are reversed) and **extinction** (in which a rewarded stimulus is no longer rewarded). Eating chocolate until it is no longer pleasant can be regarded as a form of extinction. Lesions in these regions in humans lead to difficulties on these tasks, and the amount of difficulty in reversal learning correlates with the level of socially inappropriate behavior of the patients (Rolls, Hornak, Wade, & McGrath, 1994). Many patients with damage to these regions are inappropriately joking or flirtatious rather than aggressive (Damasio et al., 1990; Grafman et al., 1996). This is consistent with the notion that they struggle to 'devalue' these positive behaviors according to social context or feedback (e.g. that other people are uncomfortable with their behavior).

Blair et al. (2006) show that activity in the OFC/VMPFC is related to the magnitude of rewards, but interestingly not the magnitude of punishments (which was more closely related to anterior cingulate activity). It was also related to the magnitude of foregone as well as chosen rewards. Thus, a choice between $500 and $400 will elicit greater activity in this region relative to $500 versus $100, even though the actual reward (i.e. $500) is the same in both. In a similar vein, Coricelli et al. (2005) argue that the OFC/VMPFC is important for the emotion of **regret** – the

Figure 4.15 The same stimulus can elicit pleasure or aversion depending on context (e.g. the person's motivational state). Chocolate is normally pleasant, but if you have just eaten two bars of it you probably do not want any more. The orbitofrontal cortex computes the current emotional status of a stimulus (i.e. whether it is *currently* desired or not), thus enabling flexible behavior. Other regions in the brain may code the long-term value of a stimulus (i.e. whether it is *normally* desired or not).

KEY TERMS

Reversal learning
Learning that the reward values of two stimuli have been swapped

Extinction
Learning that a previously rewarded stimulus is no longer rewarded

Regret
The emotion that occurs when an outcome is worse than one would have experienced if one had made a different choice

emotion that occurs when an outcome is worse than one would have experienced if one had made a different choice. Participants were asked to make a monetary gamble by choosing one of two options. In one condition, they were given feedback about what would have happened if they had chosen the other option (resulting in feelings of regret on some trials) and in another condition they were not given feedback (with no opportunity for regret). The ventral striatum was active when participants won money, but the medial OFC was only active when they were given feedback and this was correlated with the amount of regret (modeled here as the difference between chosen and not-chosen gains).

Activity in the OFC has been linked to participants' subjective reports of pleasantness to stimuli such as taste (McClure et al., 2004) and music (Blood & Zatorre, 2001). Importantly, these ratings of pleasantness are not just affected by the stimulus itself but also by the participants' beliefs about the product (which can be construed as a motivational bias). Being told the price of a wine affects the ratings of pleasantness upon tasting it – more expensive wines taste nicer – and perceived pleasantness was again related to activity in the medial part of the OFC (Plassmann, O'Doherty, Shiv, & Rangel, 2008). Of course, the experimenters administered some of the same wines twice and gave the participants different prices, so the stimuli were physically identical, but their beliefs about the quality of the wine were not identical. Presumably our beliefs about the valence of people and social situations may operate along similar principles to that documented for taste.

THE SOMATIC MARKER HYPOTHESIS: HOW FEELINGS GUIDE DECISIONS

The **Somatic Marker Hypothesis** is a theory of how feeling states have a direct role in controlling ongoing behavior in both the social (e.g. interacting with others) and non-social (e.g. risk-taking) domain (Damasio, 1994, 1996). Somatic markers form the link between previous situations stored throughout the cortex (i.e. various forms of memory) and the 'feeling' of those situations that are stored in regions of the brain dedicated to emotion (e.g. the amygdala) and the representation of body states (e.g. the insula). The somatic markers are assumed to be stored in the ventromedial frontal cortex (including parts of the orbital surface). It is worth contrasting this hypothesis with that of the James–Lange theory. First, the James–Lange theory is only concerned with conscious attribution of bodily states, whereas the somatic marker hypothesis also concerns unconscious biases. Second, the body proper is not critical to the somatic marker hypothesis as the body can be represented by an 'as if' procedure in the brain. Thus, the brain simulates what the body might feel like in the absence of a literal experience of what the body does feel like. Finally, the somatic markers are assumed to play a direct and causal role in decision-making, for example, those involving risks or social situations.

The original motivation behind the theory was driven by observations of humans with brain damage in the VMPFC/OFC region who could remember

KEY TERM

Somatic marker hypothesis
A proposal that emotional and bodily states associated with previous behaviors are used to influence decision-making

emotional events (e.g. loss of a loved one) but appeared unconcerned about them (i.e. lacked feeling states), and also exhibited problems in day-to-day life in terms of managing finances and managing social relationships. To investigate this experimentally, Damasio and colleagues devised the ***Iowa Gambling Task*** (Bechara, Damasio, Damasio, & Anderson, 1994). Players are given four decks of cards (A to D), a 'loan' of $2000 in fake bank notes, and are instructed to play so that they win the most and lose the least – see Figure 4.16. On turning each card, the player receives either a monetary penalty or gain. Playing mostly from packs A and B leads to a net loss, whereas playing mostly from packs C and D will lead to a net gain. Control participants, without a brain lesion, learn to choose from C and D

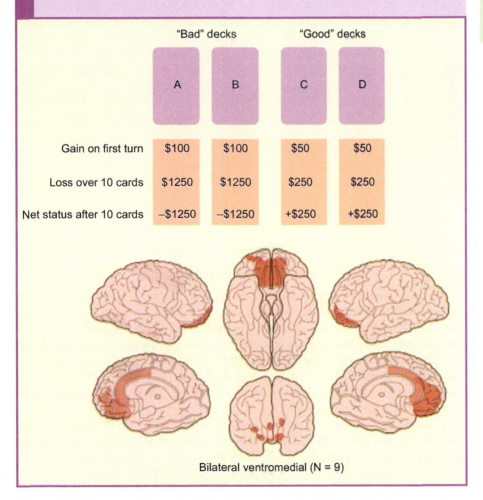

Bilateral ventromedial (N = 9)

Figure 4.16 Players receive $2000 and must choose hidden cards from one of four packs, A to D. Playing preferentially from packs A and B will result in loss, whereas playing preferentially from packs C and D will result in gain. Players are not informed of this contingency. Will they learn to avoid A and B? Patients with damage to ventromedial frontal lobes are impaired on this task. From Bechara et al. (1998). Copyright © 1998 by the Society for Neuroscience.

and to avoid A and B. Patients with lesions to the ventromedial frontal cortex do not (Bechara et al., 1994). Moreover, control participants generate an anticipatory skin conductance response (SCR) before making a selection from a risky pile (A and B), whereas these patients do not (suggesting the patients cannot use affective states to regulate behavior). Patients with lesions of the amygdala also perform poorly on the task and fail to show the anticipatory SCR (Bechara, Damasio, Damasio, & Lee, 1999). However, whilst those with ventromedial frontal cortex damage have an SCR to rewards and losses per se (winning and losing money) the patients with amygdala lesion did not. On a similar gambling task in which the odds of winning or losing were known, patients with lesions to the insula were also noted to be impaired but showed a different profile from patients with ventromedial frontal cortex damage (Clark et al., 2008). Those with damage to VMPFC/ OFC showed increased betting per se, but those with insula damage showed a failure to adjust their bets according to the chances of winning.

A somewhat different explanation of the results from the Iowa Gambling Task is that it reflects a failure of reversal learning (Maia & McClelland, 2004). This is because cards from bad decks A and B are rewarded with $100 dollars on the first turn, and cards from the good decks C and D are rewarded with only $50. Thus, patients must have to learn to avoid the previously advantageous decks, A and B. If there is initially no larger reward on the first trial of the bad decks, then patients with ventromedial frontal lesions perform normally (Fellows & Farah, 2003).

The anterior cingulate cortex: cognitive and affective evaluation of responses

The anterior cingulate is typically divided into two sections – a dorsal and ventral region – serving different functions (Bush, Luu, & Posner, 2000). This is shown in Figure 4.17. The ventral region is considered an 'affective division'. Lesions in the *ventral* anterior cingulate in humans can produce symptoms comparable to that found after orbitofrontal lesions, including reduced subjective frequency of emotional experiences and changes in social behavior (Hornak et al., 2003). Individual differences in gray matter density in this more ventral region (from within the normal population) correlate with subjective social status on a 'social ladder' even when potential confounds such as age, sex, income, and education level are taken into consideration – see Figure 4.18 (Gianaros et al., 2007). The question of whether the *dorsal* anterior cingulate is purely cognitive (e.g. Bush et al., 2000) or is involved in reward/punishment (Bush et al., 2002; Rushworth, Behrens, Rudebeck, & Walton, 2007) is discussed below.

The dorsal anterior cingulate is regarded as being particularly important in the detection of errors and in the monitoring of responses in which errors are likely to occur, for instance, when a stimulus is compatible with several responses (e.g. Botvinick, Braver, Barch, Carter, & Cohen, 2001). In support of this, anterior cingulate activity, measured with fMRI, is greater on errorful trials than non-errorful trials (Kerns et al., 2004) and there is an ERP deflection, called error-related negativity, occurring within 100 ms of an error (Gehring, Goss, Coles, Meyer, & Donchin, 1993) that has its origins in the anterior cingulate (Dehaene, Posner, & Tucker, 1994). The classic example of **response conflict**

KEY TERMS

Response conflict
Situations in which the desired response is not the easiest response

Stroop Test
A task in which participants must name the color of the ink and ignore reading the word (which also happens to be a color name)

is provided by the **Stroop Test** (Stroop, 1935). In this task, participants must name the color of the ink and ignore reading the word (which also happens to be a color name). The standard explanation for the response conflict generated by this task is that reading of the word occurs automatically and can generate a response that is incompatible with that required (e.g. MacLoed & MacDonald, 2000). Both functional imaging studies (e.g. Carter et al., 2000) and lesion studies (Stuss, Floden, Alexander, Levine, & Katz, 2001) highlight the role of the anterior cingulate. One widely accepted model of the anterior cingulate, based on these and similar findings, is that it generates a 'warning signal' when responses are likely to err, and that other regions of the brain (e.g. in the lateral prefrontal cortex) act on this signal by, for example, being slower and more cautious (e.g. Botvinick et al., 2001).

There is, however, evidence that the anterior cingulate is involved in evaluating social and emotional stimuli. Rushworth et al. (2007) argue that the function of the anterior cingulate is to assess the value of responses (i.e. whether an *action* is likely to elicit a reward or punishment). This may differ from the function of the OFC, which computes whether a given *stimulus* is currently rewarded or punished. Rats with anterior cingulate lesions are more likely to choose the 'laziest' of two options when given a small reward in a nearby chamber or a larger reward in a chamber that requires jumping over a wall – they choose the nearer response with least effort (Walton, Bannerman, Alterescu, & Rushworth, 2003). If two responses

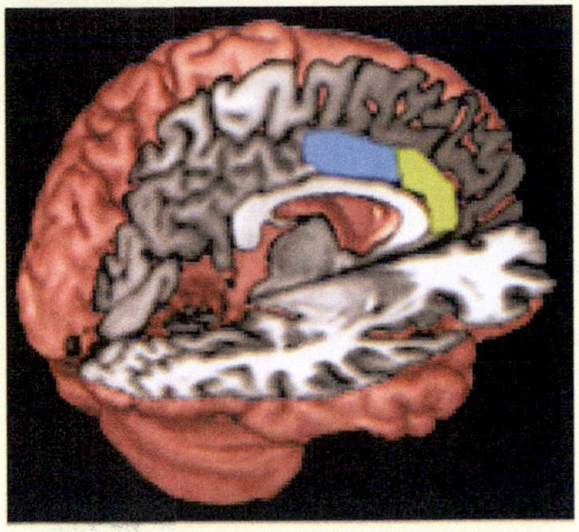

Figure 4.17 The anterior cingulate cortex lies above the corpus callosum on the medial surface of each hemisphere. It has been suggested that there are two broad divisions: a dorsal region (blue) and a ventral region (green). The ventral region is generally agreed to be involved in the processing of affective/social stimuli. But is the function of the dorsal region purely 'cognitive', or is it involved in evaluating rewards and punishments too?

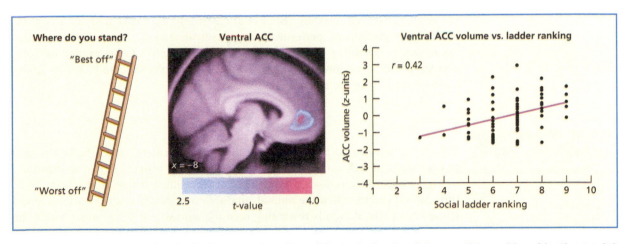

Figure 4.18 Gray matter density in the ventral portion of the anterior cingulate correlates with subjective social standing on a 'social ladder', even after more objective measures of social standing (income, education level) are controlled for. Decreased density may reflect a reduced control mechanism for social aspects of stress. From Gianaros et al. (2007). Reproduced with permission from the authors.

Figure 4.19 The Stroop Test involves naming the color of the ink and ignoring the written color name (i.e. the correct response is 'red, green, yellow, blue, yellow, white').

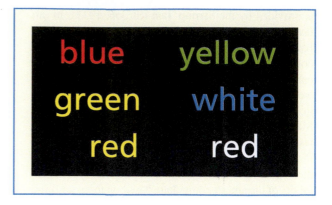

require the same effort but one requires waiting longer for a larger reward, then the rats with anterior cingulate lesions will wait for the larger reward whereas rats with orbitofrontal lesions behave impulsively by favoring the quick reward over the delayed larger reward (Rudebeck, Walton, Smyth, Bannerman, & Rushworth, 2006). Male monkeys with anterior cingulate lesions fail to adjust their responses when reaching for food when simultaneously shown a dominant male or a female in oestrus, whereas most control monkeys will pay close attention to these social stimuli, and hence take longer to respond to the food (Rudebeck, Buckley, Walton, & Rushworth, 2006).

Are the cognitive-based accounts (based on errors and response conflict) and the action-value account (based on evaluating the reward/punishment value of a response) compatible with each other? Potentially yes, given that both accounts emphasize a role in *response evaluation* but differ in the extent to which they emphasize social versus purely cognitive situations. Other theories of the anterior cingulate link it explicitly with motivation (Kouneiher, Charron, & Koechlin, 2009; Stuss & Alexander, 2007). Errors are motivationally salient events (that people work to avoid) as are rewards and punishments. The anterior cingulate also responds to the latter (e.g. monetary rewards or losses) even when there is no conflict between a habitual and non-habitual response (Blair et al., 2006). In the fMRI study of Kouneiher et al. (2009), participants performed a cognitively demanding study involving different monetary incentives: some blocks had a high incentive (more money for being correct) and others a lower incentive. High-incentive blocks were linked to greater sustained activity of the anterior cingulate.

The ventral striatum and reward

One important method for discovering the 'reward centers' of the brain has been to use electrical **self-stimulation** in animals (Olds & Milner, 1954), as illustrated in Figure 4.20. Electrodes are implanted at a specific location in the brain and a tiny current occurs when the animal produces a response, such as a lever press. If the site of stimulation is rewarding then the animal will carry on repeating this action. (Recall that the behavioral definition of a reward is a stimulus that an animal is willing to work to obtain; a non-behavioral definition could be stimuli that an animal considers pleasant.) If the stimulation site is not rewarding (i.e. neutral or punishing) then the animal does not self-stimulate. Self-stimulation does not occur for electrodes implanted in most parts of the brain, including sensory and motor cortices, but it is found in areas such as the OFC, the lateral hypothalamus, the amygdala, and – the area considered in this section – the nucleus accumbens located in the **ventral striatum** (for an overview see Rolls, 2005). Whereas the OFC and amygdala are important for learning about the emotional value of stimuli, and changes to the value of stimuli, the ventral striatum is concerned with learning the emotional value of an action, such as a lever press that delivers food or some other reward (Cardinal, Parkinson, Hall, & Everitt, 2002), and also for learning the reward value of a decision (e.g. Hare, O'Doherty, Camerer, Schultz, & Rangel, 2008).

The ventral striatum is part of the basal ganglia. These consist of subcortical gray matter structures, including the caudate nucleus, the putamen, and the globus pallidus. The caudate nucleus and putamen are collectively known as the **striatum**. The dorsal region of the striatum has more sensorimotor properties (e.g. involved in **habit** formation) whereas the ventral region may be more specialized for emotions, although the distinction is relative and not absolute (Voorn, Vanderschuren, Groenewegen, Robbins, & Pennartz, 2004). There are several loops that connect regions within the frontal cortex to the basal ganglia and on to the thalamus before returning to the frontal cortex. Each loop targets different regions of the frontal cortex and passes through different structures within the basal ganglia and thalamus (e.g. Alexander & Crutcher, 1990). The

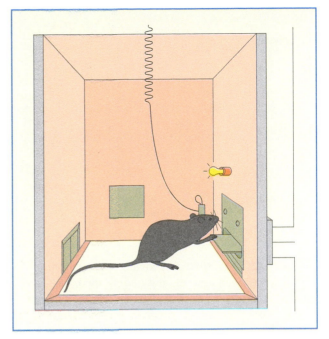

Figure 4.20 The rat is pressing a lever in order to obtain brain stimulation. This method has been used to identify regions of the brain that are associated with rewards (i.e. stimuli that the animal is motivated to work to obtain). These regions can also be considered the 'pleasure centers of the brain'. The nature of the reward differs according to the region stimulated and the current motivational state of the animal (e.g. whether it is hungry, lonely, etc.). From Olds (1956). Copyright © 1956 Scientific American, Inc. All rights reserved. Reproduced with permission.

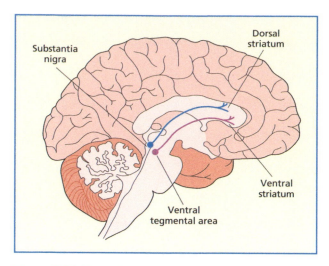

Figure 4.21 The striatum is located bilaterally in the basal ganglia, towards the front of the brain. The dorsal striatum and ventral striatum receive different dopaminergic inputs from mid-brain regions, namely the substantia nigra (implicated in Parkinson's disease) and the **ventral tegmental area**, respectively.

KEY TERM

Ventral tegmental area
A mid-brain structure in which the reward-related dopaminergic system originates

loops modulate brain activity within these frontal structures and, hence, increase or decrease the probability of a particular behavior. For example, there is a loop concerned with voluntary movement ('the motor circuit'), starting and ending in the supplementary motor area, and damage to various portions of this loop passing through the basal ganglia can lead to motor deficits such as Parkinson's disease (associated with rigidity and lack of voluntary movement) or Huntington's disease (associated with uncontrolled movements). However, the loop that is of particular relevance to reward-based learning (the 'limbisc circuit') starts and ends in the OFC and limbic regions (amygdala, hippocampus, anterior cingulate) and connects to the ventral striatum (including the nucleus accumbens and head of caudate nucleus), the ventral globus pallidus (also part of the basal ganglia) and the mediodorsal nucleus of the thalamus. As such, the ventral striatum has been described as a 'limbic–motor interface' (Mogenson, Jones, & Yim, 1980).

Neurons containing the neurotransmitter dopamine project from the mid-brain to the nucleus accumbens (Figure 4.21), and psychomotor stimulants such as amphetamine and cocaine may exert their effects on action (i.e. by increased drug taking) via this system (Koob, 1992). Initial drug taking, driven by their pleasant/rewarding nature, may transform itself into compulsive, habit-based drug taking (in which the rewards are less immediately apparent) and this may be associated with shift from the ventral striatum (emotion based) to dorsal striatum (sensorimotor based) (Everitt & Robbins, 2005). Other rewarding stimuli activate this region. Dopamine release in the nucleus accumbens of male rats increases when a female is introduced to the cage, and increases further if they have sex (Pfaus et al., 1990). Secondary reinforcers, previously associated with food, increase the release of dopamine in the nucleus accumbens of rats (e.g. Robbins, Cador, Taylor, & Everitt, 1989). In humans, an fMRI study shows that the greater the monetary reward that could be obtained in a task, the larger the activity in the ventral striatum (Knutson, Adams, Fong, & Hommer, 2001), and to give a more social example, activity in the ventral striatum correlates with male participants' desire to give a punishment to someone who has cheated (Singer et al., 2006).

One contemporary idea is that these dopaminergic neurons are not encoding reward per se but the difference between the *predicted* reward and actual reward (e.g. Schultz, Dayan, & Montague, 1997). After training to perform an action when presented with a light or tone cue, dopaminergic neurons in monkeys eventually respond to the conditioned cue itself rather than the subsequent reward, as shown in Figure 4.22 (Ljungberg, Apicella, & Schultz, 1992; Schultz, Apicella, Scarnati, & Ljungberg, 1992). If no subsequent reward appears then their activity drops below baseline, indicating that a reward was expected. fMRI studies of decision-making in humans also suggest that activity in the ventral striatum is greater when a reward is better than expected, rather than when a reward is high per se (Hare et al., 2008). In that study, the orbitofrontal cortex responded maximally according to how much a

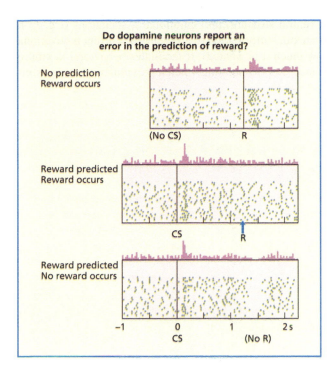

Figure 4.22 Single-cell recordings of dopamine neurons in the ventral striatum of monkeys show that the neuron responds when an unexpected reward of fruit juice is given (top), but if the reward is predicted by a cue (the conditioned stimulus) then the neuron responds to the cue and not the reward (middle). If an expected reward is omitted (bottom) the firing of the neuron falls below baseline. The results suggest that these neurons code the difference between the predicted reward and actual reward, rather than reward itself. From Schultz et al. (1997). Copyright © 1997 American Association for the Advancement of Science. Reproduced with permission.

goal was valued (i.e. a motivational 'wanting' response) whereas the ventral striatum activity was linked more to reward prediction.

Baez-Mendoza and Schulz (2013) discuss how evidence from reward processing interacts with social behavior. Of course social stimuli, i.e. the presence of others (particularly friends and loved ones), is a reward in itself and the ventral striatum is responsive to these kinds of stimuli. In human fMRI studies, viewing an image of one's mother or romantic partner will activate this region (Bartels & Zeki, 2000, 2004). In primates, one can estimate the subjective value of social stimuli (e.g. viewing an image of another monkey) by contrasting it against different amounts of a non-social reward such as fruit juice – a kind of monkey 'pay per view' (Deaner, Khera, & Platt, 2005). For instance, a male monkey may choose to have 0.2 ml of juice plus an image of a high-ranking monkey rather than 1 ml of juice and no image. They will also sacrifice juice to look at images of females in oestrus but will not trade juice for low-ranking monkeys or non-social images. The striatum (Klein & Platt, 2013) and orbitofrontal cortex (Watson & Platt, 2012) respond to both kinds of social and non-social rewards although different sets of neurons tend to tune into one or other kind of stimulus.

Evaluation

This section on motivation has highlighted the importance of three brain regions that are inter-connected: the ventral striatum, the OFC (and ventromedial frontal cortex), and the anterior cingulate cortex. The dorsal section of the anterior cingulate is important for assessing the rewards and risks of an action/response. The ventral striatum responds to rewards and the anticipation of rewards (e.g. responding to cues that may predict a subsequent reward). The OFC integrates reward-related information (e.g. from the ventral striatum) and emotional associations (e.g. from the amygdala) with other contextual information to ascertain the current motivational status of a stimulus (i.e. how much it is wanted). It facilitates behavioral flexibility enabling people to, for instance, stop responding when a previously rewarded stimulus becomes aversive.

PERSONALITY DIFFERENCES IN NEURAL NETWORKS FOR EMOTION AND MOTIVATION

Our personality can be defined as patterns of thinking and behaving that are relatively stable within individuals, but that differ from person to person. Our personality is constructed out of a number of different dimensions (traits) such as our degree of **extraversion** (seeking stimulation, both social and non-social), our degree of **neuroticism** (emotion reactivity and stress vulnerability), and so on. These are considered in Chapter 5. Whilst this describes the *structure* of personality, there is a long tradition of considering the *mechanisms* of personality in terms of individual differences in the functioning of processes relating to emotion and motivation (for a summary see Read et al., 2010). For instance, the model of Gray and McNaughton (2000) describes two fundamental mechanisms termed the Behavioural Inhibition System (BIS) and Behavioural Approach System (BAS). The BAS governs response to rewarding stimuli and parallels the trait of extraversion, whereas the BIS governs response to punishment and aversive stimuli and parallels the trait of neuroticism. These systems regulate other motivations relating to social affiliation, fear of social rejection, and so on. Davis and Panksepp (2011) have a somewhat different model in which human personality is founded upon the operation of six sub-cortical mechanisms in affective neuroscience (termed SEEKING, RAGE, FEAR, CARE, GRIEF, and PLAY). Although the details of these models are beyond the scope of the present discussion, they serve to illustrate how personality can potentially be explained in terms of differences in emotional and motivational processes. This opens up the possibility of explaining personality both at the level of neural architecture and at the level of gene-environment interactions (Hamann & Canli, 2004).

To give some concrete examples from the literature, consider the personality traits of extraversion and neuroticism. One feature of these traits is that whilst extraversion tends to be linked to positive mood, neuroticism is linked to negative mood. Canli et al. (2001) conducted an fMRI study that contrasted positive and negative affective images. Extraverts tended to activate the amygdala more when viewing positive images relative to negative ones, but the reverse was true for introverts. Neuroticism was linked to increased activity for viewing negative relative to positive images in a left prefrontal region. Canli et al. (2002) also examined the link between extraversion and the perception of fearful, happy, and neutral faces. There was a general trend for fearful faces to activate the amygdala more than happy faces (consistent with previous findings). However, the degree of amygdala activation to happy faces (not

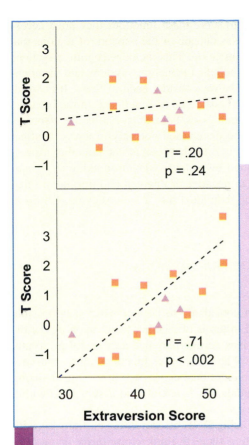

Figure 4.23 Amygdala activation to happy faces (bottom) correlates with extraversion, whereas activation to fearful faces (top) does not. Both are contrasted against a neutral face baseline. From Canli et al. (2002). Squares represent females, and triangles represent males.

fearful faces) was correlated with extraversion – see Figure 4.23. A happy face may be a more motivating to an extravert than to an introvert, whereas a fearful face has a similar motivating status for both. In terms of structural differences in the brain (assessed using VBM, greater extraversion is associated with more gray matter volume in the right amygdala whereas greater neuroticism is associated with less gray matter volume in the left amygdala (Omura, Constable, & Canli, 2005). Calder, Ewbank, and Passamonti (2011) provide a general overview of personality differences in the processing of facial expressions).

Candidate genes have been identified that are linked to both extraversion and neuroticism. In the case of extraversion, it has been linked to polymorphisms in the dopamine D4 receptor (Munafo, Yalcin, Willis-Owen, & Flint, 2008). Depue and Collins (1999) present a detailed neurobiological model of extraversion in terms of a dopaminergic behavioral approach system (BAS) linking the motivational circuits discussed in this chapter. In the case of neuroticism, the genetic evidence is much stronger and has centered, to date, on a serotonin transporter gene (Lesch et al., 1996). A short-form variation of the gene is linked to less efficient uptake or serotonin and increased neuroticism and anxiety. Participants with one or two short forms of the gene exhibit a greater amygdala response to fearful faces than do participants with two long forms (Hariri et al., 2002).

HOW EMOTIONS INTERFACE WITH OTHER COGNITIVE PROCESSES

This chapter started by introducing the idea that emotions are functional: they guide social and non-social behavior through the effects they have on memory, decision-making, and attention. In order to serve these functions they must necessarily be linked to other cognitive processes outside of the network of regions that are primarily involved in emotion and motivation. This includes cognitive systems involved in representing our thoughts and beliefs, the memory system, and the attention and perceptual system. The sections below consider each of these. It is worth noting that although these systems can be considered as functional, in the broadest sense, they can become maladaptive in certain cases. For instance, affective disorders such as depression and anxiety need to be understood not only in terms of affect (negative mood) but also in terms of the interplay between affect and other aspects of cognition. They tend to be accompanied by certain styles of thinking (e.g. catastrophizing) and attentional biases (e.g. to negative or threatening information) that maintain the negative affect and lock the individual into a vicious cycle (Gotlib & Joormann, 2010).

Thoughts, attitudes, and beliefs: emotion regulation

Although emotional processing is often considered to be automatic, it is possible nevertheless to exert a degree of control in terms of whether to act upon this information or how to interpret it. This involves an interplay between lateral prefrontal cortex (PFC) involved in cognitive control and 'executive functions', the ventromedial/orbital parts of the PFC (involved in emotional experience and contextualizing emotions), and regions such as the amygdala. This general field is subsumed within the topic of **emotion regulation**.

This can be studied experimentally using fMRI by presenting participants with an affective stimulus and then instructing them to think consciously about it in different ways. For instance, they may try to suppress it or put a negative/positive spin on it. The latter mechanism also termed *reappraisal* (for a review see Ochsner, Silvers, & Buhle, 2012). Ochsner, Bunge, Gross, and Gabrieli (2002) presented negative images (e.g. of someone in traction in a hospital) to participants in one of two conditions: either passively viewing them or a 'cognitive' condition in which they were instructed to reappraise each image 'so that it no longer elicited a negative response'. Their analysis revealed a trade-off between activity in the lateral PFC (high when reappraising) and the ventromedial PFC and amygdala (high during passive looking). When participants are asked to reappraise the stimulus negatively (i.e. making it worse than it looks), then this also engenders a similar network in the lateral PFC but tends not to dampen activity in the ventromedial PFC and amygdala (Ochsner et al., 2004). Similar results are found when comparing passive viewing of negative images and explicitly describing/labeling the images (Hariri, Bookheimer, & Mazziotta, 2000; Lieberman et al., 2007). The EEG-based event-related potential (ERP) method enables the time course of emotion regulation to be studied. Using this method it is generally found that regulatory effects emerge from around 300 ms onwards, i.e. after the perceptual stage of processing (Hajcak, MacNamara, & Olvet, 2010). In terms of brain stimulation, excitatory (anodal) tDCS over left lateral PFC reduced the rated emotionality of negative pictures (Pena-Gomez, Vidal-Pineiro, Clemente, Pascual-Leone, & Bartres-Faz, 2011). This was interpreted as increased emotion regulation induced by PFC stimulation.

KEY TERM

Emotion regulation
Deliberate attempts to control feelings through reappraisal, suppression, or other strategies

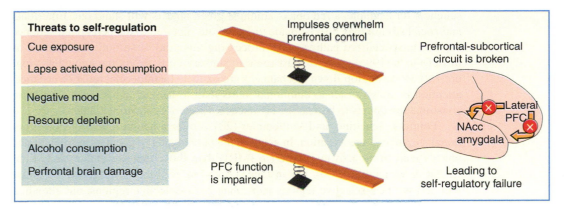

Figure 4.24 The lateral PFC represents our deliberate attempts at appraising emotional stimuli (e.g. catastrophizing them, neutering them) and also our longer-term goals and values (e.g. avoid prejudice, don't binge eat). Attempts at emotion regulation can fail either when the lateral PFC is compromised (green and blue) or when the emotional stimuli are very salient (yellow). From Heatherton and Wagner (2011).

Heatherton and Wagner (2011) propose a model of emotion regulation and its failures. They construe it as a balance between top-down processes (e.g. in the lateral PFC) and affective processes linked to reward (such as nucleus accumbens) and emotional salience (such as the amygdala). This is illustrated in Figure 4.24. In this context, the lateral PFC may act to maintain one's long-term goals (do not take drugs, stick to the diet) whereas other regions may signal contradictory behavior (take me, eat me). Self-regulatory failure takes place either when the prefrontal system is compromised (e.g. mental fatigue or 'resource depletion') or the external cues are particularly strong (e.g. after a period of abstinence or 'lapse activated consumption'). For instance, when cocaine addicts attempt to suppress their craving this results in reduced activity in orbitofrontal cortex and nucleus accumbens and this reduction is correlated with increased fMRI activity in lateral PFC (Volkow et al., 2010). The model also applies to social stimuli: for instance, when one tries to avoid acting on racist stereotypes, or to suppress (or reappraise) one's anger. In one study, fMRI activity in the lateral PFC was found to be inversely correlated with amygdala activity to viewing racial outgroup members, but only when the stimuli were presented for long enough to elicit reappraisal (Cunningham et al., 2004).

Attention and perception

Attention is the process by which certain information is selected for further processing and other information is discarded. Attention is needed to avoid sensory overload. The brain does not have the capacity to process fully all the information it receives. Nor would it be efficient for it to do so. Modern theories of attention emphasize that selection can take place at multiple levels in the cognitive system (Desimone & Duncan, 1995). At the top level, selection may occur due to task-relevance and goals (i.e. current priorities). At the bottom level, selection may occur due to perceptual salience: for example, a red object amongst green objects will stand out. Emotional and social stimuli may also be made to 'stand out' not due to the perceptual salience but due to specialized pathways for processing this kind of stimuli; for instance as in LeDoux's (1996) proposed fast route to sensory cortices via the amygdala (Figure 4.9). However, top-down mechanisms are likely to be important too. For instance, attention is drawn faster to emotional items than to neutral items when these also constitute the targets to be searched (Frischen, Eastwood, & Smilek, 2008).

Pourtois, Schettino, and Vuilleumier (2013) offer a neural model of how emotional stimuli capture attention. In general, the perceptual representations of attended objects tend to be activated more (in fMRI) than those that are unattended (Kanwisher & Wojciulik, 2000). For instance, attending to a face and ignoring a house will activate face-selective regions more than place-selective regions, but attending to a house will reverse that pattern. This boost of activation may relate to increased awareness of the attended stimuli and prioritize it for further processing. Emotional stimuli (e.g. facial expressions relative to neutral ones) tend to be linked to greater activity in perceptual representations whether attended or unattended (Vuilleumier, Armony, Driver, & Dolan, 2001). Pourtois et al. (2013) argue that this reflects an additional boost from the amygdala that serves to prioritize these stimuli.

The amygdala may not be the only region that exerts a facilitating effect on attention to emotional and social stimuli. Menon and Uddin (2010) propose an involvement of the insula in attention orienting (e.g. by initiating autonomic reactivity) and in facilitating access to the anterior cingulate which is involved in directing attention to motor responses. Finally, the medial prefrontal cortex may boost attention specifically to self-relevant stimuli. Attention towards a geometric shape (e.g. a square or triangle) can be boosted by training participants to associate it with the self (relative

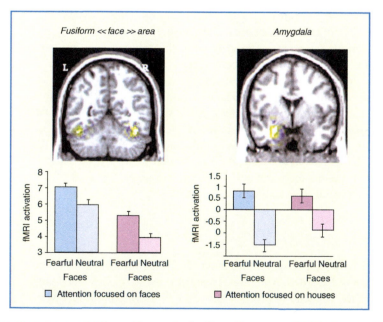

Figure 4.25 The perceptual representation of faces (in fusiform face area, FFA) is boosted for fearful faces, both when attended (blue) and unattended (orange). This boost is likely to come from the amygdala and this biases emotional stimuli for awareness and for other kinds of cognitive processing (e.g. memory, motor responding). From Vuilleumier et al., (2001).

to a known other person, or an untrained shape) and this reflects greater coupling between the ventro-medial PFC and the temporo-parietal junction (Sui, Rotshtein, & Humphreys, 2013).

Episodic memory

Memory is not a single system in the brain but is rather a constellation of different mechanisms. When the lay public use the term 'memory' they are generally referring to what psychologists term *episodic memory*: i.e. they are memories of specific events that occurred in a particular time and place. They are consciously reportable memories (and hence fall under the umbrella terms of declarative memory or explicit memory) and are often imbued with affective features and visuo-spatial imagery. It can be contrasted with other forms of memory such as semantic memory – this is information that is consciously known but is not re-experienced as an event (e.g. the capital of France). In addition, there are various unconscious forms of remembering including procedural memory (e.g. skills such as riding a bike) and also various forms of conditioned associations (e.g. tone-shock associations) that have been discussed already.

One consistent finding in the literature is that episodic memories for emotional stimuli (e.g. affective images presented experimentally, or events from one's own life) tend to be better remembered than neutral ones. This has been examined in imaging experiments using the subsequent memory paradigm. The participant is shown a series of stimuli (e.g. images) whilst scanned. They are then asked which stimuli they remember, and the experimenter can then go back and determine what, in the brain, distinguished remembered from forgotten trials when they were initially encountered. The amygdala has been consistently implicated in the memory advantage for emotional stimuli. Activation of the amygdala and hippocampus is correlated during learning of emotional scenes that are subsequently better remembered (Dolcos, LaBar, & Cabeza, 2004). In addition to coupling between the amygdala and hippocampus, there may also be a boost to object-based representations in the visual ventral stream that is consistent with its known role in attention (Kensinger, Garoff-Eaton, & Schacter, 2007). Reappraisal of negative stimuli at encoding also boosts their subsequent memory, relative to suppression or passive viewing, and this boost is linked to the amygdala, hippocampus, and lateral PFC (Hayes et al., 2010). Patients with amygdala lesions show impaired memory for the emotional details of scenes but not for other details of scenes (Adolphs, Tranel, & Buchanan, 2005).

The studies cited above use emotional stimuli but didn't contrast social (i.e. involving other people) and non-social (e.g. a snake, a gun). When social stimuli are used there tends to be a greater involvement of the ventro-medial and orbital PFC. Mitchell, Macrae, and Banaji (2004) asked participants to encode memories of people either socially (form an impression of them) or non-socially (remember the order). Subsequent memory affects were found in the medial PFC for social encoding and in the hippocampus for non-social encoding. Tsukiura and Cabeza (2008) compared memory for face–name associations with smiling or neutral faces. Smiling face–name pairs were better remembered and this was linked to increased coupling between orbitofrontal cortex and hippocampus during successful encoding and retrieval. Finally, Summerfield, Hassabis, and Maguire (2009) contrasted memories for events that happened to themselves versus other people and found medial prefrontal regions discriminated self from other during memory recall.

KEY TERM

Episodic memory
Memories of specific events that occurred in a particular time and place

SUMMARY AND KEY POINTS OF THE CHAPTER

- Emotions are states associated with stimuli that are rewarding (i.e. that we seek) or punishing (i.e. that we avoid) and they are multi-faceted in nature, consisting of emotional expressions, internal bodily responses, and subjective liking/disliking. They enable certain stimuli in the environment to be prioritized.

- Ekman has argued that there are six basic emotions that are defined on the basis of their having innate facial expressions, dedicated neural substrates, and that have evolved to deal with specific situations (e.g. disgust related to contamination). One problem with this approach is that many emotions appear to be 'basic' in one sense but not another.

- The amygdala has been specifically linked to the processing of fear (e.g. fear conditioning, recognition of fear on faces). The amygdala also has a role to play for processing other emotions (e.g. appetitive conditioning) but its role in other emotions is not necessarily the same as that for fear.

- A motivational state is one in which a goal is *desired*, whereas an emotional state is elicited when a goal is *obtained* (or not).

- The orbitofrontal cortex is involved in computing the motivational value of a stimulus (i.e. how much it is currently desired). It links emotion with current context in order to guide behavior.

- The dorsal anterior cingulate cortex is involved in computing whether an action is rewarded or punished (including detecting errors).

- The ventral striatum connects to the orbitofrontal cortex and is involved in reward-based learning. It is involved in the prediction and anticipation of reward rather than reward per se.

- "Core regions of the emotional brain interface with other brain regions to influence many aspects of cognition. The lateral prefrontal cortex interacts with emotion-related regions during appraisal of emotional stimuli. The perceptual and memory representations of emotional stimuli are boosted in their activation levels through interactions with the social and emotional brain leading to increased attention and increased remembering."

EXAMPLE ESSAY QUESTIONS

- What are 'basic emotions' and does current evidence support their existence?
- Is the amygdala the fear center of the brain?
- Contrast the roles of the orbitofrontal cortex and anterior cingulate cortex in emotions and decision-making.
- Does damage to the emotional circuitry of the brain lead to impaired social functioning?

RECOMMENDED FURTHER READING

- Fox, E. (2008). *Emotion Science.* New York: Palgrave Macmillan. Clear, comprehensive, and up-to-date.

- LeDoux, J. (1998). *The Emotional Brain.* New York: Simon & Schuster, and LeDoux, J. (2012). Rethinking the emotional brain. *Neuron, 73*(4):653–76. Although this book is now showing its age, it is still a very clear and accessible overview and a good place to start. The more recent paper brings his views up to date.

ONLINE RESOURCES

- References to key papers and readings
- Videos demonstrating tests of classical conditioning and the Iowa Gambling Task
- Interviews and lectures featuring Paul Ekman, Antonio Damasio, Elizabeth Phelps, Joseph LeDoux, and others
- Recorded lecture given by textbook author, Jamie Ward
- Multiple choice questions and interactive flashcards to test your knowledge
- Downloadable glossary

CHAPTER 5

CONTENTS

Reading faces and bodies

Our social interactions exist between other members of our species, so-called **conspecifics**. As such we need an effective system of keeping track of who is who. We need to remember what people look like and what their typical behaviors are. Facial and bodily appearances provide only superficial clues as to a person's inner state. But given that we cannot directly observe inner states but we can observe faces and bodies, there is a strong incentive to extract whatever information we can from a face or body. We need to know whether someone is likely to cooperate or cheat. Skilled basketball players, for instance, learn to detect fake passes from body language alone (Sebanz & Shiffrar, 2009). We need to know whether someone is happy or sad, or angry and likely to use force. Faces and bodies (together with voice cues) provide an important source of such information.

Recognizing someone's expression involves making inferences about someone's *current* state: they are smiling therefore they are happy. However, there is a natural tendency to go beyond this. Many people believe that we can read character traits, such as trustworthiness and aggression, from faces even when they have neutral facial expressions (Hassin & Trope, 2000). Indeed, people tend to vote for political candidates whose faces are judged to be associated with greater competency. Todorov, Mandisodza, Goren, and Hall (2005) presented participants with pairs of photographs of faces of the winner and runner-up of seats in the US congressional election. Participants were not informed about how the stimuli were constructed and were asked to note if they recognized any of the faces in the pair. (These pairs were then discarded and only those judged unfamiliar were analyzed.) Judging which of the faces was more competent predicted the overall result of the elections better than chance (68.8% relative to 50% chance), and was correlated with the margin of victory as shown in Figure 5.1.

This chapter starts by considering the basic mechanisms of recognizing a face from both a cognitive and neural perspective. Particular consideration is given to the issue of whether or not recognizing an emotional expression involves different mechanisms from recognizing familiar faces, or reading other dynamic cues in a face (such as gaze direction). Recent research in the less-studied area of body perception is then evaluated. The second half of the chapter considers how perceivers go beyond the raw information provided in order to infer other peoples' intentions from faces and bodies, and to infer their stable personality traits.

PERCEIVING FACES

Face perception has several goals. In some instances, the goal may be to recognize a particular individual (e.g. 'that is my wife'). In other instances, the goal may be to extract other types of socially relevant information such as whether the person is happy, attractive, old, where they are looking, etc. To some degree, these different aspects of face perception reflect different cognitive and neural mechanisms.

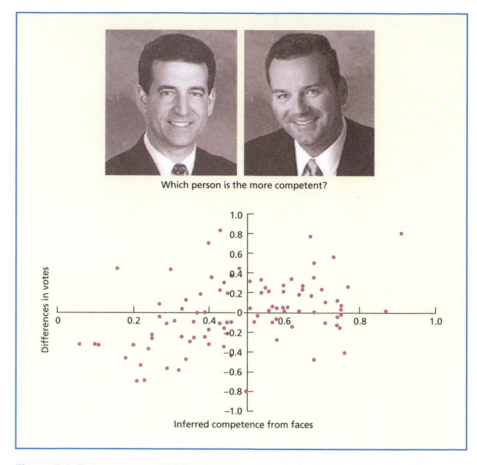

Which person is the more competent?

Figure 5.1 Todorov et al. (2005) asked participants to judge which of two faces looked more competent. The faces were unknown to the participants making the rating, but both had in fact been contenders in an election. Faces rated as more competent were likely to win with an increased majority of votes. From Todorov et al. (2005). Copyright © 2005 American Association for the Advancement of Science. Reproduced with permission.

A cognitive model

Bruce and Young (1986) proposed a cognitive model of face recognition that has largely stood the test of time and is illustrated in Figure 5.2. They assume that the earliest level of processing involves structural encoding of the face by detecting shading and curvature of surfaces and detection of edges. Following this, a distinction is made between the processing of familiar and unfamiliar faces. Recognizing familiar faces is assumed to involve matching the visual description of the face with a stored memory representation of a face. There is assumed to be a store of all known faces, and each face is said to have its own **face recognition unit**. In neural terms, single-cell recordings of primates suggest that there is a class of neurons that respond to faces but not objects and, moreover, that these neurons respond to certain faces more than other faces (e.g. Rolls & Tovee, 1995). However, neurons that respond to one (and only one) face tend not to be found (Quiroga et al., 2005). As such, face recognition units are likely to be coded in terms of the activity of a set of neurons rather than having a single neuron responding to each known face. For familiar faces, once the face has been recognized then other information may become available, such as their occupation or name. The semantic level of description (which relates to conceptual knowledge of people, rather than the face per se) is termed the **person identity node**. The process of familiar face recognition involves matching a face seen from one particular viewpoint and one particular lighting condition to a memory representation that stores the three-dimensional structure of the face (enabling it to be recognized from any view). A separate route (termed *directed visual processing*) was postulated to deal with unfamiliar faces, for example in order to match them across different views or lighting conditions. Aside from the mechanism for recognizing familiar faces, there are a number of other pathways on the Bruce and Young (1986) model that apply equally to familiar and unfamiliar faces. These include detecting the emotional expression of faces, and also using lip-reading cues from faces.

Evidence from patients with acquired neurological impairments lends support to this basic model. Impairments of face processing that do not reflect difficulties in early visual analysis are termed **prosopagnosia** (Bodamer, 1947). The term prosopagnosia is also sometimes used specifically to refer to an inability to recognize previously familiar faces. As such, care must be taken to describe the putative cognitive mechanism that is impaired rather than relying on simple labeling. The case study reported by De Renzi (1986) had profound difficulties in recognizing the faces of people close to him, including his family, but could recognize them by their voices or other non-facial information. He once remarked to his wife, 'Are you [wife's name]? I guess you are my wife because there are no other women at home, but I want to be reassured.' The patient's ability to recognize and name other objects was spared. Within the Bruce and Young (1986) model his deficit would be located at the face recognition unit stage. Patients with acquired prosopagnosia still retain the ability to recognize other socially salient information from (familiar and unfamiliar) faces, including sex, age, and emotional expressions (Tranel, Damasio, & Damasio, 1988). They can also use lip-reading cues to aid in speech perception (Campbell, Landis, & Regard, 1986). More recently, congenital prosopagnosia has been documented, which has similar characteristics to acquired prosopagnosia but is present throughout the lifespan with no known external cause (Duchaine & Nakayama, 2006).

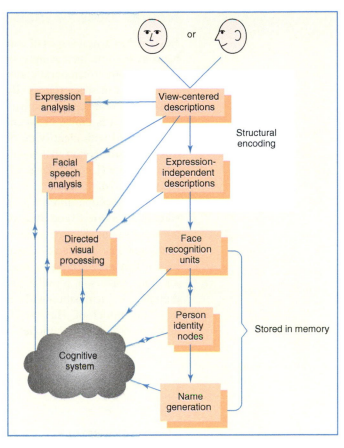

Figure 5.2 The Bruce and Young (1986) model of face processing postulates a number of separate mechanisms that reflect the different goals of the perceiver, including: recognizing a familiar face; recognizing a facial expression; and matching two images of an unfamiliar face.

KEY TERMS

Person identity node
A hypothetical entity in models of face processing that links together semantic and perceptual information about a particular individual

Prosopagnosia
Impairments of face processing that do not reflect difficulties in early visual analysis

Ventral visual stream
Runs from the occipital to the temporal lobes and is concerned with identifying objects

Dorsal visual stream
Runs from the occipital to the parietal lobes and is concerned with locating and acting on objects

Occipital Face Area (OFA)
A region in the occipital cortex that responds to faces more than objects but does not process facial identity

Fusiform Face Area (FFA)
A region in the fusiform cortex that responds to faces more than objects and is responsive to facial identity

Neural basis of face perception

Beyond the earliest stages of cortical visual processing (in the primary visual cortex), neurons become increasingly specialized in their response properties. In primates, cells in the inferotemporal cortex respond to specific shapes but not where the shape is presented in space (Gross, Rocha-Miranda, & Bender, 1972). This suggests that these neurons are concerned with the appearance of objects rather than their location. A distinction is drawn between two visual streams – a **ventral visual stream** that is concerned with identifying objects, largely irrespective of where they are, and a **dorsal visual stream** that is concerned with locating objects, largely irrespective of what they are (Ungerleider & Mishkin, 1982). Face perception depends primarily on the ventral visual stream.

In this section, three regions of the human brain are considered in detail: the so-called **occipital face area (OFA)**, **fusiform face area (FFA)**, and the superior temporal sulcus (STS). The approximate locations are shown in Figure 5.3. The label 'face area' has attracted controversy due to the claim that it may be a domain-specific region that processes only one kind of information – namely faces (e.g. Gauthier, Tarr, Anderson, Skudlarski, & Gore, 1999). The evidence for and against this domain-specific position will be considered.

The model of Haxby, Hoffman, and Gobbini (2000) presents a neuroanatomically inspired model of face perception that contrasts with the purely

---

KEY CHARACTERISTICS OF DIFFERENT FACE PROCESSING REGIONS OF THE CORE SYSTEM

Occipital face area (OFA)
- Relatively specialized for faces (not bodies or objects)
- Codes the physical aspects of facial stimuli

Fusiform face area (FFA)
- Relatively specialized for faces (not bodies or objects)
- Important for computing an invariant facial identity

Superior temporal sulcus (STS)
- Responds to faces and bodies
- Important for action perception and dynamic stimuli (e.g. lip movements)
- Integrates visual and auditory information

Figure 5.3 Approximate locations of face- and body-sensitive visual regions, shown here in the right hemisphere. The fusiform body and face areas are located on the underside of the brain, but are shown here projected onto a side view. Other regions are located laterally. EBA = extrastriate body area; FBA = fusiform body area.

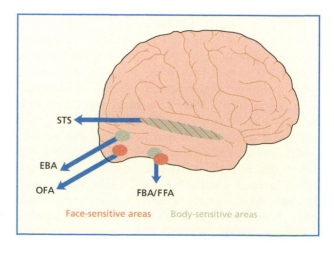

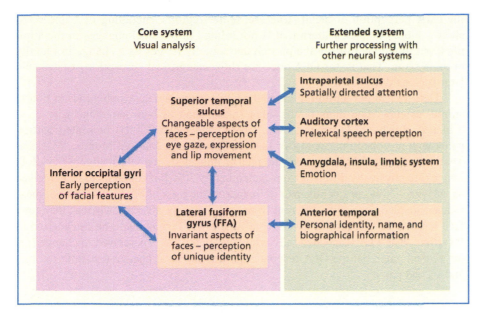

Figure 5.4 The model of Haxby et al. (2000) divides the neural substrates of face processing into a number of core mechanisms (relatively specialized for faces) and an extended system in which face processing makes contact with more general cognitive mechanisms (e.g. concerning emotion, language, action).

cognitive account offered by Bruce and Young (1986). In their model, shown in Figure 5.4, Haxby et al. (2000) consider the core regions involved in face perception to lie in the fusiform gyrus in humans (corresponding to the inferotemporal cortex identified in primates) and the STS. They also identify an 'extended system' to denote other areas of the brain that receive inputs from the core face perception system but are not essential for face perception (e.g. regions supporting semantic knowledge of people).

Occipital face area

The OFA is located in the inferior occipital gyrus. It is considered to be an early stage in perceptual analysis of faces that sends inputs to fusiform and superior temporal regions (e.g. Haxby et al., 2000). Like the 'fusiform face area' it is defined on the basis of showing a greater fMRI BOLD response to faces relative to other categories. However, it differs from the FFA in a number of key respects. For example, the OFA responds to both upright and inverted faces whereas the FFA responds more to upright faces (Yovel & Kanwisher, 2005). Fox, Moon, Iaria, and Barton (2009) used an fMRI adaptation paradigm to compare activity in the OFA with other face-sensitive regions. This method relies on the fact that the BOLD signal is reduced when the same stimulus is presented twice. However, one can vary how 'same' is defined to reveal different properties of different regions (e.g. physically same, same person, same expression). The OFA activity is sensitive to any physical change in the face stimulus. This is consistent with a role in the early perception of facial structure. Other regions (including FFA and STS) show a more complex pattern that is not related to physical changes in the stimulus, but to whether or not the participant actually perceives a change in identity or expression. That is, the OFA does not show categorical perception of faces, but later regions do.

Categorical perception refers to the tendency to perceive ambiguous or hybrid stimuli as either one thing or the other (rather than as both simultaneously or as a blend). Rotshtein, Henson, Treves, Driver, and Dolan (2005) presented participants with morphed images of two famous people (e.g. Marilyn Monroe, Margaret Thatcher) during fMRI. Consider the pair of images on the left of Figure 5.5. Both images are physically different from each other by 30% but both tend to be perceived as 'Maggie'. Now consider the pair of images on the right of Figure 5.5. Both of these images are also physically different from each other by 30% but whereas the upper face tends to be categorized as 'Maggie', the lower face tends to be categorized as 'Marilyn'. The FFA responded only when the identity was perceived to change (i.e. the second pair), but the OFA responded to any physical change (i.e. both the first and second pair).

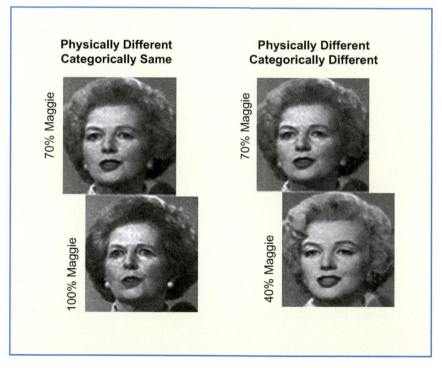

Figure 5.5 Morphing from Maggie to Marilyn! Both the left and the right pair of images differ from each other by 30%, but a change in identity is perceived to occur in the right pair only. The FFA is sensitive to changes in identity whereas the OFA is sensitive to the degree of physical change. From Rotshtein et al. (2005).

In summary, the OFA is an early (pre-categorical) perceptual region that processes the physical properties of faces.

Fusiform face area

The FFA responds to faces more than other stimuli, including bodies, and may be particularly important for recognizing known faces (Kanwisher, McDermott, & Chun, 1997; Kanwisher & Yovel, 2006). The FFA is found bilaterally, with a generally more robust BOLD response on the right. The degree of left/right asymmetry differs between individuals and is related to individual differences in visual field asymmetry for identifying faces (Yovel, Tambini, & Brandman, 2008). Thus, a greater ability at identifying a face in the left visual field is associated with greater activity in the right FFA relative to left FFA. Functional imaging has found that the FFA shows fMRI adaptation when the same face is repeated even if physical aspects of the images change (see Kanwisher & Yovel, 2006). Although the lesion location of acquired prosopagnosia varies from case to case, the evidence is broadly consistent with damage in or around the FFA (Barton, 2008).

The main rival claim to the suggestion that the FFA is specialized for faces is that this region is sensitive to expert within-category visual discriminations but not to faces per se (e.g. Gauthier et al., 1999). These accounts have two key elements: that faces require discrimination within a category (between one face and another), whereas most other object recognition requires a superordinate level of discrimination (e.g. between a cup and comb); and consequently that we become 'visual experts' at making these fine within-category distinctions through prolonged experience with thousands of exemplars. The evidence for this theory comes from training participants to become visual experts at making within-category discriminations of non-face objects, called 'Greebles' (see Figure 5.6). As participants become experts they move from part-based to holistic processing, as has often been proposed for faces (Gauthier & Tarr, 1997). In addition, they have shown that Greeble experts activate the FFA (Gauthier et al., 1999) and similar findings have been reported for experts on natural categories such as birds and cars (Gauthier, Skudlarski, Gore, & Anderson, 2000).

There is some evidence from prosopagnosia that supports the face specificity account against the visual expertise alternative. Sergent and Signoret (1992) reported a prosopagnosic patient, RM, who had a collection of over 5000 miniature cars. He was unable to identify any of 300 famous faces, or the face of himself or his wife, or match unfamiliar faces across viewpoints. Nevertheless, when shown 210 pictures of miniature cars he was able to give the company name, and for 172 he could give the model and approximate year of manufacture. This suggests that face perception and visual within-category expertise are not necessarily the same thing (see also McNeil & Warrington, 1993).

In summary, the FFA is primarily concerned with discriminating between facial identities and may act

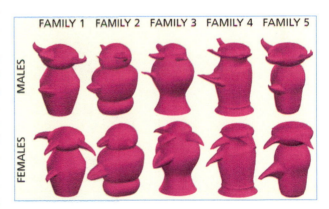

Figure 5.6 Greebles can be grouped into two genders and come from various families. To what extent does discriminating amongst Greebles resemble discriminating across faces? Images provided courtesy of Michael J. Tarr (Carnegie Mellon University, Pittsburgh, PA; see www.tarrlab.org).

as a store of known faces. The extent to which it is also involved in non-facial perception remains debated.

Superior temporal sulcus

According to the model of Haxby et al. (2000), the STS responds to the changeable aspects of a face (e.g. particular poses, gaze directions) whereas the FFA responds to the stable aspects of a face (i.e. the person's identity). The changeable aspects of a face are particularly important for extracting social cues that are likely to be fleeting (Allison, Puce, & McCarthy, 2000). In support of the distinction between STS/FFA made by Haxby et al. (2000), functional imaging studies show that when participants are asked to make judgments about eye gaze (deciding whether the face is looking in the same direction as the last face) then activity is increased in the STS but not in

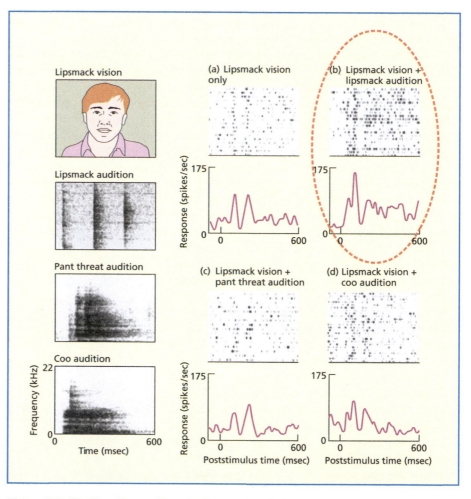

Figure 5.7 Single-cell recordings in the monkey STS show increased firing when the same vocalization is both seen and heard. This suggests that the region is not a purely visual one, but integrates across hearing and vision. From Barraclough et al. (2005). Copyright © 2005 by the Massachusetts Institute of Technology. Reproduced with permission.

the FFA (Hoffman & Haxby, 2000). In contrast, when participants are asked to make judgments about face identity (deciding whether the face is the same as the last one presented) then activity is increased in the FFA but not in the STS.

The STS region responds to bodies as well as faces, such as when observing people walk (e.g. Grossman et al., 2000). The STS may also be important for linking the ventral visual stream with the dorsal stream. This enables dynamic visual percepts of faces and bodies (in the STS) to be processed motorically (by parietal and frontal lobe systems) – a mirroring of the others' action in one's own motor system. The STS receives greater multi-sensory inputs than the FFA. Functional imaging studies show that it responds to both seen speech (i.e. facial lip-reading) and heard speech, and that the response is greater when the two correspond in terms of content and timing (Calvert, Hansen, Iversen, & Brammer, 2001). Single-cell recordings from monkeys show neurons that respond to both the sight and sound of certain actions such as lip-smacking or threat noises (Barraclough, Xiao, Baker, Oram, & Perrett, 2005). This is shown in Figure 5.7.

The role of the STS in detecting biological motion is discussed in the section on body perception, and the role of STS in initiating joint attention is considered later.

SOCIAL DISPLAY RULES

Although basic emotions are considered to be universal across cultures, culture can exert an effect in terms of social **display rules** – the extent to which one regulates emotional expressions in the presence of others. Ekman (1972) found that when both Japanese and American participants watched a stressful film in isolation there were no cultural differences in facial expression. However, in the presence of a high status experimenter the Japanese participants were more inclined to hide a display of negative emotion with a smile. Matsumoto et al. (2008) asked participants in 32 different cultures questions about their attitudes towards displays of emotion. The results revealed that individualistic cultures (e.g. USA, Australia) were more likely to endorse the display of happiness than collectivist cultures (e.g. Indonesia and Hong Kong). Moreover, participants from all cultures reported a greater emotional display towards their ingroup than an outgroup (considered here in terms of their nationality).

There are also cross-cultural differences in face-to-face looking behavior. In social interactions, Westerners consider it more important to try to maintain eye contact than East Asians for whom avoidance of eye contact can be seen as a sign of respect and deference (Argyle, Henderson, Bond, Iizuka, & Contarello, 1986). However, when trying to decode facial expressions there is a greater tendency for East Asians to fixate on the eyes whereas Westerners are more likely to scan both the eyes and mouth (Jack, Blais, Scheepers, Schyns, & Caldara, 2009). It is interesting to consider whether this scanning behavior is also reflected in cross-cultural variations in use of emoticons. Western emoticons tend to emphasize the mouth :-) whereas East Asian emoticons tend to emphasize the eyes ^_^ (Park, Baek, & Cha, 2014).

KEY TERM

Display rules
The extent to which one regulates emotional expressions in the presence of others

Perceiving emotion from faces

This section will consider two different issues. Firstly, the question of whether there is a specialized route for recognizing facial emotions relative to recognizing familiar faces (as in Bruce & Young, 1986) or whether recognizing facial expressions is part of a more general system for detecting dynamic changes in faces (as in Haxby et al., 2000). Secondly, it will conside the question of whether familiar faces (even with a neutral facial expression) have an emotional signature and how this might go awry in a fascinating symptom called Capgras delusion (reporting familiar people as imposters).

How does facial expression recognition relate to other aspects of face perception?

Both the cognitive model of Bruce and Young (1986) and the neural model by Haxby et al. (2000) make a distinction between recognizing familiar faces (facial identity) and recognizing emotional expressions in faces. Whilst there is good evidence for this basic idea, Calder and Young (2005) argue that neither model offers a satisfactory account of the evidence. Although prosopagnosic patients have been reported who are better at recognizing facial expressions than facial identity (e.g. Tranel et al., 1988), Calder and Young (2005) argue that the dissociation is not absolute. Both facial expression and facial identity tend to be impaired in prosopagnosia relative to controls, even though recognition of facial expressions is less impaired. They suggest that the structural encoding stage is important for both expressions and identity (including both the OFA and FFA) but recognizing identity fares worse after brain damage because it is more demanding. There is evidence from fMRI that the FFA responds to changes in expression as well as identity (Ganel, Valyear, Goshen-Gottstein, & Goodale, 2005).

The Bruce and Young (1986) model assumes that there is a separate route for analyzing facial expressions. If this is so, then one might be able to find brain-damaged patients who are unable to recognize facial expressions but can recognize facial identity. Insofar as such patients exist, they seem to be associated with orbital and ventromedial frontal lesions (Heberlein, Padon, Gillihan, Farah, & Fellows, 2008; Hornak, Rolls, & Wade, 1996) or somatosensory regions (Adolphs, Damasio, Tranel, Cooper, & Damasio, 2000) but *not* the STS. However, rather than postulating a single route for dealing with all emotions, Calder and Young (2005) favor the idea that each emotion is dealt with separately (e.g. the amygdala for fear, the insula for disgust) and is part of the extended system, to borrow the terminology of Haxby et al. (2000), rather than the core system of face processing.

Although not specifically discussed by Calder and Young (2005) or Haxby et al. (2000), there is one candidate mechanism that could serve as a general system for recognizing expressions but not identity – namely in terms of sensorimotor simulation (e.g. Heberlein & Adolphs, 2007). **Simulation theory** will be encountered many times during this book; it actually consists of a collection of somewhat different theories based around a unifying idea – namely that we come to understand others (their emotions, actions, mental states) by vicariously producing their current state on ourselves. This is considered at length in Chapter 6. With regard to emotions, the claim is that when we see someone smiling we also activate our own affective pathways for happiness. Moreover, we may activate the motor program needed to make us smile (this may make us smile back, or it may prepare a smile response) and we may simulate what this might feel like in terms of its sensory consequences (e.g. muscle stretch and tactile sensations on the face). As such, one could possibly recognize emotions such as happiness, fear, and disgust not just in terms of their visual appearance but in terms of the way that they activate the sensorimotor program of the perceiver.

There is evidence from electromyographic studies that viewing a facial expression produces corresponding tiny changes in our own facial musculature, even if the face is viewed briefly so as to be unconsciously perceived (Dimberg et al., 2000). However, this does not necessarily imply that this is used to recognize expressions. To address this, Oberman, Winkielman, and Ramachandran (2007) report that biting a pen lengthways uses many of the same muscles involved in smiling. They subsequently showed that the bite task selectively disrupts the recognition of happiness (see Figure 5.8). Lesion studies also suggest a direct contribution of simulation mechanisms to recognizing emotional expressions. Adolphs et al. (2000) tested the critical lesion sites in 108 patients asked to identify facial expressions from the Ekman categories. Damage to sensorimotor areas, including the right somatosensory cortex and Broca's area, were found to predict poor performance. Pitcher, Garrido, Walsh, and Duchaine (2008) applied TMS to the right somatosensory cortex of healthy participants and noted that recognition of facial expressions, but not facial identity, was disrupted by TMS in this region. The effects occurred relatively early (when TMS was applied within 170 ms of stimulus onset), suggesting that simulation pathways may be activated in parallel with visual mechanisms of expression recognition.

In summary, there is good evidence that recognition of facial emotions is (at least partially) dissociable from facial identity recognition. The evidence suggests that brain regions involved in emotional experience and sensorimotor simulation have an important role in expression recognition, but the role of the STS in expression recognition remains less clear.

Do familiar faces have an emotional signature?

A rather different issue concerns the extent to which a familiar face (with a neutral expression) generates an emotional response. In several studies, the emotional response has been measured in terms of the skin conductance response (SCR). This response occurs within seconds of an appropriate stimulus and is considered to be

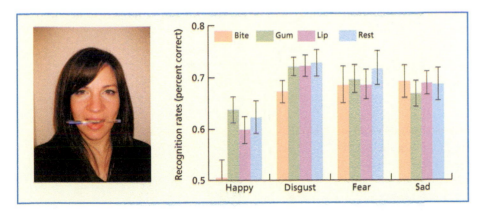

Figure 5.8 Placing a pen in the mouth horizontally and holding it with the teeth uses many of the same muscles as smiling. Performing this task can also disrupt recognition of facial expressions of happiness. Data from Oberman et al. (2007).

automatic (insofar as participants cannot control its absence or presence) and unconscious (insofar as participants are generally unaware of it).

In the neurologically typical population, familiar faces generate a greater SCR
than unfamiliar faces (Tranel, Fowles, & Damasio, 1985). Intriguingly, patients with
acquired prosopagnosia show evidence of the same trend (e.g. Tranel, Damasio, &
Damasio, 1995) – that is, they generate a greater SCR to familiar faces than unfamiliar ones despite not knowing who the familiar people are or, indeed, being unable
to classify them as familiar. There are several possible explanations for this. One is
that both the conscious recognition of familiar faces and the unconscious (emotional)
recognition of familiar faces use the same damaged mechanism (e.g. visual processes
in the fusiform). These mechanisms are presumably degraded rather than absent, but
obtaining an emotional signature is less demanding and can still occur with partial
damage (Barton, 2008). Another explanation is that there are separate processes at
early visual stages that are linked to these different outcomes: generating an emotional response versus consciously identifying the face. This would be an interesting
question to explore with functional imaging.

A dissociation between the emotional content of faces and the recognition of
facial identity has been put forward to account for a delusional symptom called **Capgras syndrome**. In the Capgras syndrome, people report that their acquaintances
(spouse, family, friends, and so on) have been replaced by 'body doubles' (Capgras & Reboul-Lachaux, 1923; Ellis & Lewis, 2001). They will acknowledge that
their husband/wife looks like their husband/wife. Indeed, they are able to pick out
their husband/wife from a line-up while maintaining that he/she is an imposter. To
account for this, Ellis and Young (1990) suggest that these patients are the opposite
of prosopagnosic. Thus, they can consciously recognize the person/face but they lack
an emotional response to them. As such, the person/face is interpreted as an imposter.
This explains why the people who are doubled are those closest to the patient, as these
would be expected to produce the largest emotional reaction. This theory 'makes
the clear prediction that Capgras patients will not show the normally appropriate
skin conductance responses to familiar faces' (Ellis & Young, 1990, p. 244). Subsequent research has confirmed this prediction as shown in Figure 5.9 (Ellis, Young,
Quayle, & DePauw, 1997). However, the findings of Tranel et al. (1995) are problematic. Their patients with damage to the ventromedial frontal lobes had abolished
skin conductance to familiar faces but reported no signs of Capgras delusion. Thus, a
lack of emotional response may well be necessary, but it is unlikely to be sufficient.

Figure 5.9 Most people
produce a greater skin
conductance response
(SCR) to personally
familiar people relative
to unfamiliar ones, but
patients with Capgras
delusion do not. From
Ellis and Lewis (2001).
Copyright © 2001
Elsevier. Reproduced
with permission.

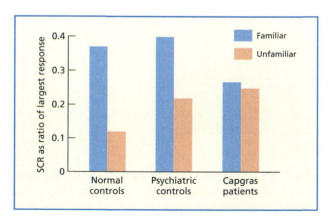

Perhaps a mechanism involved in decision-making is compromised in addition to facial emotional processing.

Evaluation

There is good evidence that face processing can be divided into different operations serving different functions and with different neural substrates. The models of Bruce and Young (1986) and Haxby et al. (2000) offer useful, and complementary, accounts of these different mechanisms. However, neither account offers a full explanation of the current evidence. For example, there appear to be multiple routes for recognizing facial expressions: one based on sensorimotor simulation that is not specialized for particular emotions, and others that make contact with core regions of emotion processing (e.g. amygdala, insula). In addition, there are conscious and unconscious correlates of emotion processing of faces – the latter may be conveniently measured by SCR.

RECOGNIZING ONE'S OWN FACE USES A SPECIAL MECHANISM

Seeing our own face activates a circuit of regions in the frontal and parietal lobes that are not found for other personally familiar faces (Uddin, Iacoboni, Lange, & Keenan, 2007). Of course, we never directly see our own faces except via mirrors or external images. Recognizing one's own reflection in a mirror requires more than just face recognition (a basic-level knowledge); it requires appreciating that the face out there corresponds to the person 'in here'. **Mirror self-recognition** (MSR) has been considered by some to be an important test of self-awareness in other species (for reviews see de Veer & Van den Bos, 1999; Schilhab, 2004). Gallup (1970) noted that, during development, both humans and chimpanzees alter their behavior towards mirrors. At early stages of development they behave as if the mirror image is another animal and show social behavior towards it. When older (18–24 months in humans) they show self-directed behaviors, such as using the mirror to explore unseen body parts. Of particular interest is their behavior on the 'mark test' (Gallup, 1970). If a red mark is placed on the forehead during anesthesia, then the animal will attempt to groom it on encountering its reflection in a mirror. This behavior is found in humans, common and pygmy chimpanzees, and orangutans, but not in gorillas or monkeys (Schilhab, 2004). Does this behavior unequivocally demonstrate self-awareness? At the very least it is consistent with *bodily* self-awareness or an ability to form a link between an external image and one's own body.

PERCEIVING BODIES

There is far less research on the visual perception of bodies compared to faces. There are no long-standing cognitive models of body perception that are equivalent to Bruce and Young's (1986) model. Whilst brain regions have been identified that

KEY TERMS

Extrastriate Body Area (EBA)
An area in visual cortex that responds more to whole bodies and body parts than faces and objects

Fusiform Body Area (FBA)
A region of inferior temporal cortex that responds relatively more to whole bodies than body parts

are important for body perception, their precise functions remain relatively under-specified relative to comparable regions in the face domain.

Visual perception of bodies

There are at least two regions of the brain that respond preferentially to the visual perception of bodies relative to other categories such as faces and objects. These have been termed the **extrastriate body area** (EBA; Downing et al., 2001) and the **fusiform body area** (FBA; Peelen & Downing, 2005) reflecting their anatomical locations, shown previously in Figure 5.3. The EBA codes an abstract description of a body plan insofar as it responds strongly not only to real photographs but also to line drawings, stick figures, and body parts (Downing et al., 2001). It shows a graded response to other types of bodies such that it has the greatest response to humans, then mammals, then fish and birds, and lastly objects – see Figure 5.10 (Downing, Chan, Peelen, Dodds, & Kanwisher, 2006). TMS over this region affects the speed of deciding whether two body parts are the same, but not an equivalent task for faces or motorcycle parts (Urgesi, Berlucchi, & Aglioti, 2004). Brain lesions in the EBA region prevent successful performance on this task (Moro et al., 2008). The FBA spatially overlaps with the fusiform face area, and is noted to respond relatively more to whole bodies than body parts whereas the EBA does not (Taylor, Wiggett, & Downing, 2007). This is consistent with the FBAs location further along the ventral stream.

What kind of function is performed by these two regions? Downing and Peelen (2011) considered the evidence to date and concluded that the EBA and FBA contain

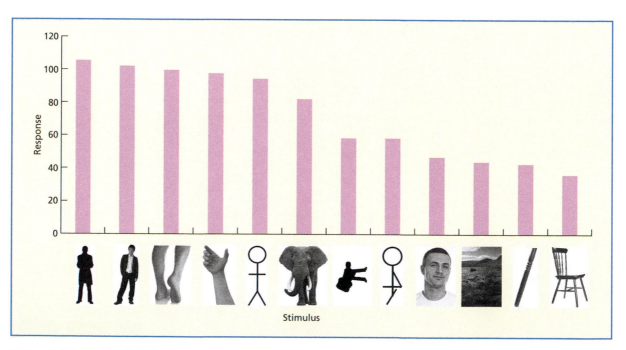

Figure 5.10 BOLD activity in the extrastriate body area (EBA) in response to different kinds of stimuli (scaled according to the response to body parts). From left to right: body silhouettes; whole bodies; various body parts; hands; stick figures; mammals; scrambled silhouettes; scrambled stick figures; faces; scenes; object parts; and whole objects. Adapted from Peelen and Downing (2007).

perceptual body representations. They represent body shape and size (e.g. fat v. thin) and posture, but are less sensitive to viewpoint. Patients with anorexia nervosa have reduced gray matter volume in the EBA, assessed using VBM, and this was correlated with errors on a body-size judgment task in which they had to match their own body to silhouettes (Suchan et al., 2010). Downing and Peelen (2011) describe the perceptual body representations as 'cognitive unelaborated' by which they mean that they are insensitive to what the body part is doing. For instance, the response of the EBA to the sight of a hand is not increased if the hand is penetrated by a needle, suggesting it does not generate an empathic response (Lamm & Decety, 2008). Similarly, it is insensitive to action context. Urgesi et al. (2007) presented static body part stimuli in a delayed matching task ('Is image 1 the same as image 2?'). The images could vary either in action or shape. TMS over EBA impaired performance on the body shape task, but not the action task, while TMS over premotor cortex (part of the mirror system) produced the opposite pattern. Kontaris et al. (2009) showed live recordings of participants' own hands in an MRI scanner and contrasted it with pre-recorded movies either with or without a concurrent hand movement. If motor information could be used to predict what is seen, then visual activity would be suppressed for live recordings. Whilst this was found in posterior STS, the EBA and FBA were activated strongly in all conditions. This does not mean that the EBA and FBA have no role at all in empathy, imitation, or action understanding but that, instead, it provides an early visual input to these processes.

The processing of body movement, so-called **biological motion,** may be more a function of the STS, whereas the processing of body configuration is a function of the EBA and FBA. The perception of biological motion is assessed by attaching light-emitting diodes (LEDs) to the joints and then recording someone walking/running in the dark as shown in Figure 5.11. When only the LEDs are viewed, most people are still able to detect bodily movement relative to a control condition in which these moving lights are presented jumbled up. A comparison of BOLD response to biological motion relative to jumbled motion reveals activity in a posterior region of the STS (Grossman et al., 2000). This region is different from the main region responsible for visual movement perception, known as **V5 (or MT)**, and there may be separate pathways into STS (for biological motion) versus V5/MT (for visual motion in general). For example, a patient with a bilateral lesion to V5/MT was unable to

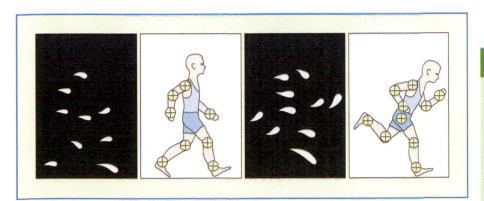

Figure 5.11 When this array of dots is set in motion, most people can distinguish between motion from biological forms (e.g. the human body) and non-biological motion (e.g. the same dots scrambled).

detect visual motion and perceived the world in terms of jerky snapshots (Zihl, Von Cramon, & Mai, 1983). Nonetheless, she was able to discriminate biological from non-biological motion (McLoed, Dittrich, Driver, Perrett, & Zihl, 1996). Single-cell recordings in monkey STS show cells that respond not only to a figure walking but also the direction of motion (towards or away from the monkey), whether the figure has his back or front to the monkey, and also whether the figure was near or far from the monkey in the testing room (Jellema, Maassen, & Perrett, 2004).

Perceiving emotion from bodies

Bodies can convey somewhat different emotional information from a face. For example, whilst a facial expression might signal fear, the body might additionally convey whether the person is likely to stay or run. Although much of the research has come from static body stimuli, there are also reliable cues to emotion from the way that people walk or move their bodies (Roether, Omlor, Christensen, & Giese, 2009).

Fearful body language elicits activation in the amygdala and fusiform gyrus relative to happy and neutral body language (Hadjikhani & de Gelder, 2003). (In these stimuli the face is blurred out so that one can be sure that it is the body itself that conveys the expression.) In order to study how emotional information from faces and bodies is combined, Meeren, van Heijnsbergen, and de Gelder (2005) created composite images in which angry/fearful faces were superimposed on angry/fearful bodies, thus generating both congruent (e.g. fear face + fear body) and incongruent (e.g. fear face + angry body) combinations. Participants asked to categorize the faces as angry or fearful were affected by the emotional status of the body, suggesting that this information is hard to ignore. This is shown in Figure 5.12. An ERP study using the same stimuli suggests that the brain detects an emotional mismatch between face and body quickly, at around 100 ms (Meeren et al., 2005).

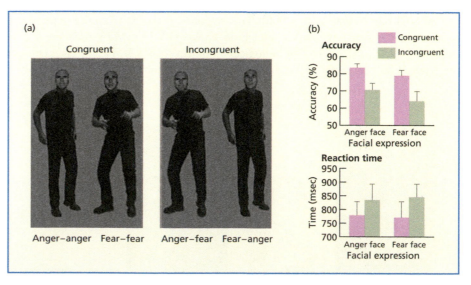

Figure 5.12 When asked to judge facial expressions of anger or fear, the emotional body expression interferes with this decision. This implies that emotional body language is recognized automatically. From de Gelder (2006). Copyright © 2006 Nature Publishing Group. Reproduced with permission.

De Gelder (2006) put forward a model of emotional body perception. The model contains two circuits and it is noteworthy that both routes have a visual perception component and a motor-based component. This raises the possibility that emotional body language, like facial expressions, could be recognized on the basis of visual appearance, or on the basis of motoric simulation of the body language, or both. Observers do not necessarily have to reproduce an action themselves, but the suggestion is that watching emotional body language prepares one for an action and that this preparatory signal may itself convey information about the other person.

1 *Reflex-like perception of emotional body language.* This involves a rapid and unconscious evaluation of the visual stimulus using mainly subcortical visual pathways (e.g. from the pulvinar and superior colliculus to the amygdala). This may initiate over-learned motor responses through a subcortical basal-ganglia motor circuit.
2 *Visuo-motor perception of emotional body language.* This involves a slower and more deliberative assessment that uses predominantly cortical structures. Visual analysis is assumed to occur in the fusiform and STS regions, which connect to emotional centers (e.g. the amygdala), and also regions containing mirror neurons in the intraparietal sulcus and premotor cortex.

With regards to the sub-cortical route there is evidence that patients with cortical blindness (who can still processes visual stimuli to a certain extent with their sub-cortical visual pathways), produce emotional contagion in the form of facial responsiveness (measured by EMG) to both facial and bodily expressions that they do not consciously perceive (Tamietto et al., 2009). Although the exact neuroanatomical route isn't known, healthy volunteers show modulation of a freezing response to the sight of fearful body postures when TMS is applied to motor cortex after only 100 ms following stimulus presentation (Borgomaneri, Vitale, Gazzola, & Avenanti, 2015).

There is good evidence of social mirroring of body language, which is a key idea in the model. Chartrand and Bargh (1999) asked two participants to cooperate in a task. However, one participant was actually an experimenter who was instructed to surreptitiously perform one of two body movements – nose rubbing or foot shaking. Different participants saw different actions so that the unperformed action in the pair served as a control. Participants who saw foot shaking were more likely to produce foot shaking than nose rubbing, and the reverse was true for those who saw nose rubbing. In a second study, Chartrand and Bargh (1999) asked the confederate to surreptitiously copy the body language of the participant (or not, as a control condition). Participants whose body language was imitated reported liking the confederate more and judged that the task was completed more smoothly. This important result suggests that mirroring of body language serves an important social function that goes beyond simulation for the sake of identifying a person's emotional state.

Evaluation

In terms of early visual processing, body perception relies on different neural substrates to face perception (the EBA is particularly important). However, beyond these earlier stages there is evidence of convergence. Bodies, like faces, activate the STS when they are dynamic (biological motion) and emotional body language activates regions of the brain specialized for emotion. Body perception is important for imitation, which promotes both social cohesion and social learning.

VOICES CONTAIN IMPORTANT SOCIAL CUES TOO

Voices, like faces and bodies, convey a large amount of socially relevant information about the people around us. It is possible to infer someone's sex, size, age, and mood from their voice. Physical changes related to sex, size, and age affect the vocal apparatus in systematic ways. Larger bodies have longer vocal tracts and this leads to greater dispersion of certain frequencies (the formants found, for example, in human vowels and dog growls are more dispersed in larger animals). Adult men have larger vocal folds (17–25 mm) than adult women (12.5–17.5 mm), resulting in a lower pitched male voice. One can also infer the current emotional state (angry, sad, etc.) from a voice even in an unfamiliar language (Scherer, Banse, & Wallbott, 2001). Individual differences in the shape and size of the vocal apparatus (teeth, lips, etc.) and resonators (e.g. nasal cavity), together with learned speaking style (e.g. accent), create a unique voice signature.

Yovel and Belin (2013) discuss the parallels between face processing and voice processing in the brain. There is evidence for a voice selective region in the STS that detects the identity of a speaker (Belin, Zatorre, Lafaille, Ahad, & Pike, 2000). This has been termed the Temporal Voice Area (TVA) and may be functionally analogous to the FFA for faces. Like the FFA it also appears to be more strongly right lateralized, which implies a closer relationship between voices and faces than between voices and speech comprehension (where the latter is more strongly left lateralized). As with faces, there is a separation between recognizing voice identity and recognizing emotion in voices as revealed through fMRI (Ethofer et al., 2012) and TMS (Banissy et al., 2010).

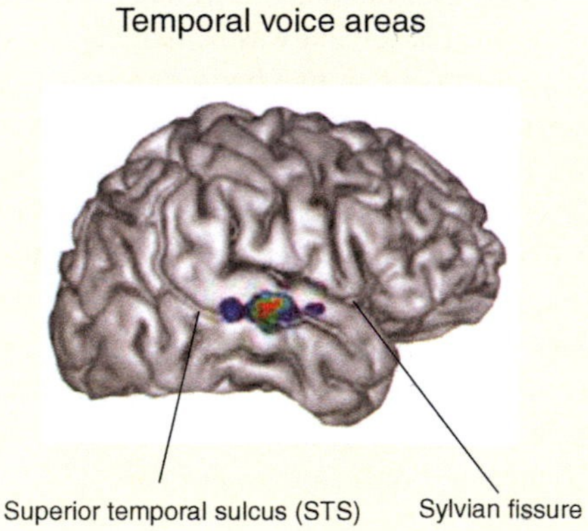

Temporal voice areas

Figure 5.13 The Temporal Voice Areas (TVA) are identified by their responsiveness to voices (whether producing speech or other vocal sounds) relative to other kinds of acoustic stimuli. A region in the right STS may be important for recognizing the identity of a speaker, and serve a broadly similar function to the FFA for faces. From Yovel and Belin (2013).

Superior temporal sulcus (STS)   Sylvian fissure

JOINT ATTENTION: FROM PERCEPTION TO INTENTION

Joint attention refers to the process by which attention is oriented to a particular object/location in response to another person's attention. Direction of eye gaze is important, but head and body orientation, or pointing, can elicit joint attention too. This may provide the foundations for making inferences about other people's intentions and actions.

Eye gaze detection

Making eye contact can be important for establishing one-to-one communication (dyadic communication), and the direction of gaze can be important for orienting attention to critical objects in the environment. Direct eye contact, in many primates, can be sufficient to initiate emotional behaviors. Macaques are more likely to show appeasement behaviors when shown a direct gaze relative to indirect or averted gazes (Perrett & Mistlin, 1990). Moreover, dominance struggles are often initiated with a mutual gaze and terminated when one animal averts its gaze (Chance, 1967). The discrimination of gaze direction in humans may be easier than in other animals due to the smaller dark region (pupil and iris) surrounded by the white sclera. One suggestion is that the white sclera evolved specifically to facilitate joint attention and enable cooperation (e.g. Tomasello, Hare, Lehmann, & Call, 2007).

Baron-Cohen argues that an eye direction detector is an innate and distinct component of human cognition (Baron-Cohen, 1995a; Baron-Cohen & Cross, 1992). Babies are able to detect eye contact from birth, suggesting that it is not a learned response (Farroni, Csibra, Simion, & Johnson, 2002). This ability is likely to be important for the development of social competence because the eyes code relational properties between objects and agents (e.g. 'mummy sees daddy', 'mummy sees the box').

The region that appears to be particularly important for eliciting joint attention from faces and bodies is the STS. Functional imaging shows that other regions such as the intraparietal sulcus are important for directing attention more generally from both social (e.g. eyes) and non-social (e.g. arrows) cues, but the STS responds particularly to social attention cues (Hooker et al., 2003). The effect is greater when the eyes look at empty space (a sort of gaze error) relative to the more usual scenario of looking at an object (Pelphrey, Singerman, Allison, & McCarthy, 2003). In humans it has been shown that more anterior portions of the STS also code for the direction of attention taking into account both the position of the head and the eyes (Carlin, Calder, Kriegeskorte, Nili, & Rowe, 2011).

The STS may be involved primarily in the *perception* of gaze, but other regions may be involved in joint attention as a gaze-based social interaction. These other regions include the medial prefrontal cortex (mPFC) and the right temporo-parietal junction (rTPJ). For instance, in one study participants were required to follow a dot with their own eyes (Williams, Waiter, Perra, Perrett, & Whiten, 2005). In one condition someone else's eyes also followed the dot (joint attention) and in another case it did not (the other person looked around randomly). The mPFC was involved in the joint attention condition. Other studies have contrasted live face-to-face interactions in the scanner during a game compared with pre-recorded ones and shown that the rTPJ, which lies adjacent to the posterior STS, is activated during the live interaction (Redcay et al., 2010). The role of these regions in mentalizing is discussed in detail in Chapter 6.

The STS of monkeys contains many cells that respond to eye direction (Perrett et al., 1985) and lesions in this area can impair the ability to detect gaze direction (Campbell, Heywood, Cowey, Regard, & Landis, 1990). Cells in the STS are not only sensitive to the direction of the eyes but also to the orientation of the head and the body. Perrett, Hietanen, Oram, and Benson (1992) report that the response rate of

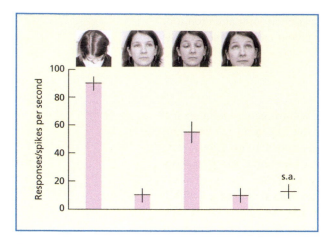

Figure 5.14 This neuron in the STS responds when gaze is oriented downwards. The activity of the neuron (spikes per second) is shown when presented with four faces and during spontaneous activity (s.a.). Adapted from Perrett et al. (1992).

Figure 5.15 In gaze cueing experiments, participants must press a button when they see a particular target (e.g. an asterisk). Performance is speeded up if gaze is directed towards the target (top) and slowed when gaze is directed away (bottom). The figure shows examples from both schematic and realistic faces, both of which elicit gaze cueing effects. From Frischen et al. (2007). Copyright © 2007 American Psychological Association. Reproduced with permission.

single cells depends mostly on eye gaze, then head orientation, followed by body posture. Some neurons may respond, for example, to a downward oriented gaze irrespective of whether the head is pointing down or whether the eyes are pointing down with the head full frontal as depicted in Figure 5.14. This suggests that these particular neurons are indeed coding the locus of *attention* rather than the orientation of particular body parts. Neurons in the lateral intra-parietal (LIP) sulcus region are involved in orienting attention by virtue of having both sensory properties (e.g. responding to visual and auditory stimuli in its receptive field) and motor properties (for generating saccades) (Bisley & Goldberg, 2010). Some neurons in this region of the macaque also show social mirroring of attention according to whether another person's gaze is oriented towards or away from its own receptive field (Shepherd, Klein, Deaner, & Platt, 2009).

In reaction time experiments, the direction of eye gaze appears to automatically orient participants to a particular location – termed **gaze cueing** (e.g. Frischen, Bayliss, & Tipper, 2007). Thus, eyes seen looking to the left will facilitate detection of a visual target presented on the left, relative to the right (see Figure 5.15). Similar effects are found when participants view head stimuli that are oriented to the left or right (Langton & Bruce, 1999). However, when head and eye cues are independently manipulated the results are more complex (Hietanen, 1999). Thus a head oriented towards the left but looking towards the right (as in Figure 5.16)

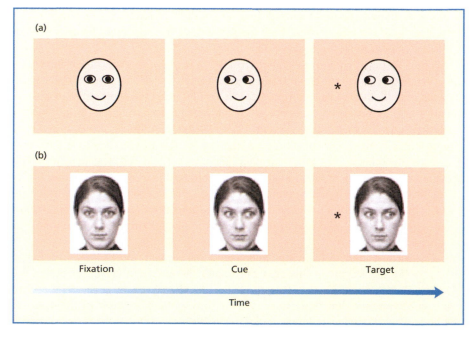

can be an effective cue to a target on the right. In this example, the person is looking towards *his* left side even though the eyes are looking directly at the observer.

Figure 5.16 Gaze cueing is modulated by head orientation. In this picture, attention is cued to our right side, even though he looks straight ahead in terms of gaze direction.

Finally, the direction of gaze and the processing of facial expression can interact with each other. A facial expression of anger is recognized faster if the gaze is direct, and a facial expression of fear is recognized faster if gaze is averted (Adams & Kleck, 2003). This is unlikely to be a perceptual level of interference between gaze and expression, but rather a more conceptual level of interference concerned with inferring behavioral intentions of approach or avoidance. Anger and direct gaze both signal approach whereas fear and averted gaze both signal avoidance, and when these two cues are mismatched the behavioral intention is more ambiguous. Mismatched gaze and expression signals are reflected in greater amygdala activity than matched signals (Adams, Gordon, Baird, Ambady, & Kleck, 2003).

WHY DO AUTISTIC PEOPLE OFTEN FAIL TO MAKE EYE CONTACT?

The social interactions of autistic people are characterized by an absence of joint attention and a failure to make direct eye contact (Sigman, Mundy, Ungerer, & Sherman, 1986). This could be due to either a difficulty in detecting where people are looking or a failure to understand the social significance of this behavior. In order to distinguish between these two possibilities, Baron-Cohen, Campbell, Karmiloff-Smith, Grant, and Walker (1995) devised a number of tests. In one test, children with autism were asked whether the eyes of another person are directed at them. They were unimpaired at this. They do, however, have difficulties in using gaze information to predict behavior or infer desire. In the four-sweets task, a cartoon face of Charlie directs his gaze to one of the sweets (Baron-Cohen et al., 1995). Children with autism are unable to decide: 'Which chocolate will Charlie take?' or 'Which one does Charlie want?'. This is shown in Figure 5.17.

Several studies have examined gaze processing difficulties in autism using neuroscientific methods. Pelphrey, Morris, and McCarthy (2005) showed people with autism two kinds of stimuli during fMRI: one in which another person looks towards an object (linked to joint attention), and a control condition in which another person looks towards empty space instead of orienting attention to the object (incongruent orienting of attention). Both people with autism and typically developing controls activated the STS when processing gaze. However, the typically developing group showed a modulation of this effect by context (such that the incongruent condition elicited more activity) but the autistic group did not. This again suggests to a problem in the social significance of gaze (rather than gaze perception itself). Whereas this study involved passive viewing of

Figure 5.17 Children with autism are able to detect which person is looking at them (top), but are unable to infer behavior or desires from eye direction (bottom). For example, they are impaired when asked 'Which chocolate will Charlie take?' or 'Which one does Charlie want?'. Top photo from Baron-Cohen and Cross (1992). Copyright © 1992 John Wiley and Sons. Reprinted with permission. Bottom panel from Baron-Cohen et al. (1995). Reproduced with permission from *British Journal of Developmental Psychology* © British Psychological Society.

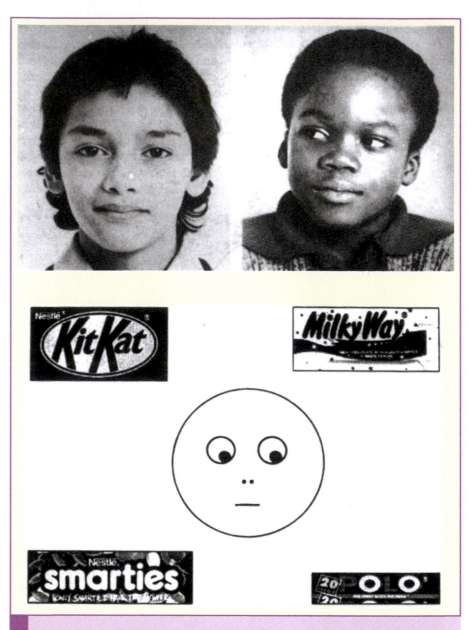

gaze, the fMRI study of Dalton et al. (2005) used eye tracking during facial discriminations. Looking at the eye region of another person (i.e. making eye contact) was linked to activation of the amygdala in both groups, but with a heightened response in the autistic group. This was interpreted as reflecting a heightened emotional response in autism to direct gaze.

Pointing and reaching

Pointing is a seemingly simple behavior, but very few animals do it. Most nonhuman animals do not understand it – they look at the fingertip, not at where it is aimed. Wild chimpanzees do not point, but captive ones raised with human contact do to a limited extent (Leavens, Hopkins, & Bard, 2005). It has even been suggested that dogs, domesticated through human contact, can outperform chimpanzees on understanding pointing (Hare, Brown, Williamson, & Tomasello, 2002).

Researchers distinguish between at least two types of pointing behavior: **proto-imperative pointing** is related to wanting (e.g. meaning 'Give me that!') whereas **proto-declarative pointing** elicits joint attention for its own sake (e.g. meaning 'Look at that!'). The latter, in particular, is considered to contain an element of 'mind reading' because it requires computing what the other person can and cannot see, whereas the former could be acquired from reward-based learning in the same way as a child learns to put his or her arms in the air to be picked up. A failure to engage in protodeclarative pointing at 18 months is an early behavioral marker of autism, several years before theory-of-mind tests are administered (Baird et al., 2000).

Materna, Dicke, and Thier (2008) asked participants to move their eyes to where the person is looking and contrasted this with the same task elicited by watching a person pointing. Functional imaging reveals that both tasks engage the same region in the posterior STS, and suggests that the region has a more general role in the social orienting of attention.

Although it is questionable whether monkeys point, or understand human pointing, their brains contain cells in STS that support the visual decoding of such behavior. Jellema, Baker, Wicker, and Perrett (2000) reported cells in the STS that respond to the sight of the arm reaching when the person directs attention to the same location (see Figure 5.18). When the same reaching movement is performed but the person

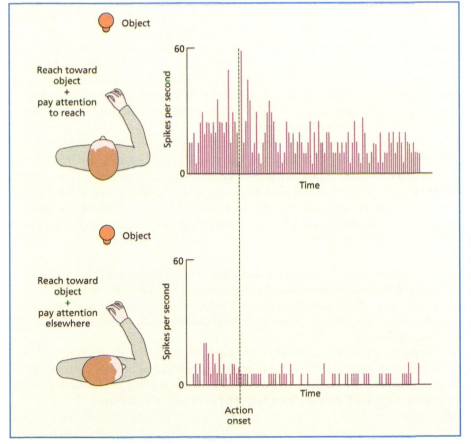

Figure 5.18 Some neurons (such as this one from the STS) respond only when the monkey watches an action performed when the actor also attends to the action. This kind of mechanism may distinguish intentional from accidental actions. From Jellema and Perrett (2005).

looks away then the response is significantly lower – the neurons did not respond strongly to head orientation per se, but only to the conjunction of head orientation and reaching. They argue that cells such as these provide the building blocks for understanding actions that are intended and goal directed, as opposed to accidental. Arm movements that are accompanied by attention are more likely to be intentional. Pointing would be one specific example of such an action.

Evaluation

Gaze detection and pointing or acting (particularly when accompanied by gaze) provide clues about the intentions of others – such as what they want, whether an action was deliberate, and whether they want to engage you in an activity. The STS is important for the visual decoding of this information, but other regions of the brain are implicated in making social inferences based on this information and in initiating joint attention.

TRAIT INFERENCES FROM FACES AND BODIES

Traits, in psychology, refer to long-term dispositions to behave or think in a particular way – for example, being an extrovert, a carer, a worrier, a risk-taker, and so on. Our **personality** can be viewed as a collection of such traits.

The suggestion that our face may be a window into our traits has a long history. The 'science' of physiognomy attempted to match facial characteristics onto traits such as criminal behavior but with no real success (Alley, 1988). However, contemporary researchers have a better handle on what the core set of traits might consist of (criminal behavior is not one of them) and rely on image-based configural techniques (e.g. that compute the relative separation of features) rather than measures of face parts such as nose length. A recent review of this field is provided by Todorov, Olivola, Dotsch, and Mende-Siedlecki (2015).

The sections below will consider evidence that people make trait inferences on the basis of superficial information and that people tend to agree (more than chance) on which traits belong with which people. Of course, proving an above-chance agreement between people does not make the association objectively true. Even if it were objectively true at the group level (i.e. in terms of statistical reliability), it does not mean that it holds true for each and every individual. In the discussion below, several possibilities should be borne in mind:

KEY TERMS

Traits
Long-term dispositions to behave or think in a particular way

Personality
A collection of traits

- *'A kernel of truth'.* It is possible that there are real associations between certain facial characteristics and certain personality traits. For example, testosterone affects masculine facial development and is also known to influence certain behaviors (i.e. correlated development of traits and physical characteristics).
- *Self-fulfilling prophecies.* People with certain facial characteristics may tend to be treated in particular ways that reinforce a particular behavioral outcome. For example, attractive people may receive more positive social interactions whereas people with facial abnormalities may tend to be shy or withdrawn.

- *Using expression cues to make trait inferences.* A smile may be used to infer friendliness, but even in neutral expressions people may infer traits from structural features that resemble expressions, such as upturning lips (Said, Sebe, & Todorov, 2009).
- *Culturally generated stereotypes with little or no objective basis*.

In all these cases, there is clearly an over-generalization. However, the accounts differ in terms of the source of the signal used for this over-generalization and the objective reliability of that signal.

Beautiful = good

The factors that make a face attractive are well understood. Rhodes (2006) lists three factors:

1 *Averageness.* Langlois and Roggman (1990) studied composite images of faces and found that as more faces were averaged together then attractiveness increased (up to a composite of about 16 faces). This factor is related to symmetry (averaging faces makes them more symmetrical) but is not identical to it. For example, averageness remains an important predictor of attractiveness when profiled faces are used (Valentine, Darling, & Donnelly, 2004).
2 *Symmetry.* In non-human animals, increased asymmetry has been related to difficulties in withstanding physical stress during development, poor nutrition, and inbreeding. Perfectly symmetrical faces created by morphing an image with its mirror image are judged to be more attractive, even when averageness is excluded (Rhodes, Sumich, & Byatt, 1999).
3 *Sexual dimorphism.* At puberty, increased testosterone in males stimulates growth of the jaw, cheekbones, brow ridges, and nose. In females, growth of these traits is inhibited by estrogen, which may also lead to the growth of fuller lips. There is evidence that females with enhanced feminine features are rated as more attractive than averaged female faces (Perrett, May, & Yoshikawa, 1994). The effects for enhanced masculine features in males tends to be weaker (see Rhodes, 2006).

In a classic study called 'What is beautiful is good', Dion, Berscheid, and Walster (1972) found that photographs of attractive faces were judged to be more likely to have socially desirable personality traits. Each person was given three photographs to rate on 27 personality traits (e.g. exciting vs. dull). The photos were of attractive, neutral, and unattractive faces (either three male or three female photos). The tendency to associate positive traits with attractive people was found irrespective of the sex of the rater or the sex of the face. This effect is related to the **halo effect** in which a person who is rated positively in one dimension tends to be rated positively in other dimensions. For example, in a mock videotaped interview featuring a potential teacher to the students, the teacher was either warm and friendly or cold and distant (Nisbett & Wilson, 1977). Participants rated the interviewer's accent and physical appearance more negatively in the cold condition even though these aspects were the same in both interviews (see Figure 5.19). Tsukiura and Cabeza (2011) argued that the link between physical attractiveness and positive personality attribution may stem from a common set of neural substrates involved in both kinds of judgment. Using fMRI, they compared facial attractiveness ratings against rating of moral goodness from short sentences ('He saved his sister from drowning.'). Both increased

KEY TERM

Halo effect
A person who is rated positively in one dimension tends to be rated positively in other dimensions.

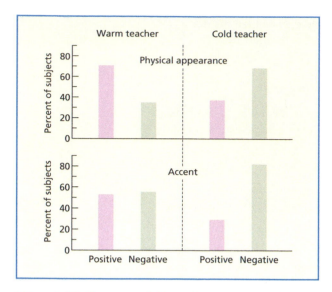

Figure 5.19 If a person behaves in a warm or cold manner then irrelevant attributes (such as appearance, accent) become judged in a positive or negative light. From Nisbett and Wilson (1977). Copyright © 1977 American Psychological Association. Reproduced with permission.

beauty and increased goodness ratings were linked to increased activity in the medial orbitofrontal cortex (linked to liking/wanting), whereas decreased beauty and decreased goodness were linked to increased insula activity (linked to aversion and disgust).

The hypothesis that 'what is beautiful is good' implies that we assign positive personality traits to attractive faces. However, some theories in evolutionary psychology would turn this on its head by arguing that traits that we consider desirable (e.g. physically strong males, youthful females) have determined what we consider as beautiful (e.g. Buss, 1989). Indeed, cultures tend to agree not only on face attractiveness but also on what long-term traits are desired in male and female partners (e.g. Buss, 1989). Nevertheless, within each culture there are individual differences in terms of what traits are desired – for example, women differ in the extent to which they desire a dominant or cooperative partner. As such, one might predict that facial attractiveness is related to the personality traits that people desire in a partner rather than (or in addition to) general factors such as symmetry and averageness. Little, Burt, and Perrett (2006) provided evidence for this prediction. For instance, women who find masculine personality traits attractive would tend to find men with masculine facial characteristics attractive, whereas those favoring more easy-going and less-assertive traits in a man would find baby-faced men attractive (see Figure 5.20). The results of this study suggest that

Figure 5.20 Which of these four faces do you find most attractive? Your answer to the question depends on the traits you desire in a partner. People who desire partners who are not easy going tend to choose (a) and those desiring easygoing partners choose (b). Those desiring nonassertive and assertive partners choose (c) and (d), respectively. From Little et al. (2006). Copyright © 2006 Elsevier. Reproduced with permission.

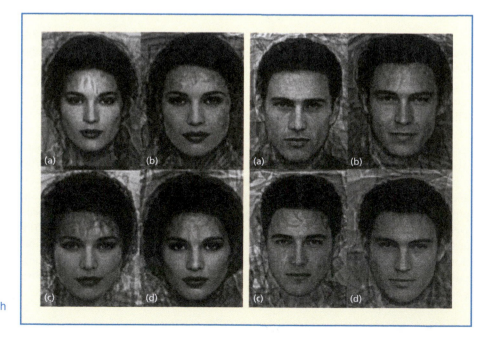

attractiveness is not just a structural property of a face but is also a projection of the traits that we value in a partner.

The 'big five' personality traits

Contemporary research into personality has attempted to address the question of how many traits are needed to describe the range of personalities in a population. According to Costa and McCrae (1985), a five-factor model offers the most satisfactory solution and the corresponding traits are known as the **Big Five**: openness to experience; conscientiousness; extraversion; agreeableness; and neuroticism (or emotional stability). Collectively, they spell the acronym OCEAN.

Penton-Voak, Pound, Little, and Perrett (2006) obtained black-and-white face photos of 294 people, and each of these people rated their own personality by questionnaire based on the Big Five. These photos were then shown to an independent group of raters who were asked to rate the photographs using the same questionnaire. Male and female faces were analyzed separately because personality is known to differ across sexes: for instance, women rate themselves as more neurotic and more agreeable (Costa, Terracciano, & McCrae, 2001). As such, raters could be expected to produce above-chance results based on gender stereotypes rather than facial characteristics. For male faces, there was a significant correlation between self-reported personality and independent ratings based on faces alone for three traits: extraversion, neuroticism, and openness to experience. For female faces, only extraversion came out as significant. In a second study, Penton-Voak et al. (2006) produced composite faces based upon the top and bottom 10% rating of self-reported traits: blending the faces of the top 10% most extraverted people to create an 'extrovert face', blending the faces of the 10% least extraverted faces to create an 'introvert face', and so on. Independent raters tended to rate the composite images as more attractive when they had been derived from socially desirable personality traits (e.g. high agreeableness, low neuroticism, etc.).

An alternative approach to producing morphed composites from real faces, is to use computer generated faces in which specific facial features (skin reflectance, lip shape, etc.) can be systematically varied and linked to participant ratings. Figure 5.21 shows examples of extrovert and introvert prototypical faces generated using this method.

KEY TERM

Big five
A collection of five personality traits that accounts for the most significant individual differences (openness to experience; conscientiousness; extraversion; agreeableness; neuroticism)

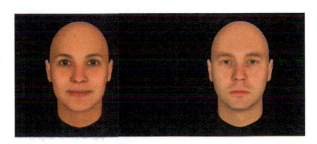

Figure 5.21 Prototypical extrovert and introvert faces (left and right, respectively) generated by computer simulations and participant ratings. From Todorov et al. (2015).

THE BIG FIVE PERSONALITY TRAITS

- *Openness to experience*: appreciation for art, emotion, curiosity, and unusual ideas and activities
- *Conscientiousness*: a tendency to show self-discipline, be meticulous, and aim for achievement
- *Extraversion*: a tendency to seek stimulation and the company of others
- *Agreeableness*: a tendency to be compassionate and cooperative towards others
- *Neuroticism*: emotionally reactive and vulnerable to stress

Dominance and facial width-to-height ratio

The concepts of dominance and aggression are related. Both entail an element of competitiveness. Whereas aggression is defined in terms of an intent to cause harm, dominance is regarded more in terms of a motivation to control others, and could employ aggressive or non-aggressive (e.g. persuasion) tactics. There has been a significant body of research that links dominant or aggressive behavior in males to a facial characteristic termed the **facial width-to-height ratio (fWHR)**. This measures the width relative to height of the internal features of the face, as illustrated in Figure 5.22. A higher fWHR in males is linked to more dominant behavior.

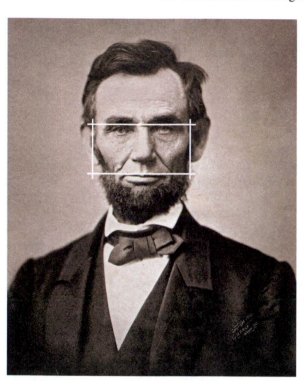

Figure 5.22 The US presidents tend to have a larger facial width-to-height ratio than average (a narrower rectangle), and this facial characteristic has been linked to dominant and aggressive behavior in males.

Carre and McCormick (2008) reported two studies: an observational study of professional ice hockey players and an experimental game-playing study. The aggressive behavior of the hockey players was measured by counting the penalty minutes that each player had accrued for behavior such as slashing, elbowing, and fighting. This objective measure of aggression correlates with fWHR as shown in Figure 5.23. In their experimental task, participants played a game against another player in which one button earns points, another button steals points from other players, and a third button protects one's own points. Aggression was operationally defined as the number of button presses that steal other players' points and this was found to correlate with fWHR of the male (but not female) players. In a different paradigm in which players could cheat to increase their chances of winning in a lottery, fWHR was found to be related to deceptive behavior (Haselhuhn & Wong, 2012). Other research using multiplayer games suggests that, in men, higher fWHR is related to a tendency to exploit someone else's trust (Stirrat & Perrett, 2010) but, also, increased cooperation to other ingroup members during an intergroup competition (Stirrat & Perrett, 2012). The fWHR is higher than expected in the US presidents (Lewis, Lefevre, & Bates, 2012) and amongst CEOs (chief executive officers) of

Figure 5.23 The facial width-to-height ratio predicts aggressive behavior in ice hockey players. Why might testosterone be a mediating factor? From Carre & McCormick (2008). Copyright © 2008 The Royal Society. Reproduced with permission.

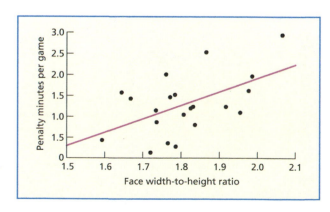

leading British companies (Alrajih & Ward, 2014). It has also been documented to be higher in the alpha males of capuchin monkeys (Lefevre et al., 2014).

What is the possible mechanism that drives this association? The original suggestion was that this was linked to variability in the amount of testosterone. This hormone is known to play a role in the development of the bone structure of the face during adolescence and has long been linked to dominant behavior. Lefevre et al. (2013) presented evidence of an association between resting salivary testosterone levels and fWHR in men. Using a measure of facial masculinity, Pound et al. (2009) found a link between facial masculinity and testosterone levels after winning a competition. The situation in women is more puzzling: the average fWHR of women isn't any lower than males, despite significantly lower testosterone levels (Lefevre et al., 2012). One possibility is that other factors (e.g. estrogen) are important for the fWHR in women.

Oosterhof and Todorov (2008) adopted a more data-driven approach to determine which traits capture most of the variability in facial differences. They obtained 14 different trait ratings for a large number of computer-generated faces and then used statistical techniques to determine what the underlying structure of the 'face space' is. They found that faces could be described in terms of two separate traits: how dominant the face appears and how positive/negative the face is judged. Trustworthiness was the trait that was most strongly linked with this positive/negative component. Examples of faces within this two-dimensional space are shown in Figure 5.24. This

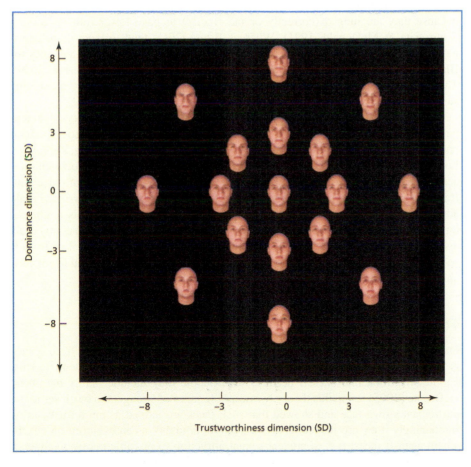

Figure 5.24 Oosterhof and Todorov (2008) found that trait ratings from faces can be described in terms of two independent dimensions: how dominant the face appears (aggressive looking vs. baby-faced) and how positive or negative the face is judged (which relates to perceived trustworthiness). From Todorov, Said, Engell, and Oosterhof (2008). Copyright © 2008 Elsevier. Reproduced with permission.

method did not consider fWHR specifically, but linked dominance to other kinds of facial characteristics relating to masculinity (which is not the same as biological sex) such as squarer jaws, anger-related structural features (e.g. prominent brows resembling a frown), and – in subsequent models – less skin reflectance (Todorov, Dotsch, Porter, Oosterhof, & Falvello, 2013). Other research has produced similar findings but added a third dimension that relates to perceived youthfulness/attractiveness (Sutherland et al., 2013).

Baby-faced adults

In the model of Oosterhof and Todorov (2008), shown in Figure 5.24, low dominance is linked to a baby-faced-like appearance of round head shape and non-prominent features (nose, jaw, brow). Several studies have asked people with these facial characteristics to rate their own traits. Berry and Brownlow (1989) found that men who were independently rated to have 'baby faces' rate themselves as being less aggressive, more approachable, and warmer than non-baby-faced men, whereas women who were independently rated to have 'baby faces' rate themselves as having low physical power and low assertiveness. Interestingly, baby-faced adolescents from low socio-economic backgrounds are more likely to enter into delinquency, which suggests that these individuals may actively react against the social stereotype of low dominance that is applied to them (Zebrowitz, Andreoletti, Collins, Lee, & Blumenthal, 1998). Baby-faced men are more likely to receive lighter sentences for crimes because they are judged (correctly or incorrectly) to have lower trait aggression (Zebrowitz & McDonald, 1991). Whereas White CEOs tend to have more dominant facial characteristics (assessed by facial width-to-height), there is a tendency for Black CEOs to have more baby-faced characteristics (Livingston & Pearce, 2009). This was explained in terms of facilitating the success of Black leaders by attenuating stereotypical perceptions that Blacks are threatening.

It has been argued that the mechanism that gives rise to trait inferences linked to baby-faced adults is that of an over-generalization of the natural instinct to respond to babies with warmth and affection, and to attribute weakness to them (e.g. Zebrowitz, Luevano, Bronstad, & Aharon, 2009). As such, this is essentially a socio-cultural account based on an erroneous cognitive bias, rather than a 'kernel of truth' account. It is certainly the case the people have a similar neural response to baby-faced male adults as they do to babies (relative to mature-faced male adults). This is characterized by increased activity in, and functional connectivity between, the amygdala and FFA (Zebrowitz et al., 2009). However, there may also be a kernel of truth if baby-facedness and low aggression are outcomes of a common biological latent variable (such as pubertal testosterone in males).

Trustworthiness

One possible metric that people may use to evaluate trustworthiness from a face is to determine whether the face looks similar to oneself. One may tend to trust people who look like oneself because such people may include family members. DeBruine (2005) tested this theory by morphing faces of strangers to be closer to the participant's own face and showed that such faces are judged as more trustworthy. Interestingly, they are judged as *less* attractive, arguably as an evolutionary mechanism against incest (i.e. to prevent sexual attraction to possible family members).

Whatever the explanation, it suggests that trustworthiness and attractiveness are not the same thing.

Although self-similarity might be one mechanism for assessing trust, it is unlikely to be the only mechanism given that different individuals tend to agree on which faces are trustworthy (e.g. Oosterhof & Todorov, 2008). In one of the earliest functional imaging studies of trustworthiness, Winston, Strange, O'Doherty, and Dolan (2002) found that activation in the amygdala increased with increasing levels of untrustworthiness. Lesions to the amygdala impair evaluation of trustworthiness, such that untrustworthy faces are deemed trustworthy (Adolphs, Tranel, & Damasio, 1998). But congenital prosopagnosics who are unable to recognize familiar faces are still able to attribute trustworthiness in similar ways to those without face recognition difficulties (Todorov & Duchaine, 2008). Similarly, people with high-functioning autism are able to evaluate trustworthiness in faces (White, Hill, Winston, & Frith, 2006). According to Todorov and Engell (2008) the role of the amygdala in trustworthiness judgments may be related to categorizing stimuli according to positive and negative valence, but is unlikely to be involved in social trait attribution per se. For instance, they found that amygdala activity in fMRI was related to trustworthiness ratings but not to dominance ratings (using the aggressive vs. baby-faced stimuli discussed above). Dominance is less easy to categorize in terms of being 'good' or 'bad' than levels of trust.

What facial characteristics are linked to perceived trustworthiness? Using human judgments of computer-generated faces (Todorov et al., 2013), it has been found that trustworthy faces have mouths resembling a smile (up-turned lips) and untrustworthy faces to have mouths resembling anger (down-turned lips). In addition, trustworthy faces tended to look more feminine and older. Thus, even though neutral facial expressions were used, participants might still use subtle differences relating to expressions and social categorical stereotypes to infer traits from individual faces.

Trait inferences from bodies

There is very little evidence concerning trait inferences from bodies, especially from a neuroscience angle. Swami et al. (2008) asked participants to make trait inferences to different images of male and female bodies that vary in body size (the body mass index, BMI). Increasing BMI was associated with the perceived trait of laziness for both male and female bodies. There was also evidence that people attribute loneliness to certain body sizes, specifically for overweight people of either sex and for underweight women. Koppensteiner (2013) examined body motion from a small set of point lights and found attributions of extraversion linked to overall activity, neuroticism to the relative velocity of movement, and openness to variation in motion direction. Kramer, Arend, and Ward (2010) extracted biological motion cues (displayed to participants using point light displays) of Barack Obama and John McCain during presidential debates. (A second experiment used UK political leaders.) Participants were told only that the movies were from people giving a public speech and they were asked to rate them on attractiveness, trustworthiness, caring, dominance, leadership, anxiety, depression, and physical health. There was some agreement across participants on trait inferences. For instance, Obama was rated as more trustworthy and dominant, whereas McCain was rated as more anxious. After the ratings were collected participants were asked who they would vote for (again they did not know who the people were). In terms of voting behavior, perceived physical health was the strongest predictor.

Evaluation

The start of this section considered various possible explanations for a link between faces and traits, and it would be interesting to return to this in light of the evidence. The strongest evidence for the 'kernel of truth' explanation comes from the link between aggression (measured through both self-report and objectively) and independently rated 'aggressive looking' faces. With regard to trustworthiness, the judgment appears to be an over-generalization of expression-relevant mechanisms to trait judgements (neutral expressions with a smile-like facial configuration are rated as trustworthy). It is unclear whether there is a kernel of truth in this. Attractive faces are rated as having attractive traits, and it is argued that this reflects our desire to seek out certain traits in potential mates.

SUMMARY AND KEY POINTS OF THE CHAPTER

- Face perception involves a number of different mechanisms with somewhat different neural substrates. The occipital face area (OFA) may compute structural properties of a face whereas the fusiform face area (FFA) computes facial identity.
- To some extent, recognizing facial expressions is separable from recognizing facial identity, and there are several candidate mechanisms for recognizing expressions: using dynamic information (e.g. STS); mapping faces onto regions specialized for emotional stimuli (e.g. amygdala, insula); or by simulating the expression motorically (i.e. the mirror system).
- The superior temporal sulcus (STS) is important for action perception for both faces and bodies, including lip reading, biological motion, eye gaze, and pointing. Eye gaze and pointing are important for establishing joint attention and inferring intentions.
- Bodies and voices provide other cues to socially relevant information. As with faces, there is evidence for some degree of separation between identifying people and recognizing their emotional state.
- There is a tendency for people to infer stable characteristics (i.e. traits) from the faces of others. Attractive faces tend to be linked to attractive traits. Similarly, faces judged to be aggressive or baby-faced tend to be linked with matching behavior (i.e. dominant vs. non-dominant behavior). These reflect a tendency to over-generalize and reflect the desire to infer the inner world of people around us based on sparse information.

EXAMPLE ESSAY QUESTIONS

- Is recognizing a facial expression different from recognizing other properties of a face?
- What kinds of social information might be conveyed in body perception and how is this related to face perception?

- What is the role of the superior temporal sulcus (STS) in face and body perception?
- Do we use facial information to infer the character of people around us? Is this information accurate?

RECOMMENDED FURTHER READING

- Bruce, V., & Young, A. W. (2012). *Face Perception*. New York: Psychology Press. An engaging book, also covering cognitive neuroscience, written by two of the most influential researchers in this field.

- Calder, A., Rhodes, G., Johnson, M., & Haxby, J. (2011). *Handbook of Face Perception*. Oxford: Oxford University Press. Extensive coverage of all aspects of face perception.

- For body perception, the following two papers are recommended:

  - Downing, P., & Peelen, M. V. (2011). The role of occipitotemporal body-selective regions in person perception. *Cognitive Neuroscience, 2*, 186–203

  - de Gelder, B., Van den Stock, J., Meeren, H. K. M., Sinke, C. B. A., Kret, M. E., & Tamietto, M. (2010). Standing up for the body. Recent progress in uncovering the networks involved in the perception of bodies and bodily expressions. *Neuroscience and Biobehavioral Reviews, 34*(4), 513–527.

ONLINE RESOURCES

- References to key papers and readings
- Videos illustrating cases of prosopagnosia and Capgras delusion
- Fun demonstrations of traits from faces and bodies, facial averaging, etc.
- TED talks by leading researchers Nancy Kanwisher and David Perrett, plus a lecture given by textbook author, Jamie Ward
- Multiple choice questions and interactive flashcards to test your knowledge
- Downloadable glossary

CHAPTER 6

CONTENTS

Understanding others

If you see someone yawning do you yawn too? Most people probably do to some extent. Some behavior, such as laughing and yawning, is socially contagious. But can any wider significance be attached to such findings? One study of contagious yawning in chimpanzees speculates that 'contagious yawning in chimpanzees provides further evidence that these apes possess advanced self-awareness and empathic abilities' (Anderson, Myowa-Yamakoshi, & Matsuzawa, 2004). Another study, this time on humans, administered tests requiring reasoning about the mental states of other people (e.g. beliefs, knowledge) as well as measuring yawn contagion, and concluded that 'contagious yawning may be associated with empathic aspects of mental state attribution' (Platek, Critton, Myers, & Gallup, 2003). Of course, there is unlikely to be anything special about yawning itself. There might be a general tendency to *simulate* the behavior of others on ourselves (internally in our minds and brains) even if we do not overtly *reproduce* it (as observable behavior on our bodies). Thus, we may understand others by creating a similar response in our brain to that found in the other person's brain. Contagious yawning, under this account, is one extreme example of this more general and, normally, more subtle tendency. This chapter will attempt to unpick these claims and place them alongside traditional concepts in social and cognitive psychology, such as empathy and theory of mind. The chapter will also consider how these processes may be disrupted after brain injury and in people with autism.

The overarching question of the chapter is how do we understand the mental states of others? **Mental states** consist of knowledge, beliefs, feelings, intentions, and desires. The process of making this inference has more generally been referred to as **mentalizing**. The term is generally used in a theory-neutral way, insofar as it

KEY TERMS

Mental states
Knowledge, beliefs, feelings, intentions, and desires

Mentalizing
The process of inferring or attributing mental states to others

Figure 6.1 It just takes one yawn to start off other yawns. How does this kind of simple contagion mechanism relate to empathy and theory of mind?

is used by researchers from a wide spectrum of views. It could be contrasted with the term 'theory of mind', which has essentially the same meaning but has tended to be adopted by those advocating a particular position, namely the notion that there is a special mechanism for inferring mental states. According to some researchers, this theory-of-mind mechanism cannot be reduced to general cognitive functions such as language and reasoning, or those involved in imitating. These arguments lie at the heart of the social neuroscience enterprise in that they raise important and divisive issues about the nature of the mental and neural mechanisms that support social behavior and the extent to which they are related to other aspects of cognition.

WHAT IS SIMULATION THEORY?

Simulation theory is not strictly a single theory but a collection of theories proposed by various individuals (e.g. Gallese, 2001; Goldman, 2006; Hurley, Clark, & Kiverstein, 2008; Preston & de Waal, 2002). However, common to them all is the basic assumption that we understand other people's behavior by recreating the mental processes on ourselves that, if carried out, would reproduce their behavior. That is, we use our own recreated (or simulated) mental states to understand, and empathically share, the mental state of others. Within this framework there are various ways in which this could occur. Gallagher (2007) broadly distinguishes between two: one could create an explicit, narrative-like simulation of another person's situation and behavior in order to understand it, or when we see someone else's behavior (e.g. their action, emotional expression) we may automatically, and perhaps unconsciously, activate the corresponding circuits for producing this behavior in our own brain. These latter versions of simulation theory tend to be intimately linked to the idea of mirror systems in which perception is tightly coupled with action.

EMPATHY AND SIMULATION THEORY

The word **empathy** is relatively modern, being little more than 100 years old. It was coined by Titchener (1909) from the German word *einfühlung* (Lipps, 1903) and originally referred to putting oneself in someone else's situation (literally 'feeling into'). This would also go under the contemporary name of **perspective taking**. This section will first consider the various different ways in which the term *empathy* is used today. This reveals potentially important distinctions that theories of empathy need to explain.

EMPATHY AS A MULTI-FACETED CONCEPT

If one starts with the working definition of empathy introduced above ('putting oneself in someone else's situation') it is clear that there are subtle, but potentially crucial, different ways in which this could be understood. Some of these are listed below and are an abridged version from Batson (2009):

1 Knowing another person's internal state, including his or her thoughts and feelings
2 Adopting the posture or matching the neural response of an observed other
3 Having an emotional reaction to someone else's situation, although it need not be the same reaction
4 Imagining how I would feel/react in that situation (i.e. given *my* personal history, traits, knowledge, beliefs)
5 Imagining how the other person would feel/react in that situation (i.e. given *their* personal history, traits, knowledge, beliefs)

The first three scenarios differ with respect to whether the knowledge/feeling is the same in self and other. Knowing about another person's internal state need not necessarily imply that the observer shares that state. This important consideration lies at the heart of some tests of theory of mind, specifically **false belief** tasks, but they are relevant to some conceptions of empathy too. The second sense in which empathy is used ('adopting the posture or matching the neural response of an observed other') is the one most closely linked with mirror systems, imitation, and contagion (emotional contagion, yawning contagion, etc.). For example, one might feel **personal distress** in response to someone else's suffering. The third sense in which the term empathy may be used differs from the second in that the person's response is not matched. For instance, one might feel a sense of **pity** to another's situation or **sympathy** or compassion towards someone who is suffering. These reactions are directed outwards (other-oriented) rather than being self-oriented (as in personal distress), and the response of the perceiver does not match that of the other person (Singer & Klimecki, 2014). The fourth and fifth notions of empathy relate more directly to the idea of perspective taking, but they differ in whether they are self-oriented versus other-oriented. The fourth scenario ('imagining how I would feel/react in that situation') could be construed as a shallow attempt to empathize, in which the level of success is dependent on self–other similarity rather than a true understanding of the other.

Given these somewhat different conceptions of empathy, it is not surprising that there is no single agreed-upon measure of empathy. Theory-of-mind tests, discussed in detail below, normally involve assessments based on linguistic reasoning of the sort: 'If X believes Y then how will he/she behave in situation Z?' Others use neural or bodily responses to seeing others in pain, for example, as a measure of empathy (e.g. Bufalari, Aprile, Avenanti, Di Russo, & Aglioti, 2007; Jackson, Meltzoff, & Decety, 2005). Of course, this presupposes a certain idea of what empathy is (i.e. that it can be measured solely in physiological ways). There are various questionnaire measures of empathy, such as the Interpersonal Reactivity Index (IRI; Davis, 1980) and the Empathy Quotient (EQ; Baron-Cohen & Wheelwright, 2004), which touch upon some of the distinctions discussed above. For example, the IRI contains separate subscales such as personal distress (items such as 'I tend to lose control during emergencies'), perspective taking (items such as 'Before criticizing somebody, I try to imagine how I would feel if I were in their place'), and empathic concern (items such as 'I often have tender, concerned feelings for people less fortunate than me'). One current trend is to incorporate questionnaire measures in functional imaging experiments. For example, watching someone drinking a pleasant or disgusting drink may activate the gustatory (taste) regions of the perceiver (Jabbi, Swart, & Keysers,

KEY TERMS

False belief
A belief that does not correspond to current reality

Personal distress
A feeling of distress in response to another person's distress or plight

Pity
A concern about someone else's situation

Sympathy
A feeling of compassion or concern for another person

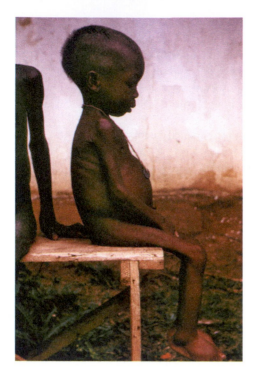

Figure 6.2 Do images of starvation evoke in you a sense of personal distress (self-focused) or a sense of pity or sympathy (other-focused)? Different individuals may have different reactions, although both can be broadly construed as empathic.

2007). Moreover, the extent to which this occurs may be greater in those people who report higher empathy on questionnaire measures (Jabbi et al., 2007). Findings such as these are often used to argue that the different concepts of empathy are related or, at least, share a common core (perhaps based upon simulation). Finally, one could potentially measure the ability to *accurately* empathize (i.e. to accurately state what another person is thinking or feeling) rather than the extent to which the person may report the motivation to empathize (i.e. most questionnaire measures) or to simulate that state themselves (which need not be linked to the ability to consciously report that state). An example of such a test is shown in Figure 6.3. The 'reading the mind in the eyes' test requires participants to match expressions in the eye region of faces to labels denoting mental states such as bored, sorry, or interested (Baron-Cohen, Wheelwright, Hill, Raste, & Plumb, 2001). Another test requires two participants to work together in a scenario that is video recorded. Each participant can then watch it back and report their own internal states as well as attempting to infer that of the other participant, thus enabling the experimenter to cross-reference the responses together in order to infer empathic accuracy (e.g. Ickes, 1993; Ickes, Gesn, & Graham, 2000). Although women tend to score higher on questionnaire measures of empathy, this gender bias is reduced (if not eliminated) when measures of empathic accuracy are used (Ickes et al., 2000). This, again, points to the need to distinguish between different conceptions of empathy such as a *disposition* to empathize (i.e. deliberately attempt to perspective take), which may be tapped by questionnaires, and an *ability* to empathize which may be tapped by performance-based measures. A recent functional imaging study based on this decoding

Figure 6.3 The extent to which people can accurately detect the mental states of others (also called empathic accuracy) may differ from the extent to which they try to empathize or take perspectives. One test along these lines is the 'reading the mind in the eyes' test. From Baron-Cohen et al. (2001). Copyright © 2001 Association for Child Psychology and Psychiatry. Reproduced with permission from Wiley-Blackwell.

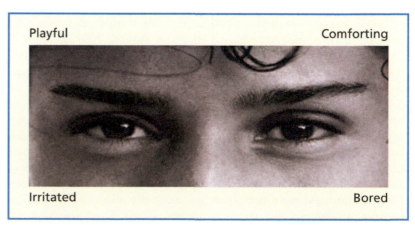

of an inter-personal exchange found that empathic accuracy was related to a network of regions including the medial prefrontal cortex, implicated in mentalizing / theory of mind (although not the temporo-parietal junction), and the premotor cortex, which has been associated with mirror systems (Zaki, Weber, Bolger, & Ochsner, 2009).

From imitation to empathy?

Most simulation theories of empathy are based on the notion of perception–action coupling, i.e. the link between seeing actions on other people and reproducing those actions on one's own motor system. In some cases the action might be literally reproduced (direct imitation), or be reproduced in a more subtle form (e.g. contractions of facial musculature that can be detected by electromyography, EMG), or reproduced solely 'in the head' of the perceiver (i.e. activation of the motor system as detected by fMRI). As discussed in Chapter 3, the candidate neural mechanism for this perception-to-action coupling is the mirror neuron system. Mirror neurons respond both when an animal performs an action and when it observes another performing the same (or similar) action. They act as a neural bridge between self and other, and it has been suggested that their capacity to support imitation provides the foundation for some aspects of social cognition such as empathy (e.g. Iacoboni, 2009). A link between imitation and empathy receives some support from social psychology in studies examining the **Chameleon Effect** in which there is a spontaneous mimicry of gestures during positive inter-personal exchanges. These studies generally use unintentional imitation in which the participant engages in a task with another person (a confederate) and the extent to which the participant imitates the confederate is measured. The participant is unaware of the true nature of the study (i.e. that his/her imitative behavior is being assessed). Participants who imitate more (based on blind scoring of their actions) whilst performing a cooperative task with a confederate tend to rate themselves as higher in trait empathy (Chartrand & Bargh, 1999). When the confederate deliberately imitates the participant in a cooperative task, then he/she is liked more by the participant than in a control condition in which imitation is avoided (Chartrand & Bargh, 1999). Van Baaren, Holland, Kawakami, and van Knippenberg (2004) showed that being imitated increases the chances of helping behavior when a confederate drops something. However, the effects are quite general. The person who has been imitated is not just more likely to help the imitator but they are more likely to help others too. It also increases the amount of money that the participant opts to donate to charity at the end of the experiment.

Carr, Iacoboni, Dubeau, Mazziotta, and Lenzi (2003) examined more directly a possible link between empathy and imitation using fMRI in humans. They showed participants emotional facial expressions under two conditions: observation versus deliberate imitation. (Note that this is different from the social psychology studies above, in which imitation was spontaneous rather than instructed.) They found increased activation for the imitation condition relative to observation in classical mirror system areas such as the premotor cortex. In addition, they found increased activation in areas involved in emotion, such as the amygdala and insula. Their claim was that imitation activates shared motor representations between self and other, but crucially, there is a second step in which this information is relayed to limbic areas via the insula. This action-to-emotion route was hypothesized to underpin empathy. Other studies have reported a positive correlation between questionnaire-based empathy scores and activation in the premotor region when observing actions

(Kaplan & Iacoboni, 2006) or listening to actions (Gazzola, Aziz-Zadeh, & Keysers, 2006).

The activity of the human motor system can be assessed using the method of **motor evoked potentials (MEPs)**. Stimulation of the brain using TMS causes the peripheral muscles to produce neuroelectrical signals known as motor evoked potentials (MEPs) – see Figure 6.4. These can be measured by electrodes attached to the skin using the principle of electromyography (EMG). TMS is applied over the primary motor cortex whilst the participant is at rest and the TMS threshold is found below which EMG responses in the muscles can no longer be reliably elicited. Asking the participant to perform a voluntary movement or simply observing the action of another person increases cortical excitability, defined as an increased MEP when this threshold level of TMS is applied (Fadiga, Fogassi, Pavesi, & Rizzolatti, 1995). This is specific to the action observed such that observing an arm action (flexing at the elbow) facilitates MEPs on the biceps but not the hand muscles and observing a hand action (writing) facilitates MEPs in the hand but not the biceps (Strafella & Paus, 2000). Observing another person in pain, such as an injection applied to the hand, results in cortical inhibition as revealed by reduced MEPs (Avenanti, Bueti, Galati, & Aglioti, 2005). Although such results may seem distantly related to the notion of empathy, they provide further evidence for the notion of mirroring. Moreover, MEPs are modulated by social factors such as questionnaire measures of empathy (Avenanti, Minio-Paluello, Bufalari, & Aglioti, 2009), the race of the hand observed (Avenanti, Sirigu, & Aglioti, 2010), and priming to think of oneself as having more or less power over others (Hogeveen, Inzlicht, & Obhi, 2014).

The model proposed by Carr et al. (2003) and Iacoboni (2009) is simple, but it is also perhaps simplistic. The assumption that limbic = emotion is an over-simplification (LeDoux, 1996), as is the claim that emotion imitation = empathy. As argued above, empathy is a broader concept than this. Recall also from Chapter 3 that the link between mirror neurons themselves and imitation is by no means uncontroversial. For example, monkeys (who possess mirror neurons) have very limited imitation abilities.

It is possible to imagine alternative scenarios to the imitation-to-empathy model within a general simulation theory framework. For example, de Vignemont and Singer (2006) suggest that it may be possible to have simulation of emotions (and empathy for emotions) without having action/motor representations as a linking step. Singer et al. (2004b) investigated empathy for pain in humans using fMRI. The brain was scanned when anticipating and watching a loved-one suffer a mild electric shock. There was an overlap between regions activated by expectancy of another

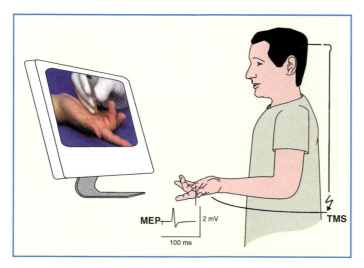

Figure 6.4 TMS applied over the primary motor cortex can trigger a neuromuscular effect in the corresponding (contralateral) body part. This can be detected using EMG (electromyography) as a motor-evoked potential (MEP). The magnitude of the MEP can be modulated upwards when seeing someone else's body part moved, or it can be modulated downwards when seeing someone else's body part in pain (reflecting cortical excitation and inhibition respectively). Does this sort of mechanism contribute towards empathy?

person's pain and experiencing pain oneself, including the anterior cingulate cortex and the insula. This provides evidence for a mirror system for pain – a system that responds to pain in self and other. However, there was little evidence that this system depends on the 'classic' mirror system for actions/goals that may support imitation. Meta-analyses of fMRI studies of empathy have also tended to center around the simulation or 'sharing' of feelings and sensations (mediated by regions such as the insula) rather than the classic action-based mirror neuron system (Fan, Duncan, de Greck, & Northoff, 2011).

Empathy beyond simulation

Some theories of empathy propose a variety of different interacting mechanisms of which simulation is only one. In such models, simulation may either be a junior or senior partner.

As noted above, watching someone in pain activates certain parts of our own pain circuitry. This offers clear support for simulation theories. However, our beliefs about the person in pain can modulate or override this mechanism. Singer et al. (2006) had participants in an fMRI scanner play a game with someone who plays fairly (a 'Goodie') and someone else who plays unfairly (a 'Baddie'). Mild electric shocks were then delivered to the Goodie and Baddie (who, of course, were only actors but the participant did not know this). Participants empathically activated their own pain regions (such as anterior insula) when watching the Goodie receive the electric shock (see Figure 6.5). However, this response was attenuated when they saw the Baddie receiving the shock. In fact, male participants often activated their pleasure and reward circuits (such as the nucleus accumbens) when watching the Baddie receive the shock, which is the exact opposite of simulation theory. This brain activity correlated with their reported desire for revenge, which suggests that although simulation may tend to operate automatically it is not protected from our higher order beliefs.

The findings of this study have implications for conditions associated with reduced empathy, such as autism and psychopathy. It suggests that there are multiple reasons why empathy might fail – because of a failure to simulate the emotions of others or because of personally or socially constructed beliefs about who is 'good' and who is 'bad'. The eminent social psychologist Bandura (2002) argues that simulation has a relatively minor role to play in empathy, arguing that if it did it would lead to emotional exhaustion, which would debilitate everyday functioning. Moreover, Bandura (2002) argues that acts of inhumanity, such as genocide, depend on our ability to self-regulate and dissociate self from other. Although genocide is an extreme example, displaying lack of empathy towards socially marginalized groups (e.g. illegal immigrants, welfare cheats) could be regarded as a typical facet of human behavior.

Other studies support this view. Although doctors may be expected to show empathy for their patients, it would be unhelpful for them to experience personal distress when performing painful procedures. Indeed acupuncturists show less activity, measured by fMRI, in the pain network (including the anterior insula and anterior cingulate) when watching needles inserted into someone, relative to controls (Cheng et al., 2007). Lamm, Batson, and Decety (2007) found that activity in these pain-related regions, induced by watching painful facial expressions induced by medical treatment, was modulated by the observer's beliefs about whether the treatment was successful or not (more activity in pain-processing regions when less successful). It was also related to whether the participants were instructed to imagine the

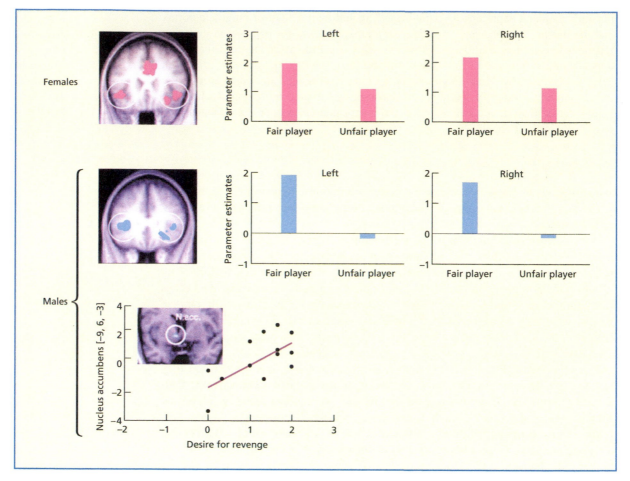

Figure 6.5 Females (pink) and males (blue) show reduced activity in brain regions that respond to pain when watching an unfair player receive a shock (shown here for the insula). In males, activity in the nucleus accumbens, measured whilst the unfair player received a shock, correlates with their self-reported desire for revenge. From Singer et al. (2006). Copyright © 2006 Nature Publishing Group. Reproduced with permission.

feelings of the patient or to imagine themselves in that situation (more activity in pain-processing regions when imagining self). This suggests that the tendency to simulate is moderated by cognitive control (e.g. based on our beliefs) and also our efforts to take different perspectives.

Studies of imitation also show that the extent to which two people imitate each other depends on the characteristics of the imitator and the person being imitated, as well as characteristics of the social situation (van Baaren, Janssen, Chartrand, & Dijksterhuis, 2009). This suggests that imitation-based simulation is flexible and context sensitive, taking into account information beyond perception–action links. For example, imitation is less likely when the confederate has a social stigma such as a facial scar or is heavily obese (Johnston, 2002). Similarly, non-deliberate imitation of facial expressions is greater for one's ethnic ingroup relative to an outgroup (Bourgeois & Hess, 2008). To return to the example of socially contagious yawning, it has been found

Figure 6.6 It may be important for doctors performing painful procedures to switch off their empathic tendencies. What kind of mechanisms in the brain might support this?

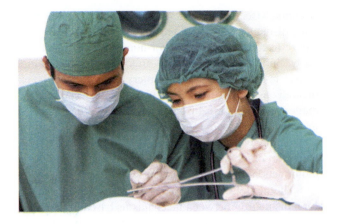

that the degree of contagion is related to social closeness: it is greatest in response to kin, then friends, then acquaintances, and lastly strangers (Norscia & Palagi, 2011).

Finally, when observing a social interaction between two people, observers use the presence of mimicry as a cue to social competence. Someone who mimics the gestures of an unfriendly person is judged as being less socially competent than someone who mimics the gestures of a friendly person (Kavanagh, Suhler, Churchland, & Winkielman, 2011). The difference can be overridden if the observer has other knowledge of the mimicker (e.g. that he is engaged in humanitarian causes). As such, simple simulation models based on perception–action coupling are insufficient to explain data obtained in more naturalistic (i.e. contextualized) social interactions.

Models of empathy

Models of empathy based *solely* on the notion of perception–action coupling or affective sharing have been shown to be lacking, but may nevertheless be one aspect of empathy. So what do alternative models of empathy look like? This section will consider three kinds of models: those that conceptualize empathy as an interaction, or trade-off, between mirroring and mentalizing; those that make a distinction between affective versus cognitive empathy; and those that make a distinction between emotion sharing versus emotion regulation.

Zaki and Ochsner (2012) consider empathy as a product of two kinds of mechanism – mirroring versus mentalizing. The extent to which one mechanism may dominate over the other is assumed to be dynamic and may depend on what the expected outcome is. Merely observing another person in a decontextualized setting may bias towards mirroring. Deciding whether to act prosocially (or otherwise) towards another person may involve some interplay between the two mechanisms, and may differ from person to person. According to this framework, in the experiment of Singer et al. (2006) the tendency to simulate another's pain would be part of the mirroring system, and the representation of the other's intentions (to play fairly or unfairly) would be part of the mentalizing (or theory of mind) system.

An alternative distinction to mirroring versus mentalizing is the proposed distinction between cognitive empathy and affective empathy (e.g. Baron-Cohen & Wheelwright, 2004; Shamay-Tsoory, Aharon-Peretz, & Perry, 2009) In theory, the cognitive/affective distinction can be regarded as separate to the mirroring/mental-izing distinction. One can attribute mental states to others that are either affective in nature (e.g. 'John is angry') or non-affective in nature (e.g. 'John thinks X'). Sim-ilarly, mirroring can either be affective in nature (emotional contagion) or non-af-fective in nature (imitation/mimicry). In practice, the extent to which the mirroring/mentalizing and affective/cognitive distinctions are related or separate remains a matter of debate. For instance in the model of Shamay-Tsoory (2011) the mirroring mechanism is assumed to be common for both actions and emotions, but separate mentalizing components are postulated for affective and non-affective mental states. However, others argue that mirror neurons should be understood solely in terms of motor actions (Kilner & Lemon, 2013).

The model of empathy proposed by Decety and Jackson (2004, 2006) argues for a distinction between mechanisms based on simulation and other types of mech-anism, but does not draw a sharp line between affective and non-affective processes. This model, particularly in its later forms (Decety & Svetlova, 2012), draws on the distinctions between emotion sharing (based on simulation), emotion understanding (linked to mentalizing) and emotion regulation (linked cognitive control and deliber-ate perspective taking). Specifically the three components of empathy postulated by Decety and Jackson (2004) are:

1 *Shared representations between self and other, based on perception–action cou-pling.* This would include mechanisms for action understanding and imitation, emotional contagion, and pain processing. However, Decety and Jackson (2004) suggest that these are widely distributed throughout the brain rather than all load-ing on some core regions (such as the premotor cortex).
2 *An awareness of self–other as similar but separate.* This is related to mechanisms of self-awareness (see Chapter 9) that enable us to attribute our own thoughts and actions as self-generated. Decety and Jackson (2004) suggest that one important brain region for this process is the right temporo-parietal junction (rTPJ). For instance, this region responds more when watching a moving dot controlled by someone else's action relative to self-generated action (Farrer & Frith, 2002) and responds more when participants are asked to imagine someone else's feelings and beliefs compared to their own (Ruby & Decety, 2004).
3 *A capacity for mental flexibility to enable shifts in perspective and self-regulation.* Decety and Jackson (2004) suggest that this is a candidate for a uniquely human component of empathy. It involves deliberate perspective taking of another's sit-uation, which may also involve inhibiting one's own beliefs and self-referential knowledge. People with high self-reported personal distress may tend to over-rely on emotional contagion rather than cognitive control. Eisenberg et al. (1994) have shown that individual differences in personal distress are related to ability to con-trol and shift attention, and Spinella (2005) reports negative correlation between behavioral measures of executive function and personal distress. Decety and Jack-son (2004) suggest that regions in the prefrontal cortex responsible for the con-trol of emotions (ventromedial and orbital regions) and the control of thought and action (lateral regions) are important. A region in the medial prefrontal cortex (con-sidered below and in Chapter 9) responds to self-referential perspective relative to other perspective.

As such, this model offers a good account of the multi-faceted nature of empathy, in terms of cognitive mechanisms, social influences, and neural substrates. It also offers one way of connecting the literature on empathy with the other main topic of this chapter: theory of mind.

Evaluation

Empathy should perhaps best be regarded as a multi-faceted concept, and is likely to be explained via several interacting mechanisms rather than a single one. One possible scenario is the interplay between mirroring (simulation) and mentalizing (theory of mind). Another postulated division is between affective versus cognitive information. Finally, some models make a distinction between bottom-up information (mirroring) and top-down information that includes mentalizing but also other kinds of cognitive control mechanisms involved in perspective taking and distinguishing self from other.

KEY TERM

Anthropomorphism
The attribution of human characteristics to non-human animals, objects, or other concepts

PROJECTING MENTAL STATES EVERYWHERE – THE ORIGINS OF ANTHROPOMORPHISM?

Anthropomorphism refers to the attribution of human characteristics to non-human animals, objects, or other concepts. This could reflect a natural tendency to attribute mental states externally, and not just to other humans who are 'like me'. Living objects are commonplace in our popular culture – think of Pixar's bouncing lamp. It has also been suggested that a belief in God is a result of the tendency to attribute mental states externally (Guthrie, 1993).

To some extent, the tendency to anthropomorphize may depend on whether something looks like us – an angry dog shows its

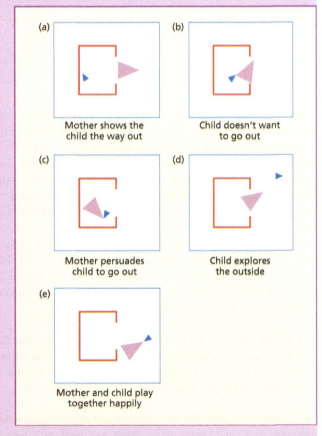

(a) Mother shows the child the way out

(b) Child doesn't want to go out

(c) Mother persuades child to go out

(d) Child explores the outside

(e) Mother and child play together happily

Figure 6.7 Mental states (e.g. want), behaviors (e.g. play), and other human characteristics (e.g. mother, child) are readily attributed to animated geometric shapes. Watching these animations, during functional imaging, activates a network of regions implicated in theory of mind. The captions were not presented in the studies, but are shown here for clarification. From Castelli et al. (2000). Copyright © 2000 Elsevier. Reproduced with permission.

teeth like an angry human. Movement as well as appearance is important. Heider and Simmel (1944) found that people readily ascribe mental states to animations of two interacting geometric objects, such as 'the blue triangle wanted to surprise the red one' (see Figure 6.7 for an example). In a functional imaging study that compared these kinds of animations with aimless movements, it was found that these moving shapes activated a network of regions that are typically activated in theory-of-mind tasks involving human agents (Castelli, Happe, Frith, & Frith, 2000). They argued that this supports the idea that intentions tend to be inferred from actions, even in situations in which participants know that the objects are not capable of having mental states.

Although anthropomorphism may be a universal tendency, some people may do it more and others may do it less. One study found that this tendency, measured in terms of mental state ratings for gadgets or terms used to describe pets, is greater in lonely people (Epley, Akalis, Waytz, & Cacioppo, 2008). Another study shows that the amount of gray matter in the left temporo-parietal junction, a key hub in the mentalizing network, is linked to the degree of anthropomorphic thinking (Cullen, Kanai, Bahrami, & Rees, 2014). People with autism use fewer mental state terms to describe the moving geometric shape stimuli and show less activity in regions linked to theory of mind when watching these animations (Castelli, Frith, Happe, & Frith, 2002).

THEORY OF MIND AND REASONING ABOUT MENTAL STATES

This section distinguishes itself from the previous one by considering in detail a certain kind of task: namely deliberate attempts to reason about mental states, and deliberate attempts to attribute mental states to others. To some extent these sorts of mechanisms are linked to those involving empathy, as discussed previously. However, the tasks used in the theory-of-mind literature are typically quite different from those considered in the section on empathy. The stimuli themselves are typically narratives or sequences of events, rather than observation of a particular state (e.g. pain). The tasks also typically require an overt response (e.g. what does Sally think or do?) whereas studies on empathy often do not (e.g. a typical measure could be the degree of imitative behavior or subtle contraction of facial muscles). We may be able to tell from someone's face or voice that they are being thoughtful, but knowing what they are thinking may involve a different computation. Consider the following two sentences (from Apperly, 2008):

(1) George likes to go to the gym in the morning, but he forgot it was closed on Mondays, so when he got there he just went straight to work.
(2) George usually goes to the gym in the morning but when he got there today it was closed, so he just went straight to work.

Both sentences describe the same observable behavior, but only the first sentence describes a mental state (i.e. an absence of knowledge due to forgetting). This gives us more insight into how George is likely to feel. He is probably annoyed that he forgot and disappointed he can't go. The first sentence requires us to think about George's mental states and, according to many researchers, this requires a special kind of mechanism that has been termed mentalizing or theory of mind.

The term 'theory of mind' derived originally from research on primate cognition. Premack and Woodruff (1978) conducted a number of studies on a chimpanzee to see if it understood an experimenter's intentions. For example, the chimp might point to a picture of a key when an experimenter was locked in a cage, the inference being 'he wants to get out'. A number of criticisms were leveled at the study. For instance, it may reflect knowledge of object associations (e.g. between lock and key) rather than mental states. In a reply to the article, Dennett (1978) suggested that one way of testing for theory of mind would be to consider false beliefs, in which someone else may hold a mental state (e.g. a belief) that differs from one's own belief and from the current state of reality. In developmental psychology, the paradigmatic false belief test is the object transfer task, such as the Sally–Anne task shown in Figure 6.8 (Baron-Cohen, Leslie, & Frith, 1985; Wimmer & Perner, 1983). Sally puts a marble in a basket so that Anne can see. Sally then leaves the room, and Anne moves the marble to a box. When Sally enters the room, the participant is asked 'Where will Sally look for the marble?' or 'Where does Sally think the marble is?' A correct answer ('In the basket') is typically taken to indicate the presence of a theory of mind. An incorrect answer is potentially more problematic to interpret. It could imply a lack of theory of mind. However, one also has to rule out other factors such as language comprehension difficulties or a failure to inhibit a more dominant response (one's own belief).

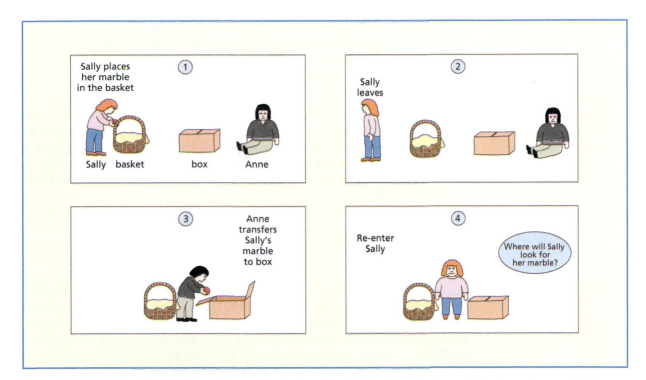

Figure 6.8 The Sally–Anne task requires an understanding of false belief and an attribution of second-order intentionality. Adapted from Wimmer and Perner (1983).

KEY TERMS

Attribution
In social psychology, the process of inferring the causes of people's behavior

Intentional stance
The tendency to explain or predict the behavior of others using intentional states (e.g. wanting, liking)

First-order intentionality
An agent possesses beliefs and desires, but not beliefs about beliefs.

Second-order intentionality
An agent possesses beliefs about other people's beliefs.

False beliefs are harder to accommodate within simulation theories because one's own belief is at odds with that attributed to the other person. This cannot be done by straightforward simulation involving shared self–other representations. It requires taking one's own mental states 'offline' and creating a hypothetical scenario different to current reality. So-called meta-representation and pretense is often regarded as a hallmark of theory-of-mind ability (Leslie, 1987).

Social psychologists use the term **attribution** to refer to the process of inferring the causes of people's behavior. The philosopher Dennett (1983) uses his own term of **intentional stance** to refer to our tendency to explain behavior in terms of mental states, which could otherwise be considered synonymous with mentalizing or theory of mind. However, Dennett (1983) has a particularly useful way of describing different levels of intentionality that might be used to account for behavior. For example, an observer might have to evoke zero-order intentionality to explain the behavior of an object, first-order intentionality to explain the behavior of some animals, and second-order intentionality to explain some human behavior.

- *Zero-order intentionality*. The assumption that an agent possesses no beliefs and desires. It responds to stimuli reflexively, such as producing a scream when frightened or running to evade a predator.
- *First-order intentionality*. The inference that an agent possesses beliefs and desires, but not beliefs about beliefs. It may produce a scream because it *believes* a predator is present or *wants* others to run away.
- *Second-order intentionality*. The inference that an agent possesses beliefs about other people's beliefs. It may produce a scream because it wants others to believe that a predator is nearby. False belief tests operate at this level (e.g. 'I think that Sally thinks that the marble is in the box').
- *Third-order intentionality*. An agent possesses beliefs about other people's beliefs concerning beliefs about other people, such as 'I think that John thinks that Sally doesn't know where the marble is'.

In this taxonomy, first-order intentionality and above would constitute 'mentalizing', taking an 'intentional stance' or theory of mind (depending on one's preferred term). Second-order intentionality does not have a special status (from a theoretical point of view), but it has acquired a special status by virtue of the fact that most tests of theory of mind operate at this level because they are more stringent and cannot be solved by stating one's own beliefs.

Mechanisms of theory of mind

Domain-specific versus domain-general accounts

Domain specificity is linked to the notion of modularity (Fodor, 1983). A cognitive mechanism, or brain region, can be said to be domain specific if it is specialized to process only one kind of information. Thus, a domain-specific theory-of-mind mechanism would be a process that is specialized for attributing mental states (Leslie, 1987). There are two dominant lines of evidence that have been brought to bear on this. First, there is the question of whether there is a specific region of the brain that responds to reasoning about mental states but not other kinds of things. It is possible that such a mechanism could be distributed in several locations, or that only one of the regions in that network is truly domain specific. Second, one can look to see if there are specific impairments in mental state attribution but not in other domains. Most evidence related to this question has come from the developmental condition of autism (e.g. Baron-Cohen, 1995b), but other lines of research have addressed this question from the perspective of acquired brain damage (e.g. Samson, 2009). These lines of evidence are considered in detail throughout the remaining part of this chapter.

Historically, explanations of theory of mind have fallen into two camps that are termed **theory-theory** and simulation theory. Theory-theory argues that we store, as explicit knowledge, a set of principles relating to mental states and how these states govern behavior (e.g. Gopnik & Wellman, 1992). In this sense, the 'theory' in theory of mind is like a mental rulebook for understanding others. This can be contrasted with simulation theory, which in one form would argue that perceptual-motor systems (rather than thinking and theorizing) are all that is needed for understanding others (e.g. Gallese & Goldman, 1998). When phrased in this way, it is reasonable to say that theory-theory makes more domain-specific assumptions whereas simulation theory can be considered a domain-general account. However, one needs to be cautious in dividing explanations into black and white dichotomies. For example, some versions of simulation theory argue that we do reason about mental states (rather than it being solely an outcome of perceptual-motor processes) but these versions are distinguished from theory-theory by making the claim that our own mental states form the foundation for understanding others (e.g. Mitchell, Banaji, & Macrae, 2005a).

Whilst the idea of a domain-specific mechanism for theory of mind is controversial, the idea that theory of mind requires basic competency in a number of domain-general mechanisms such as executive functions is not controversial, and a basic competency in language is required for many tasks. Language ability in typically developing children predicts success on a false belief task independently of age (Dunn & Brophy, 2005), and deaf children whose parents are non-native signers are delayed in passing such a task (Peterson & Siegal, 1995). This suggests that language is important for the development of theory of mind. Language may serve several functions: both a social, communicative role and also the acquisition of semantic knowledge of mental state words such as 'want' and 'think'. For example, children have to learn that these words denote concepts that are privately held (Wellman & Lagattuta, 2000). However, once a normal theory of mind is established it may not be dependent solely on language. Evidence for this assertion comes from brain-damaged patients with acquired **aphasia**. Apperly, Samson, Carroll, Hussain, and Humphreys (2006) report a single case study of a man with left hemisphere stroke who was impaired in many aspects of language, including syntax comprehension, but showed no impairments on non-verbal tests of theory of mind, including second-order inferences (X thinks that Y thinks).

Fractionating theory of mind

There is a middle ground between domain-specific accounts, which tend to treat theory of mind as a single entity, and domain-general accounts, for which theory of mind is nothing more than the interaction of domain-general resources (Apperly, 2011;

Schaafsma, Pfaff, Spunt, & Adolphs, 2015). These accounts typically assume that theory of mind can be fractionated in some way.

One of these accounts has already been mentioned briefly. Specifically, the model of Shamay-Tsoory (2011) proposes that theory of mind can be split in to affective and non-affective components (i.e. a split according to the type of information being processed), and this division occurs above and beyond the basic split between theory of mind (mentalizing) and simulation (mirroring). Patients with acquired brain damage to the orbital and ventromedial prefrontal cortex have difficulties in recognizing emotions in others (Hornak et al., 1996) and these patients may fail tests of theory-of-mind based on affective information but not on non-affective information (Shamay-Tsoory, Tibi-Elhanany, & Aharon-Peretz, 2006).

Another account of a fractionated theory of mind is offered by Apperly (2011) and Butterfill and Apperly (2013). They argue for a two-system model of theory of mind. One system is a low-level system that is assumed to work fast, automatically, and inflexibly. This may give rise to an intuitive and implicit theory of mind. This may be based on limited kinds of inputs (e.g. actions, gaze) and learned knowledge (e.g. that people are likely to act towards an object they are looking at). The low-level system includes some seemingly sophisticated abilities that underpin performance on false belief such as an appreciation of the fact that different people see different things (see Figure 6.9) as opposed to simple egocentric thinking (others see what I see). Moreover, the low-level system would enable actions to be predicted based on what other people see but, supposedly, without any explicit representation of their beliefs. The second system is a high-level system that is assumed to work more slowly but is flexible. It is based on reasoning and an explicit propositional representation of mental states. Both systems may enable false belief tests to be passed but in different ways: the low-level system may guide looking behavior whereas the high-level system may guide explicit reasoning about mental states. They assume

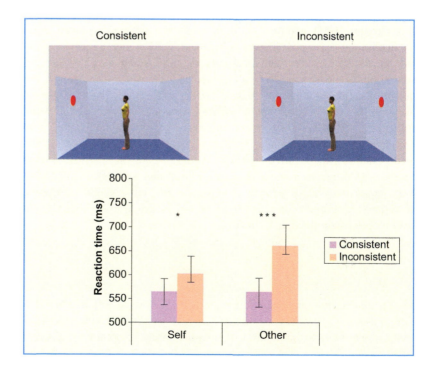

Figure 6.9 This task used by Samson et al., (2010) is taken as evidence that we automatically process the viewpoint of others. The task is to state the number of dots that can be seen from your own perspective ('self' condition) or the other person's perspective ('other' condition). On inconsistent trials, there is a mismatch between what is seen (e.g. you see two dots, the other person sees one dot) and this slows performance even when we are instructed to take an egocentric perspective.

that non-human animals and human children before the age of ~4 years rely solely on the low-level system.

The next section will go on to consider the neural substrates for theory of mind as evidenced from functional imaging (of neurologically normal adults) and neuropsychology (of brain-damaged adults). The following section will then consider autism in detail. The extent to which the evidence supports domain-specific processes, domain-general processes, or two-system models of theory of mind are considered. Developmental issues will be covered specifically in Chapter 11.

Neural substrates of theory of mind

Evidence for the neural basis of theory of mind has come from two main sources: functional imaging studies of normal participants and behavioral studies of patients with brain lesions. Numerous tasks have been used, including directly inferring mental states from stories (e.g. Fletcher et al., 1995), from cartoons (e.g. Gallagher et al., 2000), or when interacting with another person (e.g. McCabe, Houser, Ryan, Smith, & Trouard, 2001a). A review and meta-analysis of the functional imaging literature was provided by Frith and Frith (2003), who identified three key regions involved in mentalizing: the temporal poles, the temporo-parietal junction, and the medial prefrontal cortex. More recent reviews reveal a similar general pattern but with some important differences depending on the nature of the theory-of-mind test that is used (Schurz, Radua, Aichhorn, Richlan, & Perner, 2014). The evidence is summarized below.

Temporal poles

This region, shown in Figure 6.10, is normally activated in tasks of language and semantic memory. Frith and Frith (2003) suggest that this region is involved with generating **schemas** that specify the current social or emotional context, as well as in semantics more generally. Zahn et al. (2007) report an fMRI study suggesting that this region responds to comparisons between social concepts (e.g. brave–honorable) more than matched non-social concepts (e.g. nutritious–useful). Also, not all the tests of mentalizing that activated this region involved linguistic stimuli. For example, one study used triangles that appeared to interact by, say, chasing or encouraging each other (Castelli et al., 2000).

Brain damage to the temporal poles is a feature of the degenerative disorder known as **semantic dementia** (Mummery et al., 2000). Patients with semantic dementia lose their conceptual knowledge of words, objects, and people and show difficulties in language comprehension and production. However, there is little evidence from these patients that social concepts are selectively impaired. In general, although the temporal poles are important for theory of mind, there is no convincing support that it is domain specific for this kind of information and they are likely to provide more domain-general input in the form of social knowledge (Olson, McCoy, Klobusicky, & Ross, 2013).

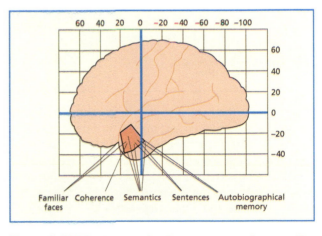

Figure 6.10 The temporal poles may support semantic knowledge, including of social concepts. Numbers indicate Talairach coordinates. Adapted from Frith and Frith (2003).

Medial Prefrontal Cortex (mPFC)

Frith and Frith (2003) reported that this region, shown in Figure 6.11, is activated in all functional imaging tasks of mentalizing to that date, and it still features prominently in a more recent meta-analysis (Schurz et al., 2014). Saxe (2006) argues that a sub-region of this area is involved in 'uniquely human' aspects of social cognition. This region lies in front of, but extends into, the ventral region of the anterior cingulate, labeled by Bush et al. (2000) as the affective division. Functional imaging studies reliably show that this region responds more to: thinking about people rather than thinking about other entities such as computers or dogs (e.g. Mitchell, Banaji, & Macrae, 2005a; Mitchell, Heatherton, & Macrae, 2002); thinking about the *minds* of people rather than thinking about their other attributes, such as their physical characteristics (Mitchell et al., 2005b); thinking about the minds of people who are similar to ourselves (Mitchell et al., 2005b); and thinking about social groups who are humanized relative to dehumanized (Harris & Fiske, 2006).

Some studies of patients with frontal lobe damage have suggested that the medial regions are necessary for theory of mind (e.g. Roca et al., 2011), but by no means all (e.g. Bird, Casteli, Malik, Frith, & Husain, 2004). This region also seems to be implicated in the pragmatics of language, such as irony ('Peter is well read. He has even heard of Shakespeare') and metaphor ('Your room is a pigsty') (Bottini et al., 1994). Interestingly, people with autism have difficulties with this aspect of language (Happe, 1995). In such instances, the speaker's *intention* must be derived from the ambiguous surface properties of the words (e.g. the room is not literally a pigsty). Functional imaging suggests that this region is involved both in theory of mind and in establishing the pragmatic coherence between ideas/sentences, including those that do not involve mentalizing (Ferstl & von Cramon, 2002).

Can a generic function be ascribed to this region? If so, how does it relate to theory of mind? Amodio and Frith (2006) argue that the function of this region is in reflecting on feelings and intentions, which they label a 'meeting of minds'. One intriguing finding concerning this region is that it can be activated when a person believes they are playing a computer game against another person relative to when they think they are playing against a computer (Rilling, Sanfey, Aronson, Nystrom, & Cohen, 2004). Even though the situation is physically identical (the participant always played the computer), the act of cooperating with another person/mind engenders activity in this region. A more recent explanation of the function of this region is similar to, but different from, that of Amodio and Devine (2006). Krueger, Barbey, and Grafman (2009) argue that the function of this region is to bind together different kinds of information (actions, agents, goals, objects, beliefs) to create what they term a 'social event'. They note that within

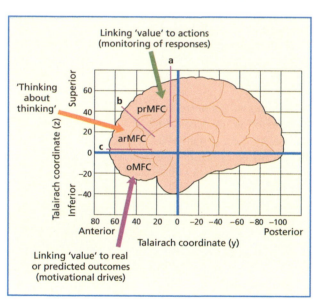

Figure 6.11 The medial frontal cortex (and adjacent regions of cingulate cortex) may contain three sub-regions with different functional specializations. Amodio and Frith (2006) regard the anterior rostral region (arMFC) as involved in 'thinking about thinking', or meta-cognition. This region is typically activated in tests of theory of mind. The orbital region (oMFC) is involved in linking value (positive or negative reinforcement) to outcomes, whereas the posterior rostral (or dorsal) region (prMFC) is involved in linking value to actions. These latter two regions are considered in Chapter 3. Adapted from Amodio and Frith (2006).

this region some sub-regions respond more when participants make judgments about themselves and also about others who are considered to be similar to themselves (this is discussed in detail in Chapter 9). This suggests that this region is not attributing mental states per se, but is considering the self in relation to others (e.g. when playing a game against a human rather than a computer). It is also consistent with some versions of simulation theory in which participants understand others by reflecting on how we would react in this scenario (e.g. Mitchell et al., 2005b). The notion of creating internal social events could also explain some of the findings of the role of this region in linking ideas in story comprehension (Ferstl & von Cramon, 2002).

Temporo-Parietal Junction (TPJ)

This region tends to be activated not only in tests of mentalizing but also in studies of the perception of biological motion, eye gaze, moving mouths, and living things in general. These skills are clearly important for detecting other agents and processing their observable actions. Some simulation theories argue that mentalizing need not involve anything over and above action perception. It is also conceivable that this region goes beyond the processing of observable actions, and is also concerned with representing mental states and perhaps even the mental states of others over and above one's own mental states. Congenitally blind people activate essentially the same network of regions identified by Frith and Frith (2003) when they perform theory-of-mind tasks (Bedny, Pascual-Leone, & Saxe, 2009). This suggests that the computations of these regions are, at least partially, independent from visual perception of agents. A more recent meta-analysis of fMRI data (neurotypical participants) on a wide range of tasks showed that whilst activity in the TPJ is found across diverse tasks, a more anterior region tends to be activated when using observable agents (e.g. 'reading the mind in the eyes') and a more posterior region for non-observed agents (e.g. verbal false belief tasks) (Schurz et al., 2014). This could conceivably support the kind of two-systems approach put forward by Apperly (2011).

The TPJ region was previously highlighted in the discussion on empathy because it responds more when participants are asked to imagine how someone else would feel relative to how they would feel (e.g. Ruby & Decety, 2004). Patients with brain lesions in this region fail theory-of-mind tasks that cannot be accounted for by difficulties in body perception (Samson, Apperly, Chiavarino, & Humphreys, 2004). Saxe and Kanwisher (2003) found activity in this region, on the right, when comparing false belief tasks (requiring mentalizing) with false photograph tasks (not requiring mentalizing but entailing a conflict with reality). A false photograph may involve taking a picture of an apple on the tree, and then the apple falling down. In this scenario, there is a conflict between reality and a representation of reality. The result was also found when the false photograph involved people and actions, consistent with a role in mentalizing beyond any role in action/person perception. The region responds to false beliefs more than false maps or signs, which differ in an important way from a false photograph in that they are designed to represent *current* reality (Perner, Aichhorn, Kronbichler, Staffen, & Ladurner, 2006).

Saxe and colleagues do not dismiss the fact that this region has a role to play in recognizing people and actions, but they claim that there may be different sub-regions within it, with one subregion specialized for the attribution of mental states as shown in Figure 6.12 (Scholz, Triantafyllou, Whitfield-Gabrieli, Brown, & Saxe, 2009). Moreover, Saxe (2006) argues that it is uniquely human in doing so. It is important to note that this region is not specialized for false belief per se. It responds to true beliefs and other

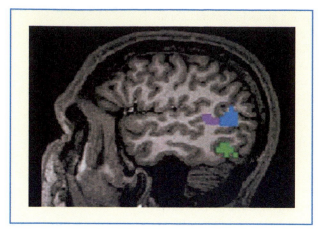

Figure 6.12 The TPJ region may contain separate sub-regions for dealing with theory of mind (shown here in blue) and recognizing actions and expressions (shown here in purple). For comparison, the position of the extrastriate body area (in green) is shown, which is involved in body perception. From Saxe (2006). Copyright © 2006 Elsevier. Reproduced with permission.

types of mental states (Saxe & Wexler, 2005). In other words, it responds to attributions of first-order intentionality as well as higher order intentionality (in Dennett's terms). Saxe and Powell (2006) have shown that this region responds to attribution of contentful mental states (such as thoughts and beliefs) rather than subjective states (such as hunger or tiredness). This suggests that it may have a role over and above 'thinking about others'. However, it is important to mention that one should be cautious in making strong claims about relative differences in BOLD signal. The differences can reflect different functional specialization (Saxe's claim), but they can also reflect the different difficulty of tasks, and the attention or strategy deployed to solve them. Other accounts propose more general functions to the TPJ that are relevant to theory of mind but without assuming it represents mental states as such. It may serve a general functioning of orienting of attention to a stimulus (Corbetta & Shulman, 2002; Mitchell, 2008) that, in social terms, may include orienting attention to other people and away from the self (Spengler, von Cramon, & Brass, 2009).

Evaluation

Functional imaging studies of the general population and, to a lesser extent, studies of people with acquired brain damage have helped to reveal the key regions involved in theory of mind and their somewhat different functions. There remains no consensus as to whether there is a domain-specific mechanism for theory of mind (i.e. a particular neural region that is dedicated to attributing mental states), but the strongest candidate region for domain specificity has shifted away from the medial prefrontal area to the TPJ region.

EXPLAINING AUTISM

He wandered about smiling, making stereotyped movements with his fingers, crossing them about in the air. He shook his head from side to side, whispering or humming the same three-note tune. He spun with great pleasure anything he could seize upon to spin. . . . When taken into a room, he completely disregarded the people and instantly went for objects, preferably those that could be spun. . . . He angrily shoved away the hand that was in his way or the foot that stepped on one of his blocks.

(This description of Donald, aged 5, was given by Leo Kanner [1943], who also coined the term autism. *The disorder was independently noted by Hans Asperger [1944], whose name now denotes a variant of autism.)*

Autism has been formally defined as 'persistent deficits in social communication and social interaction across multiple contexts' (DSM-V, American Psychiatric Association, 2013). It is a severe developmental condition that is evident before 3 years of age and lasts throughout life. There are a number of difficulties in diagnosing autism. First, it is defined according to behavior because no *specific* biological markers are known, although there are some known associations (see 'Biological markers for autism' box). Second, the profile and severity may be modified during the course of development. It can be influenced by external factors (e.g. education, temperament) and may be accompanied by other disorders (e.g. attention deficit and hyperactivity disorder, psychiatric disorders). As such, autism is now viewed as a spectrum of conditions spanning all degrees of severity. It is currently believed to affect 1.2% of the childhood population, and is three times as common in males (Baird et al., 2006). **Asperger's syndrome** falls within this spectrum, and is often considered a special sub-group. The diagnosis of Asperger's syndrome requires that there is no significant delay in early language and cognitive development, although the term is also used to denote people with autism who fall within the normal range of intelligence. Learning disability, defined as an IQ lower than 70, is present in around half of all cases of autism (Baird et al., 2006).

Much of the behavioral data has been obtained from high-functioning individuals in an attempt to isolate a specific core of deficits. On a purely theoretical level, one reason why researchers have been interested in the study of autism is the belief that it might reveal something fundamental about social interactions more generally.

> **KEY TERM**
>
> **Asperger's syndrome**
> A sub-type of autism associated with less profound non-social impairments

BIOLOGICAL MARKERS FOR AUTISM

Although there is no single biomarker that predicts the presence of autism, there are known biological differences at both the genetic and brain level. Whether a single biomarker will ever be found is uncertain. Autism may ultimately be an outcome of multiple causes including some interactions with the environment.

1 Genetic differences

- Twin studies show that autism has high heritability with 60% concordance between identical twins using a strict diagnosis and 92% concordance when using a more liberal definition (Bailey et al., 1995). The remaining influence is largely due to shared environment between twins (Hallmayer et al., 2011).
- Multiple genes convey susceptibility to autism (Ma et al., 2009), including some genetic variations known to be linked to social cognition such as the gene for the oxytocin receptor (Wu et al., 2005). Although these genetic variants are more common in people with autism they are not specific to autism.

2 Structural brain differences

- There is accelerated brain growth in early development, peaking at 2–4 years, such that people with autism have a greater brain volume (Redcay & Courchesne, 2005). This reflects both white matter and gray matter differences.
- Although increased cortical thickness (i.e. gray matter) is apparent in children with autism it may be linked to less thickness in adulthood (Raznahan et al., 2010).

Figure 6.13 Although there are some regions of the brain that show greater functional connectivity in people with autism (lower panel, red lines), the autistic brain is generally characterized by a large-scale reduction in functional connectivity (upper panel, blue lines). For the analysis, the brain was parcellated into 112 different regions (shown here as various colored nodes).

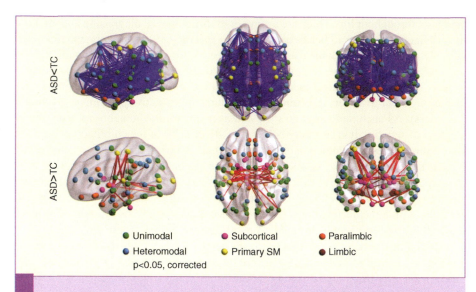

- Diffusion tensor imaging (DTI) measures the coherence of white matter connectivity (not the amount of white matter per se) and there is less coherent connectivity in the autistic brain (Anagnostou & Taylor, 2011).

3 Functional connectivity differences

- Although there are some regions of the brain that show greater functional connectivity in people with autism, the autistic brain is generally characterized by a large-scale reduction in functional connectivity as shown in Figure 6.13. This particular study used an fMRI resting state paradigm (see Chapter 2) from a consortium of researchers who provided more than 1000 datasets (Di Martino et al., 2014).

Autism as mind blindness

One candidate deficit is the ability to represent mental states, or theory of mind (e.g. Baron-Cohen, 1995b; Fodor, 1992). The first empirical evidence in favor of this hypothesis came with the development of a test of false belief devised by Wimmer and Perner (1983) and tested on autistic children by Baron-Cohen et al. (1985) as the Sally–Anne task (described above). Autistic children tend to fail the task whereas normally developing children (from 4 years on) pass the test, as do control participants with learning disability matched in IQ to the autistic children. The erroneous reply is not due to a failure of memory, because the children can remember the initial location. It is as if they fail to understand that Sally has a belief that differs from physical reality – that is, a failure to represent mental states. This has also been called 'mind-blindness' (Baron-Cohen, 1995b). Autistic children are still impaired when the false belief was initially their own. For example, in one task, the child initially expects to find candy in a candy packet and is surprised to find a pencil, but when asked what other people will think is in the packet the child replies 'pencil' (Perner, Frith, Leslie, & Leekam, 1989). This is shown in Figure 6.14.

Figure 6.14 The child initially expects to find candy in a tube of Smarties and is surprised to find a pencil. When asked what other people will think is in the packet, autistic children reply 'pencil' whereas typically developing children reply 'candy'.

Passing false belief tasks requires the ability to form meta-representations (i.e. representations of representations: in this instance, beliefs about beliefs). It was originally suggested that a failure of meta-representation may account for impaired theory of mind in autism (Baron-Cohen et al., 1985). However, other studies suggest that autistic people can form meta-representations in order to reason about false photographs in which the information depicted on the photograph differs from current reality (Leekam & Perner, 1991). If their deficit really is related to mental state representations rather than physical representations, then this offers support for the domain-specific account. A number of other studies have pointed to selective difficulties in mentalizing compared to carefully controlled conditions. For example, people with autism can sequence behavioral pictures but not mentalistic pictures (Baron-Cohen, Leslie, & Frith, 1986); they are good at sabotage but not deception – they tend to think that everyone tells the truth (Sodian & Frith, 1992); and they tend to use desire and emotion words but not belief and idea words (Tager-Flusberg, 1992). In all instances, the performance of people with autism is compared to mental-age controls to establish that the effects are related to autism and not to general level of functioning.

Functional imaging studies of autistic people carrying out theory-of-mind tasks (Happe et al., 1996; Castelli et al., 2002) have shown reduced activity in the network of regions commonly activated by controls.

Finally, it may be necessary to make a distinction between implicit mentalizing (intuitive, reflexive) and more explicit forms of mentalizing (based on reasoning), as occurs in some two-system models of theory of mind (e.g. Apperly, 2011). Whilst explicit theory of mind tends to be measured by overt predictions of behavior (as in the Sally–Anne task), the former may be measured by non-declarative means (e.g. monitoring of eye movements). For example, some high-functioning people with autism pass standard theory-of-mind measures but may still lack an intuitive understanding of others and may still show abnormal performance on other measures (e.g. eye movements to a location consistent with a false belief; Senju, Southgate, White, & Frith, 2009). By contrast, children under the age of 4 years show some implicit understanding of false beliefs (based on the same measure) despite failing on explicit measures (Onishi &

Baillargeon, 2005). This suggests a dissociation between implicit and explicit forms of theory of mind (as assessed by eye-movements and reasoning respectively). Implicit/intuitive forms may emerge early developmentally but be impaired in autism, whereas explicit forms emerge later in development and may be present or absent in autism (perhaps linked to age and level of intellectual functioning).

Empathy and feelings in autism

As discussed throughout the chapter, empathy can be measured in many ways and can almost certainly be broken down in to different kinds of cognitive and neural mechanisms. As such, we need to think carefully about how we answer a question such as 'Do people with autism lack empathy?'. On questionnaire measures of empathy, people with autism tend to score lower, implying that they do have less empathy (e.g. Baron-Cohen & Wheelwright, 2004). However, most of these measures ask about high-level aspects of empathy (e.g. such as thinking about the feelings of others), that most likely tap the mentalizing network, rather than more feeling-based aspects of empathy. Frith (2012) argues that an impairment of mentalizing need not reflect an inability to resonate with the feelings of others. When observing other people in pain, people with autism show a similar brain response in the pain matrix (Hadjikhani et al., 2014). They show normal spontaneous facial mimicry of emotional expressions measured with EMG (Deschamps, Coppes, Kenemans, Schutter, & Matthys, 2015), and normal modulation of motor-evoked potentials (MEPs) in response to TMS when viewing hand actions, including during social situations (Enticott et al., 2013; but see Theoret et al., 2005).

Many people with autism do, however, have problems in labeling their own feelings – a symptom that is known as **alexithymia** (literally 'no words for feelings') (Hill, Berthoz, & Frith, 2004). For instance, people with alexithymia tend to agree with statements such as 'I don't know if I am feeling tired or angry'. This could be interpreted as a difficulty in reflecting on one's own mental states (an emotional feeling is a mental state) as well as those of other people. Several studies have contrasted brain responses to emotional or social stimuli according to whether participants have autism, alexithymia, both autism and alexithymia, or neither. Silani et al. (2008) showed images with different levels of emotional content, and participants rated how unpleasant it made them feel or how colorful the images were. All participants showed activity in the amygdala linked to their level of emotional arousal (i.e. no evidence of impairment in basic emotional intensity). Whereas alexithymia was linked to reduced activity in the anterior insula (assumed to reflect poor awareness of specific bodily based feelings), autism was linked to reduced activity in regions implicated in mentalizing when rating their emotions. In another fMRI study, alexithymia was linked to reduced anterior insula activity during empathy for pain but autism per se was not linked to these differences, as shown in Figure 6.15 (Bird et al., 2010).

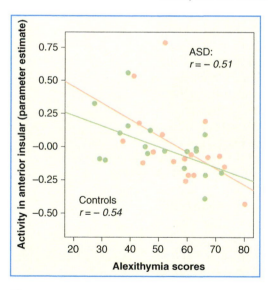

Figure 6.15 Empathy for pain is associated with activity in the anterior insula. This measure of empathy does not differ between controls and people with autism spectrum disorder (ASD), but is related to the degree of alexithymia (ability to label feelings) in the two groups (Bird et al., 2010).

Autism as executive dysfunction

The mentalizing or theory-of-mind account of autism has not been without its critics. These criticisms generally take two forms: that other explanations can account for the data without postulating a difficulty in mentalizing (e.g. Russell, 1997); or that a difficulty with mentalizing is necessary but insufficient to explain all of the available evidence (e.g. Frith, 1989). A number of studies have argued that the primary deficit in autism is one of executive functioning (Hughes, Russell, & Robbins, 1994; Ozonoff, Pennington, & Rogers, 1991; Russell, 1997). **Executive functions** refer to control processes that are needed to coordinate the operation of more specialized components of the brain, thus enabling us to switch attention from one task to another, to give priority to certain kinds of information, or to develop novel solutions, which would include inhibiting familiar solutions (e.g. Goldberg, 2001). For example, the incorrect answer might be chosen on false belief tasks because of a failure to suppress the strongly activated 'physical reality' alternative. Some patients with brain damage in prefrontal regions do this when given false belief tasks (Samson, 2009). However, it is not clear that this explanation can account for all the studies relating to mentalizing (e.g. picture sequencing). Moreover, high-functioning autistic people often have normal executive functions (e.g. Baron-Cohen, Wheelwright, Stone, & Rutherford, 1999) and early brain lesions can selectively disrupt theory-of-mind abilities without impairing executive functions (e.g. Fine, Lumsden, & Blair, 2001).

Rather than difficulties in executive function explaining impairment on theory-of-mind tasks, Baron-Cohen (2009) speculates that the opposite could be true – namely, autistic people may develop, and stick to, their own rule system rather than the 'correct' one as determined by another person, the experimenter. An experiment is, in effect, a social contract. One study found that autistic people show the greatest impairment on open-ended tasks of executive function (in which participants may induce their own rules), rather than those that require the following of simple, stated rules (White, Burgess, & Hill, 2009). On some tests of executive function autistic people show differences in the medial prefrontal region, which is implicated in mentalizing (Gilbert, Bird, Brindley, Frith, & Burgess, 2008). This again suggests that difficulties on some aspects of executive functions could be related to their social difficulties.

Autism as weak central coherence

One difficulty with the theory-of-mind explanation is that it fails to account for cognitive strengths as well as weaknesses. One popular notion of autistic people is that they have unusual gifts or 'savant' skills, as in the film *Rain Man*. In reality, these skills are found only in around 10% of the autistic population (Hill & Frith, 2003). Nevertheless, some account of them is needed for a full explanation of autism. The unusual skills of some autistic people may be partly an outcome of their limited range of interests. Perhaps one reason why some individuals are good at memorizing dates is that they practice it almost all the time. However, there is also evidence for more basic differences in processing style. For example, people on the autistic spectrum are superior at detecting embedded figures (Shah & Frith, 1983) – see Figure 6.16 – and searching for a target in an array of objects (for a review see Mitchell & Ropar, 2004).

One explanation for this is in terms of 'weak central coherence' (Frith, 1989; Happe, 1999). This is a cognitive style, assumed to be present in autism, in which processing of parts (or local features) takes precedence over processing of wholes

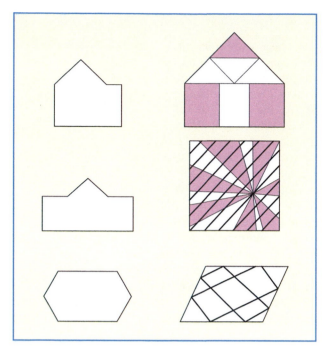

Figure 6.16 People with autism may be faster at spotting embedded figures such as the ones shown here (the figures on the left are embedded within those on the right).

(or global features). However, it is also possible that differences in social cognition in autism cause differences in the style of perceptual processing, rather than vice versa. For example, cultures that regard themselves as socially inter-dependent (i.e. strongly connected with the people around them in terms of shared goals and identity) show more global processing than those who construe themselves more socially independent (Davidoff, Fonteneau, & Fagot, 2008; Lin & Han, 2009; Nisbett, Peng, Choi, & Norenzayan, 2001). People with autism could be regarded as lying at one extreme end of this normal scale. What would cause such a pattern? Weak central coherence may be linked to differences in brain size and connectivity. There is convincing evidence the structural connectivity is less organized in the brains of people with autism, and that there is a general hypo-connectivity in terms of brain functioning (see earlier). Some theories argue that autism can be thought of as a disconnection syndrome particularly between frontal cortex and other regions (Geschwind & Levitt, 2007), and this could give rise to weak central coherence. One study found that children with autism with larger heads/brains showed a stronger local processing bias than other children with autism or typically developing controls (White, O'Reilly, & Frith, 2009).

Autism as an extreme form of the male brain

Baron-Cohen (2002, 2009) argues that the characteristics of all individuals can be classified according to two dimensions: 'empathizing' and 'systemizing'. Empathizing allows one to predict a person's behavior and to care about how others feel. Systemizing requires an understanding of lawful, rule-based systems and requires an attention to detail. Males tend to have a brain type that is biased towards systemizing (S > E) and females tend to have a brain type that is biased towards empathizing (E > S). However, not all men and women have the 'male type' and 'female type', respectively. Autistic people appear to have an extreme male type (S >> E), characterized by a lack of empathizing (which would account for the mentalizing difficulties) and a high degree of systemizing (which would account for their preserved abilities and unusual interests). Questionnaire studies suggest that these distinctions hold true (Baron-Cohen, Richler, Bisarya, Gurunathan, & Wheelwright, 2003; Baron-Cohen & Wheelwright, 2004). However, these distinctions are very broad, and it should be borne in mind that, for example, people with autism may only be impaired on particular kinds of measures relating to empathy. How does the extreme male brain hypothesis relate to other theories of autism? Baron-Cohen (2002, 2009) regards this explanation as an extension of the earlier mind-blindness theory, which has the advantage of being able to incorporate additional data. Specifically, it accounts for some of the non-social differences found in autism, and it offers an explanation for why autism is more common in men (i.e. because men are more likely to have S > E type brains). However, there are at least two ways in which these different ideas (mind blindness vs extreme male brain) could be related: that an inability to engage with others (due to a theory-of-mind deficit) leads to systemizing as a kind of compensatory strategy; or that an unusual interest or ability in systemizing leads to a lack of interest and understanding of social behavior. A third possibility is that both are true – that whatever it is that causes high systemizing also causes low empathizing. Possible mechanisms include fetal testosterone levels (e.g. Auyeung et al., 2009) or sex-related genetic differences (e.g. Creswell & Skuse, 1999). Although the extreme male brain theory predicts an autistic advantage for understanding systems, it differs from the weak central coherence theory by not making predictions about a difference between local versus global information. Finally, some research has tried to suggest a

SYSTEMIZING IN CLASSIC AUTISM AND/OR ASPERGER'S SYNDROME

Type of systemizing	Classic autism	Asperger's syndrome
sensory systemizing	tapping surfaces or letting sand run through one's fingers	insisting on the same foods each day
motoric systemizing	spinning round and round, or rocking back and forth	learning knitting patterns or a tennis technique
collectible systemizing	collecting leaves or football stickers	making lists and catalogues
numerical systemizing	obsessions with calendars or train timetables	solving math problems
motion systemizing	watching washing machines spin round and round	analyzing exactly when a specific event occurs in a repeating cycle
spatial systemizing	obsessions with routes	developing drawing techniques
environmental systemizing	insisting on toy bricks being lined up in an invariant order	insisting that nothing is moved from its usual position in the room
social systemizing	saying the first half of a phrase or sentence and waiting for the other person to complete it	insisting on playing the same game whenever a child comes to play
natural systemizing	asking over and over again what the weather will be today	learning the Latin names of every plant and their optimal growing conditions
mechanical systemizing	learning to operate the VCR	fixing bicycles or taking apart gadgets and reassembling them
vocal/auditory/verbal systemizing	echoing sounds	collecting words and word meanings
systemizing action sequences	watching the same video over and over again	analyzing dance techniques

Source: Baron-Cohen et al. (2009). Copyright © 2009 The Royal Society. Reproduced with permission.

KEY TERMS

Mu suppression
The tendency for fewer mu waves (in EEG) to be present during the execution of an action

Broken mirror theory
An account of autism in which the social difficulties are considered as a consequence of mirror system dysfunction

Mu waves
EEG oscillations at a particular frequency (8–13 Hz) that are greatest when participants are at rest

link between the extreme male brain theory and the broken mirror theory (discussed below), noting that there are sex differences (within the non-autistic population) in white/gray matter density in regions associated with the mirror system, with females showing greater density (Cheng et al., 2009). An EEG signature linked to functioning of the mirror system, termed **mu suppression**, also shows a sex difference, with females showing greater suppression (Cheng et al., 2008).

The broken mirror theory of autism

The **broken mirror theory** of autism argues that the social difficulties linked to autism are a consequence of mirror system dysfunction (Iacoboni & Dapretto, 2006; Oberman & Ramachandran, 2007; Ramachandran & Oberman, 2006; Rizzolatti & Fabbri-Destro, 2010). Hadjikhani, Joseph, Snyder, and Tager-Flusberg (2006) examined, using structural MRI, the anatomical differences between the brains of autistic individuals and matched controls. The autistic individuals had reduced gray matter in several regions linked to the mirror system, including the inferior frontal gyrus (Broca's region), the inferior parietal lobule, and the superior temporal sulcus. Although these were not the only regions where differences were found, the degree of thinning in these regions correlated with autistic symptom severity.

EEG and fMRI data also suggest differences in mirror system functioning during certain tasks. Oberman et al. (2005) used EEG to record mu waves over the motor cortex of high-functioning autistic children and controls. **Mu waves** occur at a particular frequency (8–13 Hz) and are greatest when participants are doing nothing. However, when they perform an action there is a decrease in the number of mu waves, a phenomenon termed mu suppression. Importantly, in typical controls mu suppression also occurs when people *observe* actions and, as such, it has been regarded by some as a measure of mirror system activity (e.g. Pineda, 2005). Oberman et al. (2005) found that the autistic children failed to show as much mu suppression as controls during action observation (watching someone else make a pincer movement) but did so in the control condition of action execution (they themselves make a pincer movement) as shown in Figure 6.17.

Similar findings have been obtained with fMRI. Dapretto et al. (2006) conducted a study in which autistic children and matched controls either observed or imitated emotional expressions. The imitation condition produced less activity in the inferior frontal gyrus of the autistic children relative to controls, and this was correlated with symptom severity. Differences in regions linked to face recognition (fusiform gyrus) and emotion recognition (amygdala) did not differ between groups.

The broken mirror theory makes some novel predictions about what people with autism might be impaired at, such as imitation and understanding the goals of others based on action observation. Boria et al. (2009) compared children with autism against typically developing controls in which actions were either consistent with typical use of a phone (e.g. making a call) or not (e.g. picking it up to move it; see Figure 6.18). They found that the autistic children were more likely to base their understanding of actions based on the object rather than action. In this example, they are more likely to answer 'making a call' when the object is being moved.

Although deficits in imitation are found in autism (Williams, Whiten, & Singh, 2004), these may be more apparent in spontaneous imitation than instructed imitation (e.g. Hamilton, Brindley, & Frith, 2007). This suggests that autistic people have a poor intuitive understanding of when and what to imitate (i.e. social rules) rather than in perceptual-motor interactions (at the heart of the broken mirror theory). The broken mirror theory has its critics (Dinstein, Thomas, Behrmann, & Heeger, 2008;

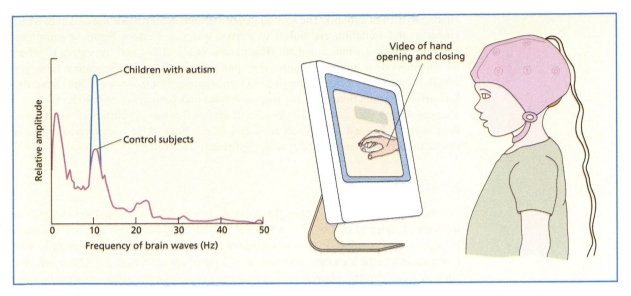

Figure 6.17 Mu waves are EEG oscillations in the 8–13 Hz range that are reduced both when performing an action and when watching someone else perform an action (relative to rest). As such, they may provide a neural signature for human mirror neurons. Autistic children show less mu suppression when watching others perform a hand action, which provides evidence in support of broken mirror theory. From Ramachandran and Oberman (2006). Reproduced with permission from Lucy Reading-Ikkanda for *Scientific American* Magazine.

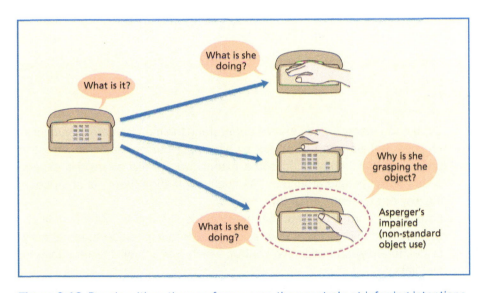

Figure 6.18 People with autism perform worse than controls at inferring intentions for nonstandard actions. In this example, they are more likely than controls to say that the person intends to make a call than to say that they are moving the phone. Figure based on Boria et al. (2009).

Southgate & Hamilton, 2008). In general, the criticism takes two forms. First, it does not account for all the unusual behavior found in autism (e.g. embedded figures; interest in systems). Defendants of the theory argue that it is not trying to explain all the features of autism (i.e. it is not a theory of autism but a theory of certain characteristics of autism). The second general criticism surrounds the extent to which empathy and imitation are linked to mirror systems. Certain forms of emotional empathy appear normal in autism (Hadjikhani et al., 2014) and the extent to which the mirror system supports emotional empathy is not clear (de Vignemont & Singer, 2006). A core deficit elsewhere (e.g. in representing mental states) could nevertheless affect the functioning of the mirror system and perhaps even lead to structural changes within that system. Heyes (2010) argues that the properties of mirror neurons may be learned as a result of social interactions. Thus, impoverished social interactions may cause mirror system dysfunction, as well as vice versa.

Evaluation

For many years the dominant explanation of autism has been a failure to represent the mental states of others. This has been termed mind blindness and has tended to be regarded as a failure to develop a theory of mind (although not necessarily with commitment to the idea that this exists as a domain-specific module). Other theories, such as weak central coherence theory and extreme male brain theory, maintain this basic idea but adopt a wider perspective in order to explain other features of autism. The most significant challenge to this idea previously came from the notion of executive dysfunction in autism, but now comes in the form of broken mirror theory. There is good evidence of mirror neuron dysfunction in autism, but it is less clear whether this dysfunction is a core feature of autism or a by-product of other deficits – given that mirror systems in general are modulated by beliefs, social knowledge, and cognitive control.

SUMMARY AND KEY POINTS OF THE CHAPTER

- Simulation theory argues that we understand the mental states (thoughts, feelings, beliefs, etc.) of others by activating our own mechanisms for producing that behavior. To some extent, we literally share the experiences of the people around us. As such, simulation theory is an appealing way of explaining empathy.
- Empathy is a broad concept that may include simulation, but it is unlikely to be limited to it. It also involves perspective taking (either automatically or deliberately) and cognitive control, which may inhibit the tendency to simulate.
- Both empathy and theory of mind (or mentalizing) involve understanding the mental states of others, but the latter is typically assessed via conscious attempts to reason about mental states, such as in false belief tasks.
- Functional imaging of the normal population reveals a network of regions that are consistently activated by tests of theory of mind, and the two regions that have provoked the most research interest are the temporo-parietal junction region and a medial prefrontal cortex region. However,

it remains controversial whether either region can be classed as domain specific for attributing mental states.

- People with autism often fail theory-of-mind tasks, leading to the theory that they have a specific impairment in representing mental states. Their difficulty is not well explained by difficulties in executive function alone or difficulties in meta-representation per se.
- There is good evidence of mirror neuron dysfunction in autism, but it is less clear whether this dysfunction is a core feature of autism or a by-product of other deficits (given that mirror systems in general are modulated by beliefs, social knowledge, and cognitive control).

EXAMPLE ESSAY QUESTIONS

- What is the evidence for and against simulation theories of empathy?
- How is empathy related to theory of mind, and in what ways are they different?
- Is there a theory-of-mind module in the human brain?
- How can the social behavior of people with autism be explained?

RECOMMENDED FURTHER READING

- Apperly, I. (2011). *Mindreaders: The Cognitive Basis of 'Theory of Mind'.* New York: Psychology Press. A thorough treatment of this topic taking into account evidence from human infants and adults, and non-human animals.

- Decety, J., & Ickes, W. (2009). *The Social Neuroscience of Empathy.* Cambridge, MA: MIT Press. An excellent collection of papers on empathy.

- Hill, E. L., & Frith, U. (2004). *Autism: Mind and Brain.* Oxford: OUP.

ONLINE RESOURCES

- References to key papers and readings
- Videos demonstrating tests of theory of mind
- Interviews and lectures featuring Rebecca Saxe, Marco Iacoboni, Uta Frith, and others
- Recorded lecture given by textbook author, Jamie Ward
- Multiple choice questions and interactive flashcards to test your knowledge
- Downloadable glossary

CHAPTER 7

CONTENTS

Interacting with others

7

Interactions have been defined as 'dyadic behavior in which the participants' actions are inter-dependent such that each actor's behavior is both a response to and a stimulus for the other participant's behavior' (Rubin, Bukowski, & Parker, 2006). This chapter is about two kinds of interaction: cooperation and competition. Cooperation entails sharing of commodities (e.g. food) and knowledge, and providing helping behavior (e.g. if someone is injured). This type of behavior is also termed **altruism**, but with the added twist that altruism is often described as 'selfless' in that no personal gain is obtained. Non-cooperation entails keeping commodities and knowledge for oneself and not providing help to others. For most people, Darwin's theory of natural selection is synonymous with competition, as exemplified in the phrase 'survival of the fittest' (a term not actually used by Darwin). When put this way, cooperation seems like a puzzle. Being social and cooperative compromises one's own time and resources. If my genes (and my traits) are to survive then they have to be of benefit to me, not to you. However, short-term costs to an individual have to be balanced against the longer-term benefits to be had through group living. Individuals working together in groups may increase chances of survival by, for instance, hunting as a group and sharing knowledge and skills.

In humans, at least, cooperative interactions between individuals are predicated upon **trust** (i.e. the belief that others will treat you fairly). Knowing who to trust and when to trust involves a complex decision-making process that will be discussed throughout the chapter. It will be dependent on the particular situation and also on prior knowledge of how others have behaved in the past. However, the basic fact that we are capable of trust at all may depend on our ability to understand that others have similar mental states to our own and on our ability to form shared goals between individuals. People who receive the benefits of cooperation but do not contribute to the group themselves are termed **freeloaders (or free riders)**, and groups typically impose sanctions on those who freeload, such as social exclusion, physical punishment, or fines. Such sanctions require norms to regulate or enforce cooperation, and these norms require consensual agreement as to what is 'fair' or 'right'.

In his book, *Why we Cooperate*, Tomasello (2009) outlines a number of different answers to that question:

1 *There is an intrinsic desire to help.* In humans at least, helping others is personally rewarding. This may relate to the capacity for empathy, which was discussed in detail in the previous chapter and will be returned to here.
2 *The benefits of reciprocity* – 'you scratch my back and I'll scratch yours'. There may be certain things that I cannot do for myself now but that you can do for me; in the future the situation may be reversed.
3 *Punishment for non-cooperation.* This has also been termed **altruistic punishment** (Fehr & Gachter, 2002). It is altruistic in the sense of being 'selfless': it has

KEY TERMS

Dyadic interactions
Face-to-face interactions

Altruism
Helping behavior that is considered 'selfless', in that no personal gain is obtained

Trust
The belief that others will behave fairly towards us

Freeloaders (or free riders)
People who receive the benefits of cooperation but do not contribute to the group themselves

Altruistic punishment
An act of punishment that has no direct benefit to the punisher but comes at a cost to the punisher

no direct benefit to the punisher and comes at a cost to the punisher (in terms of time and effort, and in terms of risk of reprisal). This factor is more likely to explain why cooperation is maintained rather than why it exists at all.

4 *A desire to conform*, by sharing in a group-level identity. Groups and identity are primarily considered in a subsequent chapter.

The evidence for these different motivations for cooperation will be outlined below. This chapter will not only consider *why* we cooperate, but also *how* we cooperate in terms of the cognitive and neural mechanisms that support this kind of behavior.

ALTRUISM AND HELPING BEHAVIOR

Evolutionary biological approaches

From an evolutionary perspective, the problem of altruism lies in explaining how it is possible to improve the survival chances (or 'fitness') of altruists, given that helping others necessarily entails a personal cost to those who help. Evolutionary

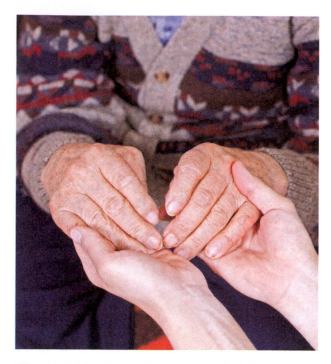

Figure 7.1 Why do people engage in altruistic acts such as helping the elderly, giving to charity, or performing favors? Given that altruism entails a cost, what kind of benefits accrue from altruism to ensure that it survives within a group?

REAL VERSUS VIRTUAL INTERACTIONS

The experimental constraints of social neuroscience place strong limitations on being able to conduct research on real-world social interactions. This has been described by some as the 'dark matter' of social neuroscience (Schilbach et al., 2013). It is certainly possible to introduce real-time interactions into an MRI scanner including across multiple scanners (Montague et al., 2002) although this poses problems for standardizing material across participants. It is also possible to have avatars that respond to the gaze direction of participants in real time (Wilms et al., 2010). The distinction between real and virtual may not be as crucial as one first assumes. Recall that one long-standing definition of social psychology includes the imagined or implied presence of others ('an attempt to understand and explain how the thoughts, feelings, and behaviors of individuals are influenced by the actual, imagined, or implied presence of others'; Allport, 1968). In general, the findings of social neuroscience support the view that participants (and their brains) respond differently when they *believe* they are interacting with a human (irrespective of whether they are or are not) relative to when they believe they are interacting with a computer program.

A separate but related issue is the cultural shift that has taken place away from face-to-face interactions to virtual ones via email and social media. Despite there being a sense of 'presence' of another social being (as opposed to, say, a Google search), these interactions do not typically occur in real time and contain little observable social information (e.g. social cues from the face, body, and voice). This 'hiding behind a screen' almost certainly shapes the tone of our interactions (Sparrow & Chatman, 2013) and may conceivably change the way that the social brain develops.

biology has come up with several main mechanisms to explain the evolution of altruism that are considered in turn below:

- Kin selection assumes that we help others who are related to us.
- Reciprocal altruism (or direct reciprocity) assumes that we provide help to others in order to obtain help from others in the future ('I'll scratch your back if you scratch mine').
- Sexual selection assumes that displays of wealth and generosity enhance the mating success of altruists.
- Indirect reciprocity assumes that one only needs to help individuals who are likely to help you (including people who you may never meet again).

Kin selection

In proposing kin selection theory, Hamilton's (1964) significant insight was to realize that it is the survival chances of a trait (in this case, helping behavior) rather than an individual that matters – it is survival of the fittest *trait*, not the fittest *person*. (This is essentially the same point made by Dawkins (1976) in his famous 'selfish gene' theory.) If an individual helps their kin, then there is a greater chance that the helping trait will survive, because there is a greater chance that their kin also carry this same trait. However, this still only applies if the benefits of helping are greater than the costs. Specifically, helping behavior can spread through a population if the cost to the organism's own reproduction (C) is offset by the benefit to the reproduction of its kin member (B) multiplied by the probability that the kin member inherits the same helping behavior (r). Thus, helping traits will tend to spread when $C < r \times B$.

There are several examples from the natural world that are consistent with kin selection. Ground squirrels are able to discriminate full siblings from half-siblings and unrelated individuals (Holmes & Sherman, 1982) and are more likely to give an

KEY TERM

Kin selection
The theory that we help others who are related to us

alarm call if close relatives are nearby (Sherman, 1977). Kinship predicts whether
vampire bats will share food with another individual (Wilkinson, 1988). It also
explains why sterile worker insects forego their own fertility for their kin (Trivers &
Hare, 1975).

Reciprocal altruism (direct reciprocity)

Kin selection does not offer an explanation of altruism towards unrelated individuals,
except as an error in detecting kin from non-kin. However, such altruism does exist.
For instance, unrelated chimpanzees will come to each other's aid when threatened
(De Waal & Luttrell, 1988). To give another example, a major predictor of whether
a vampire bat will share food with a non-relative is whether the non-relative shared
with that individual in the past (Wilkinson, 1988).

Williams (1966) and Trivers (1971) proposed the alternative mechanism of
reciprocal altruism to account for this. Reciprocal altruism is based on the economic
concept of trade (e.g. exchanging food for protection). The exchange is most often
delayed, so a cost to an individual initially will be a benefit to the individual at some
later point. What kinds of cognitive processes are needed for reciprocal altruism?
Minimally, it requires an ability to distinguish between conspecifics and to remember
their previous behavior. However, the decision is often more complex than that. Just
because someone has helped you, there is no necessary reason why you should help
them back (the freeloader problem). If you help someone else, how do you know that
they will help you in the future?

To illustrate this problem, Axelrod and Hamilton (1981) generated computer
simulations of the evolution of cooperation. They simulated populations of agents
with different kinds of behavior: some agents always cooperate, some never cooperate (freeloaders) and some employ various mixed strategies (sometimes cooperating and sometimes not). Each agent is given a 'fitness score' that determines
their chances of reproducing and passing their behavior on to subsequent generations. Pairs of agents were then selected at random to interact according to their
pre-defined behavior (i.e. both agents simultaneously decide whether to cooperate
or not). The key aspect of the simulation is that being on the receiving end of cooperation increases their fitness score (the receiving agent has benefitted from some
goods or service), but providing cooperation reduces their fitness score by some
small amount (the giving agent has incurred a cost either materially or in terms of
time and effort). But the gain from mutual cooperation is set to be greater than the
loss for mutual non-cooperation; essentially the gains of mutual cooperation are
greater than the sum of the parts. (This dyadic interaction is based on an economic
game called the Prisoner's Dilemma discussed later in the chapter in terms of real-
world social decision-making). The agents are then 'bred' (using a mathematical
formula) such that those with the highest fitness scores are more likely to generate
new agents who replicate that behavior. This process is repeated over multiple
generations.

What did Axelrod and Hamilton (1981) find? If one sets up a simulation consisting of some agents who always cooperate and some agents who always freeload,
then over time, one ends up with a population in which cooperation is bred out and
everyone freeloads. This occurs even if the population initially consists of lots of
cooperators and only a tiny number of freeloaders. Thus non-cooperation is an evolutionary stable solution of this model and, indeed, lots of species employ this tactic

by being solitary. So how does cooperation emerge as a stable solution in the model? Cooperation emerges as an alternative stable solution by having agents who *selectively* cooperate (i.e. cooperate sometimes but not always). Many different mathematical strategies were tried but the one that worked the best was called **tit-for-tat**. The strategy is very simple. Initially you agree to cooperate (i.e. you trust). After that, you pay back in kind: if someone cooperates with you then you cooperate, back on the next round (direct reciprocity), and if someone refuses to cooperate, then you refuse to cooperate until they reverse their strategy. In order for this to work then the agents need a memory that tells them about how other agents have previously treated them. To put a social spin on this, tit-for-tat is trusting (it always cooperates until someone else refuses to cooperate) and it is forgiving (if the other player switches back to cooperation there is no retaliation). This solution maximizes the benefits of cooperation but without running a high risk of exploitation. Vindictive strategies that strongly punish non-cooperation work less well. Since the original paper, tit-for-tat has been outperformed by a similar strategy based on win–stay and lose–shift (Nowak & Sigmund, 1993).

Sexual selection

There are other biologically based explanations of altruism aside from kin selection and reciprocal altruism. Zahavi (1975, 1995) argues that altruism arises as a form of **sexual selection**, analogous to the peacock's tail. Peahens' preference for a larger

Figure 7.2 Does altruism arise out of the same kind of evolutionary pressures as the peacock's tail? Altruism could be regarded as a display of wealth, or other positive attributes that are regarded as attractive, hence increasing the reproductive success of altruists.

plumage leads to peacocks with larger tails having a reproductive advantage. This sets up an evolutionary arms race in which tails become bigger and bigger despite not having a function (aside from reproductive success), and if anything they act as a handicap to the peacock. Turning the analogy back to altruism, being generous and helpful may serve the function of enhancing one's reputation (and mating success) by demonstrating wealth and the ability to provide. According to Zahavi (1995), this can occur without expectation of reciprocity: both the costs and benefits lie with the altruist. There is some evidence that humans (particularly females) rate altruistic traits as desirable (Barclay, 2010).

Indirect reciprocity

Direct reciprocity is based on returning help to someone who has helped you, but indirect reciprocity is based on helping people who you have never interacted with before and may never interact with again (thus preventing a favor from being directly returned). Figure 7.3 outlines two scenarios of indirect reciprocity (Nowak & Sigmund, 2005). In the first case, person A helps person B, but then person C helps person A (helping the helper). For such a strategy to be beneficial it needs to be based on selective cooperation rather than indiscriminate cooperation, just as it is in Axelrod and Hamilton's (1981) demonstrations of direct reciprocity. But how can people choose who to selectively cooperate with if they haven't interacted with them before? As with Zahavi's sexual selection theory, the concept of reputation is central to the idea of indirect reciprocity. In their mathematical simulations, Nowak and Sigmund (2005) assume that helping others increases one's reputation (termed 'image score') as well as incurring a cost (the cost to oneself in terms of foregone time, effort, etc.). Selective cooperation can occur between strangers provided their reputation is known and simulations show that this is a stable solution that can buffer against the effects of freeloading. How people judge the reputations of strangers in real-world interactions is less clear: it may depend on observing how others behave towards each other, gossip, or may be based on whether they belong to the ingroup.

In the second case of indirect reciprocity, person B is helped by person A, but person B then passes help on to person C (rather than back to person A). This appears to be a case of misplaced reciprocity that has proved harder to model mathematically but does occur in real-world behavior and experimental set-ups (Fowler & Christakis, 2010).

Altruism in humans

To many reading this, the examples of vampire bats sharing blood and insects sacrificing their fertility may sound like a far cry from human altruism. However, the evolutionary biological approaches described above are essentially mathematical models of those situations in which cooperation predominates over pure competition (based on costs and benefits accrued over generations). In that sense, the models do apply to humans

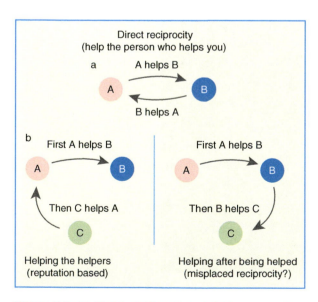

Figure 7.3 (a) Direct reciprocity and (b) two examples of indirect reciprocity that involve 'helping the helpers' (left) and 'helping as a result of being helped' (right). From Nowak (2006).

as much as to any other species. The models make no strong claims about whether the actual mechanisms in the brain are conserved across such diverse species. For instance, Nowak (2006) argues that only humans are fully capable of *indirect* reciprocity because the decision to help (or not) involves consideration of the reputations of the individuals involved, and this knowledge may be obtained via language and social norms rather than personal history of exchanges with that person. Reputation also involves thinking about what others think of us, which is by no means simple (see Chapter 6).

Where humans do appear to differ from most other species is that they can reflect on their helping behavior and they appear to have some degree of control over their decision to help or not. However, even this point is up for debate. Whilst we have a sense that our decision to help or not is a rational free choice, much of our actual decision-making process could still be based on automatic and largely unconscious biases. For example, one study measured contributions of money towards milk for tea and coffee available in a psychology department common room (Bateson, Nettle, & Roberts, 2006). The contributions were based on an honesty system where people were expected to leave money in a pot when they got a drink, but there were no sanctions against freeloading. More money was left when the pot happened to display a picture of eyes than when it displayed a picture of flowers as shown in Figure 7.4. In another study, people were likely to donate more to charity after their mannerisms were subtly imitated (van Baaren et al., 2004), but these effects depend on participants not noticing that they are being deliberately mimicked (Lakin & Chartrand, 2003). Thus, it is not true to argue that all human altruistic behavior is an outcome of conscious decision-making. Frith and Frith (2008) argue that unconscious processes in social cognition may tend to favor altruistic motives whereas conscious processes may tend to bias towards selfishness.

Whereas evolutionary theories provide a framework to understand the mechanisms by which a *population* engages in helping behavior, at the *individual* level there may be inbuilt motivational mechanisms that drive pro-social behavior. Whatever our true motivations, there is a deep-rooted sense that we help others because we care about their welfare rather than to spread our genes or receive favors in the future. As such, many theories of altruism suggest that empathy is a key component (e.g. Batson, 1991; de Waal, 2008; Piliavin, Dovidio, Gaertner, & Clark, 1981). In the model of Piliavin et al. (1981), summarized in Figure 7.5, witnessing a distressing situation leads to physiological arousal (e.g. increased heart rate) and emotional contagion (i.e. the observer also experiences high personal distress). In this model, the motivation to help is egoistic (to reduce one's own distress) rather than altruistic. However, whether or not a person helps is assumed to be determined by a cognitive appraisal (a cost–benefit analysis) rather than level of empathy per se. Helping someone may have a low cost to oneself (e.g. helping an elderly person who has fallen over) or a higher cost (e.g. intervening in a

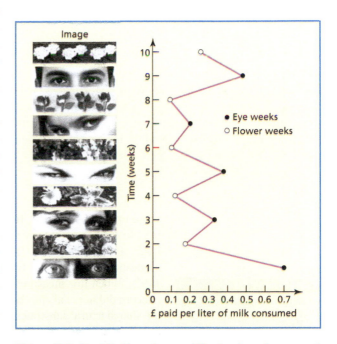

Figure 7.4 Contributions to a public good are increased when eyes are displayed on a donation tin rather than flowers. From Bateson, Nettle, & Roberts (2006). Copyright © 2006 The Royal Society. Reproduced with permission.

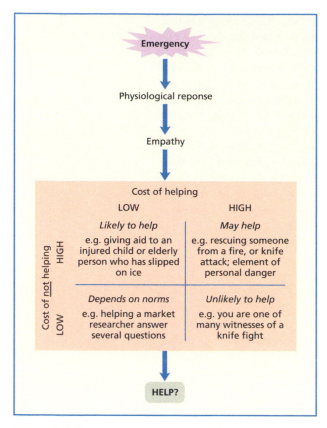

Figure 7.5 The model of altruism by Piliavin et al. (1981) includes several stages. The first stages are a physiological arousal response and an empathic (distress) response. The final stage is a cognitive appraisal in which the costs of helping and the costs of not helping are weighed. Although empathy provides the motivation for altruism, whether or not somebody acts depends on the perceived costs.

KEY TERM

Empathy-altruism model
The theory that the motivation to help is based on empathic concern for others

knife fight). The costs of not helping also need to be considered, and these may include negative emotions (e.g. guilt).

The **empathy-altruism model** of Batson (1991) extends the earlier model of Piliavin et al. (1981) but with an important difference – the motivation to help is considered to be primarily other-oriented (based on empathic concern for others) rather than self-oriented (to relieve one's own personal distress). In a series of studies, Batson and colleagues sought to distinguish between egoistic motives (self-oriented) and 'true' altruism (other-oriented) by describing in detail a particular scenario (e.g. relating to an eviction) and asking participants to state the level of help they would offer. In addition, the researchers would take self-report measures of their personal feelings (e.g. distress, fear of social disapproval) and their feelings towards the person in need of help (e.g. empathic concern). Empathic concern was found to predict helping behavior even when taking into account selfish motives such as a desire to escape aversive feelings (Batson, Duncan, Ackerman, Buckley, & Birch, 1981), social disapproval (Fultz, Batson, Fortenbach, McCarthy, & Varney, 1986), guilt (Batson et al., 1988), shame (Batson et al., 1988), and sadness (Batson et al., 1989).

The empathy-altruism model has been criticized on a number of grounds. Methodologically, the evidence is based on the assumption that people can accurately reflect on their own feelings and actions (or predict them) in imagined scenarios. Theoretically, the model hinges on the assumption that there are clear distinctions between self and other. Once that distinction becomes blurred, then so does the distinction between egoistic and altruistic motives. Batson and Shaw (1991) doubted that a merging of self and other genuinely occurs 'except perhaps in some mystical states' (p. 161). However, this claim has been challenged by research in both social psychology and, later, in social neuroscience. Cialdini, Brown, Lewis, Luce, and Neuberg (1997) used similar experimental designs to Batson but added a questionnaire measure of 'including others in the self' (Aron, Aron, & Smollan, 1992). When this measure was included it predicted helping behavior, but empathic concern did not. This has been extended by social neuroscience investigations showing shared neural substrates for processing self and other and its relevance to empathy (e.g. Decety & Chaminade, 2003). Specific examples of this include mirror systems (Rizzolatti & Craighero, 2004) but also regions such as the medial prefrontal cortex that responds to other people only when they are considered similar to one's self (e.g. Krueger et al., 2009). This research does not rule out the role of empathy in altruism – quite the opposite – but suggests that the ideas of 'true' altruism and 'true' empathy (in which 'true' = selfless) are not very meaningful.

Several social neuroscience studies have investigated altruistic decision-making. Moll et al. (2006) conducted an fMRI study of donating to charity – see Figure 7.6. Participants were shown charity names and mission statements (e.g. 'Death with dignity: Allowing euthanasia for the terminally ill') and then given donation options that either incurred a cost to the participants (e.g. You = –$2, Charity = +$5) or no cost (e.g. You = $0, Charity = +$5). They could either choose to donate or not. In a separate condition, they were given a pure reward (e.g. You = +$5, Charity = $0). Choosing to donate to a charity activated some of the same neural circuitry (including the ventral and dorsal striatum) as receiving a pure reward – even when giving incurs a cost to oneself. The authors interpret this as a 'joy of giving' effect. However, other regions did differentiate these conditions. A region in the ventromedial prefrontal cortex responded when participants decided to donate (but not to pure reward) and a region in the lateral orbitofrontal cortex was activated by decisions *not* to donate.

Harbaugh, Mayr, and Burghart (2007) used a somewhat similar paradigm to compare voluntary donations to a good cause (i.e. charity) against involuntary donations (i.e. taxation) to the same good cause, in this instance a food bank. (Note: the study design

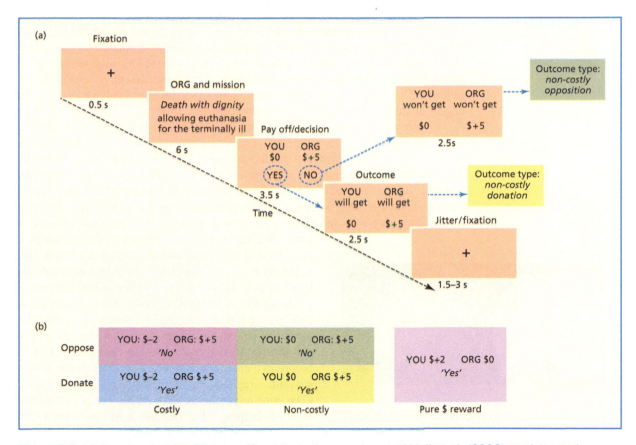

Figure 7.6 (a) An example trial, (b) a payoff matrix. In the experiment of Moll et al. (2006) participants have the option of donating to a charity or not, and that donation may come at a personal cost to them or not. In an additional condition, the participant receives a reward (without making any choice). Giving $5 to charity is similar – in neural terms – to receiving a gift of $5, which has been interpreted as a 'joy of giving' effect. From Moll et al. (2006). Copyright © 2006 National Academy of Sciences, USA. Reproduced with permission.

of Moll et al. 2006 didn't contain the taxation-like condition of forced donation). Both forms of giving were linked to activity in regions such as the striatum and insula, as was a condition in which the participant received a monetary payoff without any giving. They also examined individual differences in this neural signature: those who had a greater response in the ventral striatum to personal gains (relative to involuntary giving) tended to be less generous in their voluntary donations. This may be partly underpinned by genetic differences. Another study involving costly, voluntary giving to charity found that those with an allele leading to more rapid turnover of post-synaptic dopamine (having at least one Val allele in the COMT gene) donated twice as much to charity (Reuter, Frenzel, Walter, Markett, & Montag, 2011).

The studies above highlight the role of regions of the brain involved in reward and emotion in altruistic giving. However, other research suggests that regions of the mentalizing (or theory of mind) network are also related to the propensity towards altruism. Two of the key regions are the temporo-parietal junction (TPJ) and the medial prefrontal cortex (mPFC). Increases in gray matter density in the right TPJ, measured using VBM (voxel-based morphometry), is linked to participants willingness to pay for an altruistic act (Morishima, Schunk, Bruhin, Ruff, & Fehr, 2012). In an fMRI study participants played a game in which they won money for themselves or a charity and, in another condition, they watched someone else play the game (Tankersley, Stowe, & Huettel, 2007). The difference in activity in the right TPJ when watching other relative to self was predictive of self-reported altruistic tendencies. Rameson, Morelli, and Lieberman (2012) also measured altruism outside of the scanner, using a diary study of helping behavior, and found that the degree of activation in the medial prefrontal cortex when viewing sad images was linked to both the degree of altruism and reports of empathic feelings.

Finally some studies have examined how the neural mechanisms of altruism differ for ingroups versus outgroups. There is a general tendency to behave more altruistically towards ingroups, and this is commonly termed **parochial altruism**. The existence of parochial altruism challenges the notion of altruism as a purely 'selfless' act. In the study of Hein, Silani, Preuschoff, Batson, and Singer (2010) participants could opt to receive a painful stimulus in order to reduce pain delivered to another person who was either a member of an ingroup or outgroup (defined in terms of allegiance to a soccer team). In this scenario both the cost of helping and the cost of not helping are relatively high. The decision to help ingroup members was linked to activity in the anterior insula, a region of the brain that is involved in emotional experience and pain perception. It was also influenced by subjective reports of empathic concern, as predicted by the empathy-altruism model. In contrast the decision not to help outgroup members was linked to activity in the ventral striatum, a region normally linked to reward processing. (The same region in the study of Moll et al. (2006) was linked to helping.) The fact that it was associated with *not helping* in this study was interpreted in line with previous studies showing activity in this region associated with pleasure in other's punishment/misfortune (Singer et al., 2006). As with altruistic giving there is evidence that individual differences in altruistic punishment (and associated activity in the striatum) is linked to genetic

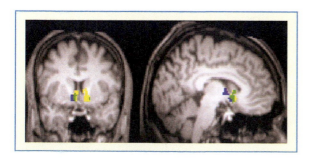

Figure 7.7 Neural response in the ventral striatum when the participant is given a monetary payoff (yellow), when making an involuntary 'taxation' to a good cause (blue), and the overlap between them (green). This is assumed to reflect a 'joy of giving' effect that occurs even when giving is mandatory. From Harbaugh et al. (2007).

differences in post-synaptic dopamine turnover (Strobel et al., 2011). Another study also reported striatal activity when punishing outgroup members, but found activity within the mentalizing network (mPFC and TPJ) when punishing an ingroup for committing the same violation (Baumgartner, Gotte, Gugler, & Fehr, 2012). This was explained in terms of participants trying to understand or justify the ingroup members' behavior.

Evaluation

The examples given in the section above, concerning charitable donations or helping people in distress, are often regarded as the clearest examples of altruism insofar as being 'selfless' acts; that is, there is no realistic chance of reciprocity. However, even in these scenarios there is a possibility for such behavior to be self-enhancing by reducing or preventing negative emotions (distress, guilt, and shame), increasing positive feelings (such as pride), and adding to one's public reputation. Although the debate about whether altruistic acts are egoistic or selfless is interesting, the dichotomy itself is rather contrived. Humans are social creatures, and our cognitive and neural apparatus cannot be easily divided into social and non-social, or self and other. Similarly, different motivations apply when considering ingroups and outgroups. The social neuroscience literature shows that charitable giving is linked to activity within the reward system, even when this represents a cost to the giver. Moreover, decisions relating to giving and individual differences in giving tendency are related to other kinds of mechanisms involving the mentalizing network and various regions of the prefrontal cortex. This dissection of a complex behavior (altruism) into different underlying mechanistic components is perhaps the main contribution of social neuroscience to the existing psychological literature. However, the astute reader will notice that the interpretation of this literature tends to rely heavily on the notion of reverse inference (e.g. we assume altruism is based on reward processing because the ventral striatum is activated; we assume altruism is based on mentalizing if the mPFC/rTPJ are activated; etc.). It is important for future research to explore these assumptions using other methods (e.g. effects of brain lesions or effects of pharmacological manipulations of reward).

GAME THEORY AND SOCIAL DECISION-MAKING

Many of the decisions we make affect the lives of others as well as ourselves, and the decisions that other people make also affect us. For example, if you decide not to help me, then how will that affect my decision to help you? Decisions to cooperate or compete typically involve complex trade-offs between costs and benefits: choosing not to cooperate may result in short-term benefits but may reduce one's reputation and the longer-term benefits of reciprocity. Social decision-making relies, to some extent, on the same mechanisms as decision-making in non-social situations. It has been claimed that there is some common currency (in neural terms) between material and social values (Ruff & Fehr, 2014). For example, lesions of the orbitofrontal cortex in mice result in them behaving impulsively – they favor small short-term gains over larger benefits accrued in the longer term (Rudebeck, Walton et al., 2006). In this instance, the reward is food. However, humans with lesions in this region

KEY TERMS

Game theory
A type of mathematical model that captures how an individual's success in making decisions is influenced by the decisions of others

Neuro-economics
Explains, using neural mechanisms, how individuals and groups make economic decisions such as assigning value to competing choices, exchange and reciprocity, and making best use of limited resources

also tend to make decisions for short-term personal gain, leading to socially non-cooperative behavior and difficulties in forming stable, long-term social relationships (e.g. Eslinger & Damasio, 1985).

Game theory is a type of mathematical model that captures how an individual's success in making decisions is influenced by the decisions of others. In *economics*, it has been typically used to find the optimal decision for an individual (i.e. the one that has the greatest benefits for the least costs), taking into account the decisions of others. This optimal decision is termed the Nash equilibrium, named after the mathematician John Nash whose life was depicted by Hollywood in the film *A Beautiful Mind*. Stated simply, David and Nick are in a Nash equilibrium if David is making the best decision he can (taking into account Nick's decision) and Nick is making the best decision he can (taking into account David's decision). In *biology*, game theory has been used to model evolution based on the concept of fitness rather than decision (e.g. Maynard Smith, 1982). Game theory has already been considered in this context to account for the population-level evolution of altruism and cooperation based on kin selection (Hamilton, 1964) and reciprocal altruism (Axelrod & Hamilton, 1981). In *psychology*, game theory is applied to 'real life' social decision-making. Real decisions do not necessarily conform to the best decision for that individual (as predicted mathematically by the Nash equilibrium). Whilst such decisions could be classed as 'irrational' or errorful they suggest, instead, that humans take into account factors other than their own personal gain when making social decisions (e.g. upholding norms of fairness even at personal cost).

These findings from psychology have been augmented by research using neuroscience methods, giving rise to the new field of **neuro-economics**. Neuro-economics

CHARACTERISTICS OF DIFFERENT GAMES COMMONLY USED IN NEURO- AND BEHAVIORAL ECONOMICS

Game	Players	Number of rounds	Principle
Prisoner's Dilemma	2	Single or multiple	Both players simultaneously decide whether to cooperate (mutual benefit) or not (either one benefits or nobody benefits).
Ultimatum Game	2	Single or multiple	One player decides how much to give; the second player decides whether to accept the offer (mutual benefit) or not (no benefits).
Public Goods Game	3+	Multiple	Each player decides how much to give to a common fund (if at all); the fund (after interest) is then distributed to all players equally.
Trust Game	2	Multiple	One player decides how much to invest (if at all); the second player then decides how much money to return to the investor (after interest).

explains, using neural mechanisms, how individuals and groups make economic decisions such as assigning value to competing choices, exchange and reciprocity, and making the best use of limited resources. For humans, the word 'value' has two meanings: it can refer to how much one is willing to give to obtain something (as in monetary value), but it is also a principle that one holds and defends (as in family values or the value of fairness). Many of the results in neuro-economics can be construed as a trade-off between these different notions of value that are tapping the same kind of neural mechanism (Ruff & Fehr, 2014).

The sections below outline various paradigms ('games') that have been used to explore social decision-making. All of the games involve a decision about whether to cooperate – leading to a mutual benefit – or not cooperate. In the case of non-cooperation, the outcome may either be neutral (maintaining or forcing a status quo), or it may involve benefitting from someone else's goodwill either passively (free-loading) or actively (exploiting). The games differ in the number of players (2 or 3+), whether it is a single round or multiple round of interactions, and in the specific outcomes of the decisions (termed payoffs).

The Prisoner's Dilemma

There is a problem with reciprocal altruism in that it is not always the optimal strategy. The optimal strategy in many situations is to receive help from you, but for me *not* to return the favor back. This is exemplified by the **Prisoner's Dilemma**. A common version of the dilemma is as follows:

Figure 7.8 Complex financial institutions are able to exist because of evolved cognitive mechanisms that permit social interactions between individuals – these include trust, norms of fairness, and responsibility.

Two suspects are arrested by the police. The police have insufficient evidence for a conviction, and, having separated both prisoners, visit each of them to offer the same deal. If one testifies for the prosecution against the other and the other remains silent, the betrayer goes free and the silent accomplice receives the full 10-year sentence. If both remain silent, both prisoners are sentenced to only six months in jail for a minor charge. If each betrays the other, each receives a five-year sentence. Each prisoner must choose to betray the other (termed 'defect') or to remain silent (termed 'cooperate'). Each one is assured that the other would not know about the betrayal before the end of the investigation. How should the prisoners act?

(Adapted from Wikipedia)

The different options can be represented by a **payoff matrix**, which lists the costs and benefits to each player based on the different independent decision options, as shown in Figure 7.9. For an *individual* the best solution (i.e. the Nash equilibrium), irrespective of what the other decides, is to defect (i.e. betray the other player). This is because defection leads to either 5 or 0 years, but cooperation leads to either 10 years or 6 months. The 'dilemma' arises because the best *collective* decision (i.e. the one that reduces the total amount of time spent in prison by both players) is for both to cooperate. The basic game can be altered by replacing the length of time in prison with other rewards or punishments, such as money. In general, participants often fail to choose the optimal solution based on maximized self-interest (i.e. always defect), suggesting they are taking into account the interests of the other player. As discussed previously, the Prisoner's Dilemma was also used in computer simulations of the evolution of cooperation by using fitness scores instead of money or prison sentences (Axelrod & Hamilton, 1981).

In the version of the Prisoner's Dilemma outlined above, the decision-making between players is effectively simultaneous. Participants have no opportunities to retaliate, reward cooperation, or develop an understanding of the other player's strategy or motives. The iterated Prisoner's Dilemma has multiple rounds with the same players, but rather than a prison scenario, a monetary scenario (or other reward/punishment) is set up. On any given trial, the greatest reward is obtained for a player who defects whilst their opponent cooperates. Thus, those who indiscriminately cooperate are easy to exploit. Those who always defect cannot be exploited, but if both players choose to defect, then they get less money than if both choose to cooperate. The dilemma here is that cooperation could lead to exploitation, but always defecting would lead to lower rewards if the other player does likewise. As discussed previously, one effective solution is tit-for-tat based on selective cooperation (cooperate initially and then copy the other player's last move).

Rilling et al. (2002) conducted an fMRI study of an iterative version of the Prisoner's Dilemma with monetary payoffs running for about 20 rounds. The behavioral results are summarized in Figure 7.10. Mutual cooperation tended to be the most

Figure 7.9 A payoff matrix for a typical Prisoner's Dilemma. If a player chooses to betray the other player they will get either 5 years or 0 years (depending on the other player's decision). If a player chooses to remain silent they will get either 10 years or 6 months in prison (depending on the other player's decision). As such, the best decision is to betray. The game can be adapted and played over repeated trials by swapping years in prison with, say, monetary rewards. In this scenario, the best decision is less obvious as a betray decision may lead to retaliation.

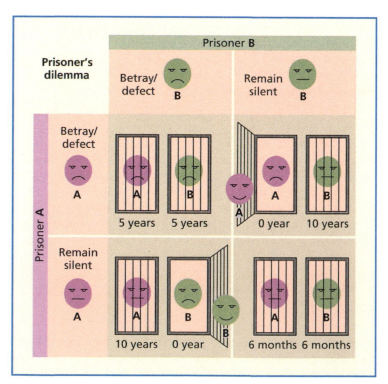

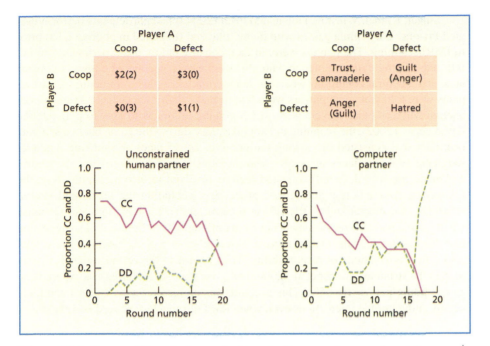

Figure 7.10 The payoff matrix for each player in the monetary version of the Prisoner's Dilemma used by Rilling et al. (2002). Player B's payoff is shown in brackets. The adjacent matrix shows the typical emotional feelings generated during the game. The bottom panels show the average pattern of results across 20 trials (CC = mutual cooperation; DD = mutual defection). When playing against another human, mutual cooperation prevails (but becomes increasingly exploitative towards the end of the game). When playing against a computer (programmed to respond using tit-for-tat) there is far less mutual cooperation and more exploitation. Adapted from Rilling et al. (2002). Copyright © 2002 Elsevier. Reproduced with permission.

common outcome (followed by mutual defection). The mutual cooperation condition had the highest activity, relative to other conditions, in regions such as the ventral striatum and the orbitofrontal/ventromedial prefrontal cortex. However, note that the mutual cooperation condition is not associated with the highest monetary reward. The highest reward condition is when the other player cooperates and you betray. Rilling et al. (2002) suggest that the reason that activity in these reward-related regions is reduced in this betrayal condition is due to negative social emotions of guilt and of fear of retaliation on future trials. They suggest that activity in these regions could have as much to do with trust and camaraderie as winning. In support of this, the striatal activity associated with the mutual cooperation condition was abolished in a separate condition in which participants played an interactive computer game (playing tit-for-tat) rather than against a human player, even though the monetary payoffs were the same. In a subsequent study, Rilling et al. (2008) studied in more detail the neural response of unreciprocated cooperation (i.e. in which a participant opts to cooperate but their partner defects). This is associated with reports of anger, irritation, and disappointment. In terms of neural correlates, unreciprocated cooperation (as opposed to mutual cooperation) engenders activity in regions such as the insula (bilaterally) and the amygdala (on the left), which are associated with emotional processing.

Singer, Kiebel, Winston, Dolan, and Frith (2004a) had participants perform iterated Prisoner's Dilemma games with many 'players' (depicted in photographs) prior to fMRI scanning. The players were, in fact, constructed by the experimenter to fall into one of three types: someone who always cheats, someone who always cooperates, and a neutral condition (where no interaction takes place). In addition, participants were told that some players were free to make their own decisions (intentional agents) whereas other players were just following orders as to how to play. Unlike Rilling et al. (2002), the scanning did not take place during the game but in a separate task later that consisted of viewing the previous set of faces and making a gender judgment. Viewing faces of people who previously cooperated (relative to neutral and/or people who defected) activated regions involved in face perception (e.g. the fusiform gyrus), emotion and reward processing (including the insula and ventral striatum), and mentalizing (TPJ). These regions tended to show a greater response when people had also previously behaved intentionally (see Figure 7.11), suggesting that these neural signals relating to trust do not reflect simple behavior or reward but also reflect attributions of responsibility. With regard to the reward-related activity (e.g. in the striatum), Singer et al. (2004a) propose that 'social fairness is experienced as rewarding per se' (p. 658). This accounts for the greater activity for intentional versus unintentional interactions even when level of monetary reward is equated.

The Ultimatum Game

In the **Ultimatum Game**, one player – the proposer – is given a pot of money and is instructed to decide what proportion of that money to give to a second player – the responder (Guth, Schmittberger, & Schwarze, 1982). The responder can then make one of two decisions. They either accept the offer, in which case it is split between both players according to the agreed amount, or they reject the offer, in which case neither player receives anything. This is illustrated in Figure 7.12. The Nash equilibrium for a non-iterative (i.e. single exchange) Ultimatum Game is for the proposer to offer the lowest amount available and for the responder to accept any offer that is given.

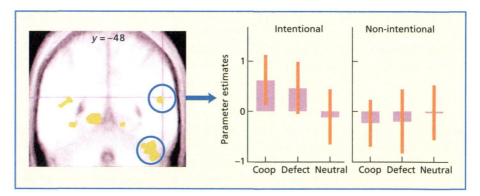

Figure 7.11 Brain activity in the right posterior STS/TPJ region (highlighted in top circle) and right fusiform (highlighted in bottom circle). These regions show greater activity when viewing faces of people previously encountered who had intentionally cooperated (i.e. free to choose their response) relative to those who were instructed how to respond. From Singer et al. (2004a). Copyright © 2004 Elsevier. Reproduced with permission.

However, responders often reject offers of money. Offers below 20% of the pot of money are reliably rejected by most participants – that is, the responder often does not make decisions in a way that maximizes their own pot of money. Why? The standard answer is that the responder considers some offers as unfair and will reject offers of money in order to punish the proposer. To put it another way, one could say that the responder values the notion of fairness and their decision reflects a trade-off between a financial value (monetary gain) and a social one (fairness). On iterative versions of the Ultimatum Game the players engage in multiple rounds in which the same players act as responder and proposer. In this instance, a rejection of a low offer has a clearer benefit to the responder because it signals to the proposer that he/she needs to increase his/her offer. In this instance, it is not necessarily the best strategy to accept any offer because doing so would invite further low offers.

Sanfey et al. (2003) conducted an fMRI study of the Ultimatum Game in which participants acted as responders to offers from many different proposers (so that there was no chance of reciprocity).

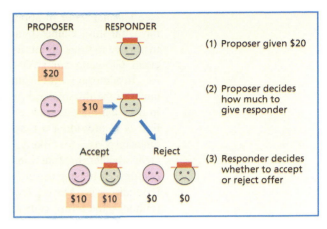

Figure 7.12 The Ultimatum Game consists of two decisions between two players. The proposer must decide how much money to offer (between $1 and $20). The responder must decide whether to accept or reject the offer. It can either be played as a one-shot game (i.e. the proposer and/or responder changes every trial, or is terminated after a single trial) or iteratively (i.e. the same proposer and responder play repeated trials).

Although not known to participants, the offers from proposers were determined in advance to be either fair (i.e. a 50/50 split of $10) or unfair (e.g. an 80/20 split of $10 in the proposer's favor). There were two other conditions: one in which participants were told that the offer came from a computer, and one in which the participant received equivalent monetary rewards without any decision-making or interaction. Participants were more likely to accept unfair offers when offered by a computer, which is consistent with the notion that rejection of unfair offers is motivated by a desire to punish. For human trials, a comparison of unfair versus fair offers showed activity in the bilateral insula, the anterior cingulate cortex, and the dorsolateral prefrontal cortex.

The insula activity was interpreted as providing an emotional signal related to unfairness (e.g. disgust or anger) that guides the behavior. Indeed, greater insula activity was found for unfair offers that were rejected relative to unfair offers that were accepted. Inducing feelings of disgust increases the rejection rates for unfair offers from a human player but not from a computer (Moretti & Di Pellegrino, 2010). Van't Wout, Kahn, Sanfey, and Aleman (2006) took a measure of emotional responsiveness (skin conductance response, SCR) during the Ultimatum Game and found a higher SCR when responders were given unfair offers relative to fair offers or relative to unfair offers from a computer. This suggests a role of emotion in the decision-making process. However, another study casts doubt on this. Civai, Corradi Dell'Acqua, Gamer, and Rumiati (2010) used a similar design but asked participants to play (as a responder) either for themselves or for a third party. The SCR to unfair offers was found only when playing for themselves, even though the behavior was equivalent in both situations. A subsequent fMRI study shows that activity in the anterior insula is equivalent in both conditions, but the 'self' condition additionally involves the medial prefrontal cortex (Corradi-Dell'Acqua, Civai, Rumiati, & Fink, 2013). This may reflect the involvement of this region in self-referential processing (see Chapter 9) or in appraising the proposer's intentions. Another fMRI study of the Ultimatum Game found an involvement of the medial prefrontal region when instructed to think about the proposer's intentions when making an unfair offer (Grecucci, Giorgetta, van't Wout, Bonini, & Sanfey, 2013).

In addition to the insula, the study of Sanfey et al. (2003) highlights the importance of the dorsolateral prefrontal cortex (DLPFC). The DLPFC has a wider role in decision-making and 'executive functions', and one general account of its function is in providing a biasing signal that makes some responses more likely and other responses less likely. However, in the context of the Ultimatum Game it is not clear whether this function lies primarily in biasing towards the more self-interested response or biasing towards responses based on fairness. To investigate this, Knoch, Pascual-Leone, Meyer, Treyer, and Fehr (2006) applied TMS over this brain region (either on the left or right) whilst the participant was acting as a responder to many different proposers (i.e. to avoid reciprocity). TMS applied over the DLPFC in the right hemisphere increases the tendency to accept unfair offers (Knoch et al., 2006) – that is, the responder behaves in a way that is more selfish / less social when this region of the brain is disrupted (see Figure 7.13). It suggests that the normal function of this region is to bias towards responses based on fairness, possibly by inhibiting a more potent urge to act in terms of self-interest. Interestingly, the participant's judgments of fairness per se were unaffected by the TMS. This suggests that this region is indeed concerned with the response itself rather than in the computation of fairness (which is possibly a function of the insula). A similar finding is found for TMS over the right DLPFC in a different game (the Trust Game, described next). TMS over the right (but not left) DLPFC reduced the amount of money returned to investors (Knoch, Schneider, Schunk, Hohmann, & Fehr, 2009). These participants behaved with more self-interest and were less concerned with building a good reputation.

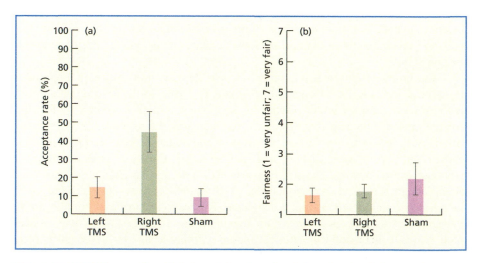

Figure 7.13 TMS over the right dorsolateral prefrontal cortex leads to people accepting more unfair offers in the Ultimatum Game (a), although their ability to decide what is fair/unfair is not compromised (b). From Knoch et al. (2006). Copyright © 2006 American Association for the Advancement of Science. Reproduced with permission.

The Public Goods Game and Trust Game

In the **Public Goods Game**, people may choose to contribute different amounts into a common pot of money, but – crucially – everybody receives the same benefits irrespective of what they put in (see Fehr & Fischbacher, 2004). The real-life analogy here is that everyone pays different amounts of taxes, but the public benefits (e.g. clean streets, national security, education) are distributed evenly. Unlike the Ultimatum Game and Prisoner's Dilemma, many players can be simultaneously involved. A typical example of such a game is that four players are initially given $20, and they have to decide how much money to donate into a public pot (between $0 and $20). The amount donated is then multiplied (e.g. doubled) and then returned equally to all four players. Let us consider three different scenarios:

1 All players contribute nothing. In this case, all players retain their initial $20 stake.
2 All players contribute everything. In this case the public pot of $80 ($20 × 4) is doubled. All players receive $40 back.
3 Three players contribute $10 and one contributes nothing. In this instance, the public pot of $30 ($10 × 3) is doubled to $60 and then divided four ways. The three players who cooperated end up with $25 in total but the person who contributed nothing gets $15 on top of their $20 stake that they held onto, making $35 in total. This is shown in Figure 7.14.

The dilemma here is that the best option, collectively, is for everyone to contribute. However, the best option for an *individual* decision maker is not to contribute (i.e. freeload) but to hope that everyone else cooperates (in this regard the Public Goods Game and Prisoner's Dilemma are variants of the same game). When such games are played, most players contribute their stake in proportion to the stake of other players – that is, if other players contribute more to the pot then these players follow the social norm and contribute more. However, as many as 30% of players will consistently freeload and contribute nothing as shown in Figure 7.15 (e.g. Fischbacher, Gachter, & Fehr, 2001).

The **Trust Game** involves an exchange between two players – an investor and a trustee (Berg, Dickhaut, & McCabe, 1995). For example, the investor may be given $20 and can choose how much of this stake to give to the trustee. The money invested is then multiplied (e.g. doubled or tripled) and the trustee decides how much to give back to the investor. This is shown in Figure 7.16. In a single exchange, the best option for the trustee (financially speaking) is to keep all the money for himself/herself. In iterative games, the trustee needs to return a fair amount of money to the investee. An unfair return would jeopardize future investment. Berg et al. (1995) report that investors send about 50% on average, although the amount sent back has a greater variance.

Both the Public Goods Game and the Trust Game depend heavily on the notion of *trust*, and in both games there is a risk of exploitation. In the Trust Game, the investor trusts the trustee to return a fair sum and not steal the money. In the Public Goods Game, an investor trusts that others will contribute to the pot. Trust may depend on social norms as to how to behave. These social norms may derive from a basic sense of right/wrong that stems from our ability to understand how others feel (e.g. 'you will not cheat me because you won't be able to cope with the guilt'). There is evidence that unconditional trust (i.e. in which players always behave fairly) is associated with greater activity in the medial prefrontal cortex than conditional

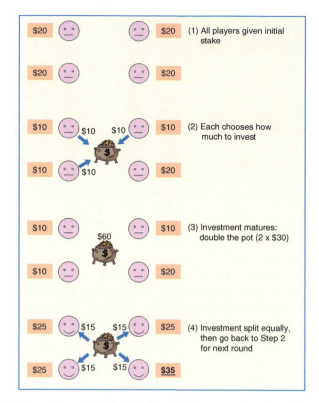

Figure 7.14 This version of the Public Goods Game can be played by two or more players and involves a voluntary donation that then increases in value (e.g. doubling or tripling) and is shared *equally* amongst all players. The equal sharing can be considered analogous to government spending that benefits all members of society (e.g. on environment, security, education, infrastructure).

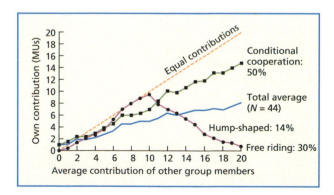

Figure 7.15 Typical results from a Public Goods Game. If other players choose to contribute more, most people decide to increase the amount that they will contribute (conformity). However, a significant proportion of people never opt to contribute anything (freeloading). From Fehr and Fischbacher (2004). Copyright © 2004 Elsevier. Reproduced with permission.

trust (Krueger et al., 2007; see also McCabe, Houser, Ryan, Smith, & Trouard, 2001b). Krueger et al. (2007) interpret this in terms of a mentalizing strategy, such that players who adopt an unconditional trust strategy are more likely to focus on the trusting intentions of others.

Trust is also likely to depend on reputation for cooperation, and merely observing cooperative individuals can evoke an extended network of brain regions involved in emotion processing and theory of mind (Singer et al., 2004a). Certain genetic differences (in monoamine oxidase A, MAOA) have been shown to make males less trusting in a Public Goods Game, but the influence of genes on behavior diminishes as participants gain more information about other players' levels of cooperativeness (Mertins, Schote, Hoffeld, Griessmair, & Meyer, 2011). Reputation may be learned via communication (e.g. gossip) as well as via direct interactions with people. Delgado, Frank, and Phelps (2005) manipulated reputation in the Trust Game by using descriptions of people's moral character, which were overall positive, neutral, or negative. In terms of their subsequent actual interactions, the three types of character were equated in terms of their level of trust (i.e. returns on investment). They found that activity in the caudate nucleus was associated with reward learning, but only for interactions with neutral characters. Thus, their reputation really did precede them (as the saying goes) and it did so by blunting the brain's mechanism for learning about rewards.

Social norms may also be enforced by punishment. In one variety of the Public Goods Game, after each round the players have an opportunity to punish any other player that they wish to (Fehr & Gachter, 2000). For example, they could spend $1 in order to remove $2 from another player's balance. In these instances, freeloaders can often expect small punishments from multiple players. Cooperation then tends to prevail on future trials. One might wonder what the motives are for punishing freeloaders. Is it motivated by social norms of fairness, or is it motivated by self-interest? Although punishment comes at a cost to the punisher, it reaps rewards in the future because more people contribute to the pot after being punished. Falk, Fehr, and Fischbacher (2005) argue that punishment is motivated by social norms rather than self-interest because players still choose to pay for punishment even on one-shot interactions in which players never re-encounter each other. Functional imaging studies show that regions relating to reward are activated when an unfair player is punished (de Quervain et al., 2004). This was interpreted as satisfaction for punishing norm violations. In contrast to this idea, Rand, Dreber, Ellingsen, Fudenberg, and Nowak (2009) argue that cooperation is more likely to be sustained via acts of selective altruism than from punishment. They set up an alternative version of the Public Goods Game in which players could give other players rewards after each round (e.g. spending $1 to give another player $2). This was found to be more beneficial than punishment in eliciting cooperation.

To examine the neural substrates of trust, King-Casas et al. (2005) conducted an iterative version of the Trust Game using two linked fMRI scanners such that the neural responses of both an investor and trustee interacting with each other could be obtained simultaneously. Behaviorally, there was strong evidence for reciprocity. If the trustee increased the returns to the investor, then the investor would increase the amount invested with the trustee in the subsequent exchange. They divided the trials into three kinds depending on whether the reciprocity was fair (i.e. like for like), malevolent (i.e. trust was betrayed), or benevolent (i.e. trust was disproportionately rewarded). An unexpected (i.e. benevolent) reward was associated with activity in the dorsal striatum (the caudate nucleus) relative to the other conditions. This fMRI signal in the trustee's brain predicted the fMRI signal in another region of the investor's brain as he/she prepared his/her response. The timing of activity in the dorsal

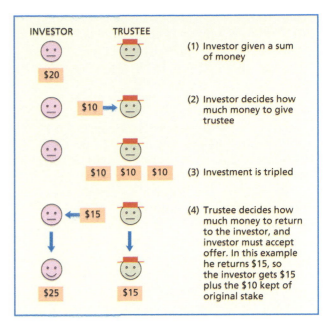

Figure 7.16 The Trust Game consists of two sequential decisions. The investor must decide how much money to invest. The sum invested then earns interest. The trustee must decide how much to return to the investor. Will the investor trust the trustee? Will the trustee keep all the money for himself/herself or return a fair amount?

Figure 7.17 The graphs show the time course of BOLD activity in a region of the striatum of the trustee playing an iterative version of the Trust Game with another human player. The purple lines show activity in the brain that precedes a decision to increase trust (i.e. repay more) and the green lines show activity in the brain that precedes a decision to decrease trust (i.e. repay less). In the early phases of the experiment (top), the decision trust-related activity (difference between green and purple) occurs when the decision of the investor is revealed. Over time (middle and bottom) the peak of this activity shifts forward in time and occurs after the investor has made a decision, but before that decision is revealed to the trustee. Thus, activity in this region shows a transition between responding to the receipt of money and responding to the expectation of money on the basis of trust. From King-Casas et al. (2005). Copyright © 2005 American Association for the Advancement of Science. Reproduced with permission.

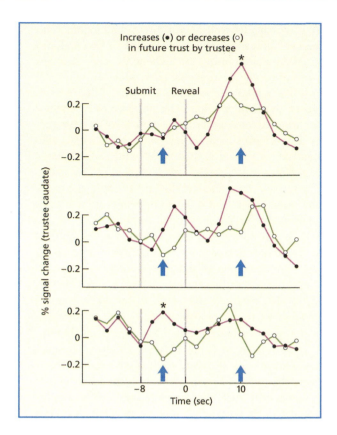

striatum changed during the course of the experiment: in early rounds it was found after the trustee learned about a favorable outcome; in later rounds it was found in the period in which the trustee was anticipating a favorable outcome (see Figure 7.17). Their explanation of this finding was that the players had learned to trust each other (i.e. by developing a reputation) and anticipated each other's responses.

CROSS-CULTURAL PERSPECTIVES ON SOCIAL NORMS OF FAIRNESS

Most of the experimental research on game theory is based on observations of university students within the Western world. Henrich et al. (2005, 2006, 2010) have attempted to extend this in 15 other cultures organized according to different principles (e.g. small-scale sedentary, semi-nomadic). The games used include the Ultimatum Game and the Public Goods Game. One of their key results was that across all cultures studied there is a tendency to not simply behave according to self-interest but to impose fairness norms in social interactions by engaging either in altruistic sharing or in punishing those who violate this norm (both of which incur a cost to the individual). The second key finding was that there is substantial variation across cultures in the extent to which these norms are enforced. In other words, all cultures are qualitatively similar but quantitatively different.

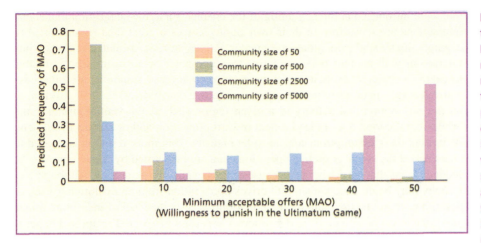

Figure 7.18 The larger the community of a human society, the more stringent are the notions of fairness. In the Ultimatum Game, people living in large communities expect a 50/50 division of the pot whereas people living in smaller communities may accept offers lower than 90/10 against them. Social norms of fairness may have a greater role when many interactions occur between strangers who are unlikely to meet again. From Henrich et al. (2010). Copyright © 2010 American Association for the Advancement of Science. Reproduced with permission.

How are these cross-cultural differences to be accounted for? Individual-level variables such as formal education level or relative wealth compared to the rest of their group did not account for the data (Henrich et al., 2005). However, some group-level variables did. The extent to which societies engage in trade with strangers and the size of the communities lived in were both positively correlated with imposition of fairness norms and punishment for unfair acts (Henrich et al., 2005, 2006, 2010), as shown in Figure 7.18 for the Ultimatum Game. This suggests that social norms for fairness may have a greater importance when interactions within a culture depend more on indirect reciprocity (with strangers who we may not meet again) than direct reciprocity (in which favors are returned to the same individual). Henrich et al. (2010) argue that the fairness norms found in large-scale societies is not just a product of our *biological* past but reflects a more recent *cultural* transition from small-group to large-group living.

Within a culture, genetic differences may account for some of the inter-individual variability for fairness preference. In the Ultimatum Game, twin studies show that responder behavior is heritable, explaining more than 40% of variation (Wallace, Cesarini, Lichtenstein, & Johannesson, 2007), and polymorphisms in a dopamine receptor gene (D4) are linked to variability in responder behavior (Zhong et al., 2010).

Evaluation

The evidence reviewed above suggests that when an individual makes a decision in a social setting (i.e. that takes into account the actual or potential decisions of others) there is a tendency not to make decisions solely on the basis of self-interest but to take into account the interests of others.

In evaluating the contribution of game theory to the social neuroscience of inter-actions, it is worthwhile returning to the four possible reasons why people cooperate that were introduced at the beginning of the chapter. First, the suggestion that social interactions are rewarding in their own right receives a great deal of support. For example, many studies demonstrate greater activity in reward-related regions when interacting with another person relative to a computer or an agent who is unable to exercise choice over their decisions, even though these situations are otherwise equivalent (e.g. in terms of monetary gain, cognitive demands). Of course, this hinges on the assumption that activity in a given region such as the ventral striatum and orbitofrontal cortex does indeed reflect reward prediction and/or reward fulfillment. At present these assumptions appear to be plausible, but more research is needed to characterize the actual cognitive operations that are performed by these regions. The second idea is that we cooperate because of the benefits of reciprocity. Although reciprocity may be the mechanism that drives cooperation from an evolutionary standpoint, it need not be the main motivational influence at the level of individual minds and brains. Are people motivated to interact because of expected reciprocal benefits or because interactions are rewarding in their own right (and the benefits of reciprocity are something of a by-product)? Evidence from game theory suggests that both punishment for non-cooperation and a desire to conform tend to increase the levels of cooperation.

Much of the evidence cited in this section has come from fMRI. Converging evidence from neuropsychology or developmental disorders (such as autism) has not been extensively brought to bear on cooperative games in the same way as it has on, say, theory of mind (but see Koenigs & Tranel, 2007; Sally & Hill, 2006). This will be an important enterprise in the future for better understanding which regions are necessary and what precise functions they perform in social and non-social contexts.

POWER AND STATUS

Power is understood simply as the capacity to influence another person or group. In other words, a person is said to have power when they are able to use certain behaviors, material objects, or information to motivate others to act in a directed fashion. Whereas power is the capacity to influence others, **status** is a perceived measure of the social worth or social rank of a person or group. Socio-economic status (SES) is the most well known metric of this. Status and power are related concepts because people of high social status are often rewarded with privileged access to resources and, hence, a capacity to influence others.

Ly, Haynes, Barter, Weinberger, and Zink (2011) asked participants to rate their own subjective SES and were then shown images of faces that, they were told, were either of higher or lower SES status than themselves. fMRI activity within the ventral striatum was linked to one's own status relative to the perceived status of others. High status people activate the ventral striatum strongly when viewing a face attributed to a higher status person, with low status people activating the same region is response to viewing someone lower in status than themselves. Thus, the neural response reflects a social comparison rather than acting as a neural signature of status.

KEY TERMS

Power
The capacity to influence another person or group

Status
A perceived measure of the social worth or social rank of a person or group

Subjectively inducing feelings of power or status has been shown to modulate various measures of empathy. One popular experimental technique is the essay priming method, in which participants are allocated randomly to one of two groups. Each group is asked to write a short description: one group (high power) is asked to write about a situation in which they have exerted power over other people, and another group (low power) is asked to write about a situation in which others have exerted power over them. Galinsky et al. (2006) found that participants primed in the high power condition were significantly worse at recognizing the feelings of others (subtle facial expressions), were more likely to take a self-oriented viewpoint, and were more likely to think that others would agree with them (when given an ambiguous statement). Similar effects are found when contrasting high and low SES, and also when participants are primed to think of themselves as high or low in status (Kraus et al., (2010). Kraus et al. (2010) suggested that the mechanism behind this biasing effect is that lower status individuals are biased to attend to social outcomes in terms of external causes (the feelings and actions of other people) whereas higher status (or empowered) individuals focus on themselves as causal agents. Other evidence has directly assessed this explanation and shown that participants primed to think of themselves as high power judge cause-and-effect as occurring closer in time when performing simple motor actions (Obhi, Swiderski, & Brubacher, 2012).

Hogeveen, Inzlicht, and Obhi (2014) used the power priming technique to examine mirroring within the motor cortex. Seeing a simple action being performed by another person (e.g. squeezing a ball), leads to activity within one's own motor cortex (a mirroring mechanism). This can be measured by externally stimulating the motor cortex with TMS. If a smaller external stimulus is needed to elicit a motor response, then it is assumed that the internal activity in the system is high (i.e. because the visual stimulus generated a mirroring response). Participants in the high power priming condition required a larger external stimulus, indicative of less mirroring and, potentially indicative of less empathy. It is only potentially indicative of less empathy because an alternative scenario is that those in high power adopt a different strategy based on mentalizing instead of mirroring (Bombari, Mast, Brosch, & Sander, 2013).

Power itself can lead to an increased tendency for dehumanization of others in that it leads to an increased tendency to treat people as objects or as a means to an end in order to achieve personal goals (Gruenfeld, Inesi, Magee, & Galinsky, 2008). People in positions of corporate power, or experimental participants who are given high power priming, tend to agree with statements such as 'this person is very useful to me' when thinking of others (Gruenfeld et al., 2008). This may also explain why high power/status is often linked to less empathy, discussed previously. Treating people as commodities in an economic game (buying/selling other people's labor) results in the kind of dehumanized neural response that typically characterizes marginalized (low-low) social groups (Harris, Lee, Capestany, & Cohen, 2014).

Evaluation

Thoughts about one's power/status affects the way that various aspects of the social brain functions. It may bias towards different kinds of information such as focusing attention either towards self-directed acts and goals versus focusing attention externally (e.g. on the wider context or the feelings and thoughts of others). This impacts upon the nature of interactions between people (e.g. the extent to which others are exploited).

SUMMARY AND KEY POINTS OF THE CHAPTER

- Altruism refers to helping others at some cost to oneself, and it may come about in a population by helping others who are likely to help us. This can be achieved via kin selection (if you are a helper then your kin may be likely to carry a helping trait) or by reciprocity based on who has helped you in the past (direct reciprocity) or who has a good reputation for helping (indirect reciprocity).
- In humans, altruism may be motivated by a concern for the welfare of others (empathy), which is facilitated by shared neural circuits for self and other.
- Experiments based on game theory suggest that humans tend not to maximize their own gain, but to act towards the collective good. Decisions take into account social norms (e.g. of fairness) and unfair decisions tend to be punished, even if the cost of punishment is personally disadvantageous.
- Social decision-making involves a network of brain regions, including those involved in emotions (e.g. the insula), cognitive control (e.g. lateral prefrontal cortex), and reward (e.g. the ventral striatum). Abstract social principles, such as trust, are starting to be understood in terms of the functioning of this network.
- Social interactions tend to be rewarding in their own right. For example, interacting with an unseen human tends to activate the brain's reward network more than interactions with a computer, even when the behavioral outcomes of those interactions are the same.

EXAMPLE ESSAY QUESTIONS

- How is human altruism similar to, and different from, that in other species?
- What is the role of empathy in altruism?
- What is 'trust' from a social neuroscience perspective?
- Do people matter more than money in human decision-making? Discuss using evidence from game theory and social neuroscience.

RECOMMENDED FURTHER READING

- Glimcher, P. W., Camerer, C., Poldrack, R. A., & Fehr, E. (2009). *Neuroeconomics: Decision Making and the Brain.* San Diego: Academic Press. An excellent, although advanced, collection of chapters.

- Rilling, J. K., & Sanfey, A. G. (2011). The neuroscience of social decision-making. *Annual Review of Psychology, 62,* 23–48. A comprehensive review of the all the main themes covered in this chapter.

- Tomasello, M. (2009). *Why We Cooperate.* Cambridge, MA: Bradford Books. An accessible book that presents evidence primarily from children and primates.

ONLINE RESOURCES

- References to key papers and readings
- Videos and websites demonstrating game theory, and how it is portrayed in popular culture
- Interviews and lectures featuring Robert Axelrod, Richard Dawkins, Alan Sanfey, Colin Camerer and others
- Recorded lecture given by textbook author, Jamie Ward
- Multiple choice questions and interactive flashcards to test your knowledge
- Downloadable glossary

CHAPTER 8

CONTENTS

Relationships

Human relationships primarily consist of friends, family, and a romantic partner. We invest a huge amount of time and effort into cultivating and maintaining these relationships. Even though most of us no longer live in close-knit communities of extended families, we find new ways of staying in touch with our inner circle. For example, undergraduates spend an average of 30 minutes a day keeping in touch with friends via Facebook alone (Pempek, Yermolayeva, & Calvert, 2009). Why? We affiliate with others because we like it, and we like it because it is good for us. It is good for us not only for the material benefits that accrue from cooperation, but also because it has protective effects on our health. Uchino, Cacioppo, and Kiecolt-Glaser (1996) reviewed 81 studies investigating the effects of social support on cardiovascular-, immune-, and endocrine-related health. Social support in terms of supportive family interactions and the presence of an intimate and confiding relationship has a protective effect against these conditions. In contrast, loneliness and lack of intimacy may have the opposite effect: for instance, being associated with greater cognitive decline in old age (Wilson et al., 2007).

At what point does a series of interactions between two people become a relationship? This is not a straightforward question to answer, but the answer almost certainly lies in their psychological state rather than the interactions themselves. For example, one might go into the same shop every day to buy groceries and, as part of this routine, chat to the shopkeeper. This does not necessarily mean that you are having a relationship with the shopkeeper! Mutually beneficial interactions (e.g. buying–selling) are not sufficient for a relationship. The appropriate test might be this: imagine you go in one day and the shopkeeper is not there because someone else is standing in. You carry out the same interactions as before (including the chatting) but there is something missing and your needs are not fully met. In this example, it would be reasonable to conclude that there was a relationship (of sorts) going on here – a **social bond**, to use another term. Social bonds are

Figure 8.1 Poets and artists have tried to capture love for millennia. Are neuroscientists going to fare any better?

KEY TERM

Social bonds
Hypothetical links between known people that induce a sense of happiness or well-being in the presence of the bonded other and a sense of wanting or longing (perhaps even distress) in their absence

characterized by a sense of happiness or well-being in the presence of the bonded other, and a sense of wanting or longing (perhaps even distress) in their absence. These bonds may develop from contact and prosocial interactions, but ultimately they need not depend upon them (e.g. the same prosocial interactions are less rewarding if the other person is substituted). In Chapter 7, we saw evidence of how the development of trust involves a shift, over time, from reward mechanisms being triggered by a positive outcome of an interaction towards reward mechanisms being triggered by the interaction itself, even before the outcome is known (King-Casas et al., 2005). This is likely to be one important neural mechanism for establishing a social bond.

An attachment is a powerful type of social bond that tends to be limited to particular kinds of relationships. Historically it was used primarily to describe infant–mother relationships (or other caregivers), but over the years it has been applied to describe romantic relationships. The emotion that is associated with being in an attachment relationship is **love**. The severing of an attachment relationship (e.g. death, divorce) can be associated with strong negative feelings of grief or distress. This chapter will start by considering love: whether it is a unitary concept and how it may be represented in the brain. This theme is continued by considering attachment more generally, focusing on infant–parent bonds and romantic bonds. Finally, the chapter will consider separation, social exclusion, and loneliness.

ALL YOU NEED IS LOVE

Love is generally not classified as a basic emotion although some researchers question this (e.g. Shaver, Morgan, & Wu, 1996). Recall that Ekman's (1992) criteria for a basic emotion include: having a universal expression; having evolved for a specific purpose; and having a distinct neurological substrate. It is true that love is not associated with a facial expression. Love is not straightforward to define and appears to be rather diverse in form: for instance, the love we have for our family feels quite different to that for a new boyfriend or girlfriend. Nevertheless, in other respects it is 'basic': it can be ascribed an evolutionary function (to form and maintain attachments, e.g. to ensure parental care), and it does appear to have its own neurobiological signature (discussed below).

KEY TERMS

Love
The emotion that is associated with being in an attachment relationship

Triangular theory of love
An explanation of love in terms of a combination of three factors (passion, intimacy, and commitment)

Types of love

Social psychologists typically define love in terms of several underlying factors. One of the most influential models along these lines is Sternberg's (1986, 1988) **triangular theory of love** shown in Figure 8.2. In this model, different types of love arise from three factors: passion (essentially sexual attraction), intimacy (feelings of warmth, closeness, and sharing), and commitment (resolve to maintain the relationship through difficulty). The factors combine to generate different kinds of love. For example: passion without commitment or intimacy is associated with infatuation (e.g. teenagers in love with pop stars); commitment and intimacy without passion are associated with companionate love; a combination of all three factors gives rise to consummate love; and so on.

Of course, the type of love experienced may change over time. Anthropological studies suggest that the passionate component of love only lasts for the first six months to three years of a relationship, which is normally enough time to ensure conception (Jankoviak & Fischer, 1992). Many relationships continue to endure for decades in which other aspects of love (e.g. intimacy) may grow. There may also be cultural differences. For example, arranged marriages begin only with a sense of commitment, but the other components (intimacy and passion) may arise through mutual respect and shared experiences (Gupta & Singh, 1982). Beyond the romantic/passionate phase of a relationship, Adams and Jones (1997) identified three factors that maintain commitment to a relationship:

1 Personal dedication: due to ongoing positive feelings towards the partner
2 Moral obligation: due to social norms or religious/cultural beliefs
3 Costs versus benefits of leaving: financial and emotional costs; availability of an alternative relationship

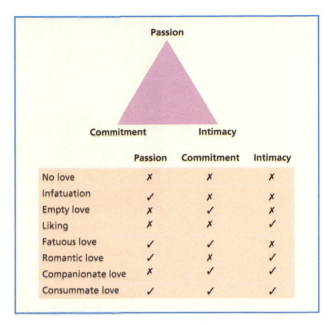

	Passion	Commitment	Intimacy
No love	X	X	X
Infatuation	✓	X	X
Empty love	X	✓	X
Liking	X	X	✓
Fatuous love	✓	✓	X
Romantic love	✓	X	✓
Companionate love	X	✓	✓
Consummate love	✓	✓	✓

Figure 8.2 The triangular theory of love aims to explain different kinds of love in terms of the presence or absence of three underlying factors: passion, commitment, and intimacy. Based on Sternberg (1988).

WHAT FACTORS DETERMINE WHOM WE FALL IN LOVE WITH?

The 'rules' of attraction have been extensively studied by psychologists. Many of these factors apply to friendships as well as romantic relationships.

- *Proximity.* This provides an opportunity to contact. The famous example of proximity is Festinger's study of friendship formation in a housing complex. People were more likely to form friends with people on the same floor relative to other floors or other buildings (Festinger, Schacter, & Back, 1950).
- *Familiarity.* People are more liked when they are seen more often, in both experimental (Jorgensen & Cervone, 1978) and naturalistic settings (Moreland & Beach, 1992).
- *Physical attractiveness.* This was considered previously in Chapter 5. However, one factor not previously considered is attractiveness of a potential partner relative to one's own perceived attractiveness. The **matching hypothesis** states that people are more likely to form long-standing relationships with those who are as equally physically attractive as they are (Walster, Aronson, Abrahams, & Rottman, 1966). In this study, male and female students were randomly paired together at a college dance (although they believed they were being paired on

KEY TERM

Matching hypothesis States that people are more likely to form long-standing relationships with those who are as equally physically attractive as they are personality). Women judged to be attractive (by independent raters) were more likely to be asked on a date, but it was similarity in relative attractiveness between male and female partners that predicted whether they would still be dating after 6 months.

- *Similarity of attitudes.* Newcomb (1961) asked students to fill in a questionnaire about their attitudes and values before moving into shared accommodation. Over the course of the semester they then noted down people who they were attracted too. Initially attraction was related to proximity but then shifted to similarity of attitudes.
- *Reciprocal liking.* If you are told that someone likes you then you are more likely to like them (Dittes & Kelley, 1956).

Neuroscience and love

Research in social neuroscience has not been based on social psychology models such as Sternberg's, but it has attempted to look at different kinds of love (for a critique see Fusar-Poli & Broome, 2007). Bartels and Zeki (2000) compared brain activity, using fMRI, in response to viewing one's romantic partner relative to a long-term friend. In a subsequent study, Bartels and Zeki (2004) contrasted the viewing of images of one's own child with another acquainted child (i.e. maternal love). In both cases, there was activity in a number of regions that the authors claim are rich in the neuro-hormones vasopressin and oxytocin (of which more later) and 'reward centers' such as the striatum. (There were also differences between maternal and romantic love.) Intriguingly, there were a number of deactivations, including the amygdala and regions that have traditionally been linked to 'mentalizing', such as temporal poles, temporo-parietal junction, and medial prefrontal cortex. A deactivation, in this context, means that the activity for the familiar acquaintance was greater than for the loved one. They argue that love involves a pull–push mechanism in the brain by which regions involved in critical social assessment (e.g. of others' intentions) are deactivated whilst regions involved in reward and attachment are activated. This may provide a basis for the unconditional nature of love that distinguishes it from context-sensitive emotional responses.

As well as being rewarding, the mere presence of a loved one may act as a buffer against stress or pain. Showing female participants pictures of a male loved one whilst the participant received a painful (but non-harmful) stimulus reduced the reported pain intensity (Master et al., 2009) and this analgesia correlates with activity in a number of regions, including the reward-related nucleus accumbens (Younger, Aron, Parke, Chatterjee, & Mackey, 2010). Similar behavioral results are found when the participant held their partner's hand (but could not see him) relative to holding the hand of a stranger or squeezing a ball (Master et al., 2009) as shown in Figure 8.3. When viewing a negative film (e.g. of a mother looking for

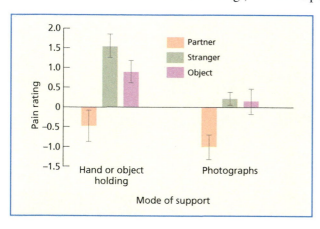

Figure 8.3 In this study, the thermal pain threshold of a group of women was determined. They were then given a number of painful stimuli during which they held their partner's hand, held a male stranger's hand, squeezed a ball, or saw images of their partner, a stranger, or anobject on a computer screen. Their pain threshold was increased in the partner conditions. From Master et al. (2009). Copyright © 2009 Association for Psychological Science. Reprinted by permission of SAGE Publications.

a lost child) then people who are in the early stages of love show less autonomic reactivity (irregular breathing) than single people, suggesting that love buffers against stress (Schneiderman, Zilberstein-Kra, Leckman, & Feldman, 2011).

The term 'falling in love' implies a lack of control and some researchers have likened the feeling to those of certain clinical disorders (e.g. Marazziti, 2009; Tallis, 2005). For example, mood elation with excess energy and sleeplessness has been likened to the manic phase of bipolar depression, and the craving and intrusive thoughts about the loved one have been likened to obsessive–compulsive disorder (OCD; Marazziti, 2009). Of course this need not reflect clinical disturbance but, instead, may be an important biological mechanism that enables us to put aside inhibitions and abandon our comfort zones (e.g. neglecting friends and family to spend time with a new partner).

There is evidence for hormonal and neurochemical changes in people who are in the early 'passionate' phase of a relationship rather than the later 'companionate' stages. Testosterone levels are lower in males and higher in females during the early stages of a romantic relationship (see Figure 8.4, top) but return to normal 12–18 months later, even when they are maintaining the same relationship (Marazziti & Canale, 2004). Plasma oxytocin levels are high in initial stages of romantic love and correlate both with signs of affection and also relationship-related worries (Schneiderman, Zagoory-Sharon, Leckman, & Feldman, 2012). The concentration of a particular serotonin transporter in the blood is lower in participants in the passionate phase of love relative to controls (either single or in longer-term relationships), and comparable to patients with OCD (Marazziti, Akiskal, Rossi, & Cassano, 1999). This is shown in Figure 8.4. The levels increase again when the same participants are tested 12–18 months later. Serotonin levels in the brain are not easy to measure in vivo in humans but the density of the serotonin transporter on blood platelets has been shown to be linked to the density of this transporter on neurons in the brain (Rausch et al., 2005). Symptoms of OCD are reduced by drugs that increase the availability of serotonin (selective serotonin reuptake inhibitors, SSRIs, such as fluoxetine/Prozac) (Bandelow et al., 2008).

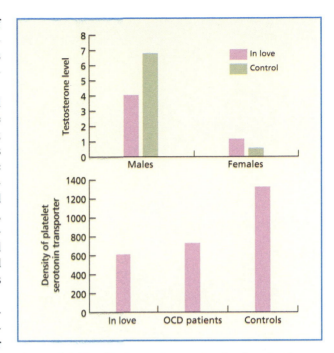

Figure 8.4 Top: The testosterone levels of males in love are lower than male controls, and those of females in love are higher than female controls. Data from Marazziti and Canale (2004). Bottom: The level of a serotonin transporter (measured in terms of a binding parameter) is lower in people in love relative to controls, and is comparable to patients with OCD. Data from Marazziti et al. (1999).

Evaluation

Love is an emotion that is elicited in the presence of (or by the thought of) another person with whom there is an attachment. There appear to be different facets to love (for instance, as articulated by Sternberg's theory) but this need not imply that love is culturally created or unworkable within a scientific framework. There is evidence for a common (and partly different) circuit for different kinds of love (e.g. maternal vs. romantic), and for major biological changes associated with the act of falling in love.

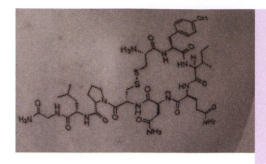

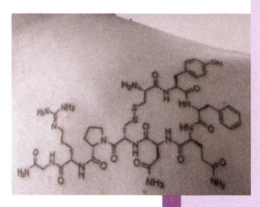

Figure 8.5 The chemical structures of oxytocin (top) and vasopressin (bottom) are similar. Some couples even opt to have them as tattoos, with the female partner having oxytocin and the male partner having vasopressin!

KEY TERMS

Oxytocin
A hormone involved in attachment formation that acts on certain receptors in the brain

Vasopressin
A similar attachment-forming hormone to oxytocin for which males have greater sensitivity than females

WHAT ARE OXYTOCIN AND VASOPRESSIN?

Both oxytocin and vasopressin are peptide hormones that are unique to mammals but have similarities with other molecules found in other species. They have a similar chemical structure (see Figure 8.5) and both are produced by the hypothalamus (supraoptic and paraventricular nuclei). They are stored in the pituitary gland from where they are released into the bloodstream. Neurons that project from the pituitary release the hormone directly into the nervous system where they may bind to specialized receptors on neurons (the oxytocin receptor, OXTR, and various vasopressin receptors). The oxytocin and vasopressin receptors are concentrated in certain regions of the brain including the amygdala and nucleus accumbens. Males are more sensitive to the effects of vasopressin than females (de Vries, 2008). Although it is generally agreed that oxytocin and vasopressin are important for relationship formation, there is doubt as to whether these hormones *specifically* serve this function or whether they have a wider and more general role in social cognition such as anxiety reduction (Churchland & Winkielman, 2012). Outside of the nervous system, oxytocin interacts with other cells in the body and is involved in uterine contraction (during labor), milk production (during breastfeeding), and both the male and female orgasm (Carter, 1998).

ATTACHMENT

An **attachment** is a long-enduring, emotionally meaningful tie to a particular individual (Schaffer, 1996). Attachments in young children provide them with comfort and security and are characterized, behaviorally, in terms of proximity seeking (i.e. maintaining close physical proximity) and separation distress when proximity is not maintained. Attachment is found in all animal species in which the young are initially in need of care and protection, as is most vividly illustrated by the case of **imprinting**. This phenomenon is found in many birds, in which newly hatched chicks follow their presumptive parent around, as shown in Figure 8.6. Research reveals that there is a narrow window of opportunity, between 15 hours and 3 days, for a gosling to imprint and the movement of a stimulus is deemed critical (Tinbergen, 1951). Once imprinted, the bird is virtually unable to learn to follow a new foster parent (although see Bolhuis, 1990).

In humans, the system is rather more flexible in terms of whom attachment relationships are formed with, the number formed, and when they may occur. Nevertheless, there are strong trends. The mother is normally the first (and strongest) attachment relationship, and this tends to emerge at around 7–8 months of age, as assessed in naturalistic settings (e.g. distress when an infant is left with a babysitter;

Figure 8.6 Research into imprinting was studied extensively by the biologist Konrad Lorenz, shown here followed by a line of goslings who, upon hatching, have apparently mistaken him for their mother. Copyright © Science Photo Library.

Schaffer & Emerson, 1964), in laboratory environments (e.g. Kotelchuck, Zelazo, Akgan, & Spelke, 1975), and across a wide range of cultures (Van Ijzendoorn & Kroonenberg, 1988). Attachment is assessed objectively in terms of some measure of separation distress (e.g. crying) or proximity-seeking behavior (e.g. attempting to follow or reach out to the adult). Shortly after a specific attachment is formed, infants tend to experience a fear of strangers, suggesting that this is a consequence of having an attachment relationship (Schaffer, 1996).

Central to biologically based accounts of attachment is the notion that interactions with attachment figures (e.g. mother) activate the reward-based mechanisms of the brain. In the behaviorist tradition, the reward was considered to be a learned association between a stimulus (e.g. mother) and the set of rewards that she provides (e.g. food, warmth). This notion was challenged, most notably by Harlow (1958) in 'The Nature of Love', by showing that newborn rhesus monkeys were selective about the objects they would seek comfort from (see Figure 8.7). For example, if given a choice between an artificial wire 'monkey' that provides milk (from a bottle) and an artificial cuddly 'monkey' (who provides no milk), they would spend more time with the latter, going over to the wire monkey only when hungry. Rather than thinking of attachment as a learned response, others such as Bowlby (1969) and Harlow (1958) argued that it is an innate response (or a primary reinforcer) that has been shaped by the evolutionary need for care and protection. In addition to triggering reward-based mechanisms, the brain's other significant reaction to attachment formation is a reduction in its stress response (e.g. Spangler & Grossmann, 1993). This may contribute to the sense of security associated with attachment relationships and also be linked to the wider benefits of secure relationships on mental and physical health. Figure 8.8 shows a simple contemporary model of secure attachment formation that emphasizes factors relating both to the child and the parent, and how these behaviors may be passed down across generations (Rilling & Young, 2014). The details of these mechanisms will be unpacked throughout the sections below.

Figure 8.7 Harlow performed a variety of studies on attachment in primates, for example, showing that infant monkeys prefer a cuddly artificial 'mother' over a wire 'mother' even if the latter provides milk. This evidence ran against the behaviorist tradition of the time, which believed that maternal love was a learned response to having needs met. Reproduced with kind permission of Harlow Primate Laboratory, University of Wisconsin.

PERCENTAGE OF INFANTS FORMING ATTACHMENTS TO PARTICULAR INDIVIDUALS

Individual	At onset of attachment	At 18 months
Mother	95	81
Father	30	75
Grandparent	11	45
Other relative	8	44
Sibling	2	24
Other child	3	14

Bowlby (1969) originally suggested that humans tend to form a single attachment relationship (to their mother), but empirical evidence suggests otherwise. By the age of 18 months, most infants have more than one attachment relationship. This may act as a 'biological insurance policy' in the event of loss of the attachment figure. Adapted from Schaffer and Emerson (1964)

KEY TERM

Medial Pre-Optic Area (MPOA)
A region of the hypothalamus that, in many mammalian species, triggers parental behaviors by responding to pregnancy-related hormonal changes

Many species, such as rats, undergo a transformative response to infants during pregnancy such that their behavior switches from attacking pups to a craving-like state in which the animal is willing to work (e.g. on a lever) to be given a pup to care for. The transition from aversion to attraction is mediated by the detection of pregnancy-related hormonal changes in a region called the **medial pre-optic area (MPOA)** of the hypothalamus which may exert its effect by decreasing amygdala responsiveness and enhancing the dopaminergic limbic pathways (Stolzenberg & Numan, 2011).

Considering human parental brain function, as noted previously, when mothers view images of their infant's face then this activates, in fMRI, a reward-related circuitry that runs from the ventral tegmental area (VTA), through the ventral striatum (including nucleus accumbens) into the medial orbitofrontal cortex (Bartels & Zeki, 2004; Leibenluft, Gobbini, Harrison, & Haxby, 2004). Although this resembles the neural response of other species, the infant-reward response in humans is almost certainly far less dependent on pregnancy-related hormone changes and the role of the MPOA in human parental behavior is presently unclear. For instance, reward-related brain responses to the sight of a human infant are found both in fathers (Mascaro, Hackett, & Rilling, 2014) and in women who have never been pregnant (Glocker et al., 2009). The latter brain response depends on their perceived 'cuteness' (accentuated babyish features). In fathers, individual differences in parenting level and style are related to circulating levels of testosterone (high testosterone linked to less involvement; e.g. Mascaro, Hackett, & Rilling, 2013).

Crying is an aversive stimulus that is linked to activity in regions such as the anterior insula and amygdala but, at least in parents, this may be down-regulated by prefrontal cortex and linked to activity in reward centers (e.g. nucleus accumbens)

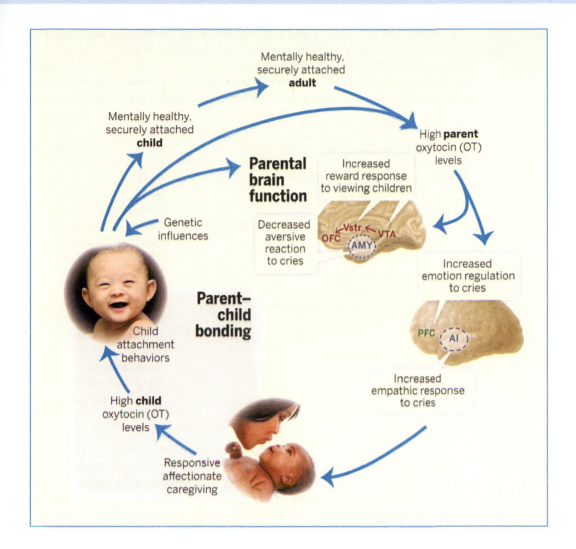

Figure 8.8 Attachment style tends to be transmitted across generations (shown by the cycling loop). This figure emphasizes key hormonal and brain responses linked to secure attachment. Oxytocin is involved in both infant-to-parent and parent-to-infant attachment. The brains of parents respond positively to the sight of infants (especially their own). They are able to emotionally regulate the sound of crying, which is perhaps a form of reappraisal (turning a negative emotion into a more positive one), enabling empathic responding (Rilling & Young, 2014).

when listening to their own infant cry (e.g. Laurent & Ablow, 2012; Lorberbaum et al., 2002). These emotion regulation mechanisms promote empathic- rather than frustration-based responses to crying.

Individual differences in parent–infant attachment style

The presence of a stable caregiver (or caregivers) during a sensitive period in the first year may be necessary for an attachment to be formed, but the quality of that attachment may depend on the quality of the interactions. The classic laboratory paradigm

KEY TERMS

Strange Situation Test
A measure of attachment during infancy in which the infant experiences separations and reunions with a stranger and with an attachment figure

Securely attached
A type of attachment characterized by proximity seeking with the attachment figure and distress at separation

Insecure/avoidant
A type of attachment characterized by ambivalence towards the attachment figure and avoidance of contact at reunion

Insecure/anxious
A type of attachment characterized by extreme distress at separation

for assessing different types of attachment in infancy is the **Strange Situation Test** developed by Ainsworth and colleagues (Ainsworth, Blehar, Waters, & Wahl, 1978). In this test, the infant is placed through a series of seven episodes involving separation from the parent, interactions with a stranger, and finally reunion with the mother. Each episode lasts around three minutes but can be curtailed if the infant is distressed.

AINSWORTH'S STRANGE SITUATION TEST

Episode	People present	Events
1	Mother, infant	Infant explores; mother watches/plays
2	Mother, infant, stranger	Stranger enters and is silent at first, then talks to mother, then interacts with infant
3	Infant, stranger	Mother leaves; stranger interacts with infant
4	Mother, infant	Mother returns and settles infant; stranger departs
5	Infant	Mother leaves; infant is alone
6	Infant, stranger	Stranger enters and interacts with infant
7	Mother, infant	Mother returns and settles infant; stranger departs

The infant's behavior is scored according to how it deals with stress and the way that it responds to the mother, particularly during the reunion (episodes 4 and 7). Using this paradigm, Ainsworth et al. (1978) identified three types of attachment style. **Securely attached** infants get moderately upset by the departure of the mother and greet her positively in reunion. Insecurely attached infants fell into two types: **insecure/avoidant** types who avoid contact with the mother, especially at reunion; and **insecure/anxious** types who show high levels of stress on separation and are difficult to console at reunion. Most infants are classed as securely attached, and this finding holds cross-culturally, although the balance between the two insecure categories varies more across cultures (Van Ijzendoorn & Kroonenberg, 1988).

Ainsworth et al. (1978) attributed the different attachment styles to the different quality of interactions between mother and infant, and to mothering styles in particular. Mothers of securely attached infants were assumed to be able to pick up on the infant's signals and respond to them appropriately and consistently. Mothers of insecurely attached infants tended to be either neglectful in their care (which was linked to insecure/avoidant infants) or inconsistent in their care (which was linked to insecure/anxious infants). Others have noted that the infant's temperament may be important for eliciting particular kinds of attachment (e.g. Belsky & Ravine, 1987).

The classification has good reliability over time and appears to have external validity in that it predicts the quality of other kinds of social behavior. Grossman (1988) found that the classification at 1 year was 87% successful at predicting the classification (based on teacher/parent ratings) at 6 years, and that securely attached 6-year-olds had better concentration and independent playing. Moreover, 4-year-olds who had previously been classified as securely attached during infancy are rated as more popular by their peers and they develop more secure friendships, particularly with friends who were also previously classified as securely attached (Park & Waters, 1989).

There are social neuroscience investigations of different attachment styles in human mothers and infants (for an overview see Swain et al., 2014). Swingler, Sweet, and Carver (2010) recorded ERPs from infants viewing either their mother's face or a stranger's face and related this to signs of distress and visual search for the mother in the Strange Situation Test. Greater distress and visual search (i.e. signs of secure attachment) during the separation phase of the test were linked to greater amplitudes and longer latencies of ERP components elicited by the mother's face for an attention-related component (termed Nc) and a P400 component that has been linked to face recognition (and may be an infant analogue of the N170 in adults). Fraedrich, Lakatos, and Spangler (2010) performed a complementary study by examining the ERP response of mothers to images of infants' faces (in various emotional expressions) and related this to attachment style (note that they did not compare images of their own infant with a stranger infant). Mothers with insecure attachments were found to have a greater amplitude to the N170 component (which is related to perceptual processing of faces), but mothers with secure attachments showed a greater P300 (which is related to more semantic aspects of face processing). ERP studies reveal that attachment style manifests itself in face perception and attention-related mechanisms, but ERP is not well suited to investigate sub-cortical contributions.

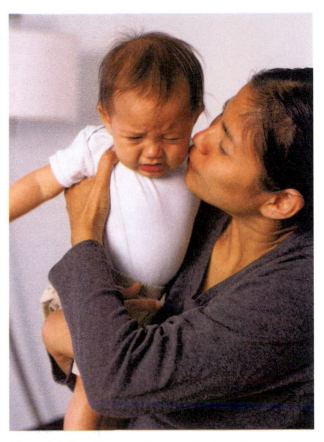

Figure 8.9 The Strange Situation Test has become the standard measure of infant attachment. The infant's behavior is scored according to how it deals with stress and the way it responds to the mother, particularly during the reunion episodes (4 and 7).

One fMRI study measured attachment security by examining the degree of synchrony (i.e. responsiveness) in mother–infant interactions and found that attachment security predicted neural activity in the nucleus accumbens when reading infant-related vignettes, whereas insecure interaction styles tended to be linked to amygdala activity (Atzil, Hendler, & Feldman, 2011). Viewing images of their own infant (relative to another infant) is linked to greater activity in the nucleus accumbens of mothers, and this activity is related both to secure attachments and to levels of oxytocin in the blood (Strathearn, Fonagy, Amico, & Montague, 2009). Mothers listening to crying activate the amygdala more for insecure attachments and activate prefrontal regions more for avoidant attachments that may be related to higher emotion regulation (Laurent & Ablow, 2012). Maternal depression, which is a large risk factor for insecure attachments, is linked to less activity within the nucleus accumbens and orbitofrontal cortex when listening to their infant cry (Laurent & Ablow, 2012).

Consistent with the model in Figure 8.8 there is some evidence to suggest that attachment styles are partially conserved from infancy. A 20-year longitudinal study that compared Strange Situation Test performance at 12 months with adult attachment style shows a 72% correspondence over this period between secure and insecure attachment styles (Waters, Merrick, Treboux, Crowell, & Albersheim, 2000). Negative life events (e.g. parental divorce, life-threatening illness) were related to changes in attachment style. A 22-year longitudinal study examined the effects of infant attachment style on adult emotion regulation using fMRI (Moutsiana et al., 2014). Participants who previously had insecure attachments as infants required more prefrontal activation (lateral and medial prefrontal cortex and anterior cingulate) to reappraise positive images. Those with previously secure attachments showed greater connectivity between these regions and the nucleus accumbens. Adult women who self-report insecure adult attachment styles to their own family have greater responsiveness in the amygdala to the sound of an infant crying, and this occurs even though they do not have children of their own (Riem, Bakermans-Kranenburg, van Ijzendoorn, Out, & Rombouts, 2012). This suggests that early attachment style has widespread effects on the social and emotional brain lasting into adulthood.

Animal models, primarily from the rat, have also explored how early maternal behavior effects later adult behavior. Maternal rats vary in the amount of caregiving (such as licking and grooming) behavior that they give (Meaney, 2001). Daughters of rats from low-caregiving mothers have an elevated response to stress (behaviorally and physiologically) and, as a result, show more promiscuous sexual behavior when mature (Cameron, 2011). This inter-generational transfer of behavior (low-care mothers, high-stressed offspring) is primarily environmentally, rather than genetically, mediated as shown from cross-fostering studies (Francis, Diorio, Liu, & Meaney, 1999). In these studies, a small number of pups from high-care mothers are given to low-care mothers for rearing, and vice versa. Importantly, the behavior of the mother doesn't change as a result of being given a new pup (i.e. she treats the biological and fostered pups the same). However, the behavior of the pups depends on the behavior of their foster mother rather than biological mother. Although the transfer across generations doesn't depend on genetic structure it is dependent on epigenetics (Weaver et al., 2004), i.e. alterations in the way that genes are expressed.

Attachment in adult relationships

It has been argued that infant attachment style carries over to adult romantic relationships (e.g. Hazan & Shaver, 1987; Simpson, 1990). Consistent with the notion that early attachment relationships provide a model for all future long-term relationships (e.g. romantic partners) is neurobiological evidence for shared substrates between parent–infant bonding and partner–partner bonding in sexual relationships (e.g. Gonzalez, Atkinson, & Fleming, 2009; Wommack, Liu, & Zuoxin, 2009). In adults, attachment styles tend to be assessed using questionnaires that ask about attitudes towards relationships. For example, the questionnaire of Hazan and Shaver (1987) was designed to classify adults into the same three categories as those proposed by Ainsworth et al. (1978). It contains statements such as 'I find it difficult to trust others completely' (insecure/avoidant) and 'I often worry that my partner doesn't really love me' (insecure/anxious). Those with an anxious adult attachment style may worry about rejection and tend to show heightened vigilance to cues of support or criticism, whereas those with an avoidant attachment style may report less need for close relationships and may tend to be distrustful of support/criticism from others. More recent questionnaire-based measures of adult attachment do not attempt to divide into categories but instead describe them according to two dimensions: avoidance (high to low) and anxiety (high to low) (e.g. Fraley, Waller, & Brennan, 2000). In this scheme, a secure attachment style would be associated with low scores on the avoidance dimension and low scores on the anxiety dimension.

Studies of adult romantic attachment styles reveal a detailed picture of the underlying neural systems (for a review see Vrticka & Vuilleumier, 2012). Gillath, Bunge, Shaver, Wendelken, and Mikulincer (2005) presented women with various relationship scenarios (e.g. break-up) during fMRI and correlated brain activity with individual attachment styles measured in terms of level of relationship anxiety and avoidance. Low levels of attachment anxiety were associated with high levels of activity in orbitofrontal cortex (given negative scenarios), whereas high levels of attachment avoidance were linked with high activity in lateral prefrontal regions (for negative and positive scenarios). This suggests different styles of control mechanism in the maintenance of relationships.

Other studies provide evidence of the role of emotion and reward processing regions in attachment styles. Lemche et al. (2006) presented participants with brief unconscious statements relating to negative attachment outcomes (e.g. 'my Mum rejects me') and compared these to a neutral condition. After these statements were presented, a probe sentence was presented that the participants had to respond to (e.g. 'other people like me', 'I trust my friends'). The degree of interference from the unconscious primes has been shown to relate to attachment insecurity and, in this study, this measure was related to activity in the amygdala as well as a skin conductance response during the task. The results were interpreted as providing evidence of the role of the amygdala in attachment *insecurity* rather than attachment per se. Priming people to feel secure in their social relationships reduces the degree of amygdala activation, in fMRI, when viewing fearful faces or threatening words (Norman, Lawrence, Iles, Benattayallah, & Karl, 2015), and amygdala volume, measured with VBM, correlates with habitual relationship anxiety (Redlich et al., 2015).

Vritcka, Anderson, Grandjean, Sander, and Vuilleumier (2008) investigated adult attachment style measured in terms of avoidance and anxiety dimensions. They used a difficult perceptual task in which they received feedback on their own performance in terms of a happy or angry facial expression. Activation of striatum and ventral tegmental area was enhanced to positive feedback signaled by a smiling face, but this was reduced in participants with avoidant attachment (see Figure 8.10). This indicates relative impassiveness to social reward in the avoidant group. Conversely, a left amygdala response was evoked by angry faces associated with negative feedback. This correlated positively with anxious attachment, suggesting an increased sensitivity to social punishment. This study differs from the others in that attachment style is entirely incidental to the task. The results suggest that attachment style affects how different individuals respond to, and interpret, social rewards (e.g. smiles) and punishments (e.g. anger). Other research suggests that those with an avoidant attachment style show less efficient emotional regulation of social stimuli (e.g. in prefrontal cortex), whereas an anxious attachment style is characterized by increased activity in the amygdala in the absence of any emotion regulation (Vrticka, Bondolfi, Sander, & Vuilleumier, 2012). These studies suggest that attachment styles have consequences even when socially interacting with strangers. In effect, attachment history may act as a schema for predicting and interpreting social interactions whether intimate or ephemeral (Vrticka & Vuilleumier, 2012).

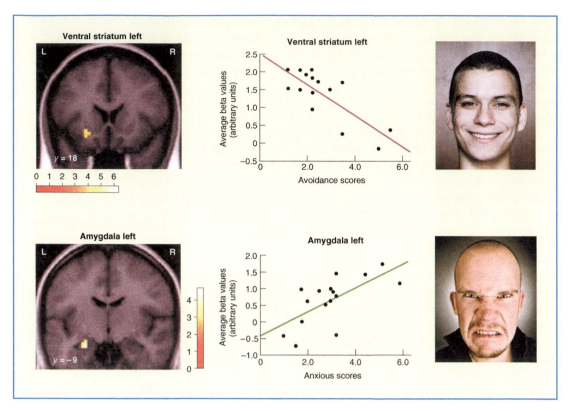

Figure 8.10 Self-reported relationship avoidance scores correlate negatively with activity in the ventral striatum when participants get feedback about a win with a smiling face (i.e. those who have an avoidant approach to relationships may find smiles less rewarding). Self-reported relationship anxiety scores are correlated with activity in the amygdala when participants get feedback about a loss with an angry face (i.e. those who are anxious about the quality of attachment may react with fear to social disapproval signaled by anger). Note that low avoidance and low anxiety scores are indicative of secure attachment. From Vritcka et al. (2008). Copyright © 2008 Vritcka et al. Photos on right are not part of the original figure.

Role of oxytocin and vasopressin in attachment

Two hormones that have been extensively studied in attachment formation are **oxytocin** and **vasopressin**. Virgin rats given an injection of oxytocin display maternal behavior (Pedersen, Ascher, Monroe, & Prange, 1982) and maternal rats given a chemical lesion that disrupts its action show reduced maternal behavior after birth (i.e. during attachment formation) but not when given several days after birth (i.e. during attachment maintenance) (Insel & Harbaugh, 1989).

Although rats do show early caregiving behavior they do not form longer-term attachments. This may reflect transient changes in the brain, such as the production of a significant number of oxytocin receptors during pregnancy (Champagne, Diorio, Sharma, & Meaney, 2001). Other species may adapt this mechanism by ensuring that the oxytocin-responsive system is more permanently available. Much of the evidence in this area has come from animal models of two mammal species: Prairie voles and montane voles (Carter et al., 1995). These two animals are similar in many respects but differ from each other in one crucial respect. **Prairie voles** form a long-enduring attachment relationship (termed **pair bonding** in this literature) with an opposite sex

Figure 8.11 The sociable and faithful prairie vole (left) and its 'love them and leave them' cousin, the montane vole (right).

partner whereas **montane voles** are promiscuous. Prairie voles share a variety of responsibilities (nest building, nest guarding, some aspects of parental care); they remain together not just in the breeding season, and on the death of a partner they rarely find a replacement (Carter et al., 1995). Montane voles do not engage in joint parental care and show little evidence of forming attachment relationships.

Pair bonding is assessed using a partner preference paradigm in which a vole is free to choose between two compartments: one containing a familiar vole and one an unfamiliar vole, both of the opposite sex. Prairie voles reliably affiliate with a familiar vole that they have lived with for 24 hours and mated with, and they avoid unfamiliar voles, whereas Montane voles show no such preference (Williams, Catania, & Carter, 1992).

The monogamous prairie vole and the promiscuous montane vole have similar oxytocin cell populations but differ substantially in the distribution of oxytocin *receptor types* in regions such as the nucleus accumbens, amygdala, and certain hypothalamic nuclei (Insel & Shapiro, 1992). The vasopressin molecule is structurally similar to oxytocin and has grossly similar behavioral effects insofar as it promotes attachment. For example, disruption of both oxytocin and vasopressin affects the ability of Prairie voles to form partner preferences under the normal conditions of 24 hours of mating contact. However, it differs from oxytocin's action in a number of ways. In prairie voles, males are more responsive to vasopressin and females to oxytocin (Cho, DeVries, Williams, & Carter, 1999). This may enhance gender-specific behavior linked to social relationships, such as the male tendency to act aggressively towards an intruder, which is associated with vasopressin (Gobrogge, Liu, Jia, & Wang, 2007). Whilst oxytocin tends to reduce anxiety, vasopressin has the opposite effect (Huber, Veinante, & Stoop, 2005).

Oxytocin's action in the central amygdala is associated with stress-reducing, anxiolytic, effects (Bale, Davis, Auger, Dorsa, & McCarthy, 2001). The

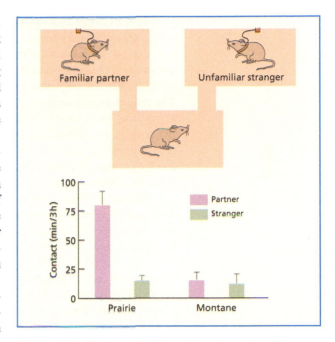

Figure 8.12 The standard experimental set-up for measuring partner preference in voles involves placing a vole in a chamber connected to two other chambers containing both a familiar vole and an unfamiliar stranger. The amount of time spent in each chamber is measured. If a prairie vole spends at least 24 hours with another vole and they are allowed to mate, then this induces a durable partner preference (an attachment). The montane vole shows no such loyalty. It actively avoids both its previous partner and the stranger. From Wommack et al. (2009).

KEY TERM

Montane voles
A species of rodent that is often studied due to its promiscuity rather than monogamy

hypothalamic-pituitary-adrenal (HPA) axis is activated during stress and produces changes in stress-related hormones such as cortisol (Erickson, Drevets, & Schulkin, 2003). Oxytocin is known to modulate the HPA by reducing the release of stress-related hormones in rodents (Neumann, Toschi, Ohl, Torner, & Kromer, 2001). Securely attached human babies secrete less cortisol than insecurely attached babies (Spangler & Grossmann, 1993), consistent with the view that secure attachment is less stress inducing. In rats, removal of the adrenal gland in late pregnancy reduces maternal behavior, but hormone injections reverse the effect (Graham, Rees, Steiner, & Fleming, 2006). In prairie voles, a stressful event has different effects on affiliative behavior in males and females (DeVries, DeVries, Taymans, & Carter, 1996). Males are more likely to affiliate with a female during stress, whereas females are less likely to affiliate with a male and more likely to choose the company of another female.

Both oxytocin and vasopressin may exert additional influences on the dopaminergic reward-based system, which originates in the ventral tegmental area and projects through the ventral striatum (including the nucleus accumbens) into the frontal cortex. Increased release of dopamine in this region is linked to sexual activity (Pfaus et al., 1990), although this is true of both monogamous and promiscuous species. The role of dopamine in attachment may be related to its interaction with hormones at key sites. For example, activation of both dopamine and oxytocin receptors within the nucleus accumbens is required for partner preference formation in prairie voles (Liu & Wang, 2003). Also, dopaminergic disruption in the medial prefrontal cortex can lead to partner preference formation in the absence of mating, suggesting that its role is not limited to sex-based reward (Smeltzer, Curtis, Aragona, & Wang, 2006).

Turning to evidence from humans, the role of oxytocin has been studied using three main techniques (note: effects of vasopressin are less studied). First, oxytocin can be administered experimentally as a nasal spray and compared against placebo. It is not clear how this form of administration exerts its effects directly on the nervous system or on other bodily functions (e.g. the heart) that might affect the nervous system indirectly (Churchland & Winkielman, 2012). Second, there are naturally occurring polymorphisms in the human oxytocin receptor (OXTR) gene that affect social behavior. Finally, one can measure the concentration of oxytocin in the body (e.g. blood plasma) and link this to experimental events or measures of individual difference. However, it is to be borne in mind that the relationship between circulating oxytocin levels in the body and levels in the brain is not fully understood (Churchland & Winkielman, 2012).

When administered nasally, oxytocin reduces the amygdala's response to fear-inducing stimuli measured using fMRI (Kirsch et al., 2005). Heinrichs, Baumgartner, Kirschbaum, and Ehlert (2003) administered either oxytocin or placebo to a group of human participants undergoing a stressful event – the Trier Social Stress Test (TSST). This involves a number of phases, including preparation and delivery of a speech to an unresponsive audience. In addition to the administration of oxytocin/placebo the participants either did or did not have social support from a best friend during the preparation phase. Both oxytocin and social support reduced the levels of cortisol associated with public speaking, and the two factors interacted with each other – that is, there were greater benefits when both occurred together relative to that expected from the effects of each alone (see Figure 8.13). This may explain the known health-related advantages of having social support for a range of physical and mental health problems that tend to increase with stress (e.g. Uchino et al., 1996).

Oxytocin may have a general social affiliative role with strangers that is linked to a reduction of stress and increased trust. Kosfeld, Heinrichs, Zak, Fischbacher, and

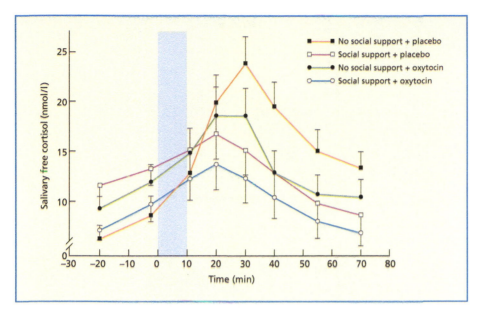

Figure 8.13 The Trier Social Stress Test (TSST) involves a number of phases, including preparation and delivery of a speech to an unresponsive audience. Levels of the stress-related hormone cortisol (measured in saliva) are lower when participants have received oxytocin and when they have received social support from a best friend in the preparation phase (and these two factors interact – the effects are significantly greater when combined). The shaded bar represents the period of public speaking. This is followed by mental arithmetic in front of a panel of evaluators. From Heinrichs et al. (2003). Copyright © 2003 Elsevier. Reproduced with permission.

Fehr (2005) administered oxytocin to human participants playing a social exchange game involving trust (the Trust Game, described in Chapter 7). Participants given oxytocin displayed higher levels of trust (in terms of the amount of money they were willing to invest), but the effects were specific to social interactions. When the same experiment was performed with a computer making random decisions (but with the same risk of winning/losing) there was no increase in trust. Within the context of group behavior, oxytocin administration increases trust and cooperation with ingroups and a defensive, but not hostile, reaction to outgroups (De Dreu et al., 2010).

Human genetic differences also shed light on the role of oxytocin. The human oxytocin receptor gene exists in two forms termed 'A' and 'G' (the A and G refer to a single difference in the DNA sequence). Given that each person has two copies of each gene, one from each parent, their genotype can be classified as AA, GG, or AG. Some studies report the prevalence of these genotypes as 25%, 25%, and 50% respectively (i.e. they are common), although this figure varies according to the ethnicity of the sample (Rodrigues, Saslow, Garcia, John, & Keltner, 2009). GG carriers are more socially sensitive. On measures of empathy, they have greater empathic accuracy (the 'reading the mind in the eyes' test), and they self-report a greater empathic disposition on questionnaire measures as illustrated in Figure 8.14 (Rodrigues et al., 2009). They are less likely to have autism (Wu et al., 2005) and show more sensitive parenting (Bakermans-Kranenburg & van Ijzendoorn, 2008). GG individuals also

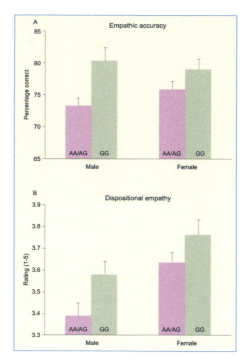

Figure 8.14 The oxytocin receptor gene exists in two main forms (A and G) such that an individual can either be AA, AG, or GG. The GG carriers score higher on a measure of empathic accuracy ('reading' the mind in the eyes) as shown in the top graph. They also self-report greater dispositional empathy as shown in the bottom graph (Rodrigues et al., 2009). Although they may be good at 'reading' social situations, GG carriers also show greater vulnerability to social stressors.

show a lower stress-related response, measured by heart rate, to startle (Rodrigues et al., 2009). These are positive influences of this genetic variant. However, GG carriers are more sensitive to previous childhood neglect (McQuaid, McInnis, Stead, Matheson, & Anisman, 2013), are more likely to develop unipolar depression and are more likely to have an adult insecure-anxious attachment type (Costa et al., 2009). Thus, social sensitivity in GG carriers makes them better able to read social cues but also less resilient to longer-term social stressors. Genetic differences in vasopressin receptors are also reported to be linked to social behavior in humans (Israel et al., 2008) and chimpanzees (Hopkins et al., 2014).

With regards to human studies examining circulating levels of oxytocin in the body, it has been found that higher levels predict generosity on the Trust Game (Zak, Kurzban, & Matzner, 2005). The findings with regards to anxiety and relationships are less straightforward. Levels of oxytocin (in women) and vasopressin (in men) in the blood plasma are *higher* in people who report relationship dissatisfaction (Taylor, Saphire-Bernstein, & Seeman, 2010). Similarly, a positive correlation between circulating levels of oxytocin and levels of social anxiety was observed in a group of patients with anxiety disorder (Hoge, Pollack, Kaufman, Zak, & Simon, 2008). Why is it that administering oxytocin (which presumably increases oxytocin levels) leads to lower stress, but higher circulating levels of oxytocin can often be linked to more stress? One possibility is that increasing oxytocin levels acts as a buffer (a biological coping mechanism) against the effects of high stress (e.g. Zelkowitz et al., 2014).

Oxytocin (and vasopressin) may signal the need to affiliate closely with others (and is therefore higher in relationship distress and social anxiety), in addition to responding to actual rewarding affiliative behavior (Taylor et al., 2010).

Evaluation

Attachment manifests itself behaviorally as distress at separation from the attached other, and comfort in the presence of the attached other. The presence of an attached other can activate the reward-based system (e.g. the nucleus accumbens) and deactivate the fear response (e.g. amygdala) and stress system (the HPA), but the initial development of an attachment may further depend on the effects of hormones such as oxytocin and vasopressin, which act on specialized receptors in the brain. Mammals that form monogamous relationships (e.g. prairie voles) tend to have far more of these receptors than mammals that do not (e.g. montane voles). Aside from attachment formation, oxytocin and vasopressin may act generally to enhance affiliation (e.g. trust) and reduce stress. In infants, different attachment styles (secure, anxious, avoidant) may reflect the quality of mother–infant interactions, but may generalize to some extent to different relationship behaviors and attitudes in childhood (peer–peer interaction) and adulthood (in terms of romantic partners). Different adult attachment styles modulate activity within the social/emotional brain both when thinking about relationships (e.g. a break-up) and in social interactions more generally (e.g. receiving positive/negative feedback from another person).

EFFECTS OF EARLY INSTITUTIONALIZATION ON ATTACHMENT

In 1990, after the collapse of the dictatorship in Romania, a series of orphanages were discovered in which abandoned children and infants had minimal care and lacked physical and mental stimulation. These children were subsequently adopted into 'good' homes, and their progress has been followed in a series of longitudinal studies. (Of course, many countries throughout the world have/had similar practices, but the Romanian model has been extensively researched). One of the most general findings of this research is that whilst their intellectual and cognitive abilities have, over years, approached the expected levels for their age, their social behavior still shows disturbances (Beckett, Castle, Rutter, & Sonuga-Barke, 2010; Kumsta et al., 2010). This includes a lack of pretend play (O'Connor, Bredenkamp, & Rutter, 1999) and social disinhibition characterized by a lack of awareness of social boundaries and an indiscriminate friendliness towards strangers (O'Connor & Rutter, 2000). The latter behavior is also found in high-quality institutions in which there is rich mental and physical stimulation but in which there is a high turnaround of staff providing care (Tizard & Rees, 1975). As such, it suggests that this aspect of social behavior – indiscriminate friendliness – is related to unavailability of a stable attachment figure rather than deprivation per se (Roy, Rutter, & Pickles, 2004). There may be a critical window for this to occur. Normal attachments begin within 6–9 months (as does a fear of strangers). Chisholm (1998) assessed attachment patterns (at the age of 4.5 years)

Figure 8.15 Many of the Romanian orphans who were discovered in 1990 in poor conditions were adopted into stable families. Their subsequent cognitive and social development has been followed into adulthood by researchers in Europe and North America. © Getty Images.

in adopted-away orphans and found that: infants adopted before 4 months of age tended to have normal attachment patterns; those adopted between 4 and 8 months had formed attachments that tended to be insecure, but they were not over-friendly to strangers; and those adopted after 8 months of age showed both insecure attachments and over-friendliness. Children adopted after the first 8 months of life also showed higher levels of stress-related salivary cortisol than those who had been institutionalized for less than 4 months in their first year of life (Gunnar, Morison, Chisholm, & Schuder, 2001). This may be a consequence of a failure to form an attachment during the most sensitive period. Children who had previously been adopted away at 1–7 years of age were found to have reduced resting levels of brain activity (measured by PET) in regions including the orbitofrontal cortex, amygdala, and hippocampus (Chugani et al., 2001).

SEPARATION, REJECTION, AND LONELINESS

The previous section on attachment considered how social bonds are formed. This section concentrates on the effects of disruption of social bonds, such as the failure to form social bonds (social exclusion), separation distress and grief (where a previous attachment is temporarily or permanently lost), and perceived isolation and lack of intimacy (loneliness). Although these are emotionally negative and stress-inducing events, they are also motivating events. We are motivated to reconnect with others (i.e. establish new social bonds or better social bonds) in order to avoid these punishing consequences.

The pain of separation and social exclusion

Separation or rejection is often described as painful (e.g. 'hurt feelings'). One suggestion is that these social aspects of pain have piggybacked onto evolutionary older mechanisms that represent physical pain (MacDonald & Leary, 2005). Regions in the anterior cingulate cortex respond to the affective and sensory components of physical pain and, to some extent, watching someone else in pain activates the same regions, particularly those that correspond to the affective component (Singer et al., 2004b). But is it associated with the 'pain' of social exclusion? Eisenberger, Lieberman, and Williams (2003) conducted an fMRI study of a cyberball game involving three players, including the one person being scanned. Players could opt to throw the ball to one of the two other players as shown in Figure 8.16. However, after a while the game was fixed such that two players consistently threw to each other, excluding the person in the scanner. There were two other conditions: one in which the player

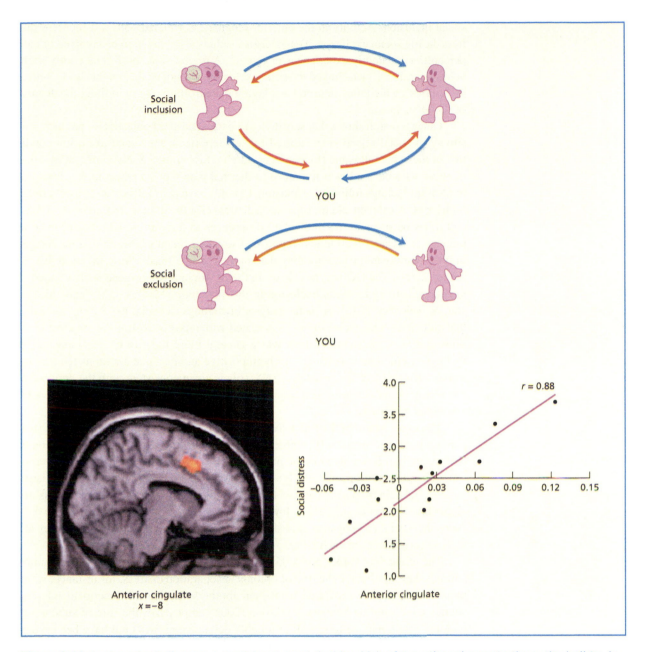

Figure 8.16 In the cyberball game a participant must decide which of two other players to throw the ball to. In a social exclusion condition, two of the players always send the ball to each other and never to the participant. In a social inclusion condition, all players get to play. Social exclusion tends to activate the anterior cingulate and this correlates with subjective levels of distress. Bottom figures from Eisenberger et al. (2003). Copyright © 2003 American Association for the Advancement of Science. Reproduced with permission.

was included, and one in which they were excluded but given the cover story of 'due to technical difficulties'. The anterior cingulate and the anterior insula were activated more by social exclusion (both real and due to technical difficulties) than social inclusion. Activity in the anterior cingulate correlated with self-reported distress during social exclusion. The anterior insula is also known to be involved in pain perception and bodily experiences in general. A region in the prefrontal cortex (right ventrolateral PFC) was linked to social exclusion, but not exclusion due to 'technical difficulties', which they interpret as playing a controlling role in limiting the distress of social exclusion.

Other research shows that sensitivity to pain (measured objectively) predicts sensitivity to social rejection (measured by self-report) on the social exclusion condition of the cyberball game (Eisenberger, Jarcho, Lieberman, & Naliboff, 2006) – that is, those who have a lower tolerance to thermal pain tend to report higher levels of unpleasant feelings following exclusion. Opioids have pain-killing (analgesic) effects both for physical pain and for separation distress (Kalin, Shelton, & Barksdale, 1988), and it has been found that individual differences in the mu-opioid receptor gene in humans are related both to self-reported distress to social exclusion and to changes in brain activity in the anterior cingulate in the cyberball task (Way, Taylor, & Eisenberger, 2009). The GG polymorphism in the oxytocin receptor gene is also linked to greater sensitivity to social exclusion in this paradigm (McQuaid, McInnis, Matheson, & Anisman, 2015). A meta-analysis of findings in social psychology revealed that although social exclusion was associated with reported feelings of negative emotions (e.g. distress, sadness) there was a general trend for it to be accompanied by feelings of emotional blunting – for both positive and negative emotions (Blackhart, Nelson, Knowles, & Baumeister, 2009). This is again consistent with the notion that there may be an analgesic response to social exclusion, which results in a general emotional numbness.

The exact nature of the link between social pain (/social exclusion) and physical pain remains controversial (Eisenberger, 2015). For instance, this cingulate region of the brain might be involved in the processing of some more general aspects of emotion rather than pain specifically. It is also the case that it is possible to reliably distinguish social pain from physical pain from fMRI data using computer-trained classifier algorithms (Wager et al., 2013). However, the theory of Eisenberger (e.g. 2015) only states that social and physical pain share some neural resources, and not that they are indistinguishable from each other.

The research of Panksepp (2005; Panksepp et al., 1980), primarily on non-human animals, has highlighted the role of opioids in separation distress, for instance, when pups and chicks are separated from their mothers. Unlike the functional imaging studies conducted on humans, Panksepp (2005) emphasizes the role of subcortical regions, which may be too small to reliably detect using fMRI but have been established as crucial in animal models. For example, the periaqueductal gray (or central gray) region is rich in opioid receptors and electrical stimulation of this region has analgesic effects (Hosobuchi, Adams, & Linchitz, 1977). However, it may also respond to 'social pain', such as when infant animals are separated from mothers (Rupniak et al., 2000), and it is also implicated in reactive aggression (Siegel et al., 1999), which often takes the form of a social threat to resources, such as threats to one's infants (Panksepp, 2005).

Finally, it should be noted that the opioid system may not be exclusively involved in separation distress and social pain, but also in reward learning and social attachment itself. In prairie voles, the mu-opioid receptor system is involved in partner-preference formation (Burkett, Spiegel, Inoue, Murphy, & Young, 2011).

Grief

Grief is an intense feeling of loss that occurs as a result of permanent separation (normally death) from a loved one. O'Connor et al. (2008) conducted an fMRI study of women who had lost a mother or sister due to breast cancer in the last five years. The grieving women consisted of two groups. One group were diagnosed with 'complicated grief' in which there was no sign of abatement of the sense of loss, yearning for the loved one, or preoccupation with thoughts of them. The other group were classified as having non-complicated grief. Both groups were shown photographs of either the loved one or a stranger with grief-related words (e.g. 'dying') or neutral words superimposed on them. Both groups showed activity in a number of regions relating to pain (dorsal anterior cingulate, insula, periaqueductal gray) when presented with images of the deceased relative to the stranger. This is consistent with an overlap between 'social pain' and physical pain found in the more artificial scenario of social exclusion in a game. However, there was some evidence that those with 'complicated grief' activated the nucleus accumbens more than those with non-complicated grief (although it was found for grief-related words and not images of the deceased). This activity was correlated with subjective reports of yearning, and was interpreted as evidence of an ongoing attachment to the deceased and an inability to fully adapt to their loss.

Loneliness

Loneliness is a related concept to social exclusion but differs from it in a number of ways. It is more akin to a *trait* (a relatively stable disposition) than a *state* (induced by a particular situation). It can also differ from social exclusion in that it need not entail an element of rejection by others. It could be an outcome of shyness or social anxiety, for instance. It could also reflect *perceived* isolation (e.g. relating to lack of intimacy) rather than actual isolation. However, loneliness and social exclusion do tend to go together because lonely people often withdraw from social situations, the net result being that others do not include them in social activities or actively reject them. Lonely people may also look for rejection cues in other people and interpret ambiguous cues threateningly. This, of course, reinforces the feeling of loneliness as shown in Figure 8.18 (Cacioppo & Hawkley, 2009).

Cacioppo, Norris, Decety, Monteleone, and Nusbaum (2009) showed images of social scenes or non-social scenes, matched for ratings of emotionality, to lonely and non-lonely people during an fMRI study. For pleasant stimuli, non-lonely people showed greater activity in the ventral striatum for

KEY TERMS

Grief
An intense feeling of loss that occurs as a result of permanent separation (normally death) from a loved one

Loneliness
A perceived social isolation and/or lack of intimacy

Figure 8.17 Queen Victoria experienced what clinical psychologists now refer to as 'complicated grief' following the early death of her husband, Prince Albert. For several years after his death she insisted that his personal effects should be laid out (e.g. hot water for shaving), and she appeared to be in a deep depression. She wore black for her remaining 40 years of reign. This photograph, taken two years after her husband's death in 1863, shows her in full mourning regalia.

Figure 8.18 Loneliness can be construed as a vicious circle in which perceived social isolation leads to hypervigilance to social threats (e.g. of not being liked), which leads to further isolation. From Cacioppo and Hawkley (2009). Copyright © 2009 Elsevier. Reproduced with permission.

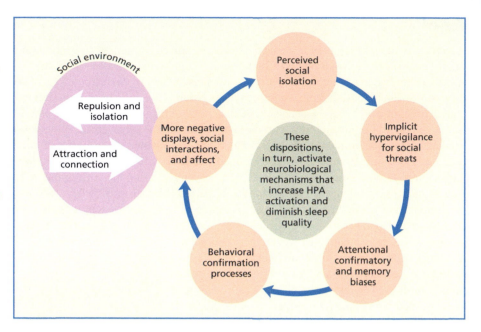

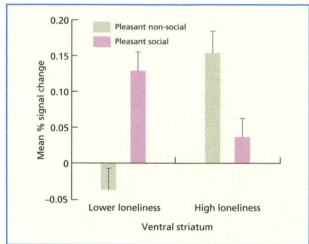

Figure 8.19 Lonely people show a different brain response to social scenes relative to non-social images. The graph shows activity in the ventral striatum and is consistent with the view that social situations are less rewarding to this group. From Cacioppo & Hawkley (2009). Copyright © 2009 by the Massachusetts Institute of Technology. Reproduced with permission.

social images than non-social images, but for lonely people the reverse was true (see Figure 8.19). This is broadly consistent with behavioral ratings, showing that lonely people judge images of social situations as less pleasant. For unpleasant stimuli, non-lonely people activated the temporo-parietal junction region (part of the mentalizing network) in response to social relative to non-social images. This pattern was absent in the lonely people, although they tended to activate their visual cortex more than non-lonely people did. Perhaps lonely people are biased to attend to the scene itself, whereas non-lonely people focus on other people.

Perceived loneliness has important health consequences. For example, perceived loneliness predicts elevated blood pressure even after statistically removing effects of objective social support and isolation (Hawkley, Masi, Berry, & Cacioppo, 2006). Cognitive decline in the elderly and risk of Alzheimer's disease are predicted by perceived loneliness even after amount of social contact and size of their social network is factored out (Wilson et al., 2007). Experimentally inducing participants to think of a lonely future has detrimental effects on subsequent performance on intelligence measures, even though their current objective levels of social support are not altered by that manipulation (Baumeister, Twenge, & Nuss, 2002). The biological mechanism by which these effects are exerted is unknown, but it is known that social isolation is associated with increased activity of the stress-related HPA axis (Adam, Hawkley, Kudielka, & Cacioppo, 2006) and affects gene expression relating to the body's defense system from infection (Cole et al., 2007).

Analogous to the suggestion that there is a relationship between physical pain and social exclusion/separation, it has been suggested that there is a link between physical warmth and feelings of social intimacy (and, conversely, between physical coldness and loneliness). This is assumed to reflect a form of 'neuronal recycling' of evolutionary older mechanisms (those relating to warmth and pain) for more recently evolved ones (i.e. relating to complex social function). Lonely people report taking baths/showers that are hotter in temperature and longer in duration that can be interpreted as a form of self-medication (Bargh & Shalev, 2012). Inagaki and Eisenberger (2013) report that holding warm objects and reading 'warm' messages from friends and family activate, in fMRI, overlapping regions such as ventral striatum and mid-insula. However, it is unclear how specific this overlap is or whether it reflects more general processes (e.g. positive affect in the broadest sense). The authors, however, did show differences between warmth and one other condition (pleasant touch).

Evaluation

Social exclusion may tend to rely on some of the same neural substrates as physical pain. This may result in temporarily blunted emotions as a counter (analgesic) response. An extreme form of this may be grief resulting from a bereavement. Lonely people have a tendency to process social stimuli differently, even when there is no element of social exclusion implied.

SUMMARY AND KEY POINTS OF THE CHAPTER

- An attachment is a long-enduring, emotionally meaningful tie to a particular individual, such as between mother and infant or between romantic partners. Love is the emotion that is elicited in the presence of (or by the thought of) another person with whom there is an attachment. Love is associated with activity in the reward circuitry of the brain (e.g. the nucleus accumbens).
- Individual differences in attachment can manifest themselves in perceptual/attention-based measures of face processing (in ERPs). In adults, attachment anxiety appears linked to the amygdala (given a negative appraisal) whereas attachment avoidance has been linked to lateral prefrontal activity and reduced reward-related responsiveness to positive feedback from others.
- Animal models based primarily on the monogamous prairie vole have been important for exploring the role of neuro-hormones such as oxytocin and vasopressin on attachment formation. This may be achieved by interfacing with the dopaminergic reward system and by reducing the activity of the stress response (e.g. in the amygdala and the HPA axis).
- Social exclusion, separation, and grief may tend to rely on some of the same neural substrates as physical pain. Lonely people have a tendency to process social stimuli differently, even when there is no element of social exclusion implied.

EXAMPLE ESSAY QUESTIONS

- Is love too subjective a concept to study scientifically?
- What have animal models contributed to our understanding of attachment?
- Do mother–infant bonding and romantic bonding share common mechanisms?
- Is 'social pain' arising from separation or rejection related to physical pain?

RECOMMENDED FURTHER READING

- De Haan, M., & Gunnar, M. R. (2009). *Handbook of Developmental Social Neuroscience*. New York: Guilford Press. Contains good up-to-date chapters on relationships and attachment (particularly animal models).
- Young, L., & Alexander, B. (2014). *The Chemistry Between Us: Love, Sex, and the Science of Attraction*. London: Penguin Group. A popular science account but written by the leading expert and up-to-date.

ONLINE RESOURCES

- References to key papers and readings
- Videos demonstrating the Strange Situation Test and Partner Preference Paradigm
- Interviews and talks given by John Cacioppo, Larry Young, Paul Zak, Naomi Eisenberger, and others
- Recorded lecture given by textbook author, Jamie Ward
- Multiple choice questions and interactive flashcards to test your knowledge
- Downloadable glossary

CHAPTER 9

CONTENTS

Groups and identity

In 1971, a group of 21 students at Stanford University were randomly assigned to be either guards or prisoners in a mock prison set-up (Zimbardo, 1972; Zimbardo, Maslach, & Haney, 1999). Being assigned to be a prisoner or a guard is to become part of a group that has certain norms and expectations. Despite receiving no detailed instructions as to how to behave, over the course of 6 days the guards became increasingly brutal in their behavior (waking the prisoners at night; enforced physical exercise; stripping naked) and the prisoners began to show signs of psychological disturbance, leading to the experiment being terminated early. The results of the Stanford Prison Experiment illustrate how easily people are able to adjust their behavior, and identity, to fit with the norms and expectations of a group – even when assignment to a group is arbitrary. A more recent replication of the study in the UK also emphasized how individuals come to identify with an arbitrarily imposed group and adjust their behavior to fit their role, although the outcome was rather different (Reicher & Haslam, 2006). In this study, there was less evidence of brutality from the guards, and they were overcome by the prisoners, who had more readily come to identify with their roles.

One might expect that social neuroscience is ill suited for addressing questions at the group level. For instance, the Stanford Prison Experiment would not lend itself well to social neuroscience investigations. However, group influences exert their pressure on individual minds (and brains) and social neuroscience, like much of social psychology, can make a distinction between individuals acting as group members (women, men, straights, gays, Blacks, Whites, etc.) and individuals acting as individuals.

Figure 9.1 What elements comprise our sense of identity? Race? Nationality? Gender? Beliefs? Our social identity is defined in terms of the various groups that we belong to.

This chapter considers several issues relating to groups and identity. First, the chapter considers the various components that are typically considered to comprise 'the self'. Some components of the self operate primarily at the level of the individual (e.g. our own personality, our sense of being in control of our actions) and others operate primarily at the level of the group (e.g. our social identity, cultural beliefs, and traditions). The second part of the chapter considers the way in which groups are assigned and evaluated, giving particular attention to the issue of prejudice.

WHAT IS A GROUP?

It would be rather too simple to describe a group as a collection of individuals, although at some level this is exactly what it is. Not all collections would be considered as a group. For example, are blue-eyed people a group? Are the collection of people who happen to be with you in the shopping mall, or on an airplane, a group? A *psychological* group would be connected by virtue of perceived relatedness, common goals, or by the way that individuals influence each other. Johnson and Johnson (1987) listed seven ways in which a group may be defined in psychology:

1. A collection of individuals who are interacting with one another
2. A social unit consisting of two or more persons who perceive themselves as belonging to a group
3. A collection of individuals who are inter-dependent
4. A collection of individuals who join together to achieve a goal
5. A collection of individuals who are trying to satisfy some need through their joint association
6. A collection of individuals whose interactions are structured by a set of roles and norms
7. A collection of individuals who influence each other

IDENTITY AND THE SELF-CONCEPT

What is it that makes me who I am? One approach to answering this question is to think about all the things that apply to me that do not apply to other people. At the most basic level, one could answer the question by saying that I occupy my own body. I cannot occupy other people's bodies and they cannot occupy mine. Similarly, I can move and control my own body (and I can create my own thoughts and ideas), but I cannot control other people, and other people cannot control me. Another thing that applies to me that distinguishes me from all others is that I have my own personal history in terms of what I have done, the places I have been, and so on. In neural terms, there are unique patterns of synaptic connections in everyone's brains that reflects their unique histories. Another approach to answering the question 'What is it that makes me who I am?' is in terms of a social identity. Instead of answering the question by listing all the things that make someone unique, one could answer the question by listing all the features that connect us with the people around us. Humans

KEY TERM

Self-awareness
Our conscious feeling of a unitary, ongoing self

are social creatures, and it would be odd if we did not define ourselves in social terms. Our social or cultural identity is determined by our membership of various groups (e.g. ethnic, religious, national, political, socio-economic) and the shared traditions (skills, beliefs, rituals) that bind groups together.

This section will begin by examining evidence that a region of the brain is specialized for thinking about the self. It will then go on to consider evidence for how the self can be related to a set of more diverse functions and mechanisms from bodily awareness to social identity.

Thinking about the self: the role of the Medial Prefrontal Cortex (mPFC)

There is a large body of neuroimaging research showing that when people are asked to evaluate information in relation to themselves (relative to others) or even attend to information that is self-related in some way then a region of the brain in the medial prefrontal cortex (mPFC) is activated.

Some of the early influential research in this domain involved participants making evaluations about personality traits (e.g. confident, talkative) either in relation to self or other. Kelley et al., (2002) found that making trait personality judgments about oneself relative to a familiar other (George W. Bush) activates a region in the medial prefrontal cortex (Kelley et al., 2002). However, making judgments of personality per se, relative to a control task (judging whether the word is in upper/lower case) activates a region implicated in semantic memory retrieval in the left lateral prefrontal cortex (Kelley et al., 2002). This suggests a possible specialized role for the medial prefrontal cortex in thinking about the self, rather than in personality judgment per se. The response of the mPFC region does not depend on whether the

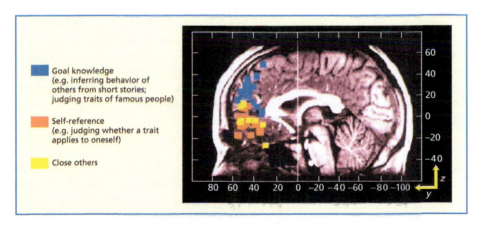

Figure 9.2 The medial prefrontal cortex (mPFC) is a crucial part of the social brain but its function is not well understood. It is activated when we are required to make judgments about other people in a wide variety of tasks, but there is some relative specialization between judgments about yourself (more ventral regions; in orange) versus other people (more dorsal regions; in blue). People who are close to us (e.g. family) or who are perceived to be similar to us tend to activate the self-related regions (in yellow). From Krueger et al. (2009). Copyright © 2009 Elsevier. Reproduced with permission.

personality traits are positive or negative but depends, instead, on the evaluation of these traits relative to the self (Moran, Macrae, Heatherton, Wyland, & Kelley, 2006).

Traits judged relative to the self (versus another) are better remembered when given a later memory test and this reflects activity in the mPFC (Macrae, Moran, Heatherton, Banfield, & Kelley, 2004). It suggests that the memory benefit is due to self-relatedness rather than semantic depth of processing (as the latter relies on a different region, in the lateral prefrontal cortex). Other research shows that recalling that a photograph was taken by the participant, rather than another person in the same campus setting, activates the mPFC (Cabeza et al., 2004). Attending to one's own emotional response, relative to the response of someone depicted in a photograph, activates the mPFC (Ochsner et al., 2004). Hearing one's own first name also activates this region (Perrin et al., 2005). It is hard to control for various confounds that might make the self 'special'. For instance, self-related stimuli are acquired early in life and occur very frequently. One experimental approach for tackling this problem is to arbitrarily associate stimuli to self or other (e.g. triangle = self, circle = a friend, square = stranger). Attending to a geometric shape that has been associated to the self also taps a network involving the mPFC that also boosts the ability to detect that shape (Sui, Rotshtein, & Humphreys, 2013).

A number of studies have shown that activity in the medial prefrontal region is not strictly specific to the self. It is also found for other people when they are judged to be similar to oneself (e.g. Mitchell, Banaji, & Macrae, 2005a), and more recent research has suggested that it is emotional closeness rather than similarity that drives this effect (Krienen, Tu, & Buckner, 2010). This kind of evidence could be taken to support the view that there aren't, in fact, any neural mechanisms in the brain that are self-specific (Gillihan & Farah, 2005). An alternative interpretation, stemming from the social psychological literature, is not to think of the self as a strict delineation between one person and everyone else but more in terms of the self as a social hub that can include close others and, to some degree, certain other social groups (Aron, Aron, & Smollan, 1992). The extent to which others are treated as an extension of self can perhaps vary according to context (current situational demands) and culture, as discussed later.

The fact that the mPFC also appears as a key region in mentalizing / theory of mind, in addition to its involvement in self-referential thought, has led to the suggestion that we use self-referential thinking in order to interpret the behavior of others (Mitchell, 2009). This is a form of simulation theory (i.e. using self to understand others) but one that is based on high-level thought rather than a lower-level form of perception–action mirroring or affective sharing.

A recent conceptualization of the functional role of the medial prefrontal cortex is in terms of binding together the different information that makes up social events – actions, agents, objects, goals, beliefs (Krueger et al., 2009). More dorsal regions of the mPFC are implicated when making inferences about the actions of others (i.e. in short stories, sentences, or single words) whereas inferences about the self in relation to one's own memories, traits, and goals are associated with activity in a more ventral region that is likely to have greater connectivity to emotion-related regions (see Figure 9.2). As such, there is evidence of a self–other continuum in this region of the brain.

Different facets of the self

The studies cited above could perhaps be best described as reflecting self-awareness. That is, the mPFC is activated by tasks in which participants explicitly process information as self-related. (How the mPFC responds to subliminally presented self-relevant information is unclear.) However, there may be more to the self than self-awareness. Indeed, there may even be multiple forms of self-awareness depending on what it is that one is aware of: awareness of a conceptual notion of self and identity, awareness of emotion, awareness of actions, and so on. Most contemporary views of the self, based on cognitive processes, regard the self as a collection of systems that operate according to different principles and draw on different kinds of knowledge. To give one example, Powell et al. (2010) show, using fMRI, a dissociation between regions involved in conceptual self processing (trait judgment tasks that activated mPFC) and agency (choosing from a selection of objects which was linked to parietal activity).

Neisser (1988) distinguished between five possible kinds of self, and Gallagher (2000) distinguished between two (these are listed below). However, in both cases these are likely to consist of several underlying mechanisms, many of which may not be unique to the self (Gillihan & Farah, 2005). One could argue that the studies examining mPFC activation to self-referential processing are primarily operating at the level of the narrative self (in Gallagher's framework) or the private/conceptual self (in Neisser's framework).

Neisser (1988) made a distinction between five kinds of self-knowledge:

1. The *ecological self* consists of one's sense of being located within the body.
2. The *interpersonal self* consists of the sense of oneself as a locus of emotion and social interaction.
3. The *extended self* consists of one's sense of existing over time.
4. The *private self* is linked more closely with self-awareness and the feeling that 'I' can reflect upon conscious experiences and feel ownership of them.
5. The *conceptual self* is linked to semantically relevant knowledge about oneself in terms of both social role (and group identities) and personal knowledge.

Gallagher (2000) broadly divides the self into two levels:

1. The *minimal self* consists of the feeling that we own our body and control our actions.
2. The *narrative self* consists of our social identity and autobiographical memory that extend through time and enable one to construct a conscious sense of unity.

Figure 9.3 provides an outline of some of the different mechanisms that contribute to the self-concept that are drawn from prominent reviews (Boyer, Robbins, & Jack, 2005; Gallagher, 2000; Gillihan & Farah, 2005; Neisser, 1988). For ease of explanation, they are divided into three different kinds of mechanism, but other divisions and groupings are possible. At one end there is the sensorimotor self, which consists of the sense that the self is located within the body (**embodiment**) and that we control our own actions and thoughts (**agency**). Although these mechanisms may be separate from each other, they can be conveniently grouped together on the principle that these self-related mechanisms are assumed to be independent of culture and shared by all – as in Gallagher's (2000) notion of a minimal self. At the other end there are culturally determined aspects of self, such as the groups that one identifies

KEY TERMS

Embodiment
The sense that the self is located within the body

Agency
The sense that we control our own actions and thoughts

Figure 9.3 The self-concept can be construed in terms of a variety of different mechanisms that draw on different kinds of information. Some self-related processes (e.g. on the right) are independent of culture, whereas others (e.g. on the left) are defined entirely by our cultural setting. Other mechanisms (in the middle) can best be construed as a mixture of culturally determined and culturally independent information.

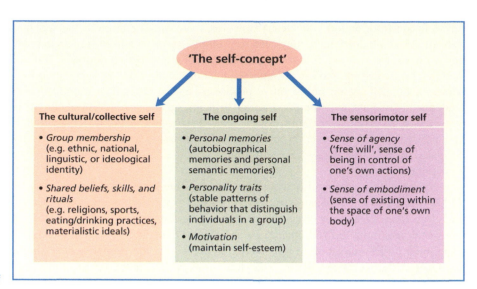

with (e.g. religious, ethnic, political) and the set of skills, beliefs, and rituals that are culturally acquired (e.g. wearing make-up, not eating pork, Protestant work ethic). Another set of mechanisms can be identified that give rise to the sense of the self as continuing over time. Key components here are our own set of personal memories, and also our personality traits and self-esteem, which are regarded as relatively stable over time and determine individual differences in behavior (e.g. wallflower vs. party animal).

The key component mechanisms are considered below in terms of possible neuroscientific mechanisms.

Sense of agency

There is a sense in which we feel in control of our own actions. We can readily distinguish situations in which we move our arm versus situations in which our arm is moved by someone else. The basic assumption is that when we ourselves generate an action, then we can use the motor program to predict what the action will feel like – for instance, the velocity, trajectory, and force on contact (Wolpert, Ghahramani, & Jordan, 1995). This is termed a **forward model**. If there is a match between predicted effects of the action and the actual action, then the action is attributed to the self; otherwise it may be attributed externally (e.g. Sato & Yasuda, 2005).

The event that generates the initial motor plan to act, and the prediction of what the action will feel like, may typically occur unconsciously. People report a conscious intention to act that is later in time than the unconscious preparation to act that can be measured in the brain. When EEG electrodes are placed over the motor cortex and participants are instructed to make a button press whenever they want to, then the brain starts to generate an electrical potential several hundred milliseconds before the person reports any intention to press the button (Libet, 1985; Libet, Gleason, Wright, & Pearl, 1983). This is shown in Figure 9.4. Thus, both the awareness of one's intentions to act and the action itself arise out of largely unconscious processes. We *infer* that our intention to act causes the action, but this is not necessarily so (Wegner, 2002). The feeling that we are in control of our actions may nevertheless be important for creating a sense of moral and social responsibility for our behavior.

KEY TERM

Forward model
The use of motor programs to infer the sensory consequences of actions

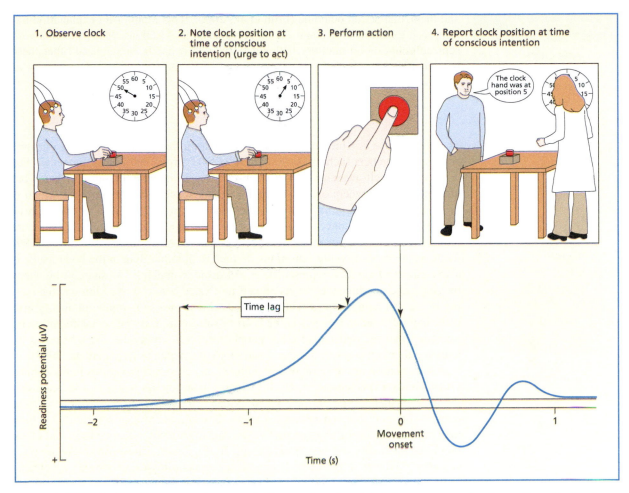

Figure 9.4 The feeling that we are in control of our actions is related to our sense of agency. Does the fact that the brain shows a change in motor cortex activity (measured by EEG) long before the participants report having any intention to act (in terms of reporting the time of the intention relative to the moving clock hand) mean that the sense of agency is an illusion with no real function? From Haggard, 2008. Copyright © 2008 Nature Publishing Group. Reproduced with permission.

Patients with schizophrenia who report **delusions of control** may have a disrupted sense of agency and, hence, a disrupted sense of (one aspect of) self (Frith, Blakemore, & Wolpert, 2000). Many patients with schizophrenia report that they feel that their actions and their thoughts are under external control – for example, 'They inserted a computer in my brain. It makes me turn to the left or right' or 'Thoughts come into my mind from outer space'. One suggestion is that there is a failure to monitor action intentions with action outcomes, thus giving rise to a feeling of external control. Frith and Done (1989) report that schizophrenic patients reporting alien control had problems in correcting errors made on a joystick game when the corrections had to be made internally (but they could correct it when given external visual feedback). This was initially explained in terms of a failure to compare conscious intentions with action outcomes, but is now interpreted in terms of the forward model comparison between an (unconscious) prediction of action outcomes

KEY TERMS

Proprioception
The sense of where
our limbs are located
in space

Interoception
The sense of the
internal state of the
body, which may
include pain, core
temperature, hunger,
heart rate, and
breathlessness

**Out-of-body
experiences**
Reports of being in a
location different from
the actual physical
location

and the actual action outcome (Frith et al., 2000). This monitoring explanation has also been extended to account for thoughts as well as actions. Many patients with schizophrenia report auditory hallucinations, and it has been suggested that these hallucinations reflect their own inner thoughts that are misattributed externally as heard speech (Frith, 1992).

Sense of embodiment

There is a sense in which our self is located within the space occupied by our own bodies. Researchers such as Damasio (1999, 2003) argue that bodily awareness lies at the core of self-awareness. Bodily senses not only include the sense of touch but also the sense of where our limbs are located in space (**proprioception**), which is given by stretch receptors in the muscles and ligaments, and also internal senses (**interoception**) that convey the internal state of the body and may include pain, core temperature, hunger, heart rate, and breathlessness. According to Damasio (1999) the reinstatement of bodily sensations (in the brain, rather than in the body itself) is a key aspect of emotion representation and decision-making. As such, in his view, the sense of embodiment accounts for self-motivated behavior in addition to the feeling of the self being located within the body. Most forms of simulation theory also place embodiment and agency at the heart of self-awareness and social cognition by assuming that we understand other people's actions, emotions, and sensations by mapping them onto our own sensory and motor mechanisms (e.g. Gallese, 2003).

A disruption of the sense of embodiment can accompany various brain lesions. **Out-of-body experiences**, in which the participant reports being in a location different from their actual physical location, can arise following damage to the right temporo-parietal junction (Blanke, Landis, Spinelli, & Seeck, 2004). This region may be involved in shifting perspective or focusing attention between self and other, including the bodily self (e.g. Blanke et al., 2005). Certain illusions can also produce something akin to an out-of-body experience in neurologically normal participants as illustrated in Figure 9.5 (Lenggenhager, Tadi, Metzinger, & Blanke, 2007). If participants see, using virtual reality, an image of someone placed in front of them and if the virtual person and the participants are stroked on the back in synchrony then the participant may report feelings such as 'I felt as if the virtual character was my body'. The participant also has difficulty in locating him/herself in space when displaced by the experimenter, and he/she gravitates towards the virtual body. The same is not found when an object rather than a body is stroked.

Memory and the self

Conway and Pleydell-Pearce (2000; also Conway, 2005) highlight the importance of memory in creating a sense of an ongoing self. Their so-called self-memory system consists of two components. The first component is autobiographical knowledge, which consists of both fact-based knowledge about one's life (e.g. our occupations and the places that we have lived) and event-based knowledge about episodes in a particular time and place. It may also contain knowledge about our general dispositions (e.g. personality). This autobiographical knowledge is assumed to be hierarchically organized, as shown in Figure 9.6, with different levels of the hierarchy linked to different neural substrates. For instance, the hippocampus may be crucial for event-based knowledge, but regions such as mPFC may act across longer timescales such as trait-based knowledge (Martinelli, Sperduti, & Piolino, 2013). Trait self-knowledge

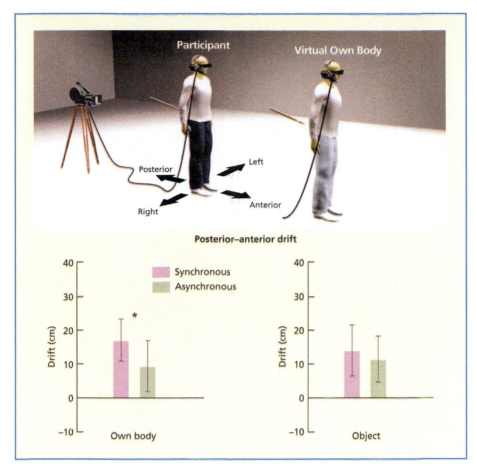

Posterior–anterior drift

Figure 9.5 Is it possible to create an experience similar to an out-of-body experience in an experimental setting? If you see an image of yourself or another person projected in front of you (using virtual reality) and if that person's back is seen to be stroked at the same time as your own back is stroked, then you tend to lose your sense of location in space and report feeling that the virtual character is your own body. The effect is not found if the stroking is out of synchrony, or if an object is seen instead of a body. Adapted from Lenggenhager et al. (2007).

is generally robust to brain damage including amnesia (Klein & Lax, 2010). The second component, which they term the 'working self', comprises the goals and motivational state of the individual. These two components interact together such that retrieving information about the past cannot be fully separated from one's current goals and aspirations. As Conway (2005) puts it, 'memory is motivated'.

According to this model, we are always viewing the past through the prism of our current goals, knowledge, and beliefs, and this may be sufficient to create a feeling of unity over time. There is evidence that people tend to remember the past in terms of their current knowledge and beliefs. Marcus (1986) asked participants to rate their attitudes on various political issues in 1973 and again in 1982. In the 1982 session they were also asked to recall the attitudes that they held in 1973. A systematic bias was found such that their previous attitudes were judged to be closer to their presently held ones. People also have a tendency to remember the past in a self-enhancing manner. Those who identify strongly with an ingroup are worse at remembering acts of historical violence by their ingroup (Sahdra & Ross, 2007) and self-enhancing memories are judged to feel more recent than (matched) memories that reflect badly on oneself (Ross & Wilson, 2002). Thinking about the present self activates the mPFC more than thinking about the past self and, in this study, the past self was equivalent to thinking about another person in the past or present (D'Argembeau et al., 2008).

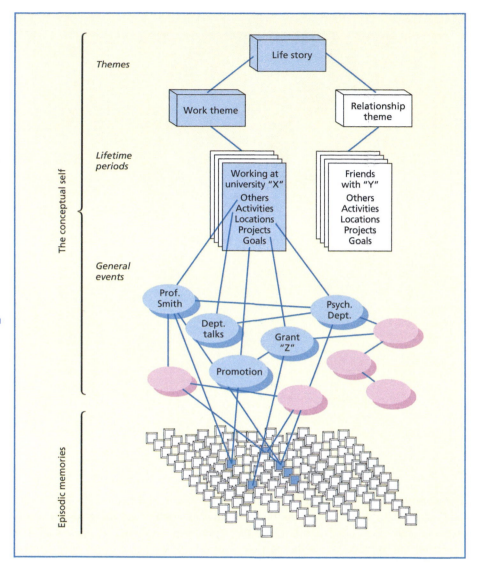

LeDoux (2002) in *Synaptic Self* also argues that memory, encoded through synaptic connectivity, is at the core of what constitutes the self. Unlike Conway and Pleydell-Pearce (2000), LeDoux (2002) does not give prominence to one particular memory system (e.g. autobiographical knowledge) but incorporates all memory systems. LeDoux's (2002) approach is inherently reductionist – it reduces psychological mechanisms (e.g. social identity, memory, personality) to a single neurobiological mechanism. (Non-reductionist approaches attempt to link together psychological and neuroscience levels of explanation but do not attempt to replace the former with the latter.) LeDoux's account enables social and cultural influences to be taken into account, as they affect synaptic learning in the same way as any other learned information. LeDoux (2002) also enables more hardwired aspects of the self (potentially including our sense of agency and embodiment, personality traits) to be incorporated in his model of the self, by arguing that synaptic connectivity is a result of genetics as well as learning. Whilst LeDoux (2002) offers one of the few unitary accounts of the self, the account fails to explain the detailed findings outlined in this section. Why do out-of-body experiences occur from disruption of those particular brain regions? Why do collectivist cultures differ from individualistic ones in the particular brain region that they do? And so on.

DO AMNESICS HAVE A DISRUPTED SENSE OF SELF?

Before beginning to answer this question it is important to stress that patients with amnesia do not have global memory impairments. Amnesia typically arises from damage to the medial temporal lobes (the hippocampus and surrounding structures). These structures appear to be particularly important for certain types of memory, but not all types of memory. They are particularly important for memories of specific episodes from one's life (Squire, 1992; Tulving, 1983), but many forms of unconscious knowledge (e.g. how to ride a bike) are spared.

One of the most striking and well-documented cases of amnesia is HM. HM began to experience epileptic seizures at the age of 10 and, by the time of leaving high school, his quality of life had deteriorated to a point where surgeons and family decided to intervene surgically. The procedure involved removing the medial temporal lobes, including the hippocampus, bilaterally (Scoville & Milner, 1957). What the surgeons did not foresee was that HM would develop one of the most profound amnesias on record. Despite acquiring few episodic memories over the intervening decades (he died in 2008), HM maintained consistent beliefs, desires, and personality traits (Corkin, 2002). He was described as altruistic, and exhibited courteous social behavior. Although he dreamed of being a neurosurgeon he attributed his failure to do so to the fact that he wore glasses (the blood would spurt on them) rather than to his memory failure. He was unsure of his age or whether he had gray hair, and he did not always recognize himself in photos. However, he did not seem shocked to look in a mirror, suggesting that he has some up-to-date knowledge (perhaps implicit) of his appearance.

Several studies have shown that amnesic patients have good insight into their own personalities when asked to decide how much a personality trait applies to them (e.g. Klein, Loftus, & Kihlstrom, 1996; Klein, Rozendal, & Cosmides, 2002). For example, Klein et al.'s (2002) amnesic patient tended to rate the same personality traits over time (correlation of .69) with similar reliability to non-amnesic controls (correlation of .74), and his scores agreed with a family member's rating of his personality (correlation of .62) in the same way as was found for controls (correlation of .64). This suggests that amnesic patients retain certain core self-knowledge in the face of severe problems in recalling specific episodes.

Cultural and social identity

Several prominent theories in social psychology make a distinction between one's personal identity (e.g. relating to one's own personality and experiences) and social identity related to membership of various groups (e.g. Tajfel & Turner, 1986). The

KEY TERMS

Inter-dependent self
A form of social identity in which individual beliefs and goals are strongly connected with the people around them

Independent self
A form of social identity in which individual beliefs and goals are viewed as independent from the people around them

extent to which one's behavior is influenced by these different identities depends on what information is salient at a given point in time, and it may, to some degree, vary across cultures.

In East Asian countries (such as China, Japan, Korea) people tend to regard themselves as having an **inter-dependent self** in which their own identity, beliefs, and goals are strongly connected with the people around them (Markus & Kitayama, 1991). The emphasis is on fitting in and attending to others. In contrast, in Western cultures (e.g. North America, Europe) people tend to perceive their own beliefs and goals as largely independent – as having an **independent self**. The emphasis is on self-discovery and personal achievement. In Japan, the success of Olympic athletes tends to be explained both in terms of personal attributes (e.g. skill, hard work) and their social background, whereas in the USA the personal attributes are given the main focus (Markus, Uchida, Omoregie, Townsend, & Kitayama, 2006). This distinction also lies at the heart of the difference between a so-called collectivist culture and an individualist culture.

Although these social differences are environmental in origin (dependent on cultural immersion), it has been claimed that they lead to more widespread changes in cognitive style outside of the social domain, and to wider differences in the way that the brain is organized and utilized (Han & Northoff, 2008). For example, Westerners tend to pay more attention to objects than background context, whereas East Asians tend to attend more to relations and contexts than objects. This manifests itself both on tests of visual perception and attention (e.g. detecting changes to objects versus backgrounds; Masuda & Nisbett, 2006) and in terms of BOLD activity in areas involved in object perception (Gutchess, Welsh, Boduroglu, & Park, 2006). Lin and Han (2009) attempted to manipulate self-construal within the same group of Chinese participants by priming them with essays containing either the inter-dependent pronoun 'we' or the independent pronoun 'I'. Priming with 'I' increases local processing over global processing, whereas priming with 'we' produces the opposite profile (Figure 9.7). ERP studies show that such changes occur in early (100 ms) components of perception (Lin, Lin, & Han, 2008).

Zhu, Zhang, Fan, and Han (2007) compared the neural substrates of self-concept in Chinese and Western participants using fMRI. In their task, they presented participants with trait adjectives (e.g. brave, childish) and asked them to judge their

Figure 9.7 Participants were asked to detect the presence of a letter (H or S) that could appear at either the local or global level (in this example, the H appears at the global level, made up of local A's). Priming with the inter-dependent pronoun 'we' increases detection of global letters over local ones, whereas the independent pronoun 'I' has the opposite effect. Graph on right taken from Lin and Han (2009).

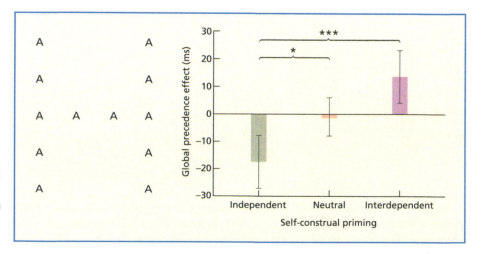

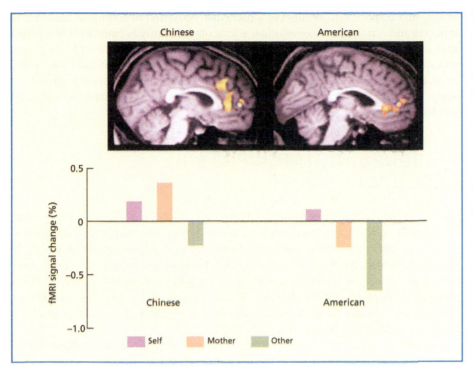

Figure 9.8 To what extent are other people, such as our mother, part of our self-concept? The answer to this question might be partly dependent on culture. This fMRI study shows that in Chinese participants (normally classed as belonging to a collectivist culture) judgments about oneself and one's mother activates the same region. In American participants (normally classed as belonging to an individualistic culture), making judgments about one's mother does not activate the same region as making judgments about oneself. It activates a separate region found when making judgments about other people in general. From Zhu et al. (2007). Copyright © 2007 Elsevier. Reproduced with permission.

appropriateness to themselves, to their mother, or to a famous person. In both cultures, a region in the medial PFC was activated when making self-referential judgments but was not activated when making trait judgments about a famous person. However, the two cultures differed when making judgments relating to their mothers. Whereas the Chinese showed similar neural responses to self and mother judgments in this region, the Westerners' responses to their mothers were similar to that of the famous person (see Figure 9.8). This is consistent with different self-concepts in which the self can be construed as inter-dependent with kin and close acquaintances in some cultures. Within a collectivist culture, the effect is more pronounced for mother relative to father or best friend (Wang et al., 2012).

The study of Zhu et al. (2007) didn't measure self-construal directly but used nationality as a proxy. Other studies have looked at endorsement of independent and inter-dependent self-construal (via questionnaire) within and between cultures. Chiao et al. (2009) reported that activity within the mPFC region was greater for those with an inter-dependent self-construal when thinking about themselves in relation to others (e.g. this personality trait applies to me when I interact with my mother), but mPFC activation was greater for those with an independent self-construal when thinking about themselves more broadly (e.g. this personality trait applies to me in general). People from a bicultural collectivist-individualistic background, such as Asian-Americans, flip between both patterns of brain activity depending on whether they are primed to think in an individualistic or collectivist way (Chiao et al., 2010).

Evaluation

Broadly speaking, the self-concept entails two different kinds of processes: *differentiation*, which is the notion that the self is a unique entity with a distinct personal history and dispositions, occupying a particular body, and in control of his/her own thoughts and actions; and *assimilation*, where one's identity is embedded in a particular social and cultural context and we identify ourselves with one or more groups and share the values and practices of those groups. The latter process may differ cross-culturally (e.g. the notion of independent versus inter-dependent selves). As such, 'the self' should not be construed as a discrete entity or linked to one brain mechanism.

INGROUPS, OUTGROUPS, AND PREJUDICE

Our social identity can be construed in terms of a collection of different group memberships, for instance relating to one's nationality, race, religion, political allegiances, and so on. Inter-group biases vary as a function of the perspective from which the groups are judged. An Asian woman may view her math ability more favorably when her ethnic identity is highlighted relative to when her gender is highlighted (Shih, Pittinsky, & Ambady, 1999). Similarly White women show evidence of a more negative attitude towards Black women when race is highlighted than when gender is highlighted (Mitchell, Nosek, & Banaji, 2003).

Although group membership operates across many dimensions, these dimensions are not always independent (or are independent but are believed to be otherwise). For example, most people who are British are also White, have English as their first language, and so on. These correlated attributes can give rise to **stereotyping**, in which a collection of attributes become linked together. Rather than considering people in terms of their unique constellation of features, there is a tendency to consider them in terms of social categories (e.g. race, gender, age). This enables a wide range of information from long-term memory to be brought to the fore (Macrae & Bodenhausen, 2000). Related to stereotyping, there is also a tendency to divide continuous dimensions into discrete categories; racial 'categories' are the most common example (e.g. Black vs. White). Stereotyping and categorization are not always bad things. They enable us to assimilate a large amount of information and make generalizations and predictions. However, they can lead to negative outcomes such as **prejudice**.

Figure 9.9 Stereotypes enable us to make quick inferences about people. But should we 'judge a book by its cover'?

Perception of ingroup and outgroup members

We tend to be better at recognizing other faces when they are a similar age to us (the **own-age effect**) and the same race as us (the **own-race effect**). These are generally considered to be expertise-related effects. We tend to have more experience with individuating people from our own age group and race (and this explains why the own-sex effect is rarely reported). Whilst most undergraduates are better at recognizing faces of people in their twenties than faces of children 8–11 years old, this does not hold true for

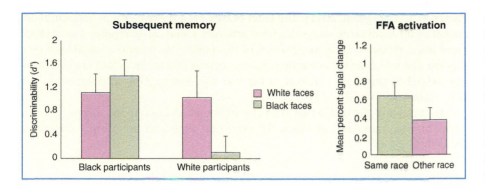

Figure 9.10 Both Black and White participants show an own-race effect in subsequent memory of faces (left figure) with the effect being more pronounced in the White participants. This own-race advantage is linked to increased activity in the fusiform face area (right) which may itself reflect an expertise-related ability to individuate conspecifics of ones own race. From Golby et al. (2001).

trainee teachers working in primary/elementary schools (Harrison & Hole, 2009). Explanations based on contact and expertise predict that it is the perceptual representations of faces (e.g. in FFA) that are being finely tuned. Indeed, one fMRI study implicated the FFA in the own-race memory advantage in both Black and White participants recognizing Black and White faces (Golby, Gabrieli, Chiao, & Eberhardt, 2001). Own-age effects have been found in the insula and medial PFC, but not the FFA, for both neutral and happy facial expressions (Ebner et al., 2013).

Studies on the own-age and own-race effect necessarily require a task of individuating (e.g. 'Have you seen this particular person before?'), but other research has examined social categorization directly by requiring participants to perform categorical gender, race, or age judgements. Ito and Urland (2003) asked participants (predominantly White) to categorize faces according to gender (ignoring race) or according to race (ignoring gender). They recorded ERPs to 'oddball' stimuli (e.g. a White face in a sequence of Black faces) that are known to elicit orienting of visual attention. Effects of race emerged as early as 100 ms (with an unexpected Black face eliciting a larger component than an unexpected White face) and gender at 200 ms. Interestingly, these early effects were independent of the task that the participants were performing, suggesting that these social categories are processed both early and automatically. Indeed the same is found in the absence of any social categorization task at all, such as when participants must detect a dot superimposed on a task-irrelevant face (Ito & Urland, 2005). These ERP components reflect visual-attention effects, but what of the N170 that is related to perceptual encoding of faces? Senholzi and Ito (2013) found that when White participants were required to individuate a face there was a larger N170 to Black than White faces but when the task was to categorize the faces than the effect was reversed (i.e. a larger N170 for White relative to Black faces). This may reflect a general bias to process Black faces categorically and White faces individually, with the N170 increasing when this bias needs to be reversed. ERP studies of the own-age effect tends to find that it occurs later than the N170 (e.g. Wiese, Wolff, Steffens, & Schweinberger, 2013).

Race and age are real-world differences that exist between individuals. However, it is possible to construct groups based on arbitrary groupings (e.g. toss of a coin, color of shirt) such that group members do not differ in terms of average appearance and are not linked to pre-existing stereotypes. This is termed the **minimal group paradigm** (Tajfel, Billig, Bundy, & Flament, 1971). Using this paradigm it has been found that there is better recognition memory for ingroup members (Bernstein, Young, & Hugenberg, 2007), and a larger N170 for ingroup members (Ratner & Amodio, 2013). There is more activity in FFA for ingroup members (relative to unfamiliar people) although FFA activity to outgroup members is not lowered (Van Bavel,

Packer, & Cunningham, 2011). The latter occurs even if the ingroups and outgroups consist of different races, suggesting that arbitrary social categorization can override racial ones. When asked to judge which of two unfamiliar faces is a member of their ingroup and which an outgroup then people tend to choose faces that are consensually judged as trustworthy as part of the ingroup (Ratner, Dotsch, Wigboldus, van Knippenberg, & Amodio, 2014).

In summary, the distinction between ingroups and outgroups affects mechanisms of perception and attention for both real-word and arbitrarily constructed group differences.

Ingroup positivity

Ingroup favoritism over outgroups can be demonstrated even when people are randomly assigned to arbitrary groups (e.g. Rabbie & Horwitz, 1969). Having graphic icons randomly assigned to oneself, relative to another, is sufficient to increase ratings of aesthetic attractiveness of those icons (Feys, 1991). One possible reason for a preference for ingroups over outgroups is that self-related attributes are more likely to be found in one's ingroup than outgroup, and there is a general tendency for self-related attributes to be regarded favorably (Greenwald et al., 2002). This general positive outlook on the self is termed **self-esteem** and is generally assumed to promote a need to belong and resilience against rejection (Baumeister & Leary, 1995).

There is evidence that people have an unconscious preference for self-related material, sometimes called **implicit self-esteem**. Nuttin (1985) found that when shown pairs of letters, people tend to prefer letters in their own name – the **name letter effect** (**NLE**). Participants are typically not aware of the origin of this bias (Nuttin, 1985), and the bias remains when one takes into consideration the fact that letters from one's own name will be more frequently encountered (Hoorens & Nuttin, 1993). Evidence from other stimuli is consistent with this. Pelham, Mirenberg, and Jones (2002) found that people are statistically more likely to choose occupations and cities to live in whose names share letters with their own first or last name (e.g. Denise the dentist from Denver). People are also more likely to donate to a hurricane relief fund if their own name shares the same initial as the hurricane name, as shown in Figure 9.11 (Chandler, Griffin, & Sorensen, 2008).

The neuroscientific basis of ingroup favoritism is not well understood, at least in comparison to outgroup negativity. Using the minimal group paradigm, the perception of ingroup members is enhanced and may rely on more enhancing strategies such as individuating (Van Bavel et al., 2011). Moreover, additional neural resources, such as the mPFC, that normally respond strongly to the self may be recruited when endorsing words belonging to an arbitrary ingroup (red or blue team; Molenberghs & Morrison, 2014). Administration of oxytocin also increases ingroup cooperation and trust (De Dreu et al., 2010). In all of these studies, there is an enhancement of ingroup processing but with a small (or non-existent) detriment to the outgroup.

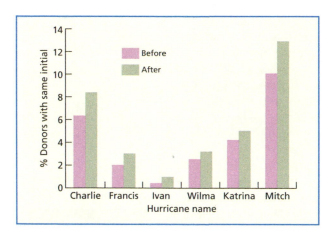

Figure 9.11 People are more likely to donate to hurricane Charlie if their name begins with the letter C (and similarly for other hurricanes). The graph compares donations before and after a hurricane. Different baseline percentages reflect different name letter frequencies (e.g. not many names begin with the letter I). This is often regarded as an implicit form of egotism or positive self-regard. Data from Chandler et al. (2008).

Outgroup negativity and prejudice

Prejudice refers to negative attitudes, emotions, or behaviors to members of a group on the basis of their membership of that group (Brown, 1995). Racism, sexism, and ageism can be regarded as specific variants of prejudice. This definition presupposes a tendency to treat individuals categorically and, unsurprisingly, stereotyping has been viewed as a key cognitive mechanism in prejudice research (Allport, 1954). The origins of the negative orientation may be attributable to historical, political, or economic reasons (e.g. that enable one group to prosper at the expense of another). However, it may also reflect a mere consequence of comparing outgroups against one's ingroups (in which the latter may be 'naturally' favored).

Measuring prejudice and outgroup negativity

One problem in conducting research on prejudice is how to measure it. Questionnaires, such as the Modern Racism Scale (McConahay, 1986), are based on self-reported racist attitudes (e.g. 'If Blacks would only try harder they could be just as well off as Whites'). However, whilst it was common in the 1950s and 1960s for people in Western countries to openly report racist attitudes, our society norms have changed such that it is no longer considered acceptable to express such views. This raises the possibility that racist beliefs may still be common but not openly endorsed. For example, scores on the Modern Racism Scale change according to the ethnicity of the person who is administering it (Fazio, Jackson, Dunton, & Williams, 1995). This social desirability problem has been addressed through the use of more implicit measures of racism, in which either the participant is unaware that his/her beliefs are relevant to

Figure 9.12 Skin color and race vary along a continuum, but we tend to make categorical judgments about them. For example, Barack Obama, 'America's first Black president' has one 'White' parent and one 'Black' parent (according to conventional labeling), but Obama is nevertheless generally categorized as Black.

the task or in which the measure that one takes is considered to be relatively automatic and immune to participants' conscious attempts to hide their beliefs. For example, electromyography (EMG) can measure muscle activity associated with smiles or frowns that are not easily visible to the naked eye. Vanman, Paul, Ito, and Miller (1997) found that with frown-related (corrugator) muscles, but not smile-related (zygomatic) muscles, EMG activity increased when White Americans were asked to imagine working in a cooperative task with a Black American, despite self-reporting potential Black partners more favorably.

One of the most commonly used implicit measures of racism is the **Implicit Association Test (IAT)** (Greenwald, McGhee, & Schwartz, 1998). This records participants' response times to categorizing words and names based on speeded button presses to two keys. For example, people may be instructed to press one button when they see names that are more likely to be given to Black people (e.g. Leroy, Aisha)

KEY TERM

Implicit Association Test (IAT)

Gives a measure of differential association of two target concepts with an attribute, by pairing different concepts and attributes to the same or different response key and another button when they see names likely to denote White people (e.g. Amanda, Matthew). At the same time, they are instructed to also press one button when they see pleasant words (e.g. friend, sunrise) and another button when they see unpleasant words (e.g. murder, vomit). The Black/White names are in upper case and the pleasant/unpleasant words are in lower case. The crux of the measure lies in how these two discriminations (Black/White vs. pleasant/unpleasant) are paired together. In the congruent condition, the response buttons are assigned according to expected prejudices – pleasant and White are responded to on the same button, and unpleasant and Black are responded to on the other button. In the incongruent condition, the response mappings are reversed – pleasant and Black are responded to on the same button, with unpleasant and White allocated to the other button. Greenwald et al. (1998) reported a difference of around 200 ms between these conditions from White participants. For example, 'murder' and 'LEROY' are responded to faster when they are paired on the same button.

The most obvious advantage of this method over self-report is that it is regarded as less susceptible to social desirability effects and may therefore be a more reliable index of true prejudicial beliefs. Another advantage is that the method can easily be adapted to explore other issues relating to stereotyping and ingroup/outgroup biases such as national identity (Devos & Banaji, 2005), political allegiances (Knutson, Wood, Spampinato, & Grafman, 2006), and stereotyped gender attitudes (Nosek, Banaji, & Greenwald, 2002b). The biggest problem with this method is ambiguity over what it is really measuring (Fiedler, Messner, & Bluemke, 2006). Does it really reflect deeply held beliefs or does it reflect cultural knowledge of stereotypes? For example, White Americans are familiar with the stereotype of a Black American (e.g. violent, athletic, good dancers) even if they, apparently, do not personally endorse them (Devine & Elliot, 1995). Another issue is how implicit measures relate to self-report measures of racist attitudes and actual discriminatory behavior. Hofmann, Gawronski, Gschwendner, Le, and Schmitt (2005) report a weak correlation of only +.20 between the IAT and various self-report measures, suggesting that they are measuring quite different things. However, the acid test is to determine which measure predicts actual discriminatory behavior. The IAT has been shown to predict various behavioral signs (e.g. speaking time, smiling, speech errors, and hesitations) in encounters with an outgroup race, more so than self-report measures (McConnell & Leibold, 2001). But does this reflect prejudice (i.e. negative evaluation of an outgroup) rather than discomfort due to lack of outgroup familiarity? Is the IAT really sensitive to long-term attitudes or to the current situational demands? For instance, the IAT score can be reduced after presenting White participants with images of famous and well-liked Black people such as Martin Luther King and Denzel Washington (Dasgupta & Greenwald, 2001).

Most social neuroscience studies relating to prejudice fall within the implicit measures approach. The methods of cognitive neuroscience (e.g. EEG, fMRI) fit well within this approach because it is possible to measure brain-related responses independently from participants' behavior and verbal reports. As such, these studies are susceptible to many of the problems of interpretation leveled at implicit behavioral measures such as the IAT. However, some genuinely new insights have emerged from these studies, including the suggestion that implicit associations may be divided between qualitatively different systems (affective vs. semantic) rather than belonging to a single network (as in Greenwald et al. 2002), and that control of prejudice need not imply *deliberate* control by the participant (for a review see Amodio, 2008).

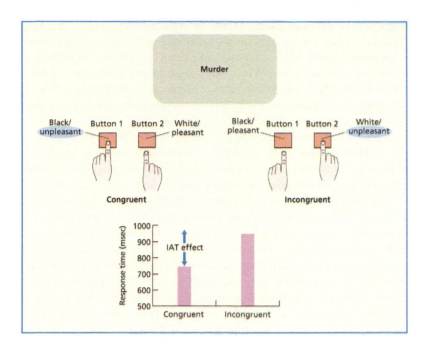

Figure 9.13 In the Implicit Association Test (IAT) people are asked to categorize names according to whether or not they denote Black or White people (e.g. Matthew, Leroy) and they are asked to categorize nouns according to whether they are pleasant or not. White participants are faster to make these judgments when the same response is required to 'Black' and 'unpleasant'. Can we use this to infer implicit racism?

The amygdala and outgroup evaluation

The neural processing of race involves a network of regions as shown in Figure 9.14 and reviewed by Kubota, Banaji, and Phelps (2012). The region that has provoked the most intense debate is the role of the amygdala.

Hart et al. (2000) showed Black and White American participants both Black and White faces during fMRI. Participants were asked to classify the gender of the face (i.e. race processing was incidental). They found that there was a slower decline in amygdala activation over the scanning session for the outgroup relative to the ingroup. This was discussed in terms of the ingroup being more familiar (i.e. unrelated to prejudiced beliefs). However, an alternative interpretation might relate amygdala activation to negative evaluation of the outgroup (i.e. relating to prejudiced beliefs) based on the assumption that the amygdala is crucial for the emotion of fear and the detection of threat (e.g. LeDoux, 2000). In support of this, Phelps et al. (2000) conducted a similar study in which White participants viewed pictures of

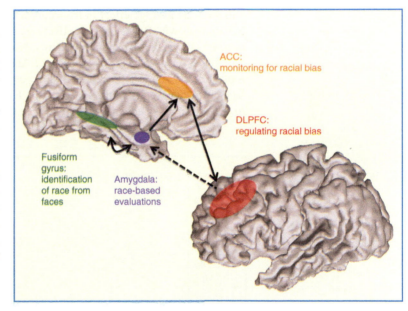

Figure 9.14 The neural processing of race is likely to begin in the fusiform gyrus, involved in face perception. Affective evaluations may be driven by the amygdala which itself may be regulated by regions such as the anterior cingulate cortex (ACC) and dorsolateral prefrontal cortex (DLPFC). From Kubota et al. (2012).

unfamiliar Black and White faces during fMRI. To maintain attention to their face, they were asked if each face was the same as the previous one. Outside of the scanner, participants completed both an implicit measure of racism (the IAT) and a self-report measure (the Modern Racism Scale). Whilst activity in the amygdala did not differ between the viewing of Black and White faces, differences in amygdala activation (Black–White) across participants correlated with the implicit, but not explicit, measure of racism as shown in Figure 9.15.

Subsequent research has confirmed amygdala activation by White participants in response to Black faces. However, the results are somewhat inconsistent and appear to depend on the nature of the task. In their second study, Phelps et al. (2000) reported that the correlation between amygdala activation and the IAT score was not found when well-liked famous Black people (e.g. Martin Luther King) were used as stimuli. Wheeler and Fiske (2005) reported greater amygdala activity when White participants made a categorical judgment about Black faces (Age over or under 21?) than an individuating judgment (Do you think he likes this vegetable?). Lieberman, Hariri, Jarcho, Eisenberger, and Bookheimer (2005) report that amygdala activation is greater when participants must match faces according to race using pictures (e.g. matching a Black face to a different Black face or a White face) rather than verbal labels (e.g. matching a Black face to either 'Caucasian' or 'African-American'). This study is noteworthy in that it was administered to separate groups of White and Black participants, enabling a direct comparison between them. Lieberman et al. (2005) found that the amygdala response was greater for Black faces, relative to White faces, in *both* Black and White participants. (Note that the earlier study of Hart et al., 2000, used participants of both races but only analyzed the data in terms of ingroup vs. outgroup by pooling across the two races.) They suggest that this reflects negative cultural stereotypes towards African-Americans rather than prejudice per se. When Black people take the IAT they still show an average effect in the same direction as White participants (i.e. negative associations with Black), but the effect size is considerably smaller, and they do report ingroup favoritism on explicit measures (Nosek, Banaji, & Greenwald, 2002a).

Using a different kind of measure, Amodio, Harmon-Jones, and Devine (2003) looked at the startle response of White participants shown White, Black, and Asian faces. A defensive eye blink (measured by EMG) is part of the normal startle response to, say, hearing a sudden loud noise. This startle response is amplified if the person

Figure 9.15 Activation of the amygdala in response to White participants viewing Black faces correlates with performance on an implicit (IAT) but not explicit measure of racism. Is the amygdala activity related to fear? From Phelps et al. (2000). Copyright © 2000 by the Massachusetts Institute of Technology. Reproduced with permission.

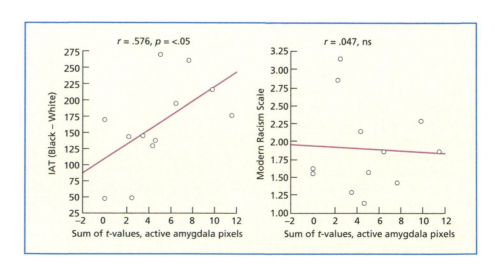

feels threatened, and this amplification is modulated by the amygdala (Davis, 1992). The White participants showed a heightened startle response to a loud noise when accompanied with Black faces relative to White and Asian faces. This is consistent with a negative evaluation of Blacks rather than relating to outgroup perception per se, given that both Black and Asian groups are outgroups to the White participants. The results of these studies converge on the conclusion that amygdala activation in response to different races does not reflect a general negative bias towards all members of racial outgroups but, rather, a more complex evaluation that takes into account familiarity with particular members of the outgroup together with negative sociocultural stereotypes.

One recent review of the literature (Chekroud, Everett, Bridge, & Hewstone, 2014) concludes that the amygdala response to Black faces is indeed related to threat-based processing, but that this is based on the negative stereotype that young, black men are violent, criminal, and dangerous. In support of this idea, they cite evidence of amygdala activity only when a Black face is using direct eye contact (Richeson, Todd, Trawalter, & Baird, 2008). Again, this suggests that it is not outgroups per se but rather a more complex evaluation, which they suggest reflects an implicit threat detection. Chekroud et al. (2014) make the interesting prediction that reports of amygdala activation may have been driven by perceiving Black *males* rather than Black females in previous research (which, to date, has not analyzed the stimuli according to gender), consistent with greater threat associated with the Black male stereotype.

Figure 9.16 Measures of implicit racism such as the IAT (a behavioral measure) and amygdala activity in fMRI (a neurophysiological measure) disappear when positive Black images, such as of Martin Luther King, are displayed. This suggests that these measures are not driven by the *perception* of an outgroup member per se.

The representation of stereotypes

Where are stereotypes represented in the brain? Amodio (2008) proposes that knowledge relevant to the IAT is represented in at least two different memory systems: a semantic memory system and an affective system of associations (based on classically conditioned associations). A person may know about group stereotypes (i.e. they represent them in their semantic memory), but they do not personally hold negative views about that group (i.e. they are not represented as a threat association). A single case study of a patient with a lesion to the amygdala found that the race IAT was unaffected (Phelps, Cannistraci, & Cunningham, 2003). This suggests that the stereotype is either not stored in the amygdala, or if it is, then it is also stored separately in some other system in the brain consistent with Amodio's (2008) theory.

Amodio and Devine (2006) devised an alternative version of the IAT in which one category of response was Black vs. White and the other category of response was changed from an affective one (pleasant vs. unpleasant) to a semantic one (physical vs. mental). According to Devine and Elliot (1995), most White Americans recognize the stereotype of Black Americans being good at physical activities (e.g. basketball, dancing) and White Americans being good at math, reading, etc. Importantly, both physical ability and mental aptitude are regarded in a positive light and so are matched for overall valence, unlike the standard race-based IAT. Amodio and Devine (2006) found that performance on the original, affective version of the IAT was uncorrelated with performance on the purely stereotypic version of the IAT. An fMRI study based on these two versions of the IAT found different neural substrates within the frontal lobe for these tasks (lateral orbitofrontal for affective, medial prefrontal for semantic) (Gilbert, Swencionis, & Amodio, 2012). They also found that the pattern of activity in the left temporal pole correlated with individual differences in the magnitude of IAT across both tasks. (Note: this study differs from previously described ones in that it is not comparing ingroups and outgroups but rather different tasks.)

The temporal pole is considered as a hub of detailed semantic knowledge and also social knowledge and person-specific knowledge (Patterson, Nestor, & Rogers, 2007). Patients who have damage to this region have semantic memory impairments but in such a way that they lose individuating knowledge and rely instead on stereotypes. For instance, if asked the color of a carrot they may answer 'green' based on stereotypical knowledge of vegetable colors (Patterson, 2007). Patients who have damage to this region show an *increase* in a gender version of the IAT, associating males with strength and females with weakness (Gozzi, Raymont, Solomon, Koenigs, & Grafman, 2009). This suggests that the temporal pole is not storing stereotypes (which would predict that the IAT would be diminished when the region is lesioned) but is perhaps storing other kinds of individuating information that guards against stereotyping.

The Stereotype Content Model argues that we tend to categorize individuals and groups along two dimensions (Fiske, Cuddy, Glick, & Xu, 2002). The first is the dimension of 'warmth', which can be construed as trustworthiness and good intentions (i.e. morally upright). There is a tendency to act cooperatively with those of high warmth and to act competitively with those of low warmth. The second dimension is termed 'competency' and is linked to status (and power). Crossing these dimensions yields four different kinds of social groups – shown in the table below – that tend to elicit specific emotions (pity, disgust, pride, and envy). The structure of this model is derived from questionnaires and is stable across cultures, albeit with more extreme stereotypes in societies with greater income inequality (Durante et al., 2013). There is also evidence for the model from neuroimaging.

	Low competence and status	High competence and status
High Warmth / Cooperative	Older, disabled [Pity]	Ingroup, allies, middle class [Pride]
Low Warmth / Competitive	Poor, immigrant, homeless [Disgust]	Rich, professionals [Envy]

Harris and Fiske (2006) showed images of people from each of the four quadrants of the model during fMRI and asked participants what emotion they felt. Activation in the mPFC was significantly lower than a fixation-cross baseline for the low-low group (addicts, homeless) but above baseline for the other social groups – see Figure 9.17. (The region was close to that reported to be activated by self-referential thought, but that comparison condition wasn't included.) The absence of mPFC activity was interpreted as a dehumanized response to these low-low outgroups.

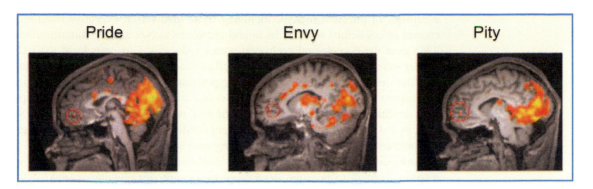

Figure 9.17 The mPFC (circled in red) was activated when shown images of different social groups and participants had to identify their own emotional response, but was not activated by a dehumanized social group (linked to feelings of disgust). Note that the activity at the posterior of the brain reflects visual cortical activity due to comparing against a fixation-cross baseline. From Harris and Fiske (2006).

Other research suggests that people make fewer mentalizing attributions to these groups (Harris & Fiske, 2011). The low-low group was, instead, linked to increased activity in the amygdala and insula and subjective disgust evaluations. This particular pattern is context dependent: when one is asked to make individuating relative to categorical judgments about the low-low social groups, then mPFC activity is observed (Harris & Fiske, 2007). In general, the results of this set of studies chimes with earlier ones that suggest that the brain's response to outgroups depends on the social knowledge and evaluation of that outgroup rather than their status as an outgroup per se.

Control mechanisms for prejudice

Whilst all members within a given culture are likely to know about common stereotypes and prejudices, people will differ in the extent to which they act upon them. Knowing about stereotypes and prejudices but not acting upon them may require greater online control. Related to this is the idea, mentioned at the outset, is the fact that people have different motives for not exhibiting prejudice. Plant and Devine (1998) attempted to classify people according to whether they had internal reasons for rejecting prejudice (e.g. endorsing statements such as 'I attempt to act in non-prejudiced ways towards Black people because it is personally important to me') versus those who report conforming to external norms (e.g. endorsing statements such as 'I attempt to appear non-prejudiced towards Black people in order to avoid disapproval from others'). It is a moot point as to whether people who report some external motivations for avoiding prejudice should be regarded as prejudiced or not. However, those who report *only* internal motivations could be reasonably described as non-prejudiced. Indeed, those reporting only internal motivations to avoid prejudice show reduced interference on the IAT test (Devine, Plant, Amodio, Harmon-Jones, & Vance, 2002) and also a reduced startle response to Black faces accompanied by a loud noise (Amodio et al., 2003).

Many of the functional imaging studies described above correlated performance on measures such as the IAT, administered outside the scanner, with performance on different tasks performed inside the scanner, typically involving presentation of Black and White faces. An alternative approach is to perform the IAT during fMRI scanning itself (e.g. Beer et al., 2008; Knutson, Mah, Manly, & Grafman, 2007). A comparison of brain activity on incongruent relative to congruent trials in these studies reveals a network of different regions, including those involved in executive functions (i.e. control of cognition) such as the lateral prefrontal cortex and the anterior cingulate. TMS over the dorsolateral prefrontal cortex reduces the magnitude of the IAT in a gender stereotype test (Cattaneo, Mattavelli, Platania, & Papagno, 2011). The anterior cingulate has been implicated in detecting potential conflicts in responses, for instance in situations in which a single stimulus could potentially give rise to two responses (e.g. the Stroop Test; see Chapter 4). The IAT is conceptually related to the Stroop Test, in that the stereotypically incongruent condition (Black + good) leads to greater response conflict than stereotypically congruent condition (Black + bad).

Amodio et al. (2004) conducted an ERP study that was designed to explore the response conflict to negative stereotypes by the anterior cingulate cortex. Although ERP is not well suited to detecting *where* neural activity is occurring, there is a particular ERP deflection termed error-related negativity (or ERN) that has a known locus within the anterior cingulate cortex (Dehaene et al., 1994). This deflection occurs when people make an error and the peak of the deflection occurs within 100 ms of the error occurring (Gehring et al., 1993). As such it is an early and automatic signature of error-ful behavior, and it occurs independently of whether the participant is aware of having made an error (O'Connell et al., 2007). Amodio et al. (2004) used a task different from the IAT in which Black or White faces are briefly presented, for 200 ms, followed by either a picture of a tool or a gun (based on Payne, 2001). The participants' task is simply to decide if a tool or gun was presented (i.e. ignoring the face). However, the combination of Black + gun is consistent with a negative Black stereotype and Black + tool is inconsistent with this stereotype. Thus, a Black face followed by a tool should elicit greater response conflict and be more error prone (relative to White face followed by tool, as a control comparison). This is found both in terms of behavioral responses (Payne, 2001) and in terms of the error-related negativity almost certainly arising from the anterior cingulate (Amodio et al., 2004). This was found for both those who were internally versus externally motivated to avoid prejudice, although a slightly later EEG component (termed the Pe) did distinguish between these different motivations (Amodio, Kubota, Harmon-Jones, & Devine, 2006). This is shown in

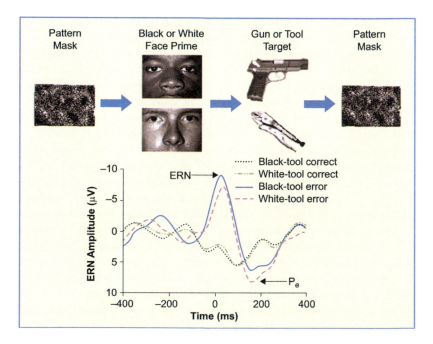

Figure 9.18 The event-related negativity (ERN), recorded using EEG, is greater when Black faces are paired with tools rather than guns because the correct response (tool) conflicts with the stereotype (i.e. association of violence with Blacks). From Amodio et al. (2004).

Figure 9.18. This later component, unlike the earlier one, is sensitive to whether one is aware/unaware of having made an error (O'Connell et al., 2007).

Evaluation

Recent research in both social psychology and social neuroscience has focused on implicit measures of prejudice. These measures are almost certainly multi-faceted; that is, different cognitive mechanisms are likely to contribute to a single behavioral measure (such as the IAT). Current research in this area suggests that there are different mechanisms involved in representing the knowledge of stereotypes (e.g. semantic memory) versus one's affective evaluation of them (e.g. in terms of potential threat). Attempts to control prejudice may engage different mechanisms depending on the extent to which one is motivated only by internal beliefs (i.e. truly non-prejudiced) or also due to an external desire to conform (i.e. potentially prejudiced).

It is to be noted that, by far, most of the social neuroscience studies relating to prejudice are based on the model of a White American ingroup and a Black American outgroup. A small number of studies have looked at the reverse (i.e. Black ingroup, White outgroup), but comparatively few have looked at other races or non-racial forms of prejudice (e.g. based on religion, sex, nationality).

COGNITIVE NEUROSCIENCE OF RELIGIOUS BELIEFS

Religious beliefs (or a lack of) typically form an important part of a person's social identity and, like race, is a prominent way of socially categorizing people. Religions are found in every human culture and consist not only in the belief of supernatural beings with human-like (e.g. desires, emotions) and non-human-like (e.g. omnipresent) abilities, but also of complex social practices and rituals. The multi-faceted nature of religion makes a single account of it, in social or cognitive terms, unlikely. However, several of the basic mechanisms discussed so far can be considered relevant. For example, the persistence of religious beliefs in the face of contradictory evidence resembles the conformity findings of Asch (1951). Others have commented on how wearing certain religious clothes (e.g. the full-length veil or chador) is a form of deindividuation in which the social norms of the group are emphasized (Jahoda, 1982).

Some aspects of religiosity may have their origins within the individual, rather than at the social, group level. For example, Persinger (1983) has argued that the temporal lobes may support certain kinds of religious and mystical experiences based upon the unusual experiences reported during epileptic seizures emanating in this area. These include dream-like states, a feeling of presence, and intense meaningfulness. However, double-blind studies in which magnetic stimulation, using TMS, is applied to the temporal lobes of non-epileptics (i.e. in which neither the participant nor the experimenter knows when the stimulation occurs) have failed to provide convincing support for this idea (Granqvist et al., 2005). Moreover, explanations along these lines fail to account for the social nature of religion. We tend to acquire the religious beliefs of those around us as a result of interacting with others rather than via individual discovery.

Figure 9.19 There are many things that a scientific account of religion must explain. One issue is how and why behavior becomes ritualized and, hence, self-perpetuating.

Some account is needed of why religious beliefs tend to be so widespread. One common socio-psychological explanation of religion is in the promotion of pro-social behavior within groups (e.g. Bloch, 2008; Bulbulia, 2004). As Bulbulia (2004) put it: 'Groups that pray together stay together, and so flourish against other groups' (p. 673). Other manifestations of religious beliefs, including rituals and monuments, would serve to identify one's own group membership and bring them together. Religions also tend to promote good deeds within the group and offer protection to the faithful against threats (real or imagined), which encourages compliance. Religious rituals may be inherently rewarding (and self-perpetuating) because they are believed to bring rewards or protect from harm. Boyer (2008) draws an analogy with non-religious rituals in people with obsessive–compulsive tendencies that are also perceived as cleansing or protecting in some way. Prayers offered by believers do indeed activate the reward circuits of the brain – that is, the belief in a reward can be rewarding in its own right (Schjødta, Stødkilde-Jørgensenb, Geertza, & Roepstorff, 2008).

A more cognitive explanation for religious beliefs views them as an extension of theory-of-mind mechanisms. Guthrie (1993), in his book *Faces in the Clouds*, argues that we have a natural tendency to project human-like mental states into the world. Similarly, Tomasello (1999) argues that the evolutionary origins of human culture lie in a tendency to infer 'hidden forces' in the world. In the social realm, the hidden forces consist of the mental states of others – people's thoughts, desires, and beliefs. But this default way of thinking may extend towards postulating mental states to inanimate objects (e.g. the Sun) and unseen or supernatural agents (as in God). We may also tend to explain unpredictable physical occurrences in terms of these hidden agents (e.g. an unexpected flood caused by God's anger). Functional imaging studies have shown that interactive prayer (relative to ritualized prayers) in believers is linked to mPFC and rTPJ regions that are implicated in theory of mind (Schjoedt, Stdkilde-Jorgensen, Geertz, & Roepstorff, 2009) and endorsing statements about religious beliefs is linked to the mPFC in believers (Harris et al., 2009). People with autism, and putative theory-of-mind deficits, are less religiously inclined (Norenzayan, Gervais, & Trzesniewski, 2012).

The pro-social behavior and the theory-of-mind explanations may explain different aspects of religiosity. The former may explain why religions spread and become ritualized, whereas the latter may explain the nature of some of the beliefs (e.g. projection of agency).

SUMMARY AND KEY POINTS OF THE CHAPTER

- Our self-awareness is constructed out of different cognitive operations and is multi-faceted (e.g. awareness of our body, our actions, of existing over time). However, many self-related mechanisms occur without us being aware of them.
- Some aspects of 'the self' do not differ across cultures (e.g. that we occupy our own bodies) whereas other aspects are embedded within our cultural context (e.g. our social identity).
- Our social identity is determined by the groups to which we consider ourselves to belong and our perceived roles within these groups (e.g. independent vs. inter-dependent self). Different social identities may activate different neural networks within the brain (including the medial prefrontal cortex), and have different consequences for cognition.
- Groups to which we belong (our ingroups) may tend to be viewed favorably because we tend to view ourselves favorably. This may also lead to negative attitudes towards outgroups.
- Social neuroscience studies of prejudice often reveal activity in the amygdala associated with perception of a racial outgroup. However, it is unclear whether this is related to negatively held cultural stereotypes or negatively held personal beliefs, but it does not seem to be an index of 'outgroup processing' per se.
- Neuroscientific and behavioral measures of cognitive control may provide an implicit measure of racist attitudes and may distinguish between different motivations for control (conformity to norms vs. personal beliefs).

EXAMPLE ESSAY QUESTIONS

- How can 'the self' be understood from a social neuroscience perspective?
- How do cross-cultural differences in social identity affect the brain and cognition?
- Does the brain process outgroups differently from ingroups?
- What has evidence from social neuroscience contributed to our understanding of prejudice?
- What effects do power and social status have on the social and emotional brain?

RECOMMENDED FURTHER READING

- Derks, B., Scheepers, D., & Ellemers, N. (2013). *Neuroscience of Prejudice and Intergroup Relations*. New York: Psychology Press. An excellent set of chapters that closely relates to many topics covered in this chapter.

- Krueger, F., & Grafman, J. (2012). *The Cognitive Neuroscience of Beliefs*. New York: Psychology Press. Contains chapters on religion, politics, stereotypes, and prejudice.

- LeDoux, J., Debiec, J., & Moss, H. (2003). *The Self: From Soul to Brain*. New York: New York Academy of Sciences. (Also published in journal form as Volume 1001 of *Annals of the New York Academy of Sciences*.) A varied collection of papers that discuss the self from a mainly cognitive-neuroscience perspective.

ONLINE RESOURCES

- References to key papers and readings
- Videos demonstrating the IAT, Libet experiment, and out-of-body illusions
- Interviews and talks given by Mahzarin Banaji, Susan Fiske, Elizabeth Phelps, and others
- Recorded lecture given by textbook author, Jamie Ward
- Multiple choice questions and interactive flashcards to test your knowledge
- Downloadable glossary

CHAPTER 10

CONTENTS

Morality and antisocial behavior

10

Morality is essentially concerned with the 'right' or 'wrong' of social behavior, and therefore, provides a system for regulating what is permissible. But on what basis do we decide what is right or wrong? Many of the social rules we follow are arbitrary: words or gestures that cause extreme offence in one country may be harmless in another; and certain forms of dress code may be normal in one culture but deemed intolerably immodest by another. Broadly speaking, we can discriminate between two different kinds of social norms: conventional norms and moral norms (Turiel, 1983). Examples of **conventional norms** include not swearing or vomiting in public, dressing neatly for a job interview, and shaking hands when being introduced. Examples of **moral norms** include not causing deliberate mental or physical harm to others. These norms may have different origins. Conventional norms may originate via consensus or authority (relating to group conformity), whereas moral norms may ultimately depend on many of the socio-cognitive processes discussed previously, such as empathy, fairness, and mentalizing. Children as young as 4 years old have an intuitive sense of the moral–conventional distinction. For instance, they realize that there are often good reasons for breaking conventional, but not moral, norms (Smetana, 1981, 1985). However, some adults – such as murderers – do sometimes fail to make this distinction (Blair, 1995), as shown in Figure 10.1.

This chapter is broadly divided into three parts. It starts by considering the kind of cognitive and neural mechanisms that support moral judgments. The second section focusses specifically on aggression and antisocial behavior. **Antisocial behavior**

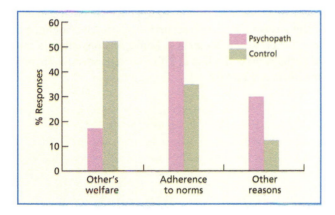

Figure 10.1 Psychopaths incarcerated for murder take a more flexible approach to moral norms. Whilst they are aware of the norms of right and wrong, they are more likely to endorse the view that these norms can justifiably be broken (e.g. for personal gain or if an authority figure is absent). In this graph, when asked to justify why a moral transgression is wrong they tend to appeal to conventional norms (e.g. 'it is illegal') rather than the welfare of others (e.g. 'it will hurt them'). Data from Blair (1995).

is defined as any behavior that violates the social norms of a particular culture, whereas aggression is defined, more specifically, as an intention to cause harm. The law can be regarded, in psychological terms, as defining those collectively agreed upon social norms that, if broken, require punishment to be meted out. As such, antisocial behavior and criminality are related but not the same thing: the latter is a subset of the former. For example, swearing and offensive language can be construed as antisocial but tend to be legal. The final section briefly considers emerging topics around neuroscience perspectives on responsibility and 'neuroethics'.

THE NEUROSCIENCE OF MORALITY

To introduce the issues faced in this section, consider the two moral dilemmas below (originally from Thomson, 1976, 1986) and illustrated in Figure 10.2.

> *The Trolley Dilemma: A trolley car (or tram) is out of control. If it continues along the current tracks it will crash into five hikers crossing the line. If it is switched to a side track it will kill only one hiker who happens to be crossing on this part of the line. You control the switch. What do you do?*

> *The Footbridge Dilemma: A trolley car (or tram) is out of control. If it continues along the current tracks it will crash into five hikers crossing the line. There is a footbridge going over the track with a fat man standing on it. If you push the fat man off the footbridge on to the line it will kill him, but it will stop the train before it gets to the five hikers, thus saving them. You are standing next to the fat man. What do you do?*

Most people, when given these scenarios, flip the switch in the Trolley Dilemma, thereby killing one, but do not push the fat man off the footbridge, thereby killing five (Petrinovich & O'Neill, 1996). These judgments normally occur intuitively and people find it hard to explain their reasoning, or to give a coherent explanation as to why they gave different answers in the two scenarios (Cushman, Young, & Hauser, 2006). Some people do give consistent answers to both questions (always saving five lives), but they would perhaps draw the line at killing one to save five people in the Organ Donor Dilemma of Hauser (2006):

> *The Organ Donor Dilemma: Five people are rushed into an ER each requiring a different organ transplant if they are to stay alive. No organs are available. However, there is another patient in the waiting room with only minor injuries; all his/her vital organs are intact. Do you kill this person in order to save the five in need of organs?*

In determining whether the implied action is permissible, people may draw on at least three different kinds of information that is, itself, underpinned by different neural mechanisms:

1. Whether it feels right or wrong to me (e.g. emotional reactions of pride or guilt)
2. Whether society deems it to be right or wrong (e.g. in terms of the law)
3. Whether the consequence of an action is likely to be net positive or net negative (a cost–benefit analysis)

Figure 10.2 Both the Trolley Dilemma and the Footbridge Dilemma involve a decision between saving five lives and killing one person. In the Trolley Dilemma, the decision concerns whether to flick a switch that sends the trolley onto a side track (killing a hiker crossing the track). In the Footbridge Dilemma, the decision is whether to push a large person off a footbridge to stop the runaway trolley from killing five people.

The third option is a so-called **utilitarian** decision, which focuses on maximizing positive outcomes (i.e. the decision focusses on the endpoint rather than the process). This kind of information biases the judgment towards saving five lives over one life in all scenarios. But, at some point (either at the Footbridge or Organ Donor Dilemma) a more intuitive gut reaction may overturn the utilitarian decision. The second source of information, relating to the consensual or legal answers to these dilemmas, tends to be largely unknown by individuals in these scenarios.

In everyday life these different sources of information should converge on the same response – that is, what feels right to me is also deemed right by society and tends to result in positive outcomes (for both me and others). In other situations they may not converge, as is the case in these dilemmas. To give other examples, some antisocial behavior may be perceived as right in the eyes of the perpetrator but wrong in terms of the law (e.g. revenge for an insult). Similarly, some politicians – even to this day – justify the use of torture on the basis that it is permissible to harm a small number of people to save a larger number.

Having considered the kinds of information people use to make moral decisions, the next section will look more closely at the mechanisms that underpin this in the brain. The section after will consider how these mechanisms differ across individuals and across cultures.

KEY TERM

Utilitarian
The moral worth of an action is determined by its outcome – maximized positive outcomes for minimized negative ones.

'Cognitive' versus emotional processes: moral emotions and moral decisions

Should a man steal a drug to pay for his wife's life? Is it okay for someone to lose their life if five other lives are saved? If someone calls you 'pig', is it okay to hit them? Examples such as these are often used in experiments in the psychology of morality and have more recently been used in social neuroscience. How are such decisions reached? Broadly speaking, two kinds of processes have been postulated: a mechanism based on emotional evaluations (or gut instincts) to these questions, or a more deliberate attempt at reasoning through the problem (e.g. considering the basis for the judgment and weighing the alternative answers). The latter is sometimes referred to as 'cognitive', although this presupposes a view that cognition and emotion are separate kinds. A better way of labeling this dichotomy is moral intuition (which tends to be more heavily emotion based, although it need not be) and moral reasoning (Haidt, 2007). Within this conceptual space, there are researchers who highlight the importance of emotions (e.g. Haidt, 2001), those that highlight the importance of reasoning (e.g. Kohlberg, Levine, & Hewer, 1983; Piaget, 1932), and those who argue for a mix of both (e.g. Greene, 2008). Others argue that the distinction itself is not valid (e.g. Moll & Schulkin, 2009).

Moral dilemmas

Greene and colleagues required participants to make decisions on moral dilemmas during functional imaging, rather than passively processing moral scenes or narratives (Greene, Nystrom, Engell, Darley, & Cohen, 2004; Greene et al., 2001).

Figure 10.3 Greene et al. (2001) gave participants different kinds of moral dilemma during fMRI scanning (and a non-moral control condition). Dilemmas involving personal harm to someone else (e.g. the Footbridge Dilemma) were more likely to engender activity in areas of the brain processing emotions, whereas dilemmas involving impersonal harm (e.g. the Trolley Dilemma) were more likely to engender activity in areas associated with working memory. (B = bilateral; L = left; R = right.) From Greene et al. (2001). Copyright © 2001 American Association for the Advancement of Science. Reproduced with permission.

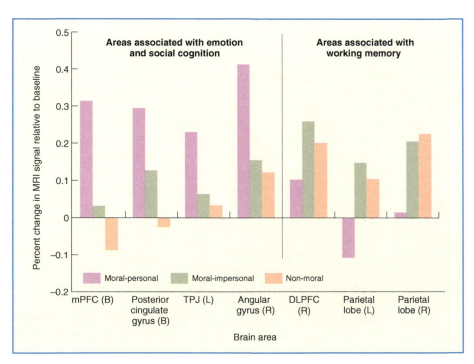

These studies were based on the Trolley and Footbridge Dilemmas (introduced above) together with other conceptually similar scenarios. Greene et al.'s (2001) explanation for the discrepancy between these scenarios is that flipping a switch in the Trolley Dilemma is *impersonal* (less emotionally based) whereas pushing the man off the footbridge in the Footbridge Dilemma is a *personal* act. Their imaging results were consistent with this. Contemplating personal scenarios such as in the Footbridge Dilemma activates regions linked to emotional processing (amygdala, posterior cingulate cortex) and social cognition (superior temporal sulcus, medial prefrontal cortex), whereas consideration of the impersonal moral dilemmas was associated with relatively greater neural activity in regions implicated in the control of behavior (dorsolateral prefrontal cortex, inferior parietal lobe) – see Figure 10.3. For those people who did opt to push the fat man off the footbridge (i.e. thereby consistently basing their judgment on cost–benefit analysis), their brain activity showed the more cognitive pattern rather than the more typical emotional pattern on these trials (Greene et al., 2004).

The theory proposed by Greene (2008) is that different senses of right/wrong are implicated in emotional versus reasoned moral decisions. Emotional judgments appeal to generic principles ('it is wrong for me to harm someone') whereas the reasoned/'cognitive' judgments are based on a cost–benefit analysis that focuses on the consequences of actions (it is better to kill one than to kill five). Other evidence is consistent with this. Increasing the cognitive demands by getting participants to perform an irrelevant task during the moral dilemma slows down decisions based on reasoning about consequences but not emotions (Greene, Morelli, Lowenberg, Nvstrom, & Cohen, 2008).

Social knowledge (e.g. of stereotypes) contributes to decision outcomes in moral dilemmas. Cikara, Farnsworth, Harris, and Fiske (2010) presented images of people from different social groups (five people from one group, and one person from another), and participants were told that a decision had been made to kill one and save five. It was deemed least acceptable to sacrifice one ingroup member (e.g. someone holding an American flag) to save five people from a low-warmth/low-competence outgroup (such as the homeless), and these particular decisions were linked to a number of prefrontal regions such as the mPFC (perhaps reflecting judgments of self–other similarity), lateral, and orbital prefrontal cortices. That is, intergroup biases provide an important context for moral decision-making.

Moral emotions

Haidt (2001, 2003) argues that most of our moral judgments are guided by our emotions, and he proposes the existence of a set of *moral emotions* that are linked to such decisions. Under this account the role of cognitive/reasoning mechanisms is to provide a post hoc justification of the decision rather than influence the decision itself. These moral emotions can be grouped into various sub-groups, or families, according to criteria such as whether the emotion is self-focused (e.g. guilt, pride) or other-focused (e.g. anger, pity) and whether it is critical (e.g. guilt) or praising (e.g. pride) – see Figure 10.5. The different kinds of emotion may be linked to (and motivate) different kinds of behavior. For instance, pity and compassion ('other-suffering' in this taxonomy) may elicit altruistic acts whereas anger and moral disgust may elicit aggressive or antisocial acts ('other-critical' in this taxonomy). Self-critical moral emotions (guilt, shame, embarrassment) may tend to protect

CATEGORIES OF MORAL EMOTIONS (SUB-DIVISIONS)

	Self-conscious		Other-conscious		
	Self-critical	Self-praising	Other-critical	Other-praising	Other-suffering
Guilt	√				
Shame	√				
Embarrassment	√				
Pride		√			
Indignation/anger			√		
Contempt/disgust			√		
Pity/compassion					√
Awe/elevation				√	
Gratitude				√	

Source: Adapted from Haidt (2003).

Figure 10.4 Moral emotions are emotions that are related to the behavior of oneself (in relation to others) or the behavior of others (in relation to oneself or others).

against antisocial acts. Feelings of anger may have a particularly important role to play in eliciting antisocial behavior. **Anger** is the emotion felt when someone else is judged to have intentionally violated a social norm. Displays of anger (in the face, body, or voice) serve a dual function of notifying the other person that they have crossed a line, and also signaling that further retributive action may be taken against them (e.g. violence). Anger is an inter-personal stop signal. **Moral disgust** involves a judgment about the moral standing of another person relative to oneself in terms of their general disposition to engage in acts that are deemed to be wrong (Tybur et al., 2009). Unlike anger, it is not necessarily associated with a triggering incident but may be based on the perceived characteristics of a person or outgroup. It is often linked to dehumanization of that person/outgroup (Fiske, Cuddy, Glick, & Xu, 2002). Acts of genocide tend to be pre-empted by characterization of the outgroup as inhumane or animal-like in their behaviors (Bandura, Barbaranelli, Caprara, & Pastorelli, 1996).

According to some researchers, moral emotions differ from basic emotions (considered in Chapter 4) in that they consist of a blend of basic emotions plus cognitive appraisals of the triggering event (e.g. Moll, de Oliveira-Souza, Zahn, & Grafman, 2008). A cognitive appraisal in this context refers to thoughts that accompany the emotion, such as what the other person is thinking, future outcomes, or the background context to the event. Moll et al. (2002) presented pictures of three kinds of emotional scenes to participants undergoing fMRI: images of moral violations (e.g. images of physical assaults, abandoned children), images of aversive scenes (e.g. dangerous animal), and pleasant images. These were matched for their self-reported arousal. The moral violation and aversive images were matched in terms of how negatively they were judged, but the moral violation images were judged as more morally unacceptable than the other affective stimuli. All affective stimuli (relative to a neutral set of images) tended to activate regions linked to emotional processing, such as the amygdala and insula, but moral emotions (relative to other affective stimuli) additionally activated regions such as the orbitofrontal gyrus, the medial prefrontal cortex (mPFC), and the right posterior superior temporal sulcus (STS). The medial PFC and right posterior STS have been linked to theory of mind / mentalizing (Amodio & Devine, 2006; Saxe, 2006), whereas the orbitofrontal cortex is implicated in the regulation of social behavior (Rolls, 1996). Similar brain regions were activated in fMRI for the moral emotions of embarrassment (Berthoz, Armony, Blair, & Dolan, 2002) and guilt (Takahashi et al., 2004) elicited by reading verbal narratives describing a norm violation relative to neutral narratives, for example, 'I left the restaurant without paying' (guilt) and 'I mistook a stranger for my friend' (embarrassment).

According to Moll and colleagues there is not a conflict between emotions and cognition that needs to be resolved by higher control, but rather moral emotions *are* an integration of emotion with cognitive appraisal (Moll et al., 2005). However, there might be situations in which there is an extra stage of 'conflict resolution' in order to decide between different courses of action. If an action has both negative and positive outcomes (e.g. hurting others to save your own reputation), then this may require an additional control mechanism to overcome more typical thinking (i.e. hurting others = wrong). Also, the distinction between basic emotions and moral emotions is not likely to be as straightforward as implied by this research. It assumes that basic emotions occur without any kind of cognitive appraisal, which may not be the case. There just might be more scope for cognitive appraisal when viewing an image of someone being hit rather than viewing an image of someone in pain. The distinction between basic emotions and moral emotions can be regarded as an extension of the broader debate as to what emotions are, when they are used, how they differ across species, and so on.

Some of the studies cited above involve violations of social norms, but they do not necessarily involve moral norms (i.e. an intentional act against a victim or victims) – or, at least, these two factors have not always been directly compared as different experimental variables (e.g. Berthoz et al., 2002; Takahashi et al. 2004). Mistaking a stranger for a friend may be embarrassing, but it is perhaps not immoral. Finger, Marsh, Kamel, Mitchell, and Blair (2006) directly contrasted these scenarios using fMRI in a 3 × 2 design. One factor was the type of transgression (moral, conventional, neutral), and the other was whether onlookers saw the transgression or not. Participants had to read the narratives silently. For example, a moral transgression may involve a narrative describing a car crash in which you killed someone (either with onlookers or not). A conventional transgression may involve a narrative about

Figure 10.5 What regions of the brain are activated when viewing (or thinking about) scenes involving moral transgressions, such as child abandonment? Is it the same pattern found when viewing other emotional stimuli that do not involve a transgression?

vomiting in a public place (either with onlookers or not). The two transgression conditions, relative to the neutral condition, were associated with activity in lateral prefrontal regions. However, whereas this pattern was found for moral transgressions irrespective of whether onlookers were present, it was only found for social transgressions in the presence of onlookers. They suggest that the activation of cognitive control mechanisms (perhaps linked to emotion regulation) may depend both on the nature of the transgression and the social context.

Beyond the reasoning-emotion dichotomy: the role of mentalizing

Although the distinction between cognitively reasoned and emotional moral judgment has some utility, not all relevant mechanisms fit easily within this dichotomy. For instance, consider the theory-of-mind or mentalizing network: Is it linked to cognitive reasoning or emotional processes? Moll et al.'s (2008) account places mentalizing on the cognitive side of moral judgment because it is part of the contextualized appraisal of emotions. However, mentalizing is very different (both neuroanatomically and cognitively) from other aspects of cognition relevant to moral reasoning such as working memory and emotion regulation. In other respects, mentalizing resembles emotional processes insofar as we tend to compute intentions and beliefs intuitively. However, it is best to consider mentalizing as making an independent contribution to moral judgment rather than force it into this emotion-cognitive dichotomy. An understanding of intentions is crucially important for moral judgment. For example, killing someone accidentally has a different moral status (and would attract different degrees of punishment) versus killing someone intentionally. This also offers a different way of thinking about the Trolley and Footbridge Dilemmas (aside from the personal/impersonal nature). Killing the lone hiker is more akin to an accidental or inevitable death – you didn't put him on the tracks – but the same can't be said of the fat man pushed off the bridge.

There is evidence that parts of the mentalizing network, such as the rTPJ, are involved in moral judgments particularly when intentionality is relevant. Using multi-voxel pattern analysis in fMRI, it has been shown that the TPJ discriminates between intentional versus accidental harm when considering moral vignettes and that differences in this activity pattern predict individual differences in how permissible the different scenarios are judged to be (Koster-Hale, Saxe, Dungan, & Young, 2013). A meta-analysis of fMRI studies that have independently looked at morality, theory of mind, or empathy finds common activation in the left and right TPJ, the dorsal part of the mPFC and right anterior temporal lobe as shown in Figure 10.6 (Bzdok et al., 2012). The next section also considers whether autism is linked to differences in moral judgment.

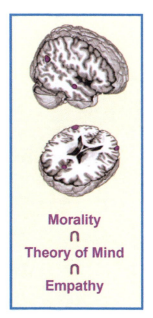

Morality
∩
Theory of Mind
∩
Empathy

Figure 10.6 A meta-analysis of fMRI studies (separately) investigating empathy, theory of mind, and morality reveal a common set of regions activated by all these tasks that center on the 'mentalizing network' of the TPJ, medial PFC, and anterior temporal lobes (Bzdok et al., 2012).

A PERSONAL ACCOUNT OF MORALITY IN AUTISM

Temple Grandin is an American professor and an activist for people with autism. Her autobiographical account of autism (Grandin, 1995, pp. 131–132) gives an illuminating description of her understanding of morals:

Many people with autism are fans of the television show Star Trek. I have been a fan since the show started. When I was in college, it greatly influenced my thinking, as each episode of the original series had a moral point. The characters had a firm set of moral principles to follow, which came from the United Federation of Planets. I strongly identified with the logical Mr. Spock, since I completely related to his way of thinking.

I vividly remember one episode because it portrayed a conflict between logic and emotion in a manner I could understand. A monster was attempting to smash the shuttle craft with rocks. A crew member had been killed. Logical Mr. Spock wanted to take off and escape before the monster wrecked the craft. The other crew members refused to leave until they had retrieved the body of the dead crew member. To Spock, it made no sense to rescue a dead body when the shuttle was being battered to pieces. But the feeling of attachment drove the others to retrieve the body so their fellow crew members could have a proper funeral. It may sound simplistic, but this episode really helped me to understand how I was different. I agreed with Spock, but I learned that emotions will often overpower logical decisions, even if these decisions prove hazardous.

Universal morals, moral variations, and absent morals

Are there any moral rules that do appear to be innate and leave little or no scope for cultural variation? The responses to moral dilemmas such as the Footbridge and Trolley Dilemmas show relatively small variability across gender, age, religion, and politics (Banerjee, Huebner, & Hauser, 2011). One of the best-documented examples of a cross-cultural/universal taboo is that against incest. For example, which of the following scenarios do you find the most reprehensible?

1 Licking ice cream off a toilet seat
2 Having consensual, contraceptive sex with your brother or sister

For most people, the second scenario is far more disgusting, but people find it hard to justify their reasoning (Haidt, 2001). One could make the claim that it is because society considers it the most reprehensible, but this would push the problem to another level: why does society consider it reprehensible? It is hard to argue that the consequences of this second case are much worse than the first case. The fact that incest feels wrong appears to be one example of a moral norm that is innate, rather than determined solely by society. Evidence for this comes from the **Westermarck effect** (Westermarck, 1891). This states that we tend not to be sexually attracted, as adults, to people who we knew in the earliest years of life, up to around 6 years. Given that we do not know (for sure) who is genetically related to us, we appear to have evolved a mechanism that applies to people who are *likely* to be our kin, namely those that we grow up with. One consequence

KEY TERM

Westermarck effect
The tendency not to be sexually attracted, as adults, to people who we knew in the earliest years of life

of this is that we are less likely to develop sexual attractions towards non-kin who we spend time with in the early years. Evidence for this comes from children reared together in a Kibbutz (Shepher, 1971), and the success/failure of Taiwanese marriages arranged at different ages of childhood (Wolf, 1995). An fMRI study contrasting disgust elicited by thoughts of incest also revealed a partially different network compared to contamination-related disgust (Borg, Lieberman, & Kiehl, 2008). The incest taboo suggests that it is theoretically possible to evolve mechanisms that determine the nature of our moral code.

However, there is variability too. There are important differences in the extent to where particular situations are cast as conventional versus moral. Cultural or inter-individual variability does not necessarily disprove the notion of an innately disposed moral disposition because innateness may specify the dimensions on which it can vary, or pose certain constraints on what can vary. As in most domains, 'nature' and 'nurture' are more likely to be collaborators rather than competitors.

There are within-culture individual differences between where the line is drawn between moral versus conventional norms. Some people suppress racist sentiments to conform to social norms (i.e. political correctness) whereas others have a strong personal belief in equality, as discussed in the last chapter (Plant & Devine, 1998). This can now be understood in terms of whether individuals represent this as a conventional or a moral norm. To give another example, Graham, Haidt, and Nosek (2009) asked participants from the USA to rate how relevant a variety of situations were to moral judgments. Across the board, people tended to rate issues relating to harm of others and fairness as being morally relevant, as would be expected (given that this is how morals are normally defined). What is more surprising is that some people apply moral judgments to issues such as respect for authority and purity (ability to control desires), as well as to norms relating to the rights and welfare of people. This was particularly true of individuals who self-declared themselves as having conservative rather than liberal moral foundations – see Figure 10.7. Graham et al. (2009) argue that definitions of morality that emphasize the rights of the individual are too narrow and that a second factor of 'maintaining social order' is needed to account for the full spectrum of what many people consider to be moral.

How is moral judgment affected by developmental or acquired conditions that affect the functioning of the brain? Adults and children with autism appear normal on many – but not all – tests relating to morality. Leslie, Mallon, and Di Corcia (2006) found that autistic children pass tests of moral reasoning. For example, they identify it as 'bad' to steal someone else's cookie when it makes them cry but that it isn't 'bad' to eat one's own cookie even if the other person (who greedily wants two cookies) starts to cry.

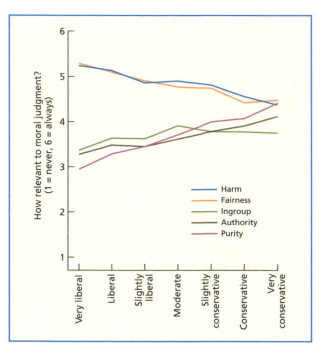

Figure 10.7 While most people regard issues relating to harm and fairness to be moral issues, people differ more in the extent to which they believe that issues relating to authority (e.g. respect for authority), group identity (e.g. group loyalty), and purity (e.g. control of desires) are morally relevant. Self-declared 'conservatives' from the USA regard authority and purity to be as morally relevant as the traditional moral areas of harm and fairness. From Haidt (2007). Copyright © 2007 American Association for the Advancement of Science. Reproduced with permission.

Blair (1996) also found that autistic children understand the moral/conventional distinction. However, other research does reveal differences particularly when there are competing sources of information. People with autism seem to rely less on emotion during moral decision-making (Brewer et al., 2015) and activate their amygdala less in these situations (Schneider et al., 2013). They tend to give more utilitarian answers to the Footbridge Dilemma, and this tendency was correlated with problems in inferring the intentions and thoughts of others (Gleichgerrcht et al., 2013).

Moran et al. (2011) used a test of moral judgment that contrasted two kinds of information in a 2 × 2 design. One kind of information was the outcome (good, bad), and the other was the intention (good, bad). The most interesting scenarios are those in which information was contradictory: good intention but bad outcome (i.e. accidental harm); or bad intention but good outcome (i.e. attempted harm). Examples of these scenarios are below:

Accidental Harm Grace and her friend are taking a tour of a chemical plant. When Grace goes over to the coffee machine to pour some coffee, Grace's friend asks for some sugar in hers. There is white powder in a container by the coffee. The white powder is a very toxic substance left behind by a scientist, and therefore deadly when ingested in any form. The container is labeled 'sugar', so Grace believes that the white powder by the coffee is sugar left out by the kitchen staff. Grace puts the substance in her friend's coffee. Her friend drinks the coffee and dies.

Putting the substance in was: Forbidden 1 2 3 4 5 6 7 Permissible

Attempted Harm Dan is giving a visitor a tour of a laboratory. Before visitors enter the testing room, all test tubes containing disease antigens must be contained in a chamber by flipping a switch. A repairman has just come to fix the switch, which had been broken. The switch has been successfully repaired, so the test tubes are quite safely contained. Thus, anybody who enters the room will be safe and unexposed. Dan believes that the switch is still broken after a conversation with the repairman, so he believes it is not safe for the visitor to enter. Dan tells the visitor to enter the testing room. The visitor does not contract any disease and is fine.

Telling the visitor to enter was: Forbidden 1 2 3 4 5 6 7 Permissible

People with autism performed normally on judgments of attempted harm (rating it as forbidden) but tended to assign moral blame to accidental harm too (they rate it as forbidden, but the controls rate it as permissible). This suggests that they rely less on the person's intentions when outcomes are negative. Disrupting the rTPJ of neurotypical participants using brain stimulation (tDCS) also affects moral judgments of accidental harm more than the other scenarios (Sellaro et al., 2015).

In terms of acquired brain lesions, the group that has been most extensively studied in relation to morality is patients with lesions to the orbitofrontal cortex and the adjoining ventromedial prefrontal cortex. Patients with lesions here often display inappropriate social behavior and, broadly speaking, the region is involved in subjective emotional experience and in integrating emotions with current context. Earlier studies with this patient group relied on tests of moral reasoning (rather than moral

Figure 10.8 Patients with damage to the ventromedial/orbitofrontal prefrontal cortex (VMPC) are more likely to endorse actions that lead to injury of another person provided they lead to a greater good (e.g. killing one person to save five). Their performance is contrasted with other brain-damaged controls (BDC) and normal controls (NC). From Koenigs et al. (2007). Copyright © 2007 Nature Publishing Group. Reproduced with permission.

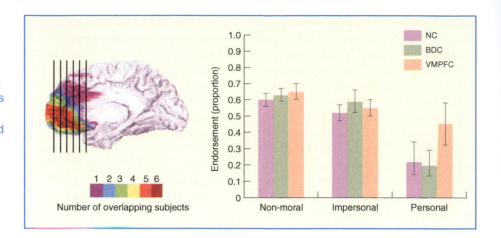

intuition) that ask about general situations (e.g. 'is it wrong to do X?') rather than personal behavior (e.g. 'how would you respond in situation X?'). These studies tend to reveal intact moral reasoning – that is, the patients know how one *ought* to behave (Bird et al., 2004; Blair & Cipolotti, 2000; Saver & Damasio, 1991). One exception to this rule may be in orbitofrontal lesions acquired during childhood. This may result in a failure to develop adequate social knowledge in the first place (Anderson, Bechara, Damasio, Tranel, & Damasio, 1999). The two cases in this study acquired their lesions before the age of 2 years, and as adults, they had a string of convictions for petty crimes. Both responded abnormally on tests of moral reasoning.

More recent studies have examined other types of moral dilemma that assess the patients' own moral intuitions, rather than their knowledge of moral norms (Ciaramelli, Muccioli, Ladavas, & di Pellegrino, 2007; Koenigs et al., 2007). These studies reveal clear differences in this domain, but only on certain kinds of moral dilemma. For example, Koenigs et al. (2007) contrasted personal dilemmas (such as the Footbridge Dilemma) with impersonal dilemmas (such as the Trolley Dilemma) and noticed that the patients performed differently on the former but not the latter. In these dilemmas, they tend to always choose to save five and kill one, irrespective of context (see Figure 10.8). In other moral dilemmas, patients with lesions in this region also tend to judge failed attempts at harming someone as permissible (Young et al., 2010). (Note: this directly contrasts to the pattern reported for people with autism where the problem lies with accidental harm rather than attempted harm). The atypical moral responses cannot be construed as 'errors' but rather as a systematic bias to respond based on outcomes rather than on the basis of emotion. Other evidence suggests they have emotional disturbances: these patients are judged by family members to exhibit low levels of empathy, embarrassment, and guilt, and on an objective assessment they show low levels of skin conductance response to emotional stimuli (Koenigs et al., 2007).

Evaluation

Moral judgments concern the rights of individuals, such as not to be harmed and to be treated fairly. These judgments activate brain regions linked to emotional processing and also regions involved in mentalizing / theory of mind, and reasoning and emotion regulation (e.g. lateral PFC). This is consistent with moral knowledge being represented at several levels: in terms of personal beliefs (or gut instincts), social norms, and reasoned pros and cons. Given that there are different sources of information that guide moral judgment it is not surprising that there are no clear cases of 'absent morals'. What exists, instead, are differences in the amount of weighting given to one kind of information over another that may occur due to normal individual differences, cross-cultural differences, or differences that arise through changes to the brain in developmental or acquired conditions. People with autism perform normally on many tasks of morality, but may perform differently in scenarios in which intentionality is crucial. Patients with damage to the orbitofrontal/ventromedial prefrontal cortex tend to base moral judgments on a rational basis (analysis of costs and benefits), which is consistent with a normal role of this region in using emotions to guide decisions.

PUNISHMENT AS SOCIALLY SANCTIONED AGGRESSION

Social norms are not only concerned with what is right/wrong and appropriate/inappropriate but also with how one should respond to a perceived wrongdoing (e.g. eye-for-eye, turn the other cheek). The severity and nature of punishment are susceptible to significant cross-cultural variation, but variability need not imply randomness. For instance, the

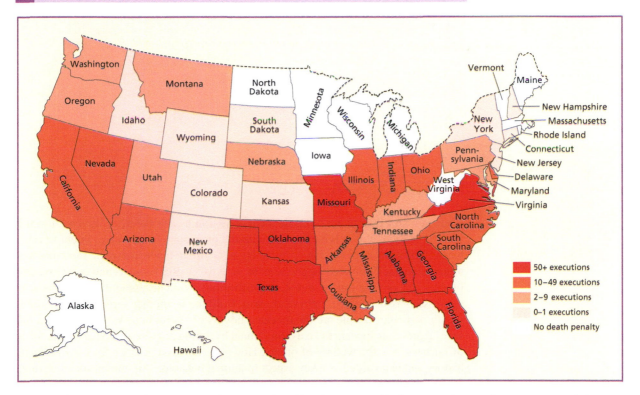

Figure 10.9 The use of capital punishment in the USA shows a divide between the northern states and the southern and western states (the latter were predominantly migrated by settlers from the South). The map shows the number of executions since 1976. Nisbett and Cohen (1996) argue that southern/western states have a 'culture of honor' in which violence is justified in order to right a wrong but not that they are associated with more violence per se.

principle that the punishment should be in proportion to the misdemeanor may be culturally universal.

The ultimate form of punishment is death, and there is significant cross-cultural variability in the extent to which it is acceptable. For instance, in some cultures families may kill one of their own family members (e.g. if they have an affair) in order to preserve the family 'honor', even though such self-initiated acts are prohibited by the law. In some cultures, the law itself permits the killing of another and this may reflect the local attitudes of the population. In the USA, for instance, there is variability in the use of (and legality of) the death penalty, with greater use in the southern and western states (see Figure 10.9). Nisbett and Cohen (1996) argue that there is a 'culture of honor' in these regions, in which violence is justified as retribution for a perceived wrong. They suggest that this reflects the cultural heritage of the immigrant populations who settled in these regions, which is perpetuated to the present day. (They note that the western states were predominantly settled by migrants from the South.) In a series of surveys they show that the presence of a 'culture of honor' attitude is independent of economic and ethnic variations in these regions. These attitudes are reflected not only by the law (e.g. in the case of capital punishment) but also in the perceived legitimacy of violence in day-to-day interactions. For instance, people from the southern states are more likely to respond with violence to someone calling them 'pig' than someone from the northern states, even when people are matched socio-economically and the questionnaire is administered in the same geographical location (Nisbett & Cohen, 1996). They are careful to note, however, that there is little evidence to suggest that there is greater acceptance of unprovoked violence (despite greater acceptance of violence as a form of retribution).

KEY TERMS

Aggression
Any behavior directed towards another individual that is carried out with the proximate (immediate) intent to cause harm

Instrumental aggression
Aggression that is self-initiated and goal-directed

Reactive aggression
Aggression that occurs in response to threat

ANGER AND AGGRESSION

Anderson and Bushman (2002) define **aggression** as any behavior directed towards another individual that is carried out with the proximate (immediate) intent to cause harm. Within this definition, a distinction is often made between **instrumental aggression** (self-initiated aggression to achieve a goal) and **reactive aggression** (aggression arising out of threat or frustration). Violence would be one form of aggression associated with physical harm. Bullying is another form of aggression in which particular people are the repeated targets of aggression. When framed in this way, aggression appears to be dysfunctional by acting against group harmony, norms of fairness, and the welfare of others. However, aggression has a long evolutionary history and is regarded as a key aspect of animal behavior. All human societies tolerate, if not openly endorse, some acts of aggression: for instance, as a means of punishment or retribution, as a means-to-an-end to overthrow an 'immoral' dictatorship, or even in certain competitive sports. Although societies tolerate some aggression, there will always be individuals (or groups of individuals) who cross the line of acceptability, and it is in this sense that aggression can be considered as pathological.

This section will describe the adaptive functions of aggression as well as pathological aspects of aggression.

Does aggression have a social function?

In non-human animals, instrumental aggression is generally considered to have a clear function: namely, social dominance (Hawley, 1999). Acts of aggression set up and maintain social hierarchies in which those at the top tend to have a leadership role (deciding what the group will do) and have privileged access to resources (such as food or mates). Those lower down the dominance hierarchy have more limited access to resources but may nevertheless benefit in other ways (e.g. receiving the protection of higher status group members). Thus, the definition of aggression outlined above can be reconsidered and expanded: the proximate (immediate) intention of an aggressive act may be to cause harm, but the ultimate (primary) intention may be to assert dominance over others (instrumental aggression) or to defend ourselves, by defending status or well-being (reactive aggression).

KEY TERM

Appeasement behaviors
Display signals that convey defeat (e.g. cowering)

This particular spin on aggression accounts for one important observation: namely, that many aggressive acts in animals do not result in any physical harm at all (Lorenz, 1966). Other cues are used to determine social dominance in order to avoid physical injury. These include physical size and aggressive displays such as posturing (e.g. to make one appear bigger), vocalizations (e.g. roaring), and facial expressions (e.g. bared teeth, direct gaze). In addition, animals have evolved other display signals to say 'back off, you win', so-called **appeasement behaviors**. These may include distress displays (e.g. fear), averting gaze, or (in humans) saying 'sorry'.

In humans and other primates, displays relating to the emotion of anger (facial, vocal, bodily) are often regarded as crucial for demonstrating aggressive intentions (e.g. Berkowitz & Harmon-Jones, 2004). Anger is linked to situations in which one's goals are unfulfilled by someone else's improper actions. It may signal the fact that somebody else has violated a social norm or has challenged their social standing (e.g. their reputation, authority). Displays of anger may differ from other emotional displays in that they often do not elicit a matched response in the perceiver. Whereas seeing happiness may elicit happiness and seeing fear may elicit fear, seeing anger does not necessarily elicit anger. An angry face may sometimes trigger a fear response – a form of appeasement – rather than a reciprocated anger response (van Honk & Schutter, 2007). To give another example, in male rats a shock may produce a fear response when alone but

Figure 10.10 There is a fine line between social dominance and aggression. In many animal species, social dominance is achieved almost exclusively through aggression (and aggressive displays). Humans and other primates employ a variety of strategies (both pro-social and aggressive) to achieve social dominance. Competitive sports, for example, are used to establish dominance hierarchies (league tables) using socially permissible levels of aggression.

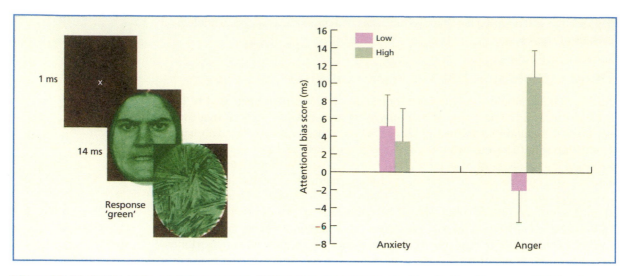

Figure 10.11 In this task, participants were divided into groups according to whether they were high/low in trait anxiety or high/low in trait angriness. The task involves presenting a colored face (subliminally) and then measuring naming times to a (supraliminal) colored display. Participants with high trait angriness are faster at naming the color after presentation of an angry face (relative to a neutral face). From van Honk et al. (2001b). Copyright © 2001 Elsevier. Reproduced with permission.

an anger/aggression response in the presence of another male (Ulrich & Azrin, 1962).

Anger as an *emotion* is a response to a situation, but angriness as a *trait* may be a stable disposition that varies across individuals (Spielberger, Jacobs, Russell, & Crane, 1983). People with high trait angriness are biased towards interpreting the intentions of others in a hostile manner (e.g. it was a wrongdoing, not an inadvertent mistake) and may opt for confrontation over appeasement. Those with high trait angriness pay more attention towards angry faces (van Honk et al., 2001b). Those with high levels of *social* anxiety (rather than general anxiety) pay less attention to angry faces and may treat them as fear inducing (Putman, Hermans, & van Honk, 2004).

One of the most influential cognitive theories of aggression is the frustration–aggression model of Dollard, Doob, Miller, Mowrer, and Sears (1939) and later revised by Berkowitz (1989, 1990). In the original version of the model, aggression was regarded as a response to having one's goals thwarted, that is as a response to frustration (Dollard et al., 1939). In the later version, Berkowitz (1990) inserted anger as a mediator – that is, frustration from having one's goals thwarted can lead to feelings of anger, and these feelings may generate aggressive acts. For Berkowitz, anger is more than an emotion; it is also an appraisal – namely, an appraisal that another person is to blame for one's unfulfilled goals through an improper or unfair action. Aggression may therefore serve a revenge motive in which one attempts to correct a perceived wrongdoing or restore fairness (Stillwell, Baumeister, & Del Priore, 2008). According to Berkowitz (1990), whether an aggressive thought is translated into an aggressive act depends on environmental cues and social norms about such behavior.

SOCIAL LEARNING OF VIOLENCE

According to Bandura (1973): 'The specific forms that aggressive behavior takes, the frequency with which it is displayed, and the specific targets selected for attack are largely determined by social learning factors'. This conclusion was reached from a large number of studies, the most influential of which involved preschool children observing aggressive attacks on a Bobo doll – a child-sized inflatable figure with a weighted bottom so that it pops back up after being hit. For example, Bandura, Ross, and Ross (1963) showed children films of an adult kicking, attacking with a hammer, and shouting at the doll (a control video showed non-aggressive behavior). Children exposed to the aggressive video tended to behave aggressively when introduced to the doll themselves. This effect was enhanced if they had seen the adult rewarded after the aggressive act but was diminished if they saw the adult punished (Bandura, 1965). Although not considered here, another important strand of evidence is the observation of violence on television (e.g. Johnson, Cohen, Smailes, Kasen, & Brook, 2002; Paik & Comstock, 1994).

Social learning theory places an emphasis on *behavior* rather than the underlying *intentions*. Children are able to make the distinction between aggression and play-fighting. Similarly, they understand from a young age that dolls and inanimate things do not experience pain and other feelings. As such, it is unclear whether aggressive acts in these studies really reflect aggressive intentions to harm. This does not mean that social learning has no role, but as a general theory of aggression it is unsatisfactory (Anderson & Bushman, 2002).

Figure 10.12 The children attack the Bobo doll after seeing an adult behave aggressively towards it. From Bandura et al. (1961). Copyright © 1961 American Psychological Association. Reproduced with permission.

KEY TERM

Displaced aggression
An aggressive act
from a higher status
person is redirected to
someone of perceived
lower status.

As in other species, there is evidence that human aggression is related to social dominance. According to Hawley, Little, and Rodkin (2007) moderate hostility and aggressive self-expression characterize highly ambitious, successful, and powerful people. However, it need not involve violence. Such people may tend to use a mix of coercive and pro-social methods to dominate the control of resources, and the use of these two strategies tends to be positively correlated (Hawley, 2002, 2003). Aggressive acts are not always directed towards the person that has elicited anger, but may be displaced towards someone perceived to be of lower social status. This is known as **displaced aggression**. Examples of this include: men who are treated with hostility by their boss, who then shout at their wife or children when they commit a minor transgression; higher degrees of aggression in workplaces where managers behave unfairly; prisoners who are bullied, who then bully others lower down the hierarchy (Marcus-Newhall, Pedersen, Carlson, & Miller, 2000). Research suggests that violent acts are more common in societies in which there is greater inequality between the richest and poorest, but the overall level of wealth is not important (Wilkinson & Pickett, 2009). Poverty does not cause anger and violence, but the perception of injustice and unfairness can. Mazur and Booth (1998) argue that antisocial actions often reflect attempts to dominate figures in authority or to prevail over a constraining environment.

Biological basis of anger and aggression

The amygdala as a modulator of the fight-or-flight response

KEY TERMS

Fight-or-flight response
The decision whether to respond to a threat aggressively or to escape from the threat

Periaqueductal gray (also called the central gray)
A mid-brain region involved in defensive rage and modulation of pain

Although the amygdala has been strongly implicated in the perception and generation of fear, it is also important in aggression. It is perhaps not surprising that aggression and fear should have partly overlapping neural substrates, given that both are linked to the **fight-or-flight response** (fight = aggression, flight = fear). Lesions of the amygdala disrupt social dominance hierarchies in primates (Rosvold et al., 1954) and can result in unusual tameness in situations in which a fight-or-flight response would be the norm (Kluver–Bucy syndrome; Kluver & Bucy, 1939). The specific role of the amygdala is in the *regulation* of aggression (making an aggressive act more or less likely). Studies of reactive aggression in cats show that a defensive rage reaction can be elicited by direct stimulation of the dorsal **periaqueductal gray** region of the mid-brain, either electrically or chemically (Siegel et al., 1999). In addition, different sub-regions of the amygdala and hypothalamus were found to influence this response through both inhibitory and excitatory mechanisms as shown in Figure 10.13 (Siegel et al., 1999). In humans, amygdala lesions affect aggressive behavior but may either increase its likelihood (van Elst, Woermann, Lemieux, Thompson, & Trimble, 2000) or decrease it (Ramamurthi, 1988). This is consistent with a regulatory role in aggression and also with the view that there are different subregions of the amygdala that affect aggression in different ways.

The ventral striatum, dopamine, and the 'warrior gene'

The perception of anger has been linked to the ventral striatum and regions of the orbitofrontal/ventromedial prefrontal cortex that it projects to. Dougherty et al. (1999) asked participants to read stories from their own life that previously made them angry during fMRI and found that these stories trigger high ratings of angry feelings again, correlated with activity in these frontal regions. However, these regions have been linked to subjective feelings of emotion in general (Hornak et al., 2003; Koenigs et al., 2007), and it is unclear whether it is anger-specific. A selective deficit in recognizing anger has been reported following damage to the ventral striatal region of the basal ganglia (Calder et al., 2004), and the dopamine system in this region has been linked to the production of aggressive displays in rats (van Erp & Miczek, 2000). Lawrence, Calder, McGowan, and Grasby (2002) found that disruption of a certain class of dopamine receptors (D2) disrupts anger recognition in humans. Given the wider role that this region plays in motivation and reward-based processing, it has been argued that the ventral striatum dopamine system may be particularly important for *instrumental* forms of aggression, which are goal based and pre-meditated rather than a response to an immediate threat (e.g. Couppis & Kennedy, 2008). The latter may be more strongly linked to the amygdala–hypothalamus–periaqueductal gray interactions described earlier.

There is evidence for at least one genetically based individual difference linked to aggression. This has been labelled as the **warrior gene**. A large Dutch family with a strong history of violence in its male members were found to have a mutation in the gene coding for monoamine oxidase A (MAOA), an enzyme involved in the metabolism of dopamine, norepinephrine, and serotonin (Brunner et al., 1993). The effect is more pronounced in men because the gene is coded on the X chromosome. Men (XY) have only a single copy of the gene whereas women (XX) have two copies, one of which can compensate for the other. Although this family had a rare mutation, within the general population there is a common polymorphism called the low activity (L) variant that is linked to less enzymes produced and, hence, fewer neurotransmitters removed. This L variant is also present in primates, and has an

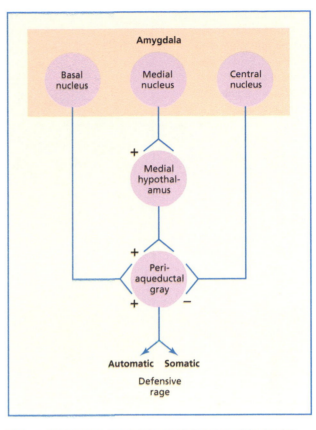

Figure 10.13 The amygdala modulates activity in the periaqueductal gray (also called central gray) region via direct excitatory (+) and inhibitory (−) connections and via the medial hypothalamus. The periaqueductal gray generates defensive autonomic and somatic ('fight') responses via brainstem nuclei. Adapted from Siegel et al. (1999).

evolutionary history dating back at least 25 million years, leading to the suggestion that it was positively selected for by evolution to create warrior-like behavior (Gibbons, 2004). Its prevalence is likely kept in balance by the fact that it would be disadvantageous for all members of a society to carry it. Children with this gene who were maltreated are more likely to show antisocial behavior as an adult – a gene X environment interaction (Caspi et al., 2002). In an experimental study of aggression, it was found that there was an interaction between MAOA genotype and presence/absence of social exclusion in the cyberball paradigm (Gallardo-Pujol, Andres-Pueyo, & Maydeu-Olivares, 2013). Specifically, those with the L variant showed more aggressive punishment behavior (taking points away from another player) following exclusion. In terms of brain structure, the L variant is linked to gray matter volume reductions in the amygdala and the cingulate cortex (bilaterally), but an increase in gray matter in the lateral part of the orbitofrontal cortex – see Figure 10.14 (Meyer-Lindenberg et al., 2006). The changes in orbitofrontal cortex were much greater for males with the L variant. In terms of functional activity using fMRI, judgments of angry and fearful faces (relative to non-face stimuli) produced more activity in the amygdala of L carriers and less activity in

Figure 10.14 People the low (L) activity variant of the MAOA gene (the so-called warrior gene) have reduced volume in the amygdala and cingulate cortices (top). They also have increased gray matter volume in the lateral orbitofrontal cortex and this region (but not other regions) interacts with sex (bottom). From (Meyer-Lindenberg et al., 2006)

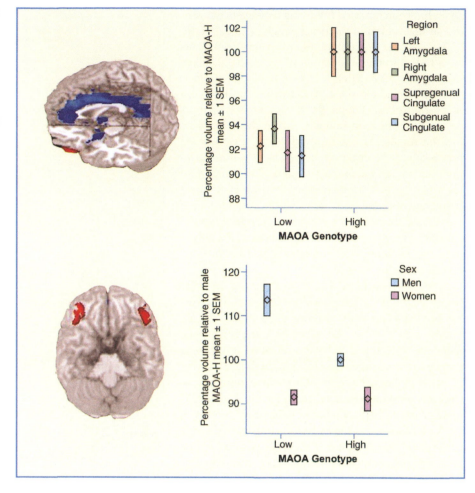

orbitofrontal cortex. This could be interpreted as a high emotional arousal signal with reduced tendency to regulate. Gender (rather than genotype) affected the connectivity between these regions during the task with males having less functional connectivity (Meyer-Lindenberg et al., 2006).

The role of testosterone

The androgen hormone **testosterone** has been linked to anger, aggression, and social dominance. Testosterone has two types of effect on the brain. First, it has an *organizing* effect during development. It stimulates the development of certain neural circuits and changes the sensitivity of others, making them more sensitive to testosterone. Testosterone is implicated in the production of predominantly male-related behaviors and may account for sex-related differences in aggression (Mazur & Booth, 1998). Women do produce testosterone (via the adrenal glands and ovaries), but men produce far more of it (via the testes), with notable peaks prenatally and at puberty. Second, testosterone has an *activating* effect throughout the lifespan, that is, it binds to specific neural receptors and directly influences the functioning of certain neural circuits.

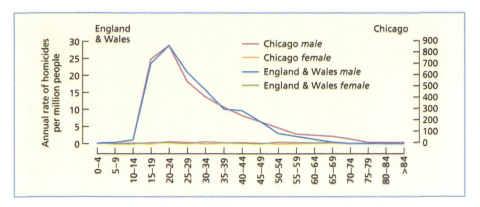

Figure 10.15 The age and sex distribution of convicted murderers shows a remarkably similar distribution between two different populations (Chicago, USA, vs. England & Wales), notwithstanding significant differences in the overall homicide rates (there is a 20-fold difference between them). What do you think is the relative contribution of biology and culture to these trends? From Cronin (1991). Copyright © 1991 Cambridge University Press. Reproduced with permission.

Participants given a testosterone injection show a greater response to angry faces, measured in terms of heart rate change, but not to neutral or happy faces, which was interpreted as an increased willingness to defend status – see Figure 10.16 (van Honk et al., 2001a). Resting levels of testosterone in men correlate with increased activity in the ventromedial prefrontal cortex when viewing angry faces (relative to neutral ones), but this negatively correlates with amygdala activity (Stanton, Wirth, Waugh, & Schultheiss, 2009). Ehrenkranz, Bliss, and Sheard (1974) studied prisoners divided into three categories: persistently physically aggressive, socially dominant but not physically aggressive, and neither physically aggressive nor socially dominant. Levels of testosterone were significantly higher in the first two groups relative to the third (but the first two groups were not different from each other). Other research links testosterone to the levels of violence of the crime committed by members of a prison population and the extent to which they violate rules in the prison (Dabbs, Carr, Frady, & Riad, 1995). In a non-prison sample of male army veterans, levels of testosterone were positively related to various self-report measures on antisocial behaviors (e.g. truancy at school and work) and violence (Dabbs & Morris, 1990). Socio-economic status (SES) was also found to be a moderating variable: testosterone had a greater influence on antisocial behavior in those from lower SES backgrounds.

Studies such as these establish a relationship between testosterone level and the level of aggression, dominance, and antisocial behavior. However, they do not establish the direction of cause and effect. Evidence suggests that the effects are bidirectional: high levels of testosterone are associated both with acts of competition and aggression *and* with the outcome of these acts. Testosterone levels increase prior to a competitive sporting engagement as a preparatory effect, but after the engagement the testosterone levels of the losers drop and those of the winners increase or remain stable (Elias, 1981). Comparable results are found for non-physical competitions such as chess (Mazur, Booth, & Dabbs, 1992). They are even found when watching our own national team win or lose in the soccer World Cup (Bernhardt, Dabbs, Fielden, & Lutter, 1998) or when our political party wins or loses a general election (Stanton, Beehner, Saini, Kuhn, & LaBar, 2009). This modulation of testosterone by

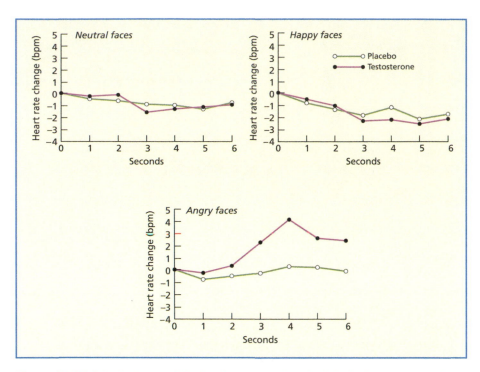

Figure 10.16 A testosterone injection increases the physiological response to the sight of angry faces. This may bias people towards a 'fight' rather than a 'flight' response. From van Honk et al. (2001a). Copyright © 2001 American Psychological Association. Reproduced with permission.

winning and losing may increase and decrease (respectively) the future motivation to take on more fights or challenges. A fMRI study shows that administration of testosterone (relative to a placebo) is associated with increased activity in the ventral striatum in a (non-social) game in which female participants played for money (Hermans et al., 2010). In social decision-making games, administration of testosterone relative to a placebo is associated with reduced levels of both trust (Bos, Terburg, & van Honk, 2010) and generosity (e.g. increases in the number of unfair offers made to others during the Ultimatum Game; Zak et al., 2009).

Pathological aspects of aggression

The *Diagnostic and Statistical Manual of Mental Disorders* (American Psychiatric Association) proposed the term **antisocial personality disorder** (**ASPD**) to replace the older terms of **psychopathy** and **sociopathy**, which have pejorative connotations. As such, in their view, these are different labels for essentially the same thing. The latest version (DSM-V) highlights the following pattern, which must be pervasive and extend over the course of development:

- Impairments in self-functioning such as ego-centrism or a failure to apply internal standards relating to lawful and ethical behavior
- Impairments of inter-personal functioning such as lack of remorse, lack of empathy, and lack of intimacy
- Antagonism such as callousness, manipulativeness, and hostility
- Disinhibition such as impulsivity and irresponsibility

Such individuals comprise around 3–4% of the male population and less than 1% of the female population (Robins, Tipp, & Przybeck, 1991). There is greater prevalence of ASPD traits in people with the low variant MAOA gene, after excluding for environmental adversity in the form of early physical abuse (Reti et al., 2011).

Others continue to use the term psychopath and measure it in different (but related) ways to the DSM-V criteria. The 'Psychopathy Checklist' (e.g. Hare, 1980) contains items that are similar to those used to diagnose ASPD (e.g. glibness and superficial charm, grandiose sense of self-worth). Perhaps unsurprisingly, a high proportion of inmates at a maximum security prison were found to meet the criteria for ASPD and scored highly on the Psychopathy Checklist (Hart & Hare, 1996). Psychopathy is also over-represented amongst corporate leaders relative to community samples (Babiak, Neumann, & Hare, 2010) which is consistent with the view that it is related to aggression, in broad terms, rather than violence/criminality specifically.

Psychopathy

Several studies have examined emotional processing in people diagnosed as psychopaths. Lykken (1957) was the first to note that these people do not show a normal fear-conditioned response to aversive stimuli (see also Flor, Birbaumer, Hermann, Ziegler, & Patrick, 2002). This is is linked to reduced amygdala activity during aversive conditioning (Veit et al., 2002). In addition they show reduced autonomic activity to fear and sadness in others (Blair et al., 1997). When presented with vignettes describing happiness, sadness, embarrassment, and guilt, a psychopathic group demonstrated particular problems with attributing guilt (Blair et al., 1995). Guilt tended to be described in terms of happiness or indifference. Blair (1995) put forward a cognitive model of psychopathy to account for this evidence in terms of an inability to respond appropriately to the distress cues of others, which normally acts as a 'violence inhibition mechanism'. Psychopaths perform normally on moral dilemmas such as the Trolley and

Figure 10.17
Psychopaths appear to be aware of social norms of right and wrong, but are willing to violate them. This may reflect an indifference to the consequences of aggression (e.g. they do not show a normal response to fear in others). Copyright © Sunset Boulevard/Corbis.

Footbridge Dilemmas (Cima, Tonnaer, & Hauser, 2010), consistent with an awareness of right/wrong but an indifference to the consequences of an aggressive act. When seeing someone else stroked or in pain there is a tendency for the observer to simulate the affective and sensory properties of the event (see Chapter 6). When similar paradigms are used in people with psychopathy they show less of this mirroring neural response, but this can be offset by instructing them to put themselves in the other person's perspective (Meffert, Gazzola, den Boer, Bartels, & Keysers, 2013).

Others have argued that prefrontal dysfunction is a core deficit in psychopathy (Gorenstein, 1982; Newman & Lorenz, 2002). The orbitofrontal cortex (OFC) and ventromedial prefrontal cortex have strong bidirectional connections with the amygdala and are implicated in making emotionally guided decisions (Bechara et al.,

1994; Bechara, Damasio, Damasio, & Lee, 1999). In a study of emotional memory, psychopathic individuals showed reduced medial OFC activity to emotional words, in addition to reduced amygdala responses (Kiehl et al., 2001). They also show less medial OFC activity during aversive conditioning (Birbaumer et al., 2005) and less medial OFC activity when making cooperation decisions relative to defect decisions in the Prisoner's Dilemma experiment (Rilling et al., 2007). The relationship between these differences and the actual symptoms of psychopathy is unclear: for example, patients with acquired lesions to these regions show a different behavioral pattern to psychopaths (see below). It is possible that a wider network of regions is disrupted in psychopathy (e.g. see Tiihonen et al., 2008). For example, functional imaging studies often show an *increase* in activity in lateral prefrontal regions, in addition to reduced orbitofrontal activity (e.g. Kiehl et al., 2001). This could be consistent with a deliberate attempt to suppress the influence of emotions, or a compensatory mechanism for a lack of context-appropriate emotional responses in the first place (Kiehl, 2008).

THE EXTRAORDINARY CASE OF PHINEAS GAGE

One of the most famous cases in the neuropsychological literature is that of Phineas Gage (Harlow, 1848/1993; Macmillan, 1986). On 13 September 1848, Gage was working on the Rutland and Burlington railroad. He was using a large metal rod (a tamping iron) to pack explosive charges into the ground when the charge accidentally exploded, pushing the tamping iron up through the top of his skull; it landed about 30 m behind him. The contemporary account noted that Gage was momentarily knocked over but that he then walked over to an ox-cart, made an entry in his time book, and went back to his hotel to wait for a doctor. He sat and waited half an hour for the doctor and greeted him with, 'Doctor, here is business enough for you!' (Macmillan, 1986).

Not only was Gage conscious after the accident, but he was able to walk and talk. Although this is striking in its own right, it is the cognitive consequences of the injury that have led to Gage's notoriety. Before the injury, Gage held a position of responsibility as a foreman and was described as shrewd and smart. After the injury, he was considered unemployable by his previous company; he was 'no longer Gage' (Harlow, 1848/1993). Gage was described as

> irreverent, indulging at times in grossest profanity . . . manifesting but little deference for his fellows, impatient of restraint or advice when it conflicts with his desires . . . devising many plans of future operation, which are no sooner arranged than they are abandoned in turn for others.
>
> (Harlow, 1848/1993)

After various temporary jobs, including a stint in Barnum's Museum, he died of epilepsy (a secondary consequence of his injury) in San Francisco, some 12 years after his accident.

Where was Phineas Gage's brain lesion? This question was answered by an MRI reconstruction of Gage's skull, which found damage restricted to the frontal lobes, particularly the left orbitofrontal/ventromedial region and the left anterior region (Damasio, Grabowski, Frank, Galaburda, & Damasio, 1994). Research suggests that this region is crucial for certain aspects of decision-making, planning, and social regulation of behavior, all of which appeared to have been disrupted in Gage. Other areas of the lateral prefrontal cortex are likely to have been spared.

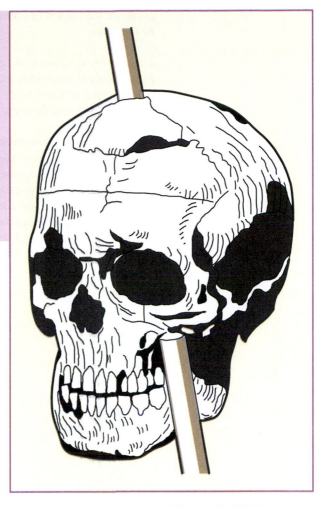

Figure 10.18 A large metal rod was blown through the skull of Phineas Gage as a result of an explosion. Remarkably, he survived. However, his social behavior was profoundly changed.

Acquired sociopathy/ASPD following damage to the orbitofrontal lobe

When testing a number of their patients with frontal lobe lesions, Damasio and colleagues (1990) made an interesting observation, namely, that many of their patients met a published American Psychiatric Association criterion for sociopathy (now called ASPD). The term *acquired sociopathy* is used to refer to those individuals who did not exhibit such symptoms prior to their brain injury. One of the earliest and most famous neurological cases in the literature is that of Phineas Gage (Macmillan, 1986). After a lesion to the left frontal lobe, Gage was noted to be 'irreverent, indulging at times in grossest profanity'. Patient MGS is described as a modern case of Phineas Gage (Dimitrov, Phipps, Zahn, & Grafman, 1999). He had been decorated for service in Vietnam, with more than 10 medals and a Purple Heart. Following a head injury, he was demoted for incompetent behavior. After an honorary discharge, he was noted to be sarcastic, lacking in tact (e.g. inappropriate disclosure of sexual history), moody, and unable to manage his own finances.

More detailed analysis has revealed that this kind of behavior arises specifically after lesions of the orbitofrontal cortex (particularly bilaterally) and also following lesions of parts of the medial surface of the frontal lobes, including the ventral region of the anterior cingulate (Hornak et al., 2003). Patients with lesions limited to the lateral prefrontal cortex do not show this socially disrupted behavior. Hornak et al. (2003) conducted a variety of assessments on patients with different lesions to the frontal cortex, including asking the patients about their own experiences of emotion (e.g. whether they feel angry more/less often) and asking a close relative about their social behavior using a questionnaire – see Figure 10.19. Patients with bilateral orbitofrontal damage reported changes to subjective emotional experience relative to patients with lateral prefrontal damage, although the direction of change was variable (some reported less emotion, others more). Similar results were obtained with this group for the ratings of social behavior given by a relative. These patients tended to be rated as:

- less likely to notice when other people were sad/angry/disgusted;
- less likely to respond to emotions in others (e.g. through comfort or reassurance);
- less cooperative and more impatient;
- less close to their family and have problems with close relationships.

Tranel, Bechara, and Denburg (2002) draw a somewhat different conclusion about the anatomical location within the prefrontal cortex needed to disrupt social functioning. They suggest that damage to the *right* ventromedial/orbitofrontal cortex, but not the left, gives rise to symptoms of acquired sociopathy. This was found on a range of objective tests (e.g. anticipatory skin conductance responses to risky decisions) as well as being apparent in assessments of daily social living.

Double dissociations between acquired sociopathy and executive functions have been reported (Bechara, Damasio, Tranel, & Anderson, 1998), and also different

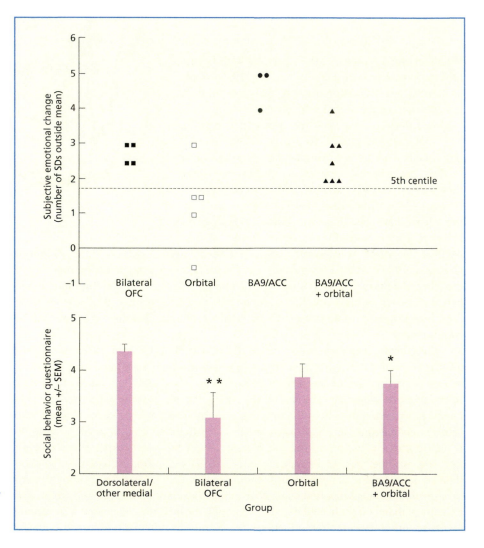

Figure 10.19 Patients with lesions to the orbitofrontal cortex (OFC) and/or the ventromedial cortex (including regions of the anterior cingulate cortex, ACC) show subjective changes in emotional intensity (top) and disturbances in social behavior as rated by a relative (bottom). Figures from Hornak et al. (2003). Copyright © 2003 Oxford University Press. Reproduced with permission.

neural substrates are implicated (orbitofrontal in sociopathy and lateral frontal in executive functions). The patients' difficulties appear to lie in the control of behavior specifically from emotional cues, rather than from a general problem in executive functions. Subjective emotional experiences of both a moral (e.g. pride, guilt) and a basic nature could be described as serving a motivational function on behavior: we are motivated to do 'right' because we want to achieve pride, joy, happiness, and we want to avoid guilt, shame, or distress. In the absence of this emotional compass the quality of social interactions will almost certainly deteriorate.

Antisocial and aggressive traits in childhood

Psychopathy and ASPD/sociopathy both require presentation in adulthood for diagnosis. However, comparable traits in childhood include a diagnosis of conduct disorder and also bullying behavior. Within these children, callous-unemotional (CU) traits (e.g. lack of guilt, absence of empathy, use of others to achieve one's goals) are linked to aggression and are often seen as an early marker of psychopathy (Frick & White, 2008).

Decety, Michalska, Akitsuki, and Lahey (2009) conducted an fMRI study of youths with aggressive conduct disorder. When watching another person in pain, both the conduct disorder and control groups activated regions of the brain linked to pain processing (including the insula and anterior cingulate cortex). However, the conduct disorder group activated it more, suggesting, paradoxically, a heightened neural empathic response. In addition, the conduct disorder group activated other regions not found in controls, including the ventral striatum. One possibility is that pain-producing acts (i.e. acts of aggression) are rewarding, and this overrides the tendency to be driven by empathy. Although both psychopathy and autism are often referred to as problems with empathy, they reflect different underlying mechanisms. People with autism tend not to be knowingly cruel. In a study that directly contrasted the two groups, boys with psychopathy reported experiencing less subjective fear and less empathy for victims of aggression but performed normally on a measure of cognitive perspective taking (Jones, Happe, Gilbert, Burnett, & Viding, 2010). Boys with autism showed the opposite profile. Non-aggressive conduct disorder wasn't linked to a distinctive cognitive profile.

Bullying is distinguished from other forms of aggressive behavior in that hostile acts are targeted towards specific individuals. Bullies appear to have good social cognition when tested formally. On tests of understanding deception, emotional reasoning, and theory of mind, bullies (aged 7–10 years) score as high as their non-bully peers and significantly higher than the victims of bullying (Sutton, Smith, & Swettenham, 1999). This suggests that many bullies have good social intelligence coupled with an ability (or willingness) to ignore the effects that their actions have on victims. A brain imaging study of boys (aged 10–16 years) with conduct disorder and callous, unemotional traits, showed normal activity within regions linked to mentalizing (mPFC and rTPJ) during a theory-of-mind task, consistent with the view that they have normal mentalizing abilities (O'Nions et al., 2014). By contrast, a group of boys with autism, in the same study, showed reduced activity in these regions.

Evaluation

Whilst it would be relatively easy to draw up lists of social and cultural factors (variations in gun law, inequality levels and SES, 'culture of honor' mentality) and biological factors (testosterone, young male predominance, genetic predispositions) relating to aggression, this would give a misleading picture of the field. The reason why it is misleading is that these factors are not separate but intertwined: for instance, individual differences in aggression in adults are affected by gene–environment interactions (Caspi et al., 2002), testosterone–SES interactions (Dabbs & Morris, 1990), and so on. One reason why they are intertwined is because aggression and displays of anger serve adaptive social functions (establishing dominance, maintaining social order). Some individuals, however, do exhibit extreme forms of aggression. In cases of adult psychopathy this is related to reward-based, goal-directed motivations for aggression rather than a lack of understanding of social norms or deficits in mentalizing or moral reasoning. This may involve inhibiting distress cues of others or an additional difficulty in learning fear-related associations (by the amygdala). In cases of acquired orbitofrontal damage, it may reflect an inability to use emotional cues to guide social behavior, resulting in antisocial trends but without high levels of goal-directed aggression.

CONTROL AND RESPONSIBILITY: 'IT WASN'T ME; IT WAS MY BRAIN'

Another key concept when considering antisocial behavior is that of **responsibility** – the extent to which someone can be held to account for his/her actions (e.g. Eastman & Campbell, 2006). This is related to the degree of control that people have over their behavior. Accidentally hitting somebody is not considered an antisocial act, but intentionally hitting someone is. Responsibility is typically deemed to vary across individuals (e.g. children vs. adults) and across situations (e.g. defensive, provoked, unprovoked). Imagine you find your lover in bed with someone else. Is it permissible to respond violently? In many countries, the law would offer a minimal punishment for such an act based on the assumption that the cheated lover had reduced control over his/her violent actions – a so-called crime of passion. These legal systems make assumptions about underlying psychological mechanisms – namely that there are some situations in which emotions drive antisocial behavior with minimal ability to override it. It is, of course, an empirical question as to whether this assumption is true. These kinds of issues are explored in the emerging field of **neuroethics** (e.g. Farah, 2005; Moreno, 2003).

The ability to control behavior is central to the commonsense notion of 'free will' (or agency): namely, that we have a sense of being able to select between different courses of action. Controlled actions are intentional rather than accidental or reflexive (e.g. such as moving one's hand away from a flame). Control and responsibility, however, are not the same thing: a person with paranoid schizophrenia who attempts to kill their neighbor for 'tuning into their thoughts' would have control over this action but have diminished responsibility; a drunk driver who kills a pedestrian would have diminished control but would still be considered responsible for this action as the potential consequences could be foreseen.

The concept of controlled behavior is most easily explained by reference to its antithesis: automatic behavior (Schneider & Shiffrin, 1977). Whereas controlled behavior is considered slow, conscious, and based on reason, automatic behavior is considered fast, often unconscious, and based on intuition. Intuition, in this context, may refer either to stereotyped behaviors that have been performed numerous times in the past (so-called schemas) or to our gut reactions (e.g. based on emotions).

Traditionally, these two sources of decision-making (controlled vs. automatic) have been regarded as being in opposition to each other, but with controlled behavior having the upper hand such that it is able to override automatic behavior (Miller & Cohen, 2001; Norman & Shallice, 1986). This controlling behavior is also referred to as the 'executive functions' of the brain and has been most closely associated with the lateral prefrontal cortex (Fuster, 1989; Goldberg, 2001; Stuss & Benson, 1986).

The distinction between automatic versus controlled is by no means clear-cut. Behavior is rarely solely one or the other. It has also been argued that much of our control over behavior occurs unconsciously (e.g. Suhler & Churchland, 2009). For example, in the social domain many of our actions are influenced by biases of which we are completely unaware. This raises important philosophical questions about whether we really are in control of, and responsible for, our actions (Doris, 1998). Consider the following findings from the social psychology literature:

- Participants are more likely to interrupt a (staged) conversation between the experimenter and another person when they have been primed, in a previous task, with words relating to rudeness than those primed with words relating to politeness or not primed at all (Bargh, Chen, & Burrows, 1996).
- Participants who have been primed by an irrelevant task (containing words relating to old age) subsequently walk more slowly to get an elevator than those primed by words relating to young age (Bargh et al., 1996).
- People are more likely to litter in a particular place when it contains graffiti than when it does not (Keizer, Lindenberg, & Steg, 2008).
- If you find a dime on the street, you are more likely to help a passerby who accidentally drops some papers (Isen & Levin, 1972).

Studies such as these raise important questions about the notion of control. Is our sense of control just a post-hoc justification for the unconscious decisions we make? Can control occur unconsciously and, if so, are we responsible for such actions? Suhler and Churchland (2009) argue that we can be considered to be in control of our actions (and responsible for them) even if we are not consciously aware of all the information that enters into the decision. Their approach is to define control and responsibility relative to a normative model of brain function, rather than attempting to link it to prescribed ideas about what controlled behavior is (i.e. that one needs to be consciously aware of the basis of one's decisions). According to them, if normative control turns out to be unconscious, then so be it. They propose that lack of control and responsibility would then be inferred by disruptions of this normative neurobiological model. This represents a philosophical rather than a pragmatic approach to the issue, given the absence of an agreed-upon normative model of control or even a definition of abnormal (statistically rare, qualitatively different, etc.).

Eastman and Campbell (2006) take a different approach to Suhler and Churchland (2009) by addressing the question of how the neuroscience of aggression can be interpreted by the law. They note that all the functional and structural imaging studies of psychopathy are essentially correlational in nature. It is not clear what, if anything, is causally related to the committing of a criminal act. Psychopaths presumably have similar brain structure both when committing an antisocial act and when behaving responsibly. What may differ between these situations is the interaction with an external influence at the time of the act.

Control over actions may wax and wane over time. De Wall, Baumeister, Stillman, and Gailliot (2007) asked participants to complete a task involving a high degree

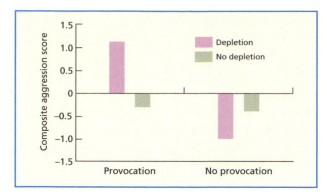

Figure 10.20 Participants who received an insult (provocation condition) *and* had previously exerted high levels of control on an unrelated task (depletion condition) chose louder and longer blasts of noise for the experimenter to receive. This suggests that our ability to exert self-control over aggression is diminished if we have previously had to exert control on a different task. From De Wall et al. (2007). Copyright © 2007 Elsevier. Reproduced with permission.

of control (watching a video but ignoring words that appeared unexpectedly at the bottom of the screen) and compared them to another group shown the same stimuli but given no control instructions. Their prediction was that having to exert sustained control over this task would reduce their ability to exert control over a subsequent aggressive impulse. After this task, they either received an insult or praise from the experimenter about a short essay they had written earlier ('This is one of the worst essays I've ever read' vs. 'Excellent! No comments'). Finally, they played a competitive task against the experimenter in which the loser would receive a blast of loud noise. The participant could choose the level and duration of noise. Participants who had previously exerted high control and were then given the insult gave louder and longer blasts of noise to the experimenter (see Figure 10.20). This suggests that our ability to control aggression is not fixed but is related to prior cognitive activity. Whether our responsibility for the aggressive act can be said to wax and wane is harder to answer.

Evaluation

The notions of control and responsibility are closely linked because both imply an element of intentionality or 'free will' over actions. They differ insofar as responsibility operates at the social level (i.e. the extent to which others hold you to account for your actions) whereas control operates at the level of the individual (Goodenough, 2004). The extent to which structural or functional brain difference could be used to argue for diminished responsibility is unclear due to several factors: establishing causation; establishing 'normality'; the extent to which normal control is itself influenced by unconscious (and uncontrollable) biases.

SUMMARY AND KEY POINTS OF THE CHAPTER

- Moral emotions (e.g. guilt, shame, moral disgust, anger) consist of emotions arising as a result of an appraisal of behavior (either one's own or somebody else's) relative to some normative standard of behaving. They tend to involve not only regions specialized for emotions but also regions implicated in higher cognition (lateral prefrontal cortex) and mentalizing.
- Moral judgments of right/wrong are underpinned by different sources of knowledge: an emotional reaction (or gut instinct; e.g. based on empathic concern for others), consensual norms (e.g. the law), and reasoned decisions (e.g. based on comparisons between costs and benefits). These

different sources of knowledge may sometimes conflict with each other (requiring effortful control), and different individuals may give different weightings to the importance of these different factors.

- Patients with lesions to the orbitofrontal cortex and/or ventromedial prefrontal cortex may exhibit antisocial tendencies. They fail to respond appropriately to emotional cues, and may eschew emotional information in favor of more rational judgments in moral dilemmas.

- Feelings of anger are associated with lack of goal attainment due to the perceived improper/unfair actions of other people. Expressions of anger signal disapproval and possible aggressive intentions.

- Aggression (normal and pathological) has been linked to brain structures such as the amygdala (involved in regulating aggression), periaqueductal gray (involved in generating a reactive 'fight' response to a threat), the ventral striatum (linked to anger), and the orbitofrontal/ventromedial prefrontal cortex (contextual modulation of emotions).

- Responsibility is related to the notion of intentional control of actions and 'free will'. Although important legally, it is unclear how it can be translated into neuroscientific theory.

EXAMPLE ESSAY QUESTIONS

- Are morals different from other kinds of social norm?
- What is the role of testosterone on aggression?
- Why does damage to the orbitofrontal cortex (and ventromedial frontal cortex) give rise to antisocial behavior?
- Evaluate the claim that both psychopathy and autism are underpinned by deficits in empathy.
- How can studies in neuroscience shed light on the legal definition of responsibility?

RECOMMENDED FURTHER READING

- Hodgins, S., Viding, E., & Plodowski, A. (2009). *The Neurobiological Basis of Violence: Science and Rehabilitation.* Oxford: Oxford University Press. Many of the chapters are also published in the *Philosophical Transactions of the Royal Society B* (12 August 2008; volume 363, issue 1503).

- Sinnott-Armstrong, W. (2008). *Moral Psychology. Volume 3: The Neuroscience of Morality, Emotion, Brain Disorders and Development.* Boston, MA: MIT Press. Includes an up-to-date overview of all topics covered in this chapter.

ONLINE RESOURCES

- References to key papers and readings
- Videos demonstrating the Trolley and Footbridge Dilemmas, 'Bobo doll' studies, and effects of ventro-medial/orbital PFC damage on emotional decisions
- Interviews and talks given by Robert Hare, Patricia Churchland, Jonathan Haidt, and others
- Recorded lecture given by textbook author, Jamie Ward
- Multiple choice questions and interactive flashcards to test your knowledge
- Downloadable glossary

CONTENTS

CHAPTER 11

Developmental social neuroscience

11

The structure and function of the brain, and the neurons within it, are not resolutely fixed at birth according to a predetermined blueprint. Being part of a socially enriched environment affects both the structural and functional development of the brain (e.g. Branchi et al., 2006; Chugani et al., 2001). Cells and pathways in the brain may whither or grow depending on the quality and quantity of social interactions during development. As such, brain-level explanations – such as those being developed in the field of developmental social neuroscience – offer an exciting opportunity for resolving the **nature–nurture debate**. The brain is the organ in which gene-based influences ('nature') and environment-based influences ('nurture') come together. Traditionally within developmental science this has been construed as a series of interactions between a child's ability to understand and engage with the environment coupled with the suitability of the environment for providing the appropriate inputs. For example, the eminent developmental psychologist Jean Piaget (1896–1980) considered development as progressing through various stages: the structure of the stages was construed as largely predetermined, but successful passage to the next stage required appropriate interactions between the child and the environment.

Within this nature–nurture debate it is important not to confuse **phylogenetic development** (i.e. of species) with **ontogenetic development** (i.e. of individuals). It is widely accepted that many of the evolutionary developments that distinguish humans from other primates lie in the domain of social intelligence, and in this sense we can say that many facets of human social behavior are innate. However, this does not mean that human infants enter the world with a mature understanding of the social world, or that innate is the same as 'present from birth' (the term for this is congenital), and nor does it mean that developing good social skills is inevitable. The question of what the initial 'start-up kit' consists of in human infants remains open for debate, as do the many complex gene–environment interactions that occur during development that lead to adult social competence.

This chapter considers social development in three broad stages: infancy (birth to 18 months), childhood (18 months to puberty), and adolescence (puberty to adulthood). The chapter will consider the relationship between early social competencies that can be considered to be innate (e.g. mimicry of tongue protrusion, preference for face-like stimuli, emotion contagion) and the development of related skills found in older children and adults (e.g. empathy, goal-based imitation, face-specific processing). Before considering different stages of development, the chapter will explore in more detail how genes and environments shape social cognition and social behavior over a lifespan.

THE NATURE AND NURTURE OF SOCIAL COGNITION

Behavioral genetics is concerned with studying the inheritance of behaviors and cognitive skills. The classic methods of behavioral genetics are twin studies and adoption studies. These provide ways of disentangling nature and nurture.

Most behaviors run in families but it is hard to know to what extent this reflects shared environment or shared genes. When a child is placed into an adopted home, he or she will effectively have two sets of relatives: biological relatives with whom the child no longer shares any environment, and adopted relatives with whom the child shares an environment but not genes. Will the child more closely resemble the biological or adoptive family, thus emphasizing a role of nature or nurture, respectively? In many cases, it is not possible to contact or test the biological relatives, but the genetic contribution can still be estimated by comparing the adopted child with non-adopted siblings in the household (i.e. both the adopted and non-adopted siblings share family environment but not genes).

Twin studies follow a similar logic. Twins are formed either when a single fertilized egg splits into two (monozygotic or **MZ twins**) or when two eggs are released at the same time and separately fertilized (dizygotic or **DZ twins**). MZ twins are genetically identical; they share 100% of their genes. DZ twins are non-identical and share only 50% of their genes (i.e. the same as non-twin siblings). Studies of twins reared apart combine the advantages of the standard twin study and adoption study.

Figure 11.2 shows the similarity (correlations) between MZ and DZ twins on a wide range of social cognitive measures that have been discussed in this book (Ebstein, Israel, Chew, Zhong, & Knafo, 2010). Higher correlations reflect higher degrees of similarity amongst twins. However, it is the relative difference between MZ and DZ twins that is indicative of genetic influences. **Heritability** is an estimate of *how much* genetics contributes to a trait. In particular, heritability is the proportion of variance in a trait, in a given population, that can be explained by genetic differences amongst individuals. If MZ twins correlate with each other by 1.00 and

Figure 11.1 Identical twins look the same, but do they think the same?

Figure 11.2 Top: correlations between identical (MZ, purple) and non-identical (DZ, orange) twins on various measures of social behavior. Bottom: estimates of the percentage of variability in each measure that can be assigned to genetic effects (purple), shared environment (orange), and unshared environment (green). From Ebstein et al., (2010)

if DZ twins correlate with each other by 0.50, then heritability is 100%. The degree to which an MZ correlation is lower than the perfect correlation of 1.0 is assumed to reflect **unshared environment** – i.e. those environmental differences that distinguish between twins (e.g. different peer groups, exposure to different pathogens).

KEY TERM

Unshared environment
The proportion of variance in a trait, in a given population, that can be accounted for by events that happen to one twin but not another, or events that affect them in different ways

The remaining portion of variance is assigned to **shared environment** – i.e. the environmental factors that are common to both twins (e.g. family SES, common parenting styles, common diets). The lower panel of Figure 11.2 shows the estimates of heritability and shared and unshared environment for this set of social traits.

Behavioral genetic studies are important for establishing that there is a genetic component to a trait and for quantifying how much variance in that trait is attributable to genetics. However, behavioral genetics is effectively silent about the underlying mechanisms and heritability statistics are often misinterpreted. For example, consider the statistic that female infidelity has a heritability of ~50% (Cherkas, Oelsner, Mak, Valdes, & Spector, 2004). What can we conclude from that? We can't conclude that there is a gene (or genes) for female infidelity, as this statistic gives no clues as to how the underlying genes operate (the genes are likely to affect many behaviors and not just infidelity). Nor could we conclude that 50% of female infidelities are caused by genetics. It is not possible to infer causality from such a statistic. The figure simply reflects how much variability in this behavior is due to (undefined) genetic influences, and these figures themselves may change from culture to culture even when the genetic make-up is similar. For instance, heritability estimates for female infidelity could be very different in a culture that punishes this behavior oppressively (e.g. death by stoning) relative to a more permissive society.

The new frontier of the nature–nurture debate is concerned with how genes and environments influence each other mechanistically. At least three broad scenarios have been identified.

- **Epigenetic influences**. Environmental influences can alter the expression of genes. Although the sequence of DNA is normally fixed in a given individual and across all cells in his or her body, the timing and the degree of functioning of genes in the DNA can be affected by the environment (see Chapter 2). For example, increased maternal nurturing by a rat affects expression of a stress-reducing gene in its offspring that persists throughout their lifetime (Weaver et al., 2004) and early maltreatment in humans is also linked to epigenetic changes (McGowan et al., 2009).
- **Gene–environment correlations** (rGE) are genetic influences on people's exposure to different environments (Plomin, DeFries, & Loehlin, 1977). For example, people will seek out different environments (e.g. drug taking and novelty seeking) depending on their genotype (Benjamin et al., 1996). Also, the environment that a parent creates for raising his or her children will depend on the parent's own dispositions (intellect, personality, mental illnesses), which are partly genetic in origin.
- **Gene X environment interactions** (G × E) occur when susceptibility to a trait depends on a particular combination of a gene and environment. The effects of the gene and environment together exceed what would be expected from the sum of the parts. Previous chapters have presented examples of gene X environment interactions with regards to the oxytocin receptor gene (Chapter 8) and the so-called warrior gene (Chapter 10). Another example is given in Figure 11.3 showing that the probability of developing Schizophrenia-related symptoms in early adulthood is linked to an interaction between genetic disposition and smoking of cannabis during adolescence (Caspi et al., 2005).

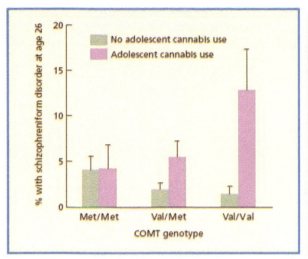

Figure 11.3 The COMT gene is involved in the metabolism of the neurotransmitter dopamine, and the gene exists in two main forms (termed Val and Met). Each person has two copies of the gene. If you have a Val copy of the gene *and* you smoke cannabis during adolescence, then there is an increased risk of displaying symptoms of schizophrenia at age 26 – a gene X environment interaction. Note that there was no evidence that people with this genotype were more likely to smoke cannabis (which would have been a gene–environment correlation). Reprinted from Caspi et al. (2005) Copyright 2005, with permission from Elsevier.

Evaluation

Behavioral genetics, and the associated methods of twin and adoption studies, play an important role in quantifying the role of genetic and environmental influences in determining the variability in a given trait. Most complex social behaviors are a result of both and three candidate mechanisms that link genes and environments are: epigenetics, gene-environment correlations, and gene X environment interactions.

BUILDING A BRAIN

The nervous system derives from a set of cells arranged in a hollow cylinder, the **neural tube**. By around 5 weeks after conception, the neural tube has organized into a set of bulges and convolutions that will go on to form various parts of the brain. Closer to the hollow of the neural tube are several proliferative zones in which neurons and glial cells are produced by division of proliferating cells (neuroblasts and glioblasts). Purves (1994) estimates that the fetal human brain must add 250,000 neurons per minute at certain periods in early development. The newly formed neurons must then migrate outwards towards the region where they will be employed in the mature brain. At birth, the head makes up around a quarter of the length of the infant. Although the brain itself is small (450 g)

KEY TERM

Neural tube
A set of cells arranged in a hollow cylinder in an embryo from which the nervous system derives

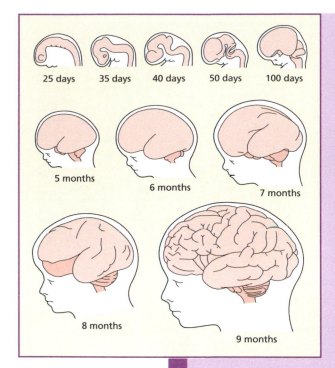

Figure 11.4 The embryonic and fetal development of the human brain

relative to adult human size (1400 g), it is large in comparison to remote human ancestors and living primates (a newborn human brain is about 75% of that of an adult chimpanzee). The vast majority of neurons are formed prior to birth, so the expansion in brain volume during postnatal development is due to factors such as the growth of synapses, dendrites, and axon bundles; the proliferation of glial cells; and the myelination of nerve fibers.

Huttenlocher and Dabholkar (1997) measured the synaptic density in various regions of human cortex at different ages (see Figure 11.5). In all cortical areas studied to date, there is a characteristic rise and then fall in synapse formation (synaptogenesis). In primary visual and primary auditory cortex the peak density is between 4 and 12 months, at which point it is 150% above adult levels, but this falls to adult levels between 2 and 4 years. In the prefrontal cortex, the peak is reached after 12 months but does not return to adult levels until 10–20 years old. Why does the number of synapses rise and then fall during the course of development? It is not necessarily the case that more synapses reflects more efficient functioning. During development, a process of fine-tuning the brain to the demands of the environment renders some connections redundant.

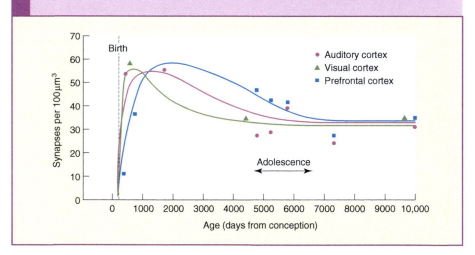

Figure 11.5 Synapse formation has a rise-and-fall pattern. It peaks soon after birth, but different cortical regions differ greatly in the time taken to fall again to adult synaptic levels. From Huttenlocher and Dabholkar (1997). Reprinted with permission of John Wiley & Sons, Inc.

SOCIAL LEARNING DURING INFANCY

Infancy is defined as the period from birth to 18 months, beyond which the child is normally capable of language production. During this period the infant's social world is dominated by its parents or other caregivers, for which the infant will establish a powerful emotional bond. (Attachment and relationships were discussed in Chapter 8.) The infant learns to recognize these key figures via sight, sound, and smell. It learns to read the social cues that they provide (e.g. emotional expressions) and to understand their significance via shared activities (e.g. joint attention, turn-taking).

Recognizing others

The infant's first vision of the social world occurs at birth. However, other senses, particularly hearing, are able to acquire socially relevant information from within the womb. The newborn comes into the world already knowing the voice of the mother (DeCasper & Fifer, 1980), and her native language (Mehler, Bertoncini, Barriere, & Jassikgerschenfeld, 1978). This is demonstrated using an experimental set-up in which the infant's sucking switches on a tape recording: it sucks more strongly to hear a familiar recording. Infants even show a preference for listening to a story that the mother had frequently read aloud during her late pregnancy relative to an unfamiliar story (DeCasper & Spence, 1986). However, this behavioral evidence does not prove that they possess adult-like voice recognition from birth. In adults, there is a region in the upper part of the superior temporal sulcus that responds more to voices than to other kinds of auditory stimuli (Belin et al., 2000). Grossmann, Oberecker, Koch, and Friederici (2010) used near infrared spectroscopy to examine the emergence of this specialization and found that the region responds selectively to voices in 7-month-olds but not 4-month-olds. Other evidence suggests that at 7 months of age infants can discriminate between happy, angry, and neutral prosody, and ERP evidence suggests that they are able to link these to the appropriate facial expressions (Grossmann, Striano, & Friederici, 2005).

Far more is known about face recognition by infants than voice recognition. Infants have an early preference for face-like stimuli relative to control stimuli. Infants within an hour of birth can show greater tracking of a moving face-like stimulus over a non-face-like stimulus – see Figure 11.6 (Johnson et al., 1991). This evidence has been used to propose an innate device for orienting attention towards stimuli that are likely to be faces (Morton & Johnson, 1991). However, this falls short of claiming that the newborn brain contains a representation of a standard face. Macchi Cassia et al. (2004) manipulated photographs of real faces by inverting the internal features (such that the eyes are below the mouth) but such that the head itself retains the normal orientation. Newborns show a preference for a 'top-heavy' face but not necessarily the correct arrangement of features, as shown in Figure 11.7. At 3 months of age, infants show preferential looking at faces of their own race relative to different races (Bar-Haim, Ziv, Lamy, & Hodes, 2006) and prefer to look at faces that are of the same sex as their primary caregiver (Quinn, Yahr, Kuhn, Slater, & Pascalis, 2002).

At what age can an infant recognize specific faces? Infants show a preference for the mother, relative to a stranger, as early as 2–3 days old, and this occurs when odor cues are eliminated and the stranger has a similar hair style (Bushnell, Sai, & Mullin, 1989). However, early face recognition of the mother appears to depend on being able to link the mother's face with the mother's voice, which was already familiar from the prenatal stage (Sai, 2005). As such the mother's face may be learned differently to other faces. Long-term recognition of familiar people from a variety of views is possible by 3 months (Pascalis, de Haan, Nelson, & de Schonen, 1998). Before 3

Figure 11.6 Newborns in the first hour of life were sat on the experimenter's lap and shown a handheld stimulus. Upon fixating the stimulus, the experimenter slowly moved it to the side (90°). The extent to which the baby tracked the stimulus (head and eye turns) was measured. From Johnson et al. (1991). Copyright © 1991 Elsevier. Reproduced with permission.

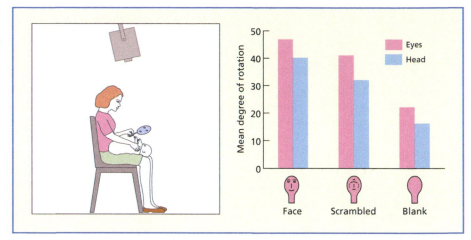

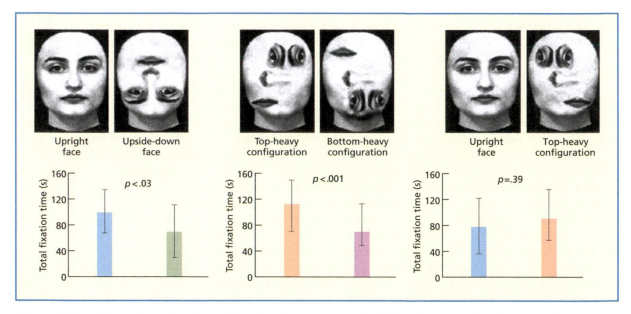

Figure 11.7 In this study, pairs of face-like stimuli are presented side by side (with side of presentation counter-balanced). The preferential looking time for one stimulus relative to another is measured. Infants 1–3 days old tend to prefer top-heavy face-like configurations, although they do not necessarily distinguish possible from impossible faces. From Macchi Cassia et al. (2004). Copyright © 2004 Association for Psychological Science. Reprinted by permission of SAGE Publications.

months, studies have tended to rely only on short-term familiarity using habituation, but nevertheless demonstrate the ability to distinguish between a recently exposed picture of a face and a completely novel face (Turati, Cassia, Simion, & Leo, 2006). Language may play a role in individuating faces even in infancy. Infants at 6 months old can discriminate different monkey faces (as assessed by habituation) but fail to do so at 9 months. However, if infants are trained to individuate them by pairing each face with a different name, then the tendency to individuate is retained at 9 months (Scott & Monesson, 2009). Pairing each face with the same category label 'monkey' leads to the normal decline in individuation.

Tzourio-Mazoyer et al. (2002) conducted a functional imaging study of face processing in infants and noted that the regions implicated in adult face processing are activated but that there are also other regions activated. For instance, regions that are functionally specialized for language in adults respond to faces in infants. Brain damage in the fusiform region at only one day post-birth can lead to severe difficulties in face recognition (Farah, Rabinowitz, Quinn, & Liu, 2000). This suggests that this region may be committed to the processing of faces prior to birth even if it takes many years to reach an adult level of specialization. Studies outside infancy also suggest a slow maturation. Golarai et al. (2007) compared face perception (relative to objects) in children (7–11 years), teenagers, and adults using fMRI. Whilst all three groups responded similarly to objects, the fusiform face area was three times larger in adults relative to children. It would be interesting to know if this reflects cumulative experience or a property of the region itself – adults know more faces than children, but our knowledge of objects does not accumulate in the same way.

At around five months of age a child is able to hold and manipulate an object. As such, their interactions with other people shift away from face-to-face interactions (**dyadic interactions**) to joint attention-based interactions in which infant and adult may engage each other in looking at other objects or people (**triadic interactions**). Kaye and Fogel (1980) recorded mother–baby sessions at several points in time. At 6 weeks 70% of the session time was based on face-to-face interactions, reducing to 33% at

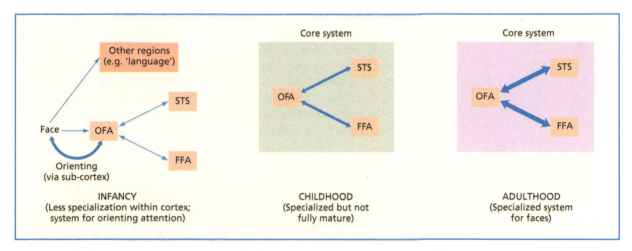

Figure 11.8 Functional imaging studies of infant face processing suggest that faces engage a wider network of regions than found in children and adults, in addition to showing activity in the 'core' system. During childhood this system becomes more specialized, but the spatial extent of the activation in regions such as the fusiform face area (FFA) is far less than in adults. OFA = occipital face area; STS = superior temporal sulcus.

about six months. The ability to recognize and respond to eye gaze cues is crucial for establishing joint attention. Newborns prefer to look at eyes, making direct contact relative to averted eyes (Farroni et al., 2002), and faces presented with a direct gaze are subsequently recognized better by 4-month-old infants (Farroni, Massaccesi, Menon, & Johnson, 2007). However, the youngest age at which an infant is capable of following a gaze to a peripheral location is around three months (Farroni, Mansfield, Lai, & Johnson, 2003). These mechanisms may enable joint attention to develop. However, merely following the gaze of another person is not sufficient for making the link between behavior and psychological states (i.e. that seeing is a source of knowing). Up until 10 months an infant will follow the head-turns of someone who has their eyes closed, but beyond this age they only follow head-turns when the eyes are open (Brooks & Meltzoff, 2005). This suggests that before this point they have little understanding of the significance of attention-orienting behavior, even though they can engage in it. A source localization study of ERP components to eye gaze in infants revealed sources of activity in occipital regions and prefrontal regions but *not* reliably in the region that is most closely linked to eye-gaze processing in adults – namely the superior temporal sulcus (Johnson et al., 2005). This suggests an early use of general processes (e.g. in visual perception, motor coordination) in gaze detection that becomes more specialized during later development. Another EEG study (Mundy, Card, & Fox, 2000) suggests a difference between brain mechanisms for *responding* to joint attention cues (more posterior) and *initiating* such cues by looking back and forth between adult and object (more anterior).

By about seven months of age, infants are generally able to distinguish between dynamic emotional expressions (Soken & Pick, 1999). However, this need not imply that they use an adult-like system to achieve this. For example, infants can be far better at discriminating certain emotions when posed by their mother rather than by a stranger (Kahana-Kalman & Walker-Andrews, 2001), suggesting that emotion recognition is not separable from face recognition in early development. Also, infant ERP studies show an influence of emotion on far later components of the ERP signal than expected from the adult literature (Leppanen, Moulson, Vogel-Farley, & Nelson, 2007; Nelson & DeHaan, 1996).

Social referencing becomes possible by linking gaze cues and facial expressions to learn about the potential emotional significance of objects (Klinnert et al., 1983). Recall from Chapter 4 that social referencing is based on a learned (classically conditioned) association between an adult's emotional response and attention directed towards an object, enabling an infant to learn whether something is 'good', 'safe', 'forbidden', etc. This kind of behavior comes online at around 12 months, although it continues to mature subsequently. For example, infants of this age respond appropriately to where the adult focuses their emotion even if the infant is attending elsewhere (Moses, Baldwin, Rosicky, & Tidball, 2001).

ADAPTING THE METHODS OF COGNITIVE NEUROSCIENCE FOR INFANTS AND CHILDREN

Methods such as fMRI and EEG are generally considered suitable for infants and children. One advantage of using these methods in younger people is that they do not necessarily require a verbal or motor response to be made.

Functional MRI and structural MRI

Gaillard, Grandon, and Xu (2001) provide an overview of some of the considerations needed. If one wants to compare across different ages, then the most significant problem is that the structural properties of the brain change during development. Although the volume of the brain is stable by about five years of age, there are differences in white- and gray-matter volumes until adulthood (Reiss, Abrams, Singer, Ross, & Denckla, 1996). The hemodynamic response function is relatively stable after 7 years of age but differs below this age (Marcar, Strassle, Loenneker, Schwarz, & Martin, 2004). Both the differences in brain structure and blood flow make it harder to compare activity in the same region across different ages. Younger children also find it harder to keep still in the scanner and this motion can disrupt the reliability of the magnetic resonance signal.

Near-infrared spectroscopy (NIRS)

One relatively new method that is now being used in developmental social neuroscience is **near-infrared spectroscopy (NIRS)** (for a summary see Lloyd-Fox, Blasi, & Elwell, 2010). This measures the amount of oxygenated blood and is – like fMRI – a hemodynamic method. Unlike fMRI it accommodates a good degree of movement and is more portable. The infant can sit upright on their parent's lap. However, it has poorer spatial resolution and does not normally permit whole-head coverage.

ERP/EEG

When working with young subjects using ERP/EEG, a limiting factor is the child's willingness to tolerate the electrodes, the task, and the time commitment required (Thomas & Casey, 2003). Children and adults can show quite different patterns of ERP (e.g. in terms of latency, amplitude, or scalp distribution), even for tasks that both groups find easy, as shown in Figure 11.9 (Thomas & Nelson, 1996). These could reflect either age-related cognitive differences (i.e. the same task can be performed in different ways at different ages) or non-cognitive differences (e.g. the effects of skull thickness, cell packing density, or myelination).

KEY TERM

Near-infrared spectroscopy (NIRS)
A hemodynamic method that measures blood oxygenation, normally in one brain region

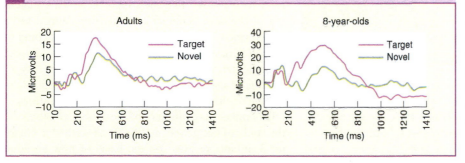

Figure 11.9 Adults and children show very different visual ERP waveforms, despite having equivalent behavioral performance. Adapted from Thomas and Nelson (1996).

Imitation, action understanding, and early communication

As noted in previous chapters, many researchers limit the use of the term imitation to those situations in which an individual reproduces the goals and intentions of another person rather than copying their motor actions. The latter, motor mimicry, appears to be present from birth, and goal-based imitation appears to emerge after the first year. Newborn infants will mimic simple actions such as tongue protrusion (Meltzoff & Moore, 1977, 1983) – that is, they demonstrate an understanding that a seen tongue being protruded corresponds to their own, unseen, motor ability to do the same. Meltzoff and Moore (1977) concluded that, 'the ability to use intermodal equivalences is an innate ability of humans' (p. 78). At 3 months of age (but not 1 month) an infant shows signs of being able to discriminate when its own actions (facial, vocal, hands) are being imitated by an adult (Striano, Henning, & Stahl, 2005). They gaze more intently at this condition relative to control conditions in which the adult acts expressively but non-imitatively.

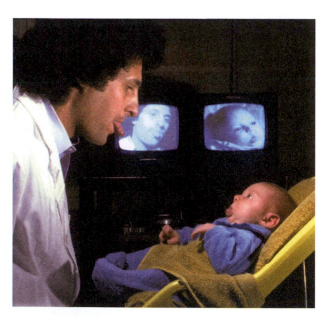

Figure 11.10 This 23-day-old infant imitates the tongue protrusion of the experimenter, suggesting an understanding of the link between seen actions of another and their own unseen actions. Photo by Andrew N. Meltzoff and E. Ferorelli. Used with permission from Andrew N. Meltzoff.

A parallel finding in the auditory–vocal domain is that (newborn) infants are susceptible to emotional contagion – that is, if other infants around them are crying then they too will be more likely to cry or show other distress cues (Simner, 1971). An infant is more likely to cry in response to the recorded cries of another infant than to a recording of their own previous crying (Dondi, Simion, & Caltran, 1999) and are more likely to do so than in response to an equally loud non-human sound (Simner, 1971).

Both motor mimicry and emotional contagion are assumed to reflect automatic reactions over which an infant has no control. Nevertheless, infants may use this behavior to learn the principle that other people are like themselves. Meltzoff and Decety (2003) refer to it as the 'like me' hypothesis. This matching of *physical* behavior (actions) is assumed to be transformed, during development, into a matching of *mental* behavior (thoughts, feelings, etc.) leading to the emergence of empathy and mind reading. By 14 months of age, infants show evidence of being able to imitate on the basis of understanding the intentions of others rather than mimicry of the actions of others. In the study of Gergely et al. (2002), described in Chapter 3, infants respond on the basis of how an adult would have responded (should their arms be free) rather than how they actually did respond. Carpenter, Akhtar, and Tomasello (1998) found that infants aged 14–18 months would imitate acts differentially according to whether they were intentional versus accidental. Each action sequence had two parts and was performed on the same object. However, one part of the action would be accompanied by the experimenter saying 'whoops!' and the other part by saying 'there!' (signaling accidental and intentional actions, respectively). The infants were more likely to imitate the intentional actions than the accidental ones, suggesting an understanding of goals as distinct from actions. At 9 months of age, EEG recordings of infants' motor systems (mu rhythms) suggests that they discriminate actions in which a goal can be inferred relative to motorically similar but ambiguous actions (Southgate, Johnson, El Karoui, & Csibra, 2010).

One topic of controversy is the age at which mirror neurons develop and, more specifically, whether they are innate or a product of learning. Cook et al. (2014) have argued that whilst the development of certain visual and motor abilities may be genetically guided, the particular patterns of association between the visual and motor domain (i.e. those that characterize mirror neurons) are a result of associative learning as a result of seeing one's own actions and through social interactions including imitation. This account is illustrated in Figure 11.11. The critical evidence to adjudicate between these positions must come specifically from newborns rather than older infants. In support of innateness, there is evidence for mimicry from birth in both humans (Meltzoff & Moore, 1977) and macaques (Ferrari et al., 2006), and evidence for mu suppression in newborn macaques when observing tongue protrusion and lip smacking (Ferrari et al., 2012). The latter is shown in Figure 11.12.

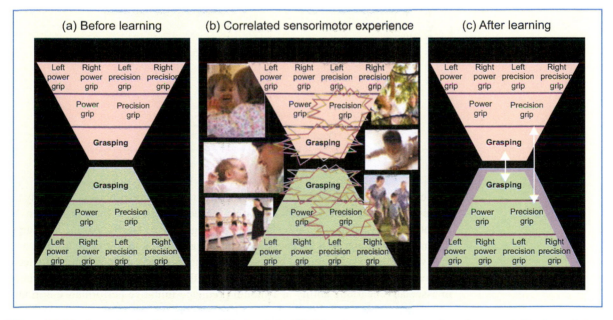

Figure 11.11 According to the theory of Cook et al. (2014), mirror neurons develop via associative learning between representations in the visual and motor systems. Over time, parts of the motor system respond to the observation of actions (i.e. they become mirror neurons) and different parts of the system may be tuned to specific or broad parts of the action (e.g. responding either to specific mechanical movements, or to any kind of grasping action). This challenges the view that mirror neurons may have evolved specifically for action understanding or social cognition.

Figure 11.12 Certain EEG oscillations are diminished (event-related desynchronizations, ERD) when the newborn macaque produces facial gestures (lip smack, LS, and tongue protrusion, TP) and when observing them produced by the experimenter. Does this prove that the mirror system is innate and functioning, in some form, from birth? From Ferrari et al. (2012).

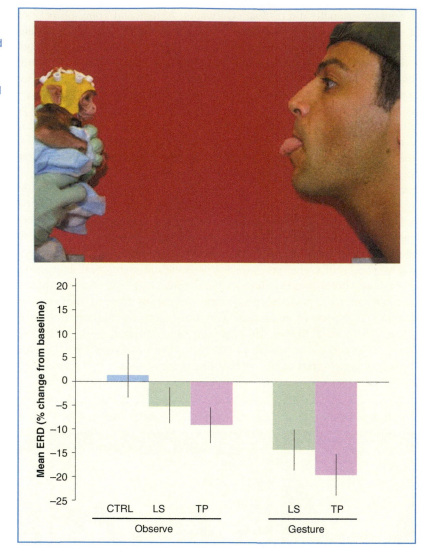

However, Cook et al. (2014) argue that this evidence is insufficient. They suggest that evidence for mimicry is limited to only one act in humans: tongue protrusion, but not other acts such as pouting or mouth opening (Ray & Heyes, 2011). They argue that mu suppression could reflect a general arousal of the motor system rather than the systematic correspondence between visual and motoric actions that characterize mirror neurons. In support of their own theory, they note that individual differences in associative learning ability at 1 month predicted imitative performance 8 months later (Reeb-Sutherland, Levitt, & Fox, 2012). Moreover, in adults, it is possible to elicit 'counter-mirror' responses such that, for instance, observing movement of the index finger can increase motor excitability of the little finger after learning to pair this action with this atypical visual stimulus (Catmur, Walsh, & Heyes, 2007).

Gesture-based communication can occur before the production of the first spoken words. Carpenter, Nagell, and Tomasello (1998) examined protodeclarative pointing during infancy. Protodeclarative pointing carries the meaning 'look at that!'. It is important, theoretically, because it implies that the infant understands that seeing is a route to knowing and that a different viewpoint leads to different knowledge. The infant and an experimenter interacted with a toy, but then a second experimenter entered with a more interesting toy. Crucially, the second experimenter could be seen by the infant but not by the first experimenter. Gestures by the infant to engage the attention of the first experimenter were seen to emerge at around 12–13 months of age. In terms of possible neural substrates, the level of EEG activity in frontal regions at 14 months of age predicts protodeclarative pointing at 18 months (Henderson, Yoder, Yale, & McDuffie, 2002). This suggests that this region may be important for initiating shared attention.

Although language development is typically studied from a 'cognitive' rather than 'social' developmental perspective, the social element is crucial given that communicative acts *are* social interactions. Note, for instance, that a difficulty in communication is one of the basic markers for autism. Kuhl (2007) has argued that normal language learning is gated by the social brain: that is, the degree to which an infant has developed an adequate understanding of social interactions will determine the pace and richness of speech acquisition. The quality of the interactions between infant and parent is also likely to be crucial, as well as the infant's own readiness. Tomasello and Todd (1983) videotaped a number of sessions of mother–child interactions between 12 and 18 months. They calculated the amount of time that the mother and child spent in activities requiring joint attention, and found that this correlates with overall vocabulary size at 18 months. Different types of attentional engagement predicted learning of particular types of words: the infants of mothers who *followed* their child's attention tended to learn more object names, whereas infants of mothers who *directed* their child's attention tended to learn more social words.

Kuhl, Tsao, and Liu (2003) studied phoneme discrimination by infants raised in English-speaking communities (see Figure 11.13). At 10–12 months of age, infants tune-in to the phonemes of their language, and this is achieved by reducing (tuning-out)

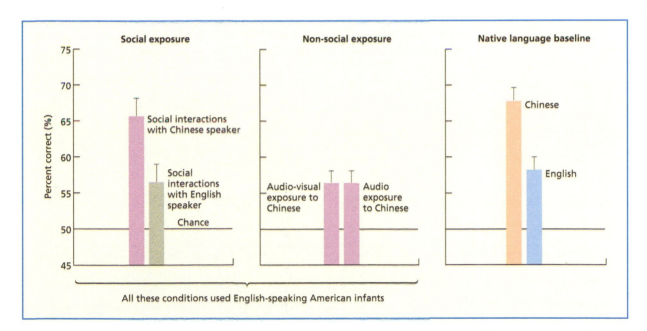

Figure 11.13 Social contact with a Chinese speaker enhances the discrimination of Chinese phonemes by English-speaking American infants (10–12 months old). The same effect is not found for comparable language exposure via CDs (i.e. audio only) or DVDs (i.e. audio-visual). Adapted from Kuhl et al. (2003).

their ability to detect the phonemes of other languages. Infants were exposed to Chinese phonemes either via social interaction (over 12 sessions of 25 minutes) with a Chinese speaker or using the equivalent material presented non-socially via DVD (i.e. audio-visual) or CD (i.e. audio only). Only those infants exposed to Chinese socially showed an ability to make Chinese phonemic distinctions (and did not, in fact, differ from infants raised in monolingual Chinese cultures). One suggestion is that this learning depends on the level of triadic interactions between infant, speaker, and object, which is found in face-to-face interactions but not from DVDs and CDs (Mills & Conboy, 2009).

Adults, across most cultures, adjust the qualities of their speech when talking to infants using so-called **motherese** or **infant-directed speech**. This has several features, including higher pitch overall, more pitch variability, and a slower tempo with elongated vowels (Garnica, 1977). Infants show a preference for this kind of speech and one possible origin of the preference is that it is related to happy-sounding prosody (Singh, Morgan, & Best, 2002). Whilst adults tend to increase the pitch of their voice both when talking to pets and when talking to infants, they only exaggerate the pronunciation of vowels when talking to their infants (Burnham, Kitamura, & Vollmer-Conna, 2002). ERP studies comparing motherese with normal speech showed that infants at 6 and 13 months had larger amplitudes of certain ERP components to motherese relative to normal speech, notably over a component at 600–800 ms, which has been linked to attentional processes rather than word recognition – see Figure 11.14 (Zangl & Mills, 2007). Preschool children with autism do not fully distinguish motherese from normal speech, either in terms of preference or electrophysiological measures (Kuhl, Coffey-Corina, Padden, & Dawson, 2005).

Evaluation

Although the infant's expressive and communicative abilities are limited, he/she is both an avid consumer and user of social cues. Within the first year of life the infant is able to recognize others and respond to affective cues in faces and voices, and respond to attention-orienting cues (e.g. eye gaze). However, the infant is not a

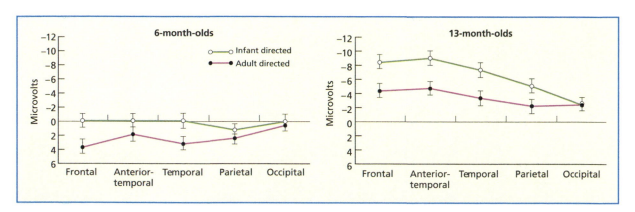

Figure 11.14 Motherese (or infant-directed speech) may increase attention to words and is associated with changes in neural processing, measured here using EEG (showing the mean amplitude of the ERP component at 600–800 ms). The infants listened to both familiar and unfamiliar words presented either in adult-directed or infant-directed speech. From Zangl and Mills (2007).

passive learner. He/she will engage in interactions (pointing, turn-taking, etc.) and is capable of manipulating the attention of others. Many of these behaviors suggest that the infant is behaving like a 'rational' being (Gergely & Csibra, 2003) in that there appears to be an element of *interpretation* of social cues (e.g. responding to head turns only when the eyes are opened). The neuroscience evidence for this period, such as it is, suggests that the neural substrates supporting social behavior are far less specialized than found in adults (or older children).

THE DEVELOPMENT OF SELF-RECOGNITION AND SELF-AWARENESS

Lewis and Brooks-Gunn (1979) conducted an important study examining how children recognize themselves. Mirrors provide multiple cues to self-recognition: the person in the mirror looks like me, and the person in the mirror follows my movements. These different cues can be pulled apart using video cameras and monitors, such that there is a live stream (equivalent to the mirror condition) or a recorded stream so that the person looks like them but does not follow the movements. Between 3 and 8 months, infants show some signs of self-recognition based on shared movement cues. Between 8 and 12 months they show clear signs of recognizing themselves using movement cues, but they do not recognize their own face in a non-live video stream and they do not pass the mark test (i.e. rubbing a lipstick mark off their forehead). This suggests that they do not fully connect the image in the mirror to themselves. Between 12 and 24 months they can recognize themselves based on appearance even if the movements are not in synchrony, but it is only by 21–24 months that children pass the mark test. This suggests that bodily self-recognition develops incrementally up to the age of 2 years.

Figure 11.15 Infants and children seem to use different cues to recognize themselves at different developmental stages. Only at 2 years do they use mirrors to rub off paint smudges on their foreheads, implying a full understanding that the images out there correspond to their own bodies.

Lewis and Carmody (2008) investigated the neural basis of self-representation by measuring structural changes (using structural MRI) in different brain regions in infants/children aged between 5 and 30 months (note: not longitudinally). They had three tasks: mirror self-recognition (based on the mark test), use of personal pronouns (e.g. me, mine), and use of pretend play (e.g. feeding a doll with a spoon). Of the regions considered, only the temporo-parietal junction showed an association with self-representation (and the largest correlation was with mirror self-recognition). As well as being part of the 'mentalizing' network (e.g. Saxe, 2006) this region is involved in putting oneself in another bodily perspective (Arzy, Thut, Mohr, Michel, & Blanke, 2006), and lesions can induce out-of-body experiences (Blanke et al., 2004). Recognizing oneself in a mirror may require a loosening of the sense of embodiment by projecting oneself onto an external location.

KEY TERM

Childhood amnesia
The inability of children and adults to recall episodes from the first few years of life

Other research suggests that children may not have a full understanding of the continuity of the self over time until 4 years. Povinelli and Simon (1998) conducted a version of the mark test in which children were video-recorded in two different locations at two time points – 1 week previously and a few minutes ago. Both 4- and 5-year-olds searched for the mark when shown the video of themselves a few minutes ago but not when shown the video taken 1 week ago. In contrast, 3-year-olds did not distinguish between these time points: they tended not to search for the mark in either condition. The authors interpret this as a difficulty in understanding that 'the self' can be duplicated in time and a struggle to represent the 'past self' and 'present self' as separate. This is consistent with other lines of evidence. As adults, our earliest recollections date from the period between 3 and 4 years old, and we remember far fewer events before the age of 6 years than would be expected from simple forgetting (Pillemer & White, 1989). The inability to recall episodes from our first few years is termed **childhood amnesia** or infantile amnesia. Perner and Ruffman (1995) claim that until 3–4 years children have *knowledge* about their past but they are unable to recall events as having been experienced by themselves (i.e. as autobiographical memories as opposed to facts). This shift is assumed to reflect changes in self-awareness (perhaps linked to theory-of-mind development) rather than changes in basic memory processes.

THE SOCIAL BRAIN IN CHILDHOOD: UNDERSTANDING SELF, UNDERSTANDING OTHERS

The childhood period from the end of infancy (at 18 months) to adolescence (at puberty) is characterized by an increasingly complex and diverse set of social interactions. The child is increasingly influenced by his/her peers and behavior becomes more guided by the norms and skills of the surrounding culture. Relating to this, the evidence from behavioral genetics emphasizes the greater role of unshared environment (e.g. peers) over shared environment (e.g. family) in explaining variability in social behavior (Ebstein et al., 2010).

The period of childhood is particularly long in humans compared to most other species. There is a correlation, in primate species, between brain size, social group size, and the length of the childhood period (Joffe, 1997). This long period of immaturity gives human children extra social learning opportunities before achieving adulthood.

Developing a theory of mind

Meltzoff and Moore (1977) argue that one of the mechanisms that is innate is a simple 'body scheme' that enables different parts of the body to be activated by vision (of others) and by touch and action (by oneself). This self–other link provides a foundation for the infants' understanding that others are 'like me', and according to Meltzoff and Decety (2003), 'infant imitation is the seed and the adult theory of mind is the fruit' (see also Meltzoff, 2007). By 18 months, the imitative behavior of toddlers reproduces the intentions of adults by following what people *meant* to do and ignoring, say, accidents (Meltzoff, 1995). At 18 months, most infants engage in pretend play and understand that when their mother uses the banana as a telephone she is not mistaken but is deliberately substituting one object for another (Leslie, 1987). Leslie (1987) argues that there has been a decoupling between ideas (what people think) and behavior (what people do). An absence of pretend play at 18 months is an early indicator of autism, along with a lack of protodeclarative pointing discussed earlier (Baird et al., 2000).

The ability to pass false belief tasks, giving an explicit verbal answer, typically emerges around 4 years of age, as shown in Figure 11.16. This poses a paradox. If skills such as pretend play and understanding intentions emerge by the end of infancy, then why does it take several more years for the 'classic' theory-of-mind tests, based on false belief, to be passed? One possibility is that aspects of non-social cognition need to catch up. These tasks require reasonable language abilities: following a narrative and remembering a sequence of events. However, from the age of 2 years onwards children use mental state words such as 'want', 'wish', and 'pretend' in an appropriate context, such as 'I thought it was an alligator. Now I know it's a crocodile' (Shatz, Wellman, & Silber, 1983). It may also be related to the particular nature of the mental state being tapped by theory-of-mind tasks (i.e. false beliefs) compared to imitation tasks (i.e. intentions). False belief tasks require second-order intentionality (e.g. beliefs about beliefs; Dennett, 1983), which could emerge later than first-order intentionality. False belief tasks also require inhibiting a strongly competing response (e.g. the child's own knowledge of the correct location) to perform correctly. Gergely and Csibra (2003) argue that from around 12 months infants understand that an action is a means to a goal (rather than the action being the goal), but they suggest that this falls short of being able to represent mental states as separate from current reality (as required in many theory-of-mind tests). Others suggest that the understanding of some mental states, such as desires, may develop earlier than others, such as beliefs (e.g. Wellman, 2002). This contrasts with the view that an understanding of mental states emerges as a result of maturation of a single overarching theory-of-mind module.

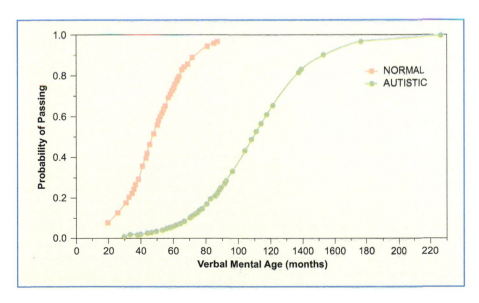

Figure 11.16 The probability of passing two different false belief tasks (Sally–Anne Task and Smarties Task) as a function of verbal mental age. Most typically developing children pass both from around 4 years (48 months), but children with autism show a significant developmental delay even when equating for mental age. From (Happe, 1995).

There is some evidence of a functioning theory of mind before the age of 4 years, even on false belief tasks, when the other demands of the task are minimized. Onishi and Baillargeon (2005) used a looking-based measure to claim that infants can represent false beliefs as early as 15 months. Their procedure is shown in Figure 11.17. In their task the adult saw a toy slice of watermelon placed in one of two boxes. A curtain was then lowered to obscure the adult's view. In the true belief condition, the object remains where it is. In the false belief condition, it is transferred to the other box. The curtain is then raised, and the adult reaches into one of the two boxes (either consistent or inconsistent with their belief), and the infant's looking time is measured. In the false belief condition, longer looking times were found when the experimenter reached into the correct physical location (inconsistent with the false belief) relative to when the experimenter reached into the incorrect location (consistent with the false belief). That is, the infants appeared to be surprised when the adult didn't act according to their beliefs, as indicated by the longer looking time. This raises the question as to whether it is the same theory-of-mind system that gives rise both to looking-based correct responses (present at late stages of infancy) and to verbally based correct responses (that emerge around 4 years), or whether there are two different systems. Evidence for the two-systems theory comes from a study of 3-year-olds showing that the same children can give contradictory answers depending on how

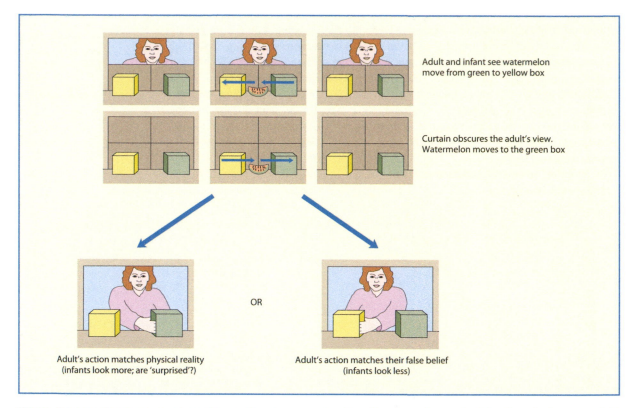

Figure 11.17 Infants as young as 15 months show surprise (i.e. they have longer looking times) when an adult reaches for an object in a box that matches its true location but does not match the adult's presumed beliefs about the location. Is this evidence for an understanding of false beliefs several years before infants can explicitly reason about such mental states? From Onishi and Baillargeon (2005). Copyright © 2005 American Association for the Advancement of Science. Reproduced with permission.

their understanding of beliefs is measured (Clements & Perner, 1994). The children may look towards the correct answer (the false belief) but verbally state the incorrect answer (their own belief). Butterfill and Apperly (2013) argue that there is an implicit form of theory of mind that emerges towards the end of infancy that supports the looking behavior reported by Onishi and Baillargeon (2005), and a later emerging explicit theory of mind that supports the verbal behavior of the traditional false belief tasks. They also argue that people with autism who pass explicit theory-of-mind tests may still lack the implicit theory-of-mind system. In one study, adults with autism were able to pass false belief tests but didn't show the characteristic looking behavior (Senju et al. 2009), this forming a double dissociation when contrasted against the pattern reported in typically developing 3-year-olds (Clements & Perner, 1994).

Evidence from neuroscience could perhaps be used to adjudicate between the one-system and two-system positions. Are the 'core' regions implicated in adult theory of mind used as early as 18 months, or is it a different network? At present there is very little evidence either way. At 4 years of age, EEG studies suggest that differences in regions such as the temporo-parietal junction and medial prefrontal cortex (both implicated in adult theory of mind) discriminate between children who fail and pass *explicit* measures of theory of mind, that is, those requiring an overt prediction of behavior (Sabbagh, Bowman, Evraire, & Ito, 2009). Between 5 and 11 years, the TPJ responds more selectively in fMRI to mental states over and above judgments of physical appearance or social relationships, and also predicted performance on higher-order belief reasoning (Gweon, Dodell-Feder, Bedny, & Saxe, 2012). In this 5–11 age range, children also become able to pass higher order belief reasoning of the sort 'John thinks that Mary thinks that . . .' (Perner & Wimmer, 1985).

Development of empathy and pro-social behavior

As noted previously, 18-month-old children are more likely to imitate intentional versus accidental actions, and this has been taken to imply that they have an understanding of the causes of actions (perhaps in terms of mental states) rather than copying motor programs. Another line of research suggests that children of this age also show pro-social helping behavior that discriminates between accidental versus intentional behavior (Warneken & Tomasello, 2006). Young children help when an experimenter accidentally drops something or if they cannot open a cabinet door properly because their arms are full. However, they are less likely to help if the experimenter deliberately drops something or deliberately bumps into the cabinet door (Warneken & Tomasello, 2006). This occurs without an external reward being offered and without the need for parental encouragement (Warneken, Hare, Melis, Hanus, & Tomasello, 2007), and in fact, giving rewards for helping may *reduce* the likelihood of future helping behavior (Warneken & Tomasello, 2008). Young children, like adults, may find the act of helping a reward in its own right. This intrinsic reward may be devalued when it is paired with an external award (i.e. the positive associations become linked with the external stimulus rather than their own internally generated behavior).

Young children may show concern and empathy for others and this may provide a motivation for helping in the absence of external rewards. Young children (18 months to 2 years) watched an adult doing a drawing and saw another adult grab the drawing and tear it up (Vaish, Carpenter, & Tomasello, 2009). In a control condition, one adult grabbed a blank piece of paper lying in front of the other adult. The children produced facial expressions of concern in the first scenario (as assessed by blind raters), and increased levels of concern predicted helping behavior towards the adult.

It is unclear from this study whether helping behavior is driven by the need to alleviate the child's own personal distress (self-oriented) or a genuine concern for the adult (other-oriented). Children, like adults, may react differently when confronted with distress. Fabes, Eisenberg, Karbon, Troyer, and Switzer (1994) measured children's (6–8 years) physiological response (heart rate increase) and behavioral response to a crying infant in another room, listened to over an intercom. Children who were better able to regulate their own emotional response (as measured by a small increase in heart rate) spontaneously engaged in comforting behavior over the intercom. Children who were less able to regulate their own emotion (large increase in heart rate) showed signs of distress themselves, and their behavior showed irritation (rather than comforting) and avoidance. For instance, these children were more likely to switch off the intercom.

Decety and Svetlova (2012) describe the development of empathy in terms of three broad stages (see Figure 11.18). Affective arousal is functioning from birth and supports behavior such as emotion contagion and attachment-related behaviors. It involves regions of the brain that support emotional processing such as the amygdala, ventral striatum, OFC, and hypothalamus. Emotion understanding develops during early childhood and is linked to the mentalizing network. Emotion regulation does not mature until adulthood and involves the maturation of lateral prefrontal cortex and its connections with those regions supporting affective arousal and emotion contagion. When participants aged from 7 to 40 years are watching another person being

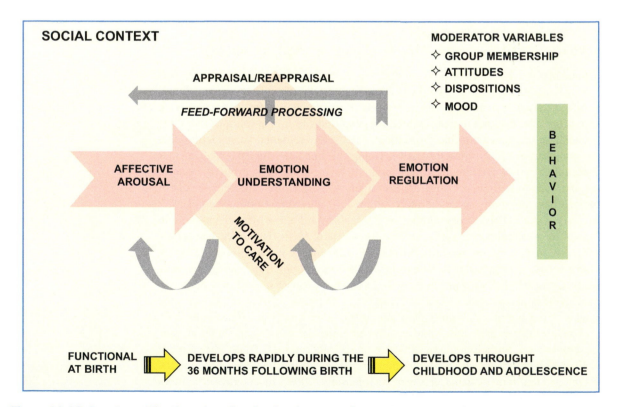

Figure 11.18 Decety and Svetlova describe the development of empathy in terms of three broad stages/ mechanisms that interact with each other: affective arousal, emotion understanding, and emotion regulation.

intentionally hurt by another person, there is a developmental shift in the degree of fMRI activity in various regions (Decety & Michalska, 2010). Younger participants tended to activate regions such as the amygdala and posterior insula more (perhaps reflecting greater affective arousal), whereas older participants tended to activate regions such as the dorsolateral PFC more (perhaps reflecting greater emotion regulation).

At 3 years of age, children show signs of discriminating who to cooperate with and who not to cooperate with. Olson and Spelke (2008) gave children a doll ('Reese') and then introduced, by way of a narrative, four more dolls. The dolls could either be family, friends, or strangers to Reese. They may have shared with Reese previously, or they may have shared with others recently. The child was then given some shells that 'Reese' had to distribute to the other dolls. When given four shells, children tended to distribute equally (i.e. one each) irrespective of status. However, when given fewer shells, then children tended to favor family and friends over strangers, those who had previously shared with them (direct reciprocity), and those who had shared with others (indirect reciprocity).

Other research based on the Ultimatum Game suggests that older children (aged 9–10 years) often apply rigid fairness norms (everything being shared equally), although adolescents and adults show more flexibility (Murnighan & Saxon, 1998). Crone and Westenberg (2009) argue that children might be basing their judgment on earlier-maturing emotion-processing regions and that overriding these responses, in favor of self-interest, may require mature functioning of the dorsolateral prefrontal cortex.

Some studies have investigated children's empathy using EEG and fMRI. Light et al. (2009) measured EEG activity in 6–10-year-old children when performing a pleasurable task involving a pop-out toy. Changes in EEG activity were then correlated with individual differences in empathy, measured by the child's response (facial, vocal, bodily) to seeing the experimenter experience pain (trapping finger) followed by relief. Children showing high empathic concern on this measure tended to show greater EEG activity over prefrontal regions during the pop-out task, although the laterality difference shifted during the course of the task (shifting from right to left). Greimel et al. (2010) used fMRI to study changes in empathy from 8 to 27 years in which participants had to infer an emotional state from a face or judge their own emotional response to a face (relative to a control task of judging the width of a face). Interestingly, accuracy was unaffected by age, but the brain regions supporting this task did shift developmentally. For instance, activity in the inferior frontal gyrus (linked to the human mirror neuron system) increased over age and was associated with differences between self and other perspectives (the left side showed an age-related increase in the self-versus-other contrast, whereas the right side showed the opposite). It is interesting to note that both of these studies suggest developmental differences in prefrontal functioning related to empathy

Figure 11.19 When 3-year-old children are asked to distribute shells, on a doll's behalf, to other dolls they show evidence of kin favoritism, direct reciprocity (giving to those who have given to you), and indirect reciprocity (giving to those who have given to others) when an equal distribution of resources is not possible. From Olson and Spelke (2008). Copyright © 2008 Elsevier. Reproduced with permission.

either across individuals of the same age (Light et al., 2009) or developmentally at different ages (Greimel et al., 2010).

Role of the social environment: family, social groups, and culture

Culture is not normally considered to influence the development of mind reading, as assessed by theory-of-mind tasks. For example, Callaghan et al. (2005) noted that passing false belief tests emerged between 3 and 5 years in a variety of cultures (rural Peru, Samoa, rural Canada, urban India, and urban Thailand). However, this does not necessarily imply equivalence in terms of underlying neural mechanisms. For instance, there is evidence, from fMRI, of different neural substrates for mentalizing in childhood (9 years) between English-speaking American children and bilingual Japanese children (Kobayashi, Glover, & Temple, 2006). It remains unclear whether these differences relate to differences in family dynamics or cultural/linguistic differences.

One aspect of the environment that *does* affect the development of mentalizing occurs at the level of the family, rather than at the level of culture (i.e. society). Children who live in families that engage in extensive talk about the feelings and thoughts of others tend to pass theory-of-mind tests at an earlier age (Dunn, Brown, Slomkowski, Tesla, & Youngblade, 1991) and are better at detecting the feeling states of others from vignettes when tested several years later at age 6, even after taking into account verbal ability (Dunn, Brown, & Beardsall, 1991). One may wonder about the direction of cause and effect in these studies; perhaps these parents have better genes for developing a social brain rather than providing their children with a richer mentalizing environment. However, the children of parents *trained* to discuss mental states show a larger improvement in mind reading tasks, including false belief, suggesting that the effect is environmental (Lohmann & Tomasello, 2003). Another important facilitating environmental factor within a family is the presence of a sibling (Jenkins & Astington, 1996).

A child's awareness of his/her social identity is formed in the childhood years. Although children can reflect on internal states (e.g. hunger, wanting) from a young age, they do not attribute stable traits to themselves (shyness, kindness, etc.) until around the age of 7 years (Eder, 1990). The tendency to describe others in terms of psychological traits rather than in terms of their behavior increases throughout childhood. However, it is not until adolescence that they make a significant number of *comparisons* of psychological traits (e.g. 'X is kinder than me') when asked to describe others (Barenboim, 1981).

The *perception* of social categories such as age, sex, and race emerges during infancy. However, an own-race *preference* emerges during the early stages of childhood (2.5 to 5 years) as demonstrated in experimental paradigms involving giving and taking toys from ingroup or outgroup strangers (Kinzler & Spelke, 2011). White American children, aged 6 and 10 years, show an equivalent level of implicit racial bias on the Implicit Association Test (IAT) as adults (Baron & Banaji, 2006). The children also showed high levels of explicit ingroup preference. Black American children, on the whole, showed no ingroup/outgroup preference on this test (Newheiser & Olson, 2012). However, Black children with an explicit preference for high status tended to have an implicit outgroup bias (i.e. associating White with positive evaluations). Although adult-like implicit race associations emerge during childhood, they are not necessarily supported by the same brain networks. The amygdala responsiveness of White participants to Black faces emerges during adolescence, not childhood, and

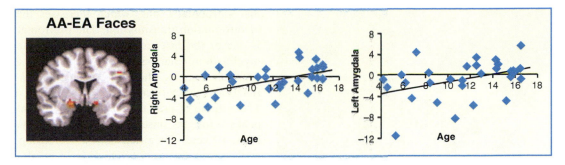

Figure 11.20 The amygdala responsiveness when viewing African American (AA) faces relative to European American (EA) faces emerges around the age of 14 years (Telzer et al., 2013).

also depends on the amount of inter-racial contact (Telzer, Humphreys, Shapiro, & Tottenham, 2013). Similarly, an fMRI study comparing children (aged 6–16 years) who had been raised in a single-race orphanage (either East Asian or Eastern Europe) showed greater amygdala activity linked to outgroup face processing, and this was related to the age of adoption (Telzer et al., 2013).

Evaluation

During the preschool and early school years, the child comes to understand that people (including themselves) are embedded in a social and cultural context comprising norms (e.g. fairness, right/wrong), beliefs, and roles (e.g. go to work, raise a family). They also become aware that different people have different traits (e.g. kindness, shyness) and that they too have a particular social identity (based on their traits, sex, beliefs).

THE ADOLESCENT BRAIN

Adolescence is formally defined as the period between onset of puberty and adulthood and is characterized by significant hormonal changes and changes to physical appearance. For adolescents an over-arching social concern involves the cultivation of their social identity (how they see themselves) and their reputation (how they are perceived by others). This involves a need for social comparison between themselves and others in terms of their social standing (e.g. popularity, dominance) but also in terms of their adherence to other social norms (e.g. Sebastian, Viding, Williams, & Blakemore, 2010). Adherence to age-appropriate social norms (e.g. fashion or music trends) is one way of appearing popular or, at least, guarding against unpopularity. Changes in social cognition go hand-in-hand with structural and functional changes that occur in the brain during adolescence.

Structural changes in the adolescent brain

The region of the brain described as undergoing the most structural change during adolescence is the prefrontal cortex. These changes occur somewhat differently in the white matter and gray matter. White matter density increases steadily during the first two decades of life, stabilizing during late adolescence (Sowell et al., 1999). This is generally attributed to **myelination** – the increase in the fatty sheath

Figure 11.21 The success of many contemporary fashion trends (from mods and rockers to punks, new wave, and Goths) is driven by teenagers. What changes in the adolescent brain and their social world might explain this?

that surrounds axons, which increases the speed of information transmission. The prefrontal cortex is one of the last areas to achieve adult levels of myelination. Changes in gray matter take a rather different course, showing an inverted-U-shaped function characterized by a peak and then a decline. The volume of gray matter in the prefrontal cortex peaks before the onset of adolescence (12 years for boys, 10 years for girls), before decreasing during adolescence to adult levels (Giedd et al., 1999). The maturation tends to occur earliest for posterior regions of the frontal lobes, and latest for anterior regions (Gogtay et al., 2004). These changes in gray matter are likely to be associated with the density of synaptic connections (Huttenlocher & Dabholkar, 1997).

Studies of animals – both rodents and non-human primates – have highlighted important changes to another brain system during adolescence, namely in terms of dopaminergic inputs to the orbitofrontal/ventromedial prefrontal cortex through the ventral striatal 'reward' system (Casey, Getz, & Galvan, 2008; Ernst & Spear, 2009). Specifically, there is a developmental increase in dopaminergic input to the frontal regions in early adolescence due to changing patterns of receptor binding (Tseng & O'Donnell, 2005), shifting towards greater predominance of dopaminergic activity within the ventral striatum in later adolescence (Ernst & Spear, 2009). Ernst, Pine, and Hardin (2006) have suggested that these changes lead to a dominance of reward-related over punishment-related motivations (see Figure 11.22). Whereas the reward-related system is linked to changes in the dopaminergic striatal–frontal pathway, the punishment-related system is linked to amygdala functioning. During adolescence, the amygdala shows a reduced response to stress, measured by stress-induced gene regulation in rats (Kellogg, Awatramani, & Piekut, 1998), and functional imaging in humans shows altered amygdala response to fear expressions relative to adults (Killgore & Yurgelun-Todd, 2010). Ernst et al. (2006) argue that this imbalance between reward/punishment-related cues in adolescence, coupled with immature control (due to late maturation of prefrontal cortex), leads to greater risk-related behavior. It may also tend to lead to greater independence as the adolescent 'follows his/her own heart' rather than norms prescribed by their parents. Casey et al. (2008) propose a similar model that highlights the different rates of maturation of reward-related regions (e.g. the ventral striatum) and prefrontal regions (although this model does not consider the amygdala in detail). The faster maturation of the ventral striatal pathway than the prefrontal cortex during adolescence is assumed to give rise to increased risk-taking and a greater propensity for rewarding social affiliations (friends, romantic partners).

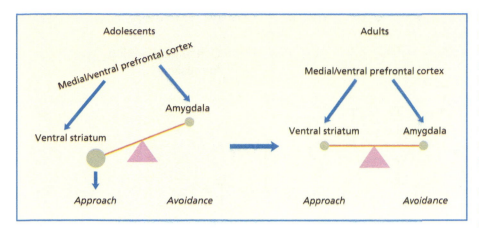

Figure 11.22 Many neuroanatomically based models explain adolescent behavior in terms of an imbalance between late-maturing regions of the prefrontal cortex and earlier-maturing regions such as the ventral striatum. This leads to increased reward-based behavior (greater sociality, thrill seeking) coupled with less cognitive control (greater risk-taking). From Ernst and Spear (2009).

Gaining control over the social world

Although many theories refer to adolescence in terms of risk-taking, it is important to clarify what is meant by this. Risk-taking isn't necessarily the same thing as impulsivity (taking of immediate rewards) which tends to be greatest in younger children. On questionnaire measures, adolescence is related to increased sensation-seeking relative to younger children and adults (Harden & Tucker-Drob, 2011). This is shown in Figure 11.23. Sensation seeking can be regarded as motivated, goal-directed behavior whereas impulsivity is more of an unchecked stimulus–response behavior. Gardner and Steinberg (2005) measured risk-taking using a computerized task in which the participant had to drive a car as close to a wall as a possible, to accumulate more points, but without crashing. Risky driving was greatest in the adolescent group relative to an older youth and adult group. Moreover, the effect was much pronounced in the adolescent group when a peer was present, suggesting that risk-taking has a strong social element.

Consistent with the structural changes occurring in prefrontal regions, several functional imaging studies have shown that adolescents show different patterns of brain activity in frontal areas in socially relevant tasks. Wang, Lee, Sigman, and Dapretto (2006) conducted an irony comprehension task on children/early adolescents (9–14 years) and adults. The participants had to decide whether a speaker was sincere or ironic using both prosodic cues (e.g. sarcastic voice) and contextual cues (integrating prior information). The younger group were more likely to activate regions of the prefrontal cortex (both medially and laterally) than the adult group on this task, which may reflect greater effort and a less intuitive response. A similar finding was obtained by Blakemore, den Ouden, Choudhury, and Frith (2007). They compared judgments about mind-reading scenarios (e.g. 'You are at the cinema and have trouble seeing the screen – do you move to another seat?') with physical scenarios (e.g. 'A huge tree suddenly comes crashing down in a forest – does it make a loud noise?'). Adults showed greater activity than adolescents in the right superior temporal sulcus (STS) region in response to mind-reading scenarios, whereas adolescents showed greater activity than adults in the medial prefrontal cortex region in response to mind-reading scenarios. This result is intriguing, given the claim that the medial prefrontal cortex may serve a general function in representing self in relation to other (Amodio & Devine, 2006), whereas the right STS region has been linked

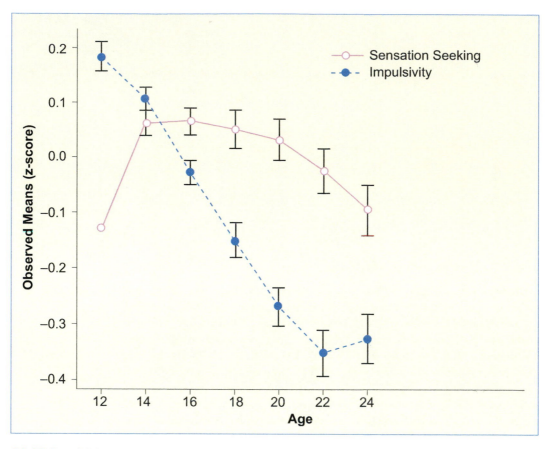

Figure 11.23 Impulsivity and sensation seeking were measured via questionnaire in different age groups. An example of an impulsivity item is 'I think that planning takes the fun out of things'. An example of a sensation seeking item is 'I enjoy new and exciting experiences, even if they are a little frightening or unusual' (Harden & Tucker-Drob, 2011).

more specifically to representing mental states (Saxe, 2006). However, differences in activation need to be interpreted carefully. An increase in activity in frontal regions by adolescents could reflect the fact that these regions work less efficiently (less efficiency generating more effort and more activity), rather than reflecting a greater importance for conducting the task. The cortical thickness in these regions of the mentalizing network shows a developmental reduction during adolescence (Mills, Lalonde, Clasen, Giedd, & Blakemore, 2014).

Although children from the age of 4 years upwards tend to pass theory-of-mind tasks, competency on some of these tasks may not reach adult levels until late in adolescence. Dumontheil et al. (2010) developed a more challenging theory-of-mind test (although based on the same principle as the Sally–Anne task) for older participants. Participants were given an array of compartments with objects on them that were all visible to the participant, but a 'director' was positioned on the other side of the array and could only see a subset of the objects due to an occlusion. The director would call out instructions to the participant (e.g. 'move the small ball to the left') and the participant would have to respond appropriately by taking into account the director's perspective (Figure 11.24). There was found to be a significant difference between older adolescents (14–17 years old) and adults (19 years and above) on this task, but

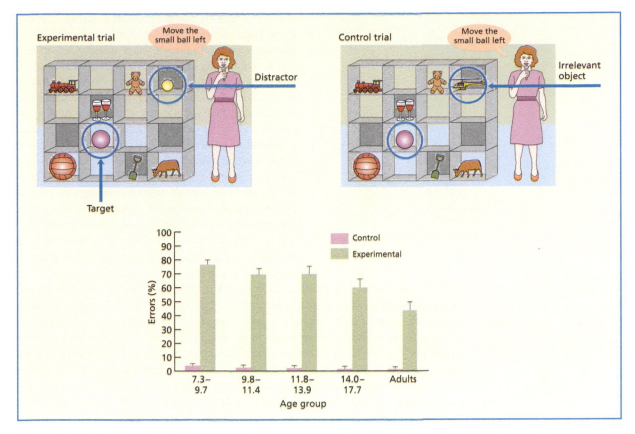

Figure 11.24 In a difficult theory-of-mind task (which requires inhibiting an alternative response), adolescents do not perform as well as adults. Thus, whilst competence in these tasks emerges at an early age (from 4 years), sophisticated performance develops more slowly and depends on domain-general resources such as executive functions. From Dumontheil et al. (2010). Copyright © 2009 the authors. Journal compilation copyright © 2010 Blackwell Publishing Ltd. Reproduced with permission from Wiley-Blackwell.

not on a control task that did not require perspective taking. The authors argue that although a basic understanding of theory of mind comes online at around 4 years of age (if not before) it may require a mature system of executive functioning in order for it to be optimized, and this may not come online until the end of adolescence. It would also be interesting to see how individual differences in social behavior may be related to developing expertise in theory of mind. For example, would an adolescent who is very preoccupied with his/her reputation (e.g. 'what does person X think of me?') perform better than someone who is less concerned but of the same age?

A similar conclusion to Dumontheil et al. (2010) was reached in a study of the Ultimatum Game (described in Chapter 7) with adolescents between 9 and 18 years old (Guroglu, van den Bos, & Crone, 2009). When the responder had the opportunity to reject an offer, there was an age-related increase during adolescence in the amount of offer that the proposer (the participant) proposed. This was interpreted as a greater developmental inclination to take on board the perspective of the other person.

Much of the adolescent concern for reputation and motivation for risk-taking may be related to the need to develop relationships with others. This is spurred on,

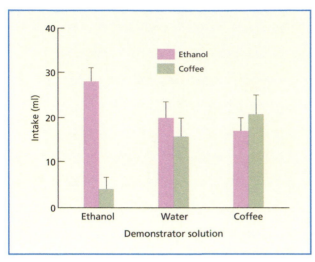

Figure 11.25 If a sober adolescent rat is exposed to an intoxicated sibling then he/she will voluntarily consume more alcohol. In this experiment, a sibling (the 'demonstrator') was administered a solution of coffee (COF), ethanol (EtOH), or water (H_2O). The demonstrator was then allowed to interact with the other rat, the 'observer' (who would be able to smell the coffee/ethanol on the demonstrator's breath). This observer was then removed and, in isolation, was given two cups containing ethanol and coffee. The intake of each (out of a maximum of 50 ml) was recorded. From Hunt et al. (2001). Copyright © 2001 John Wiley & Sons, Inc. All rights reserved. Reproduced with permission.

at least in part, by reaching sexual maturity. Chronological age and the onset of puberty, assessed via hormonal measures, in females have independent effects on the functioning of the brain when appraising social emotions (Goddings, Heyes, Bird, Viner, & Blakemore, 2012). The presence of intimate and confiding relationships has a positive influence on health and stress, and the absence of them during adolescence is related to vulnerability to mental health problems (Pine, Cohen, Gurley, Brook, & Ma, 1998). Social rejection, in the cyberball game (discussed in Chapter 8), leads to greater negative feelings in adolescence relative to adulthood (Sebastian et al., 2010). Experimentation with alcohol and drugs by this age group is typically a 'social' activity insofar as it normally occurs in social gatherings rather than via individual discovery. Alcohol reduces anxiety, which may facilitate certain social interactions (e.g. Varlinskaya & Spear, 2002). Of course, many recreational drugs stimulate the reward systems of the brain (e.g. Koob, 1992), as do social interactions themselves. By pairing the two together, the social interaction itself may be perceived as being more rewarding than in the absence of such a stimulant. It would be easy to dismiss the interaction between adolescence and drug/alcohol use as being related to social norms rather than developmental changes in the brain. However, adolescent rats show comparable effects in the absence of any social norms for alcohol use (see Figure 11.25). In adolescent rats, but not adult rats, low doses of alcohol facilitate social interactions (Varlinskaya & Spear, 2002). If a sober adolescent rat is exposed to an intoxicated sibling, then alcohol drinking by the sober rat is enhanced, and in a dose-dependent manner – the more the sibling has consumed, the more he/she will consume (Hunt, Holloway, & Scordalakes, 2001).

Evaluation

Adolescence is no longer considered simplistically as half-adult/half-child but rather as a distinct phase in development that is qualitatively different from both childhood and adulthood. The prefrontal cortex reaches maturity during adolescence, but other structures relating to reward (e.g. ventral striatum) and punishment (e.g. amygdala) may also mature during this period and – crucially – not necessarily at the same rate.

SUMMARY AND KEY POINTS OF THE CHAPTER

- Infants show an innate preference for face-like stimuli (e.g. for a top-heavy configuration), but this falls short of innate knowledge of the exact layout of a face. These stimuli do not selectively engage regions that, in adults, have been proposed to be face specific. This suggests that specialization emerges slowly over time.

- Between 12 and 18 months, infants show evidence of being able to understand the relationship between actions and intentions (e.g. imitating intentional actions more than accidental ones) and to understand the relationship between seeing and knowing (e.g. by directing an adult's attention to an object that the infant can see but the adult cannot). It is unclear whether this is evidence of a theory of mind at this early age or whether these skills are precursors to the later development of this ability (as measured on *explicit* judgments of mental states).
- Evidence from EEG and fMRI suggests that the neural substrates of social cognition (e.g. on tasks of empathy, face processing, or min-dreading) differ between children, adolescents, and adults. This can even occur when behavioral performance across these age groups does not differ; that is, seemingly mature levels of performance are not necessarily supported by a mature neural architecture.
- The preschool and early school years could be considered as a process of enculturation in which the child learns about social norms (e.g. fairness, reciprocity), engages in peer-based interactions, and develops an understanding of his/her own social identity.
- Structural changes in the brain during adolescence consist of late maturation of the prefrontal cortex (important for the control of behavior) and changes in the reward-based systems of the brain. This combination may give rise to increased risk-taking by this age group. Adolescents also show increased concern with reputation (how others perceive them) and reputation management (adherence to peer-endorsed fashions), which may relate to development changes in the functioning of certain prefrontal regions (e.g. medial prefrontal cortex).

EXAMPLE ESSAY QUESTIONS

- How is face recognition in infancy and childhood related to that found in adults?
- Does the development of early language and gestural communication depend on knowledge of the social world?
- What kind of basic abilities and social interactions act as developmental precursors to the emergence of mature forms of empathy and theory of mind?
- What physical changes occur in the brain during adolescence and how might this relate to changes in social functioning at this age?

RECOMMENDED FURTHER READING

- De Haan, M., & Gunnar, M. R. (2009). *Handbook of Developmental Social Neuroscience*. New York: Guilford Press. Up-to-date but not all areas are covered in detail (e.g. theory of mind, social norms).

- Happe, F., & Frith, U. (2014). Towards a developmental neuroscience of atypical social cognition. *Journal of Child Psychology and Psychiatry, 55*, 553–577. A comprehensive review paper of the field.

- Zelazo, P. D., Chandler, M., & Crone, E. (2010). *Developmental Social Cognitive Neuroscience*. New York: Psychology Press. An excellent collection of papers.

ONLINE RESOURCES

- References to key papers and readings
- Videos demonstrating neonatal imitation, face recognition in infants, and altruism in children
- Interviews and talks given by Andrew Meltzoff, Sarah-Jayne Blakemore, Renee Baillargeon, and others
- Multiple choice questions and interactive flashcards to test your knowledge
- Downloadable glossary

References

Adam, E. K., Hawkley, L. C., Kudielka, B. M., & Cacioppo, J. T. (2006). Day-to-day dynamics of experience-cortisol associations in a population-based sample of older adults. *Proceedings of the National Academy of Sciences of the United States of America, 103*(45), 17058–17063.

Adams, J. M., & Jones, W. H. (1997). The conceptualisation of marriage commitment: An integrative analysis. *Journal of Social and Personal Relationships, 11*, 1177–1196.

Adams, R. B., Gordon, H. L., Baird, A. A., Ambady, N., & Kleck, R. E. (2003). Effects of gaze on amygdala sensitivity to anger and fear faces. *Science, 300*(5625), 1536–1536.

Adams, R. B., & Kleck, R. E. (2003). Perceived gaze direction and the processing of facial displays of emotion. *Psychological Science, 14*(6), 644–647.

Adolphs, R. (1999). Social cognition and the human brain. *Trends in Cognitive Sciences, 3*, 469–479.

Adolphs, R. (2002). Neural systems for recognizing emotion. *Current Opinion in Neurobiology, 12*, 169–177.

Adolphs, R., Damasio, H., Tranel, D., Cooper, G., & Damasio, A. R. (2000). A role for somatosensory cortices in the visual recognition of emotion as revealed by three-dimensional lesion mapping. *Journal of Neuroscience, 20*(7), 2683–2690.

Adolphs, R., Gosselin, F., Buchanan, T. W., Tranel, D., Schyns, P., & Damasio, A. R. (2005). A mechanism for impaired fear recognition after amygdala damage. *Nature, 433*, 68–72.

Adolphs, R., Tranel, D., & Buchanan, T. W. (2005). Amygdala damage impairs emotional memory for gist but not details of complex stimuli. *Nature Neuroscience, 8*(4), 512–518.

Adolphs, R., Tranel, D., & Damasio, A. R. (1998). The human amygdala in social judgment. *Nature, 393*(6684), 470–474.

Adolphs, R., Tranel, D., Damasio, H., & Damasio, A. (1994). Impaired recognition of emotion in facial expressions following bilateral damage to the human amygdala. *Nature, 372*, 669–672.

Aguirre, G. K., Zarahn, E., & D'Esposito, M. (1998). The variability of human BOLD hemodynamic response. *NeuroImage, 8*, 360–369.

Ainsworth, M. D. S., Blehar, M. C., Waters, E., & Wahl, S. (1978). *Patterns of Attachment*. Hillsdale, NJ: Erlbaum.

Alexander, G. E., & Crutcher, M. D. (1990). Functional architecture of basal ganglia circuits: Neural substrates of parallel processing. *Trends in Neurosciences, 13*, 266–271.

Alley, T. R. (1988). Physiognomy and social perception. In T. R. Alley (Ed.), *Social and Applied Aspects of Perceiving Faces*. Hillsdale, NJ: Earlbaum.

Allison, T., Puce, A., & McCarthy, G. (2000). Social perception from visual cues: Role of the STS region. *Trends in Cognitive Sciences, 4*, 267–278.

Allport, G. W. (1954). *The Nature of Prejudice*. Reading, MA: Addison-Wesley.

Allport, G. W. (1968). The historical background of modern social psychology. In G. Lindzey & E. Aronson (Eds.), *Handbook of Social Psychology*. New York: McGraw Hill.

Alrajih, S., & Ward, J. (2014). Increased facial width-to-height ratio and perceived dominance in the faces of the UK's leading business leaders. *British Journal of Psychology, 105*(2), 153–161.

Amaral, D. G., Price, J. L., Pitkanen, A., & Carmichael, S. T. (1992). Anatomical organization of the primate amygdaloid complex. In J. P. Aggleton (Ed.), *The Amygdala*. New York: Wiley-Liss.

Amodio, D. M. (2008). The social neuroscience of intergroup relations. *European Review of Social Psychology, 19*, 1–54.

Amodio, D. M., & Devine, P. G. (2006). Stereotyping and evaluation in implicit race bias: Evidence for independent constructs and unique effects on behavior. *Journal of Personality and Social Psychology, 91*(4), 652–661.

Amodio, D. M., & Frith, C. D. (2006). Meeting of minds: The medial frontal cortex and social cognition. *Nature Reviews Neuroscience, 7*(4), 268–277.

Amodio, D. M., Harmon-Jones, E., & Devine, P. G. (2003). Individual differences in the activation and control of affective race bias as assessed by startle eyeblink response and self-report. *Journal of Personality and Social Psychology, 84*(4), 738–753.

Amodio, D. M., Harmon-Jones, E., Devine, P. G., Curtin, J. J., Hartley, S. L., & Covert, A. E. (2004). Neural signals for the detection of unintentional race bias. *Psychological Science, 15*(2), 88–93.

Amodio, D. M., Kubota, J. T., Harmon-Jones, E., & Devine, P. G. (2006). Alternative mechanisms for regulating racial responses according to internal vs. external cues. *Social Cognitive and Affective Neuroscience, 1*, 26–36.

Anagnostou, E., & Taylor, M. J. (2011). Review of neuroimaging in autism spectrum disorders: What have we learned and where we go from here. *Molecular Autism, 2*.

Anders, S., Eippert, F., Weiskopf, N., & Veit, R. (2008). The human amygdala is sensitive to the valence of pictures and sounds irrespective of arousal: An fMRI study. *Social Cognitive and Affective Neuroscience, 3*(3), 233–243.

Anderson, C. A., & Bushman, B. J. (2002). Human aggression. *Annual Review of Psychology, 53*, 27–51.

Anderson, J. R., Myowa-Yamakoshi, M., & Matsuzawa, T. (2004). Contagious yawning in chimpanzees. *Proceedings of the Royal Society of London Series B-Biological Sciences, 271*, S468–S470.

Anderson, S. W., Bechara, A., Damasio, H., Tranel, D., & Damasio, A. R. (1999). Impairment of social and moral behavior related to early damage in human prefrontal cortex. *Nature Neuroscience, 2*, 1032–1037.

Apperly, I. (2011). *Mindreaders: The Cognitive Basis of Theory of Mind*. New York: Psychology Press.

Apperly, I. A. (2008). Beyond simulation-theory and theory-theory: Why social cognitive neuroscience should use its own concepts to study "theory of mind". *Cognition, 107*(1), 266–283.

Apperly, I. A., Samson, D., Carroll, N., Hussain, S., & Humphreys, G. (2006). Intact first- and second-order false belief reasoning in a patient with severely impaired grammar. *Social Neuroscience, 1*(3–4), 334–348.

Argyle, M., Henderson, M., Bond, M., Iizuka, Y., & Contarello, A. (1986). Cross-cultural variations in relationship rules. *International Journal of Psychology, 21*(3), 287–315.

Aron, A., Aron, E. N., & Smollan, D. (1992). Inclusion of other in the self scale and the structure of interpersonal closeness. *Journal of Personality and Social Psychology, 63*(4), 596–612.

Arviv, O., Goldstein, A., Weeting, J. C., Becker, E. S., Lange, W. G., & Gilboa-Schechtman, E. (2015). Brain response during the M170 time interval is sensitive to socially relevant information. *Neuropsychologia, 78*, 18–28.

Arzy, S., Thut, G., Mohr, C., Michel, C. M., & Blanke, O. (2006). Neural basis of embodiment: Distinct contributions of temporoparietal junction and extrastriate body area. *Journal of Neuroscience, 26*, 8074–8081.

Asch, S. E. (1951). Effects of group pressure upon the modification and distortion of judgements. In H. Guetzkow (Ed.), *Groups, Leadership and Men*. Pittsburgh: Carnegie Press.

Ashburner, J., & Friston, K. J. (2000). Voxel-based morphometry: The methods. *NeuroImage, 11*, 805–821.

Asperger, H. (1944). 'Autistic psychopathy' in childhood. In U. Frith (Ed.), *Autism and Asperger Syndrome*. Cambridge: Cambridge University Press.

Attwell, D., & Iadecola, C. (2002). The neural basis of functional brain imaging signals. *Trends in Neurosciences, 25*, 621–625.

Atzil, S., Hendler, T., & Feldman, R. (2011). Specifying the neurobiological basis of human attachment: Brain, hormones, and behavior in synchronous and intrusive mothers. *Neuropsychopharmacology, 36*(13), 2603–2615.

Auyeung, B., Baron-Cohen, S., Ashwin, E., Knickmeyer, R., Taylor, K., & Hackett, G. (2009). Fetal testosterone and autistic traits. *British Journal of Psychology, 100*, 1–22.

Avenanti, A., Bueti, D., Galati, G., & Aglioti, S. M. (2005). Transcranial magnetic stimulation highlights the sensorimotor side of empathy for pain. *Nature Neuroscience, 8*, 955–960.

Avenanti, A., Minio-Paluello, I., Bufalari, I., & Aglioti, S. M. (2009). The pain of a model in the personality of the onlooker: Influence of state-reactivity and personality traits on embodied empathy for pain. *NeuroImage, 44*, 275–283.

Avenanti, A., Sirigu, A., & Aglioti, S. M. (2010). Racial bias reduces empathic sensorimotor resonance with other-race pain. *Current Biology, 20*(11), 1018–1022.

Axelrod, R., & Hamilton, W. D. (1981). The evolution of cooperation. *Science, 211*, 1390–1396.

Babiak, P., Neumann, C. S., & Hare, R. D. (2010). Corporate psychopathy: Talking the walk. *Behavioral Sciences & the Law, 28*(2), 174–193.

Baez-Mendoza, R., & Schultz, W. (2013). The role of the striatum in social behavior. *Frontiers in Neuroscience, 7*.

Bailey, A., Lecouteur, A., Gottesman, I., Bolton, P., Simonoff, E., Yuzda, E., . . . Rutter, M. (1995). Autism as a strongly genetic disorder – evidence from a british twin study. *Psychological Medicine, 25*(1), 63–77.

Bailey, A. A., & Hurd, P. L. (2005). Finger length ratio (2D : 4D) correlates with physical aggression in men but not in women. *Biological Psychology, 68*(3), 215–222.

Baird, G., Charman, T., Baron-Cohen, S., Cox, A., Swettenham, J., Wheelwright, S., & Drew, A. (2000). A screening instrument for autism at 18 months of age: A 6-year follow-up study. *Journal of the American Academy of Child Adolescent Psychiatry, 39*, 694–702.

Baird, G., Simonoff, E., Pickles, A., Chandler, S., Loucas, T., Meldrum, D., & Charman, T. (2006). Prevalence of disorders of the autism spectrum in a population cohort of children in South Thames: The Special Needs and Autism Project (SNAP). *Lancet, 368*(9531), 210–215.

Bakermans-Kranenburg, M. J., & van Ijzendoorn, M. H. (2008). Oxytocin receptor (OXTR) and serotonin transporter (5-HTT) genes associated with observed parenting. *Social Cognitive and Affective Neuroscience, 3*(2), 128–134.

Bale, T. L., Davis, A. M., Auger, A. P., Dorsa, D. M., & McCarthy, M. M. (2001). CNS region-specific oxytocin receptor expression: Importance in regulation of anxiety and sex behavior. *Journal of Neuroscience, 21*(7), 2546–2552.

Ballantyne, A. O., Spilkin, A. M., Hesselink, J., & Trauner, D. A. (2008). Plasticity in the developing brain: Intellectual, language and academic functions in children with ischaemic perinatal stroke. *Brain, 131*, 2975–2985.

Bandelow, B., Zohar, J., Hollander, E., Kasper, S., Moeller, H.-J., Allgulander, C., . . . Guideli, W. T. F. T. (2008). World federation of societies of biological psychiatry (WFSBP) guidelines for the pharmacological treatment of anxiety, obsessive-compulsive and post-traumatic stress disorders – first revision. *World Journal of Biological Psychiatry, 9*(4), 248–312.

Bandura, A. (1965). Influence of models reinforcement contingencies on the acquisition of imitative responses. *Journal of Personality and Social Psychology, 1*(6), 589–595.

Bandura, A. (1973). *Aggression: A Social Learning Analysis.* Englewood Cliffs, NJ: Prentice-Hall.

Bandura, A. (2002). Reflexive empathy: On predicting more than has ever been observed. *Behavioral and Brain Sciences, 25*(1), 24.

Bandura, A., Barbaranelli, C., Caprara, G. V., & Pastorelli, C. (1996). Mechanisms of moral disengagement in the exercise of moral agency. *Journal of Personality and Social Psychology, 71*, 364–374.

Bandura, A., Ross, D., & Ross, S. A. (1961). Transmission of aggression through imitation of aggressive models. *Journal of Abnormal and Social Psychology, 63*, 575–582.

Bandura, A., Ross, S. A., & Ross, D. (1963). Imitation of film-mediated aggressive models. *Journal of Abnormal Psychology, 66*(1), 3–11.

Banerjee, K., Huebner, B., & Hauser, M. D. (2011). Intuitive moral judgments are robust across demographic variation in gender, education, politics, and religion: A large-scale web-based study. *Journal of Cognition and Culture, 10*, 253–281.

Banissy, M.J., Kanai, R., Walsh, V., & Rees, G. (2012). Inter-individual differences in empathy are reflected in human brain structure. *NeuroImage, 62*(3), 2034–2039.

Banissy, M.J., Sauter, D., Ward, J., Warren, J.E., Walsh, V., & S. K., S. (2010). Suppressing sensory-motor activity modulates the discrimination of auditory emotions but not speaker identity. *Journal of Neuroscience, 30*, 13552–13557.

Barclay, P. (2010). Altruism as a courtship display: Some effects of third-party generosity on audience perceptions. *British Journal of Psychology, 101*, 123–135.

Barenboim, C. (1981). The development of person perception in childhood and adolescence – from behavioral-comparisons to psychological constructs to psychological comparisons. *Child Development, 52*(1), 129–144.

Bargh, J. A., Chen, M., & Burrows, L. (1996). Automaticity of social behavior: Direct effects of trait construct and stereotype activation on action. *Journal of Personality and Social Psychology, 71*(2), 230–244.

Bargh, J.A., & Shalev, I. (2012). The substitutability of physical and social warmth in daily life. *Emotion, 12*(1), 154–162.

Bar-Haim, Y., Ziv, T., Lamy, D., & Hodes, R. M. (2006). Nature and nurture in own-race face processing. *Psychological Science, 17*, 159–163.

Barker, A. T., Jalinous, R., & Freeston, I. L. (1985). Non-invasive magnetic stimulation of human motor cortex. *Lancet, 1*, 1106–1107.

Baron, A.S., & Banaji, M.R. (2006). The development of implicit attitudes – Evidence of race evaluations from ages 6 and 10 and adulthood. *Psychological Science, 17*(1), 53–58.

Baron-Cohen, S. (1995a). The Eye-Direction Detector (EDD) and the Shared Attention Mechanism (SAM): Two cases for evolutionary psychology. In C. Moore & P. Dunham (Eds.), *The Role of Joint Attention in Development.* Hillsdale, NJ: Lawrence Earlbaum Associates.

Baron-Cohen, S. (1995b). *Mindblindness: An Essay on Autism and Theory of Mind.* Cambridge, MA: MIT Press.

Baron-Cohen, S. (2002). The extreme male brain theory of autism. *Trends in Cognitive Sciences, 6*, 248–254.

Baron-Cohen, S. (2009). Autism: The Empathizing-Systemizing (E-S) theory. In *Year in Cognitive Neuroscience 2009* (Vol. 1156, pp. 68–80). Oxford: Blackwell Publishing.

Baron-Cohen, S., Ashwin, E., Ashwin, C., Tavassoli, T., & Chakrabarti, B. (2009). Talent in autism: Hyper-systemizing, hyper-attention to detail and sensory hyper-sensitivity. *Philosophical Transactions of the Royal Society of London, Series B, 364*(1522), 1377–1383.

Baron-Cohen, S., Campbell, R., Karmiloff-Smith, A., Grant, J., & Walker, J. (1995). Are children with autism blind to the mentalistic significance of eyes? *British Journal of Developmental Psychology, 13*, 379–398.

Baron-Cohen, S., & Cross, P. (1992). Reading the eyes: Evidence for the role of perception in the development of theory of mind. *Mind and Language, 6*, 166–180.

Baron-Cohen, S., Leslie, A. M., & Frith, U. (1985). Does the autistic child have a 'theory of mind'? *Cognition, 21*, 37–46.

Baron-Cohen, S., Leslie, A. M., & Frith, U. (1986). Mechanical, behavioral and intentional understanding of picture stories in autistic-children. *British Journal of Developmental Psychology, 4*, 113–125.

Baron-Cohen, S., Richler, J., Bisarya, D., Gurunathan, N., & Wheelwright, S. (2003). The systemizing quotient: An investigation of adults with Asperger syndrome or high-functioning autism, and normal sex differences. *Philosophical Transactions of the Royal Society of London B, 358*, 361–374.

Baron-Cohen, S., & Wheelwright, S. (2004). The empathy quotient: An investigation of adults with Asperger syndrome or high functioning autism and normal sex differences. *Journal of Autism and Developmental Disorders, 34*, 163–175.

Baron-Cohen, S., Wheelwright, S., Hill, J., Raste, Y., & Plumb, I. (2001). The "Reading the Mind in the Eyes" test revised version: A study with normal adults, and adults with Asperger syndrome or high-functioning autism. *Journal of Child Psychology and Psychiatry and Allied Disciplines, 42*(2), 241–251.

Baron-Cohen, S., Wheelwright, S., Stone, V., & Rutherford, M. (1999). A mathematician, a physicist and a computer scientist with Asperger syndrome: Performance on psychology and folk physics tests. *Neurocase, 5*, 475–483.

Barraclough, N. E., Xiao, D., Baker, C. I., Oram, M. W., & Perrett, D. I. (2005). Integration of visual and auditory information by superior temporal sulcus neurons responsive to the sight of actions. *Journal of Cognitive Neuroscience, 17*, 377–391.

Barrett, L. F. (2006). Are emotions natural kinds? *Perspectives on Psychological Science, 1*, 28–58.

Barrett, L.F., & Satpute, A. B. (2013). Large-scale brain networks in affective and social neuroscience: Towards an integrative functional architecture of the brain. *Current Opinion in Neurobiology, 23*(3), 361–372.

Barrett, L.F., & Wager, T.D. (2006). The structure of emotion – evidence from neuroimaging studies. *Current Directions in Psychological Science, 15*(2), 79–83.

Bartels, A., & Zeki, S. (2000). The neural basis of romantic love. *NeuroReport, 11*, 3829–3834.

Bartels, A., & Zeki, S. (2004). The neural correlates of maternal and romantic love. *NeuroImage, 21*(3), 1155–1166.

Barton, J.J.S. (2008). Structure and function in acquired prosopagnosia: Lessons from a series of 10 patients with brain damage. *Journal of Neuropsychology, 2*, 197–225.

Bateson, M., Nettle, D., & Roberts, G. (2006). Cues of being watched enhance cooperation in a real-world setting. *Biology Letters, 2*, 412–414.

Batson, C., & Shaw, L. L. (1991). Encouraging words concerning the evidence for altruism. *Psychological Inquiry, 2*, 159–168.

Batson, C. D. (1991). *The Altruism Question: Toward a Social-psychological Answer*. Hillsdale, NJ: Erlbaum.

Batson, C. D. (2009). These things called empathy: Eight related but distinct phenomena. In J. Decety & W. J. Ickes (Eds.), *The Social Neuroscience of Empathy*. Cambridge, MA: MIT Press.

Batson, C. D., Batson, J. G., Griffitt, C. A., Barrientos, S., Brandt, J. R., Sprengelmeyer, P., & Bayly, M. J. (1989). Negative-state relief and the empathy altruism hypothesis. *Journal of Personality and Social Psychology, 56*(6), 922–933.

Batson, C. D., Duncan, B. D., Ackerman, P., Buckley, T., & Birch, K. (1981). Is empathic emotion a source of altruistic motivation? *Journal of Personality and Social Psychology, 40*(2), 290–302.

Batson, C. D., Dyck, J. L., Brandt, J. R., Batson, J. G., Powell, A. L., McMaster, M. R., & Griffitt, C. (1988). 5 studies testing 2 new egoistic alternatives to the empathy altruism hypothesis. *Journal of Personality and Social Psychology, 55*(1), 52–77.

Baumeister, R.F., & Leary, M.R. (1995). The need to belong – desire for interpersonal attachments as a fundamental human-motivation. *Psychological Bulletin, 117*(3), 497–529.

Baumeister, R. F., Twenge, J. M., & Nuss, C. K. (2002). Effects of social exclusion on cognitive processes: Anticipated aloneness reduces intelligent thought. *Journal of Personality and Social Psychology, 83*(4), 817–827.

Baumgartner, T., Gotte, L., Gugler, R., & Fehr, E. (2012). The mentalizing network orchestrates the impact of parochial altruism on social norm enforcement. *Human Brain Mapping, 33*(6), 1452–1469.

Baxter, M. G., & Murray, E. A. (2002). The amygdala and reward. *Nature Reviews Neuroscience, 3*(7), 563–573.

Bechara, A., Damasio, A. R., Damasio, H., & Anderson, S. W. (1994). Insensitivity to future consequences following damage to human prefrontal cortex. *Cognition, 50*, 7–15.

Bechara, A., Damasio, H., Damasio, A. R., & Lee, G. P. (1999). Different contributions to the human amygdala and ventromedial prefrontal cortex to decision making. *Journal of Neuroscience, 19*, 5437–5481.

Bechara, A., Damasio, H., Tranel, D., & Anderson, S. W. (1998). Dissociation of working memory from decision making within the human prefrontal cortex. *Journal of Neuroscience, 18*, 428–437.

Bechara, A., Tranel, D., Damasio, H., Adolphs, R., Rockland, C., & Damasio, A. R. (1995). Double dissociation of conditioning and declarative knowledge relative to the amygdala and hippocampus in humans. *Science, 269*(5227), 1115–1118.

Beckett, C., Castle, J., Rutter, M., & Sonuga-Barke, E. J. (2010). VI. Institutional deprivation, specific cognitive functions, and scholastic achievement: English and Romanian adoptee (ERA) study FINDINGS. *Monographs of the Society for Research in Child Development, 75*(1), 125–142.

Bedny, M., Pascual-Leone, A., & Saxe, R. R. (2009). Growing up blind does not change the neural bases of Theory of Mind. Proceedings of the National Academy of Sciences of the United States of America, 106(27), 11312–11317

Beer, J. S., Stallen, M., Lombardo, M. V., Gonsalkorale, K., Cunningham, W. A., & Sherman, J. W. (2008). The Quadruple Process model approach to examining the neural underpinnings of prejudice. *Neuroimage, 43*(4), 775–783.

Bedny, M., Pascual-Leone, A., & Saxe, R. R. (2009). Growing up blind does not change the neural bases of Theory of Mind. *Proceedings of the National Academy of Sciences of the United States of America, 106*(27), 11312–11317.

Belsky, J., Jonassaint, C., Pluess, M., Stanton, M., Brummett, B., & Williams, R. (2009). Vulnerability genes or plasticity genes? *Molecular Psychiatry, 14*(8), 746–754.

Belsky, J., & Ravine, M. (1987). Temperament and attachment security in the strange situation: An empirical rapproachment. *Child Development, 58*, 787–795.

Benjamin, J., Li, L., Patterson, C., Greenberg, B. D., Murphy, D.L., & Hamer, D.H. (1996). Population and familial association between the D4 dopamine receptor gene and measures of novelty seeking. *Nature Genetics, 12*, 81–84.

Bernstein, M.J., Young, S.G., & Hugenberg, K. (2007). The cross-category effect – Mere social categorization is sufficient to elicit an own-group bias in face recognition. *Psychological Science, 18*(8), 706–712.

Bentin, S., Allison, T., Puce, A., Perez, E., & McCarthy, G. (1996). Electrophysiological studies of face perception in humans. *Journal of Cognitive Neuroscience, 8*, 551–565.

Bentin, S., & Deouell, L. Y. (2000). Structural encoding and identification in face processing: ERP evidence for separate mechanisms. *Cognitive Neuropsychology, 17*, 35–54.

Berg, J., Dickhaut, J., & McCabe, K. (1995). Trust, reciprocity and social history. *Games and Economic Behavior, 10*, 122–142.

Berkowitz, L. (1989). Frustration aggression hypothesis – examination and reformulation. *Psychological Bulletin, 106*(1), 59–73.

Berkowitz, L. (1990). On the formation and regulation of anger and aggression – a cognitive-neoassociationistic analysis. *American Psychologist, 45*(4), 494–503.

Berkowitz, L., & Harmon-Jones, E. (2004). Toward an understanding of the determinants of anger. *Emotion, 4*(2), 107–130.

Bernhardt, P. C., Dabbs, J. M., Fielden, J. A., & Lutter, C. D. (1998). Testosterone changes during vicarious experiences of winning and losing among fans at sporting events. *Physiology & Behavior, 65*, 59–62.

Berry, D. S., & Brownlow, S. (1989). Were the physiognomists right – personality-correlates of facial babyishness. *Personality and Social Psychology Bulletin, 15*(2), 266–279.

Berry Mendes, W. (2009). Assessing autonomic nervous system activity. In E. Harmon-Jones & J. S. Beer (Eds.), *Methods in Social Neuroscience*. New York: The Guilford Press.

Berthoz, S., Armony, J. L., Blair, R. J. R., & Dolan, R. J. (2002). An fMRI study of intentional and unintentional (embarrassing) violations of social norms. *Brain, 125*, 1696–1708.

Bickart, K.C., Wright, C.I., Dautoff, R.J., Dickerson, B.C., & Barrett, L. F. (2011). Amygdala volume and social network size in humans. *Nature Neuroscience, 14*(2), 163–164.

Birbaumer, N., Viet, R., Lotze, M., Erb, M., Hermann, C., Grodd, W., & Flor, H. (2005). Deficient fear conditioning in psychopathy – A functional magnetic resonance imaging study. *Archives of General Psychiatry, 62*(7), 799–805.

Bird, C. M., Casteli, F., Malik, O., Frith, U., & Husain, M. (2004). The impact of extensive medial frontal lbe damage on 'theory of mind' and cognition. *Brain, 127*, 914–928.

Bird, G., Silani, G., Brindley, R., White, S., Frith, U., & Singer, T. (2010). Empathic brain responses in insula are modulated by levels of alexithymia but not autism. *Brain, 133*(5), 1515–1525.

Bisley, J. W., & Goldberg, M. E. (2010). Attention, intention, and priority in the parietal lobe. *Annual Review of Neuroscience, 33*, 1–21.

Bjornebekk, A., Westlye, L.T., Fjell, A. M., Grydeland, H., & Walhovd, K.B. (2012). Social reward dependence and brain white matter microstructure. *Cerebral Cortex, 22*(11), 2672–2679.

Blackhart, G. C., Nelson, B. C., Knowles, M. L., & Baumeister, R. F. (2009). Rejection elicits emotional reactions but neither causes immediate distress nor lowers self-esteem: A meta-analytic review of 192 studies on social exclusion. *Personality and Social Psychology Review, 13*, 269–309.

Blackmore, S. (1999). *The Meme Machine*. Oxford: OUP.

Blair, K., Marsh, A. A., Morton, J., Vythilingam, M., Jones, M., Mondillo, K., Pine, D. C., Drevets, W. C., & Blair, J. R. (2006). Choosing the lesser of two evils, the better of two goods: Specifying the roles of ventromedial prefrontal cortex and dorsal anterior cingulate in object choice. *Journal of Neuroscience, 26*(44), 11379–11386.

Blair, R. J. R. (1995). A cognitive developmental approach to morality: Investigating the psychopath. *Cognition, 57*, 1–29.

Blair, R. J. R. (1996). Morality in the autistic child. *Journal of Autism and Developmental Disorders, 26*, 571–579.

Blair, R. J. R., & Cipolotti, L. (2000). Impaired social response reversal: A case of acquired 'sociopathy'. *Brain, 123*, 1122–1141.

Blair, R. J. R., Jones, L., Clark, F., & Smith, M. (1997). The psychopathic individual: A lack of responsiveness to distress cues? *Psychophysiology, 34*, 192–198.

Blair, R. J. R., Sellars, C., Strickland, I., Clark, F., Williams, A. O., Smith, M., & Jones, L. (1995). Emotion attributions in the psychopath. *Personality and Individual Differences, 19*(4), 431–437.

Blakemore, S. J., den Ouden, H., Choudhury, S., & Frith, C. (2007). Adolescent development of the neural circuitry for thinking about intentions. *Social Cognitive and Affective Neuroscience, 2*(2), 130–139.

Blanke, O., Landis, T., Spinelli, L., & Seeck, M. (2004). Out-of-body experience and autoscopy of neurological origin. *Brain, 127*, 243–258.

Blanke, O., Mohr, C., Michel, C. M., Pascual-Leone, A., Brugger, P., Seeck, M., Landis, T., & Thut, G. (2005). Linking out-of-body experience and self processing to mental own-body imagery at the temporoparietal junction. *Journal of Neuroscience, 25*, 550–557.

Bloch, M. (2008). Why religion is nothing special but is central. *Philosophical Transactions of the Royal Society B, 363*, 2055–2061.

Blood, A. J., & Zatorre, R. J. (2001). Intensely pleasurable responses to music correlate with activity in brain regions implicated in reward and emotion. *Proceedings of the National Academy of Science, USA, 98*, 11818–11823.

Bodamer, J. (1947). Die prosopagnosie. *Archiv fur Psychiatrie und Zeitschrift fur Neurologie, 179*, 6–54.

Bombari, D., Mast, M. S., Brosch, T., & Sander, D. (2013). How interpersonal power affects empathic accuracy: Differential roles of mentalizing vs. mirroring? *Frontiers in Human Neuroscience, 7*.

Bonini, L., Rozzi, S., Serventi, F.U., Simone, L., Ferrari, P.F., & Fogassi, L. (2010). Ventral premotor and inferior parietal cortices make distinct contribution to action organization and intention understanding. *Cerebral Cortex, 20*(6), 1372–1385.

Borgomaneri, S., Vitale, F., Gazzola, V., & Avenanti, A. (2015). Seeing fearful body language rapidly freezes the observer's motor cortex. *Cortex, 65*, 232–245.

Boucsein, W., Fowles, D.C., Grimnes, S., Ben-Shakhar, G., Roth, W.T., Dawson, M.E., . . . Soc Psychophysiological Res Ad, H. (2012). Publication recommendations for electrodermal measurements. *Psychophysiology, 49*(8), 1017–1034.

Boyd, R., & Richerson, P. J. (1985). *Culture and the Evolutionary Process*. Chicago: University of Chicago Press.

Bolger, D. J., Perfetti, C. A., & Schneider, W. (2005). Cross-cultural effect on the brain revisited: Universal structures plus writing system variation. *Human Brain Mapping, 25*, 92–104.

Bolhuis, J. J. (1990). Mechanisms of avian imprinting: A review. *Biological Reviews, 66*, 303–345.

Borg, J. S., Lieberman, D., & Kiehl, K. A. (2008). Infection, incest, and iniquity: Investigating the neural correlates of disgust and morality. *Journal of Cognitive Neuroscience, 20*(9), 1529–1546.

Boria, S., Fabbri-Destro, M., Cattaneo, L., Sparaci, L., Sinigaglia, C., Santelli, E., Cossu, G., & Rizzolatti, G. (2009). Intention understanding in autism. *Plos One, 4*(5).

Bos, P. A., Terburg, D., & van Honk, J. (2010). Testosterone decreases trust in socially naive humans. *Proceedings of the National Academy of Science, USA, 107*, 9991–9995

Bottini, G., Corcoran, R., Sterzi, R., Paulesu, E., Schenone, P., Scarpa, P., Frackowiak, R. S. J., & Frith, C. D. (1994). The role of the right hemisphere in the interpretation of figurative aspects of language: A positron emission tomography activation study. *Brain, 117*, 1241–1253.

Botvinick, M. M., Braver, T. S., Barch, D. M., Carter, C. S., & Cohen, J. D. (2001). Conflict monitoring and cognitive control. *Psychological Review, 108*(3), 624–652.

Bourgeois, P., & Hess, U. (2008). The impact of social context on mimicry. *Biological Psychology, 77*(3), 343–352.

Bowlby, J. (1969). *Attachment and Loss: Volume 1—Attachment*. London: Hogarth Press.

Boyer, P. (2008). Religion: Bound to believe? *Nature, 455*, 1038–1039.

Boyer, P., Robbins, P., & Jack, A. I. (2005). Varieties of self-systems worth having. *Consciousness and Cognition, 14*(4), 647–660.

Branchi, I., D'Andrea, I., Fiore, M., Di Fausto, V., Aloe, L., & Alleva, E. (2006). Early social enrichment shapes social behavior and nerve growth factor and brain-derived neurotrophic factor levels in the adult mouse brain. *Biological Psychiatry, 60*(7), 690–696.

Brent, L. J. N., Chang, S. W. C., Gariepy, J.-F., & Platt, M. L. (2014). The neuroethology of friendship. *Year in Cognitive Neuroscience, 1316*, 1–17.

Brewer, R., Marsh, A. A., Catmur, C., Cardinale, E. M., Stoycos, S., Cook, R., & Bird, G. (2015). The impact of autism spectrum disorder and alexithymia on judgments of moral acceptability. *Journal of Abnormal Psychology, 124*(3), 589–595.

Brunner, H. G., Nelen, M. R., Vanzandvoort, P., Abeling, N., Vangennip, A. H., Wolters, E. C., . . . Vanoost, B. A. (1993). X-linked borderline mental-retardation with prominent behavioral disturbance – phenotype, genetic localization, and evidence for disturbed monoamine metabolism. *American Journal of Human Genetics, 52*(6), 1032–1039.

Brooks, R., & Meltzoff, A. N. (2005). The development of gaze following and its relation to language. *Developmental Science, 8*, 535–543.

Brown, R. (1995). *Prejudice*. Oxford: Blackwell.

Bruce, V., & Young, A. W. (1986). Understanding face recognition. *British Journal of Psychology, 77*, 305–327.

Buccino, G., Lui, F., Canessa, N., Patteri, I., Lagravinese, G., Benuzzi, F., Porro, C. A., & Rizzolatti, G. (2004). Neural circuits involved in the recognition of actions performed by nonconspecifics: An fMRI study. *Journal of Cognitive Neuroscience, 16*, 114–126.

Buckner, R. L., Andrews-Hanna, J. R., & Schacter, D. L. (2008). The brain's default network: Anatomy, function, and relevance to disease. *Annals of the New York Academy of Sciences, 1124*, 1–38.

Bufalari, I., Aprile, T., Avenanti, A., Di Russo, F., & Aglioti, S. M. (2007). Empathy for pain and touch in the human somatosensory cortex. *Cerebral Cortex, 17*(11), 2553–2561.

Bulbulia, J. (2004). The cognitive and evolutionary psychology of religion. *Biology and Philosophy, 19*, 655–686.

Burkett, J. P., Spiegel, L. L., Inoue, K., Murphy, A. Z., & Young, L. J. (2011). Activation of mu-opioid receptors in the dorsal striatum is necessary for adult social attachment in monogamous prairie voles. *Neuropsychopharmacology, 36*(11), 2200–2210.

Burnham, D., Kitamura, C., & Vollmer-Conna, U. (2002). What's new, pussycat? On talking to babies and animals. *Science, 296*(5572), 1435–1435.

Bush, G., Luu, P., & Posner, M. I. (2000). Cognitive and emotional influences in anterior cingulate cortex. *Trends in Cognitive Sciences, 4*, 215–222.

Bush, G., Vogt, B. A., Holmes, J., Dale, A. M., Greve, D., Jenike, M. A., & Rosen, B. R. (2002). Dorsal anterior cingulate cortex: A role in reward-based decision making. *Proceedings of the National Academy of Sciences of the United States of America, 99*(1), 523–528.

Bushnell, I. W. R., Sai, F., & Mullin, J. T. (1989). Neonatal recognition of the mothers face. *British Journal of Developmental Psychology, 7*, 3–15.

Buss, D. M. (1989). Sex-differences in human mate preferences – evolutionary hypothesis tested in 37 cultures. *Behavioral and Brain Sciences, 12*(1), 1–14.

Butterfill, S. A., & Apperly, I. A. (2013). How to construct a minimal theory of mind. *Mind & Language, 28*(5), 606–637.

Buttelmann, D., Carpenter, M., Call, J., & Tomasello, M. (2007). Encultured chimpanzees imitate rationally. *Developmental Science, 10*, F31–F38.

Byrne, R. W., & Corp, N. (2004). Neocrotex size predicts deception rates in primates. *Proceedings of the Royal Society of London B, 271*, 1693–1699.

Bzdok, D., Schilbach, L., Vogeley, K., Schneider, K., Laird, A. R., Langner, R., & Eickhoff, S. B. (2012). Parsing the neural correlates of moral cognition: ALE meta-analysis on morality, theory of mind, and empathy. *Brain Structure & Function, 217*(4), 783–796.

Cabeza, R., Prince, S. E., Daselaar, S. M., Greenberg, D. L., Budde, M., Dolcos, F., . . . Rubin, D. C. (2004). Brain activity during episodic retrieval of autobiographical and laboratory events: An fMRI study using a novel photo paradigm. *Journal of Cognitive Neuroscience, 16*(9), 1583–1594.

Cacioppo, J. T., & Berntson, G. G. (1992). Social psychological contributions to the decade of the brain: Doctrine of multi-level analysis. *American Psychologist, 47,* 1019–1028.

Cacioppo, J. T., & Hawkley, L. C. (2009). Perceived social isolation and cognition. *Trends in Cognitive Sciences, 13,* 447–454.

Cacioppo, J. T., Norris, C. J., Decety, J., Monteleone, G., & Nusbaum, H. (2009). In the eye of the beholder: Individual differences in perceived social isolation predict regional brain activation to social stimuli. *Journal of Cognitive Neuroscience, 21*(1), 83–92.

Cahil, L., Prins, B., Weber, M., & McGaugh, J. L. (1994). Beta-adrenergic activation and memory for emotional events. *Nature, 371,* 702–704.

Cahill, L., Weinberger, N. M., Roozendaal, B., & McGaugh, J. L. (1999). Is the amygdala a locus of 'conditioned fear'? Some questions and caveats. *Neuron, 23,* 227–228.

Calder, A. J., Ewbank, M., & Passamonti, L. (2011). Personality influences the neural responses to viewing facial expressions of emotion. *Philosophical Transactions of the Royal Society B-Biological Sciences, 366*(1571), 1684–1701.

Calder, A. J., Keane, J., Lawrence, A. D., & Manes, F. (2004). Impaired recognition of anger following damage to the ventral striatum. *Brain, 127,* 1958–1969.

Calder, A. J., Keane, J., Manes, F., Antoun, N., & Young, A. W. (2000). Impaired recognition and experience of disgust following brain injury. *Nature Neuroscience, 3*(11), 1077–1078.

Calder, A. J., & Young, A. W. (2005). Understanding the recognition of facial identity and facial expression. *Nature Reviews Neuroscience, 6*(8), 641–651.

Calder, A. J., Young, A. W., Rowland, D., Perrett, D. I., Hodges, J. R., & Etcoff, N. L. (1996). Facial emotion recognition after bilateral amygdala damage: Differentially severe impairment of fear. *Cognitive Neuropsychology, 13,* 699–745.

Caldwell, C. A., & Millen, A. E. (2008). Experimental models for testing hypotheses about cumulative cultural evolution. *Evolution and Human Behavior, 29*(3), 165–171.

Call, J., & Tomasello, M. (2008). Does the chimpanzee have a theory of mind? 30 years later. *Trends in Cognitive Sciences, 12,* 187–192.

Callaghan, T., Rochat, P., Lillard, A., Claux, M. L., Odden, H., Itakura, S., Tapanya, S., & Singh, S. (2005). Synchrony in the onset of mental state reasoning: Evidence from five cultures. *Psychological Science, 16,* 378–384.

Calvert, G. A., Hansen, P. C., Iversen, S. D., & Brammer, M. J. (2001). Detection of audio-visual integration sites in humans by application of electrophysiological criteria to the BOLD effect. *NeuroImage, 14,* 427–438.

Cameron, N. M. (2011). Maternal programming of reproductive function and behavior in the female rat. *Frontiers in evolutionary neuroscience, 3,* 10.

Campbell, R., Heywood, C., Cowey, A., Regard, M., & Landis, T. (1990). Sensitivity to eye gaze in prosopagnosic patients and monkeys with superior temporal sulcus ablation. *Neuropsychologia, 28,* 1123–1142.

Campbell, R., Landis, T., & Regard, M. (1986). Face recognition and lip reading: A neurological dissociation. *Brain, 109,* 509–521.

Canli, T., Sivers, H., Whitfield, S. L., Gotlib, I. H., & Gabrieli, J. D. E. (2002). Amygdala response to happy faces as a function of extraversion. *Science, 296*(5576), 2191–2191.

Canli, T., Zhao, Z., Desmond, J. E., Kang, E. J., Gross, J., & Gabrieli, J. D. E. (2001). An fMRI study of personality influences on brain reactivity to emotional stimuli. *Behavioral Neuroscience, 115*(1), 33–42.

Cannon, W. B. (1927). The James-Lange theory of emotions: A critical examination and an alternative theory. *American Journal of Psychology, 39,* 106–124.

Capgras, J., & Reboul-Lachaux, J. (1923). L'illusion des sosies dans un delire sytematise chronique. *Bulletin de la Societe Clinique do Medecine Mentale, 2,* 6–16.

Cardinal, R. N., Parkinson, J. A., Hall, J., & Everitt, B. J. (2002). Emotion and motivation: The role of the amygdala, ventral striatum, and prefrontal cortex. *Neuroscience and Biobehavioral Reviews, 26*(3), 321–352.

Carlin, J. D., Calder, A. J., Kriegeskorte, N., Nili, H., & Rowe, J. B. (2011). A head view-invariant representation of gaze direction in anterior superior temporal sulcus. *Current Biology, 21*(21), 1817–1821.

Carpenter, M., Akhtar, N., & Tomasello, M. (1998). Fourteen- through 18-month-old infants differentially imitate intentional and accidental actions. *Infant Behavior and Development, 21,* 315–330.

Carpenter, M., Nagell, K., & Tomasello, M. (1998). Social cognition, joint attention, and communicative competence from 9 to 15 months of age. *Monographs of the Society for Research in Child Development, 63*(4).

Carr, L., Iacoboni, M., Dubeau, M. C., Mazziotta, J. C., & Lenzi, G. L. (2003). Neural mechanisms of empathy in humans: A relay from neural systems for imitation to limbic areas. *Proceedings of the National Academy of Sciences of the United States of America, 100*(9), 5497–5502.

Carre, J. M., & McCormick, C. M. (2008). In your face: Facial metrics predict aggressive behaviour in the laboratory and in varsity and professional hockey players.

Proceedings of the Royal Society B-Biological Sciences, 275(1651), 2651–2656.

Carter, C. S. (1998). Neuroendocrine perspectives on social attachment and love. *Psychoneuroendocrinology, 23,* 779–818.

Carter, C. S., DeVries, A. C., & Getz, L. L. (1995). Physiological substrates of mammalian monogamy: The prairie vole model. *Neuroscience and Biobehavioral Review, 19,* 303–314.

Carter, C. S., MacDonald, A. M., Botvinick, M., Ross, L. L., Stenger, V. A., Noll, D., & Cohen, J. D. (2000). Parsing executive processes: Strategic vs. evaluative functions of the anterior cingulate cortex. *Proceedings of the National Academy of Science, USA, 97,* 1944–1948.

Casey, B. J., Getz, S., & Galvan, A. (2008). The adolescent brain. *Developmental Review, 28*(1), 62–77.

Caspi, A., McClay, J., Moffitt, T. E., Mill, J., Martin, J., Craig, I. W., Taylor, A., & Poulton, R. (2002). Role of genotype in the cycle of violence in maltreated children. *Science, 297*(5582), 851–854.

Caspi, A., Moffitt, T. E., Cannon, M., McClay, J., Murray, R. M., Harrington, H. L., . . . Craig, I. W. (2005). Moderation of the effect of adolescent-onset cannabis use on adult osychosis by a functional polymorphism in the catechol-o-methyltransferase gene: Longitudinal evidence for a gene X environment interaction. *Biological Psychiatry, 57,* 1117–1127.

Caspi, A., Sugden, K., Moffitt, T. E., Taylor, A., Craig, I. W., Harrington, H., . . . Poulton, R. (2003). Influence of life stress on depression: Moderation by a polymorphism in the 5-HTT gene. *Science, 301*(5631), 386–389.

Castelli, F., Frith, C., Happe, F., & Frith, U. (2002). Autism, Asperger syndrome and brain mechanisms for the attribution of mental states to animated shapes. *Brain, 125,* 1839–1849.

Castelli, F., Happe, F., Frith, U., & Frith, C. D. (2000). Movement and mind: A functional imaging study of perception and interpretation of complex intentional movements. *NeuroImage, 12,* 314–325.

Catmur, C., Walsh, V., & Heyes, C. (2007). Sensorimotor learning configures the human mirror system. *Current Biology, 17*(17), 1527–1531.

Cattaneo, Z., Mattavelli, G., Platania, E., & Papagno, C. (2011). The role of the prefrontal cortex in controlling gender-stereotypical associations: A TMS investigation. *NeuroImage, 56*(3), 1839–1846.

Cavada, C., Company, T., Tejedor, J., Cruz-Rizzolo, R. J., & Reinoso-Suarez, F. (2000). The anatomical connections of the macaque monkey orbitofrontal cortex. A review. *Cerebral Cortex, 10*(3), 220–242.

Chance, M. (1967). The interpretation of some agonistic postures: The role of "cut-off" acts and postures. *Symposium of the Zoological Society of London, 8,* 71–89.

Champagne, F., Diorio, J., Sharma, S., & Meaney, M. J. (2001). Naturally occurring variations in maternal behavior in the rat are associated with differences in estrogen-inducible central oxytocin receptors. *Proceedings of the National Academy of Sciences of the United States of America, 98*(22), 12736–12741.

Chandler, J., Griffin, T. M., & Sorensen, N. (2008). In the "I" of the storm: Shared initials increase disaster donations. *Judgment and Decision Making Journal, 3*(5), 404–410.

Chapman, H. A., Kim, D. A., Susskind, J. M., & Anderson, A. K. (2009). In bad taste: Evidence for the oral origins of moral disgust. *Science, 323*(5918), 1222–1226.

Chartrand, T. L., & Bargh, J. A. (1999). The Chameleon effect: The perception-behavior link and social interaction. *Journal of Personality and Social Psychology, 76*(6), 893–910.

Cheesman, J., & Merikle, P. M. (1984). Priming with and without awareness. *Perception & Psychophysics, 36*(4), 387–395.

Cheetham, M., Pedroni, A. F., Antley, A., Slater, M., & Jancke, L. (2009). Virtual milgram: Empathic concern or personal distress? Evidence from functional MRI and dispositional measures. *Frontiers in Human Neuroscience, 3,* 13.

Chekroud, A. M., Everett, J. A. C., Bridge, H., & Hewstone, M. (2014). A review of neuroimaging studies of race-related prejudice: Does amygdala response reflect threat? *Frontiers in Human Neuroscience, 8.*

Cheng, Y., Chou, K. H., Decety, J., Chen, I. Y., Hung, D., Tzeng, O. J. L., & Lin, C. P. (2009). Sex differences in the neuroanatomy of human mirror-neuron system: A voxel-based morphometric investigation. *Neuroscience, 158*(2), 713–720.

Cheng, Y. W., Lee, P. L., Yang, C. Y., Lin, C. P., Hung, D., & Decety, J. (2008). Gender differences in the Mu rhythm of the human mirror-neuron system. *Plos One, 3*(5).

Cheng, Y. W., Lin, C. P., Liu, H. L., Hsu, Y. Y., Lims, K. E., Hung, D., & Decety, J. (2007). Expertise modulates the perception of pain in others. *Current Biology, 17*(19), 1708–1713.

Cherkas, L. F., Oelsner, E. C., Mak, Y. T., Valdes, A., & Spector, T. D. (2004). Genetic influences on female infidelity and number of sexual partners in humans: A linkage and association study of the role of the vasopressin receptor gene (AVPR1A). *Twin Research, 7*(6), 649–658.

Chiao, J. Y. (2010). At the frontier of cultural neuroscience: Introduction to the special issue. *Social Cognitive and Affective Neuroscience, 5*(2–3), 109–110.

Chiao, J. Y., & Blizinsky, K. D. (2010). Culture-gene coevolution of individualism-collectivism and the serotonin transporter gene. *Proceedings of the Royal Society B-Biological Sciences, 277*(1681), 529–537.

Chiao, J. Y., Harada, T., Komeda, H., Li, Z., Mano, Y., Saito, D., . . . Iidaka, T. (2009). Neural basis of individualistic and collectivistic views of self. *Human Brain Mapping, 30*(9), 2813–2820.

Chiao, J. Y., Harada, T., Komeda, H., Li, Z., Mano, Y., Saito, D., . . . Iidaka, T. (2010). Dynamic cultural influences on neural representations of the self. *Journal of Cognitive Neuroscience, 22*(1), 1–11.

Chisholm, K. (1998). A three year follow-up of attachment and indiscriminate friendliness in children adopted from Romanian orphanages. *Child Development, 69*(4), 1092–1106.

Chittka, L., & Niven, J. (2009). Are bigger brains better? *Current Biology, 19*(21), R995–R1008.

Cho, M. M., DeVries, A. C., Williams, J. R., & Carter, C. S. (1999). The effects of oxytocin and vasopressin on partner preferences in male and female prairie voles (Microtus ochrogaster). *Behavioral Neuroscience, 113*(5), 1071–1079.

Chomsky, N. (1980). Rules and representations. *Behavioral and Brain Sciences, 3*, 1–61.

Chugani, H. T., Behen, M. E., Muzik, O., Juhasz, C., Nagy, F., & Chugani, D. C. (2001). Local brain functional activity following early deprivation: A study of postinstitutionalized Romanian orphans. *Neuroimage, 14*(6), 1290–1301.

Churchland, P. S., & Sejnowski, T. J. (1988). Perspectives on cognitive neuroscience. *Science, 242*, 741–745.

Churchland, P. S., & Winkielman, P. (2012). Modulating social behavior with oxytocin: How does it work? What does it mean? *Hormones and Behavior, 61*(3), 392–399.

Cialdini, R. B., Brown, S. L., Lewis, B. P., Luce, C., & Neuberg, S. L. (1997). Reinterpreting the empathy-altruism relationship: When one into one equals oneness. *Journal of Personality and Social Psychology, 73*(3), 481–494.

Ciaramelli, E., Muccioli, M., Ladavas, E., & di Pellegrino, G. (2007). Selective deficit in personal moral judgment following damage to ventromedial prefrontal cortex. *Social Cognitive and Affective Neuroscience, 2*(2), 84–92.

Cikara, M., Farnsworth, R. A., Harris, L. T., & Fiske, S. T. (2010). On the wrong side of the trolley track: Neural correlates of relative social valuation. *Social Cognitive and Affective Neuroscience, 5*(4), 404–413.

Cima, M., Tonnaer, F., & Hauser, M. D. (2010). Psychopaths know right from wrong but don't care. *Social Cognitive and Affective Neuroscience, 5*(1), 59–67.

Civai, C., Corradi-Dell'Acqua, C., Gamer, M., & Rumiati, R. I. (2010). Are irrational reactions to unfairness truly emotionally-driven? Dissociated behavioural and emotional responses in the Ultimatum Game task. *Cognition, 114*, 89–95.

Clark, A. (2003). *Natural-born Cyborgs: Minds, Technologies, and the Future of Human Intelligence.* Oxford: Oxford University Press.

Clark, A. (2008). *Supersizing the Mind: Embodiment, Action and Cognitive Extension.* Oxford: Oxford University Press.

Clark, L., Bechara, A., Damasio, H., Aitken, M. R. F., Sahakian, B. J., & Robbins, T. W. (2008). Differential effects of insular and ventromedial prefrontal cortex lesions on risky decision-making. *Brain, 131*, 1311–1322.

Clements, W. A., & Perner, J. (1994). Implicit understanding of belief. *Cognitive Development, 9*, 377–395.

Clore, G. L., & Ortony, A. (2000). Cognition in emotion: Always, sometimes or never? In R. D. Lane & L. Nadel (Eds.), *Cognitive Neuroscience of Emotion.* Oxford: Oxford University Press.

Cohen, L., Lehericy, S., Chochon, F., Lemer, C., Rivaud, S., & Dehaene, S. (2002). Language-specific tuning of visual cortex functional properties of the visual word form area. *Brain, 125*, 1054–1069.

Cole, S. W., Hawkley, L. C., Arevalo, J. M., Sung, C. Y., Rose, R. M., & Cacioppo, J. T. (2007). Social regulation of gene expression in human leukocytes. *Genome Biology, 8*(9), 13.

Collins, D., Neelin, P., Peters, T., & Evans, A. (1994). Automatic 3D intersubject registration of MR volumetric data in standardized Talaraich space. *Journal of Computer Assisted Tomography, 18*, 192–205.

Connor, R. C. (2007). Dolphin social intelligence: Complex alliance relationships in bottlenose dolphons and a consideration of selective environments for extreme brain size evolution in mammals. *Philosophical Transactions of the Royal Society B, 362*, 587–602.

Connor, R. C., Wells, R., Mann, J., & Read, A. (2000). The bottlenose dolphin: Social relationships in a fission-fusion society. In J. Mann, R. Connor, P. Tyack & H. Whitehead (Eds.), *Cetacean Societies: Field studies of Whales and Dolphins.* Chicago: Chicago University Press.

Conway, M. A. (2005). Memory and the self. *Journal of Memory and Language, 53*(4), 594–628.

Conway, M. A., & Pleydell-Pearce, C. W. (2000). The construction of autobiographical memories in the self-memory system. *Psychological Review, 107*, 261–288.

Cook, R., Bird, G., Catmur, C., Press, C., & Heyes, C. (2014). Mirror neurons: From origin to function. *Behavioral and Brain Sciences, 37*(2), 177–192.

Corbetta, M., & Shulman, G. L. (2002). Control of goal-directed and stimulus-driven attention in the brain. *Nature Reviews Neuroscience, 3*(3), 201–215.

Corradi-Dell'Acqua, C., Civai, C., Rumiati, R. I., & Fink, G. R. (2013). Disentangling self- and fairness-related neural mechanisms involved in the ultimatum game: An fMRI study. *Social Cognitive and Affective Neuroscience, 8*(4), 424–431.

Coricelli, G., Critchley, H. D., Joffily, M., O'Doherty, J. P., Sirigu, A., & Dolan, R. J. (2005). Regret and its avoidance: A neuroimaging study of choice behavior. *Nature Neuroscience, 8*(9), 1255–1262.

Corkin, S. (2002). What's new with the amnesic patient HM?. *Nature Reviews Neuroscience, 3*, 153–160

Cosmides, L. (1989). The logic of social exchange: Has natural selection shaped how humans reason? Studies with the Wason selection task. *Cognition, 31*, 187–276.

Costa, B., Pini, S., Gabelloni, P., Abelli, M., Lari, L., Cardini, A., . . . Martini, C. (2009). Oxytocin receptor polymorphisms and adult attachment style in patients with depression. *Psychoneuroendocrinology, 34*(10), 1506–1514.

Costa, P. T., & McCrae, R. R. (1985). *The NEO Personality Inventory Manual*. Odessa, FL: Psychological Assessment Resources.

Costa, P. T., Terracciano, A., & McCrae, R. R. (2001, February). *Gender differences in personality traits across cultures: Robust and surprising findings*. Paper presented at the 2nd Annual Meeting of the Society-for-Personality-and-Social-Psychology, San Antonio, Texas.

Couppis, M. H., & Kennedy, C. H. (2008). The rewarding effect of aggression is reduced by nucleus accumbens dopamine receptor antagonism in mice. *Psychopharmacology, 197*(3), 449–456.

Craig, A. D. (2009). How do you feel – now? The anterior insula and human awareness. *Nature Reviews Neuroscience, 10*(1), 59–70.

Creswell, C. S., & Skuse, D. H. (1999). Autism in association with Turner syndrome: Genetic implications for male vulnerability to pervasive developmental disorders. *Neurocase, 5*, 511–518.

Critchley, H. D., Elliott, R., Mathias, C. J., & Dolan, R. J. (2000). Neural activity relating to generation and representation of galvanic skin conductance responses: A functional magnetic resonance imaging study. *Journal of Neuroscience, 20*, 3033–3040.

Critchley, H. D., Mathias, C. J., Josephs, O., O'Doherty, J., Zanini, S., Dewar, B. K., Cipolotti, L., Shallice, T., & Dolan, R. J. (2003). Human cingulate cortex and autonomic control: Converging neuroimaging and clinical evidence. *Brain, 126*, 2139–2152.

Critchley, H. D., Wiens, S., Rotshtein, P., Ohman, A., & Dolan, R. J. (2004). Neural systems supporting interoceptive awareness. *Nature Neuroscience, 7*(2), 189–195.

Crone, E. A., & Westenberg, P. M. (2009). A brain-based account of developmental changes in social decision making. In M. D. Haan & M. R. Gunnar (Eds.), *Handbook of Developmental Social Neuroscience*. New York: The Guilford Press.

Cronin, H. (1991). *The Ant and the Peacock*. Cambridge, UK: Cambridge University Press.

Cullen, H., Kanai, R., Bahrami, B., & Rees, G. (2014). Individual differences in anthropomorphic attributions and human brain structure. *Social Cognitive and Affective Neuroscience, 9*(9), 1276–1280.

Cunningham, W. A., Johnson, M. K., Raye, C. L., Gatenby, J. C., Gore, J. C., & Banaji, M. R. (2004). Separable neural components in the processing of black and white faces. *Psychological Science, 15*(12), 806–813.

Cunningham, W. A., Raye, C. L., & Johnson, M. K. (2004). Implicit and explicit evaluation: FMRI correlates of valence, emotional intensity, and control in the processing of attitudes. *Journal of Cognitive Neuroscience, 16*(10), 1717–1729.

Cushman, F., Young, L., & Hauser, M. (2006). The role of conscious reasoning and intuition in moral judgment: Testing three principles of harm. *Psychological Science, 17*(12), 1082–1089.

Custance, D. M., Whiten, A., & Bard, K. A. (1995). Can young chimpanzees (Pan troglodytes) imitate arbitrary actions? Hayes and Hayes (1952) revisited. *Behaviour, 132*, 837–859.

D'Argembeau, A., Feyers, D., Majerus, S., Collette, F., Van der Linden, M., Maquet, P., & Salmon, E. (2008). Self-reflection across time: Cortical midline structures differentiate between present and past selves. *Social Cognitive and Affective Neuroscience, 3*(3), 244–252.

Dabbs, J. M., Carr, T. S., Frady, R. L., & Riad, J. K. (1995). Testosterone, crime, and misbehavior among 692 male prison-inmates. *Personality and Individual Differences, 18*(5), 627–633.

Dabbs, J. M., & Morris, R. (1990). Testosterone, social-class, and antisocial-behavior in a sample of 4,462 men. *Psychological Science, 1*(3), 209–211.

Dalton, K. M., Nacewicz, B. M., Johnstone, T., Schaefer, H. S., Gernsbacher, M. A., Goldsmith, H. H., . . . Davidson, R. J. (2005). Gaze fixation and the neural circuitry of face processing in autism. *Nature Neuroscience, 8*(4), 519–526.

Damasio, A. R. (1994). *Descartes' Error: Emotion, Reason and the Human Brain*. New York: G.P. Putnam & Sons.

Damasio, A. R. (1996). The somatic marker hypothesis and the possible functions of the prefrontal cortex. *Philosophical Transactions of the Royal Society of London B, 351*, 1413–1420.

Damasio, A. R. (1999). *The Feeling of what Happens: Body and Emotion in the Making of Consciousness*. New York: Harcourt.

Damasio, A. R. (2003). Feelings of emotion and the self. In J. LeDoux, J. Debiec & H. Moss (Eds.), *Self: From Soul to Brain* (Vol. 1001, pp. 253–261). New York: New York Acad Sciences.

Damasio, A. R., Grabowski, T. J., Bechara, A., Damasio, H., Ponto, L. L. B., Parvizi, J., & Hichwa, R. D. (2000). Subcortical and cortical brain activity during the feeling of self-generated emotions. *Nature Neuroscience, 3*(10), 1049–1056.

Damasio, H., Grabowski, T., Frank, R., Galaburda, A. M., & Damasio, A. R. (1994). The return of Phineas Gage: Clues about the brain from the skull of a famous patient. *Science, 264*, 1102–1105.

Damasio, A. R., Tranel, D., & Damasio, H. (1990). Individuals with sociopathic behavior caused by frontal damage fail to respond autonomically to social stimuli. *Behavioral Brain Research, 41*, 81–94.

Damoiseaux, J. S., Rombouts, S. A. R. B., Barkhof, F., Scheltens, P., Stam, C. J., Smith, S. M., . . . Beckmann, C. F. (2006). Consistent resting-state networks across healthy subjects. *Proceedings of the National Academy of Sciences of the United States of America, 103*(37), 13848–13853.

Dapretto, M., Davies, M. S., Pfeifer, J. H., Scott, A. A., Sigman, M., Bookheimer, S. Y., & Iacoboni, M. (2006). Understanding emotions in others: Mirror neuron dysfunction in children with autism spectrum disorders. *Nature Neuroscience, 9*, 28–30.

Darwin, C. (1872/1965). *The Expression of the Emotions in Man and Animals*. Chicago: University of Chicago Press.

Darwin, C. J. (1871). *The Descent of Man and Selection in Relation to Sex*. London: John Murray.

Dasgupta, N., & Greenwald, A. G. (2001). On the malleability of automatic attitudes: Combating automatic prejudice with images of admired and disliked individuals. *Journal of Personality and Social Psychology, 81*(5), 800–814.

Davidoff, J., Fonteneau, E., & Fagot, J. (2008). Local and global processing: Observations from a remote culture. *Cognition, 108*(3), 702–709.

Davis, K.L., & Panksepp, J. (2011). The brain's emotional foundations of human personality and the affective neuroscience personality scales. *Neuroscience and Biobehavioral Reviews, 35*(9), 1946–1958.

Davis, M. H. (1980). A multi-dimensional approach to individual differences in empathy. *JCAS Catalog of Selected Documents in Psychology, 75*, 989–1015.

Davis, M. H. (1992). The role of the amygdala in fear-potentiated startle – implications for animal-models of anxiety. *Trends in Pharmacological Sciences, 13*(1), 35–41.

Dawkins, R. (1976). *The Selfish Gene*. Oxford: OUP.

De Dreu, C.K.W., Greer, L.L., Handgraaf, M.J.J., Shalvi, S., Van Kleef, G.A., Baas, M., . . . Feith, S.W.W. (2010). The neuropeptide oxytocin regulates parochial altruism in intergroup conflict among humans. *Science, 328*(5984), 1408–1411.

de Gelder, B. (2006). Towards the neurobiology of emotional body language. *Nature Reviews Neuroscience, 7*(3), 242–249.

de Quervain, D. J. F., Fischbacher, U., Treyer, V., Schelthammer, M., Schnyder, U., Buck, A., & Fehr, E. (2004). The neural basis of altruistic punishment. *Science, 305*(5688), 1254–1258.

De Renzi, E. (1986). Prosopagnosia in two patients with CT scan evidence of damage confined to the right hemisphere. *Neuropsychologia, 24*, 385–389.

de Sousa, A., & Cunha, E. (2012). Hominins and the emergence of the modern human brain. In M. A. Hofman & D. Falk (Eds.), *Evolution of the Primate Brain: From Neuron to Behavior* (Vol. 195, pp. 293–322). Elsevier: New York.

de Veer, M. W., & Van den Bos, R. (1999). A critical review of methodology and interpretation of mirror self-recognition research in nonhuman primates. *Animal Behaviour, 58*, 459–468.

de Vignemont, F., & Singer, T. (2006). The empathic brain: How, when and why? *Trends in Cognitive Sciences, 10*(10), 435–441.

de Vries, G.J. (2008). Sex differences in vasopressin and oxytocin innervation of the brain. *Advances in Vasopressin and Oxytocin: From Genes to Behaviour to Disease, 170*, 17–27.

de Waal, F. B. M. (2008). Putting the altruism back into altruism: The evolution of empathy. *Annual Review of Psychology, 59*, 279–300.

de Waal, F. B. M., & Luttrell, L. M. (1988). Mechanisms of social reciprocity in 3 primate species – symmetrical relationship characteristics or cognition. *Ethology and Sociobiology, 9*(2–4), 101–118.

De Wall, C. N., Baumeister, R. F., Stillman, T. F., & Gailliot, M. T. (2007). Violence restrained: Effects of self-regulation and its depletion on aggression. *Journal of Experimental Social Psychology, 43*(1), 62–76.

Dean, L.G., Kendal, R.L., Schapiro, S.J., Thierry, B., & Laland, K.N. (2012). Identification of the social and cognitive processes underlying human cumulative culture. *Science, 335*(6072), 1114–1118.

Dean, L.G., Vale, G.L., Laland, K.N., Flynn, E., & Kendal, R.L. (2014). Human cumulative culture: A comparative perspective. *Biological Reviews, 89*(2), 284–301.

Deaner, R.O., Khera, A.V., & Platt, M.L. (2005). Monkeys pay per view: Adaptive valuation of social images by rhesus macaques. *Current Biology, 15*(6), 543–548.

DeBruine, L. M. (2005). Trustworthy but not lust-worthy: Context-specific effects of facial resemblance. *Proceedings of the Royal Society B-Biological Sciences, 272*(1566), 919–922.

DeCasper, A. J., & Fifer, W. P. (1980). Of human bonding – newborns prefer their mothers voices. *Science, 208*(4448), 1174–1176.

DeCasper, A. J., & Spence, M. J. (1986). Prenatal maternal speech influences newborns' perception of speech sounds. *Infant Behavior and Development, 9*, 133–150.

Decety, J., & Chaminade, T. (2003). When the self represents the other: A new cognitive neuroscience view on psychological identification. *Consciousness and Cognition, 12*, 577–596.

Decety, J., & Jackson, P. J. (2004). The functional architecture of human empathy. *Behavioral and Cognitive Neuroscience Reviews, 3*, 71–100.

Decety, J., & Jackson, P. L. (2006). A social-neuroscience perspective on empathy. *Current Directions in Psychological Science, 15*(2), 54–58.

Decety, J., & Michalska, K.J. (2010). Neurodevelopmental changes in the circuits underlying empathy and sympathy from childhood to adulthood. *Developmental Science, 13*(6), 886–899.

Decety, J., Michalska, K. J., Akitsuki, Y., & Lahey, B. B. (2009). Atypical empathic responses in adolescents with aggressive conduct disorder: A functional MRI investigation. *Biological Psychology, 80*(2), 203–211.

Decety, J., & Svetlova, M. (2012). Putting together phylogenetic and ontogenetic perspectives on empathy. *Developmental Cognitive Neuroscience, 2*(1), 1–24.

Dehaene, S., & Cohen, L. (2007). Cultural recycling of cortical maps. *Neuron, 56*, 384–398.

Dehaene, S., Dehaene-Lambertz, G., & Cohen, L. (1998). Abstract representations of numbers in the animal and human brain. *Trends in Neurosciences, 21*, 355–361.

Dehaene, S., Posner, M. I., & Tucker, D. M. (1994). Localisation of a neural system for error detection and compensation. *Psychological Science, 5*, 303–305.

Delgado, M. R., Frank, R. H., & Phelps, E. A. (2005). Perceptions of moral character modulate the neural systems of reward during the trust game. *Nature Neuroscience, 8*, 1611–1618.

Dennett, D. C. (1978). Beliefs about beliefs. *Behavioral and Brain Sciences, 1*, 568–570.

Dennett, D. C. (1983). Intentional systems in cognitive ethology – the panglossian paradigm defended. *Behavioral and Brain Sciences, 6*(3), 343–355.

Denkhaus, R., & Boes, M. (2012). How cultural is 'cultural neuroscience'? Some comments on an emerging research paradigm. *Biosocieties, 7*(4), 433–458.

Denson, T. F., Pedersen, W. C., Ronquillo, J., & Nandy, A. S. (2009). The angry brain: Neural correlates of anger, angry rumination, and aggressive personality. *Journal of Cognitive Neuroscience, 21*(4), 734–744.

Depue, R. A., & Collins, P. F. (1999). Neurobiology of the structure of personality: Dopamine, facilitation of incentive motivation, and extraversion. *Behavioral and Brain Sciences, 22*(3), 491–517.

Deschamps, P. K. H., Coppes, L., Kenemans, J. L., Schutter, D., & Matthys, W. (2015). Electromyographic responses to emotional facial expressions in 6–7 year olds with autism spectrum disorders. *Journal of Autism and Developmental Disorders, 45*(2), 354–362.

DeScioli, P., & Kurzban, R. (2009). The alliance hypothesis for human friendship. *PLoS One, 4*(6).

Desimone, R., & Duncan, J. (1995). Neural mechanisms of selective visual-attention. *Annual Review of Neuroscience, 18*, 193–222.

Devine, P. G., & Elliot, A. J. (1995). Are racial stereotypes really fading – the Princeton trilogy revisited. *Personality and Social Psychology Bulletin, 21*(11), 1139–1150.

Devine, P. G., Plant, E. A., Amodio, D. M., Harmon-Jones, E., & Vance, S. L. (2002). The regulation of explicit and implicit race bias: The role of motivations to respond without prejudice. *Journal of Personality and Social Psychology, 82*(5), 835–848.

Devos, T., & Banaji, M. R. (2005). American = white? *Journal of Personality and Social Psychology, 88*(3), 447–466.

DeVries, A. C., DeVries, M. B., Taymans, S. E., & Carter, C. S. (1996). The effects of stress on social preferences are sexually dimorphic in prairie voles. *Proceedings of the National Academy of Sciences of the United States of America, 93*(21), 11980–11984.

Di Martino, A., Yan, C. G., Li, Q., Denio, E., Castellanos, F. X., Alaerts, K., . . . Milham, M. P. (2014). The autism brain imaging data exchange: Towards a large-scale evaluation of the intrinsic brain architecture in autism. *Molecular Psychiatry, 19*(6), 659–667.

di Pellegrino, G., Fadiga, L., Fogassi, L., Gallese, V., & Rizzoloatti, G. (1992). Understanding motor events: A neurophysiological study. *Experimental Brain Research, 91*, 176–180.

Diamond, J. (1997). *Guns, Germs and Steel*. London: Cape.

Dijksterhuis, A. (2005). Why we are social animals: The high road to imitation as social glue. In S. Hurley & N. Chater (Eds.), *Perspectives on Imitation: From Neuroscience to Social Science* (Vol. 1). Cambridge, MA: MIT Press.

Dimberg, U., & Petterson, M. (2000). Facial reactions to happy and angry facial expressions: Evidence for right hemisphere dominance. *Psychophysiology, 37*, 693–696.

Dimberg, U., Thunberg, M., & Elmehed, K. (2000). Unconscious facial reactions to emotional facial expressions. *Psychological Science, 11*(1), 86–89.

Dimitrov, M., Phipps, M., Zahn, T. P., & Grafman, J. (1999). A thoroughly modern Gage. *Neurocase, 5*, 345–354.

Dinstein, I., Thomas, C., Behrmann, M., & Heeger, D. J. (2008). A mirror up to nature. *Current Biology, 18*(1), R13–R18.

Dion, K., Berscheid, E., & Walster, E. (1972). What is beautiful is good. *Journal of Personality and Social Psychology, 24*(3), 285–290.

Dittes, J. E., & Kelley, H. H. (1956). Effects of different conditions of acceptance upon conformity to group norms. *Journal of Abnormal and Social Psychology, 53*, 100–107.

Dolan, R. J. (2007). The human amygdala and orbitofrontal cortex in behavioural regulation. *Philosophical Transactions of the Royal Society of London Series B, 362*, 787–799.

Dolcos, F., LaBar, K. S., & Cabeza, R. (2004). Interaction between the amygdala and the medial temporal lobe memory system predicts better memory for emotional events. *Neuron, 42*(5), 855–863.

Dollard, J., Doob, L., Miller, N., Mowrer, O., & Sears, R. (1939). *Frustration and Aggression*. New Haven, CT: Yale University Press.

Dondi, M., Simion, F., & Caltran, G. (1999). Can newborns discriminate between their own cry and the cry of another newborn infant? *Developmental Psychology, 35*, 418–426.

Doris, J. M. (1998). Persons, situations, and virtue ethics (Moral psychology). *Nous, 32*(4), 504–530.

Dorus, S., Vallender, E. J., Evans, P. D., Anderson, J. R., Gilbert, S. L., Mahowald, M., . . . Lahn, B. T. (2004). Accelerated evolution of nervous system genes in the origin of Homo sapiens. *Cell, 119*(7), 1027–1040.

Dougherty, D. D., Shin, L. M., Alpert, N. M., Pitman, R. K., Orr, S. P., Lasko, M., Macklin, M. L., Fischman, A. J., & Rauch, S. L. (1999). Anger in healthy men: A PET study using script-driven imagery. *Biological Psychiatry, 46*(4), 466–472.

Downing, P. E., Chan, A. W., Peelen, M. V., Dodds, C. M., & Kanwisher, N. (2006). Domain specificity in visual cortex. *Cerebral Cortex, 16*, 1453–1461.

Downing, P. E., Jiang, Y. H., Shuman, M., & Kanwisher, N. (2001). A cortical area selective for visual processing of the human body. *Science, 293*, 2470–2473.

Downing, P. E., & Peelen, M. V. (2011). The role of occipitotemporal body-selective regions in person perception. *Cognitive Neuroscience, 2*(3–4), 186–203.

Duchaine, B. C., & Nakayama, K. (2006). Developmental prosopagnosia: A window to content-specific face processing. *Current Opinion in Neurobiology, 16*(2), 166–173.

Dumontheil, I., Apperly, I. A., & Blakemore, S. J. (2010). Online usage of theory of mind continues to develop in late adolescence. *Developmental Science, 13*, 331–338.

Dunbar, R. I. M. (1992). Neocortex size as a constraint on group size in primates. *Journal of Human Evolution, 20*, 469–493.

Dunbar, R. I. M. (1998). The social brain hypothesis. *Evolutionary Anthropology, 6*, 178–190.

Dunbar, R. I. M. (2004). Gossip in evolutionary perspective. *Review of General Psychology, 5*, 100–110.

Dunbar, R. I. M., & Shultz, S. (2010). Bondedness and sociality. *Behaviour, 147*(7), 775–803.

Dunn, J., & Brophy, M. (2005). Communication, relationships, and individual differences in children's understanding of mind. In J. W. Astington & J. A. Baird (Eds.), *Why Language Matters for Theory of Mind*. Oxford: Oxford University Press.

Dunn, J., Brown, J., & Beardsall, L. (1991). Family talk about feeling states and children's later understanding of others' emotions. *Developmental Psychology, 27*, 448–455.

Dunn, J., Brown, J., Slomkowski, C., Tesla, C., & Youngblade, L. (1991). Young children's understanding of other people's feelings and beliefs: Individual differences and their antecedents. *Child Development, 62*, 1352–1366.

Durante, F., Fiske, S. T., Kervyn, N., Cuddy, A. J. C., Akande, A., Adetoun, B. E., . . . Storari, C. C. (2013). Nations' income inequality predicts ambivalence in stereotype content: How societies mind the gap. *British Journal of Social Psychology, 52*(4), 726–746.

Dushanova, J., & Donoghue, J. (2010). Neurons in primary motor cortex engaged during action observation. *European Journal of Neuroscience, 31*, 386–398.

Eastman, N., & Campbell, C. (2006). Neuroscience and legal determination of criminal responsibility. *Nature Reviews Neuroscience, 7*(4), 311–318.

Ebner, N. C., Johnson, M. R., Rieckmann, A., Durbin, K. A., Johnson, M. K., & Fischer, H. (2013). Processing own-age vs. other-age faces: Neuro-behavioral correlates and effects of emotion. *NeuroImage, 78*, 363–371.

Ebstein, R. P., Israel, S., Chew, S. H., Zhong, S., & Knafo, A. (2010). Genetics of human social behavior. *Neuron, 65*(6), 831–844.

Eder, R. A. (1990). Uncovering young childrens psychological selves – individual and developmental differences. *Child Development, 61*(3), 849–863.

Ehrenkranz, J., Bliss, E., & Sheard, M. H. (1974). Plasma testosterone – correlation with aggressive-behavior and social dominance in man. *Psychosomatic Medicine, 36*(6), 469–475.

Eisenberg, N., Fabes, R. A., Murphy, B., Karbon, M., Maszk, P., Smith, M., Oboyle, C., & Suh, K. (1994). The relations of emotionality and regulation to dispositional and situational empathy-related responding. *Journal of Personality and Social Psychology, 66*(4), 776–797.

Eisenberger, N. I. (2015). Social pain and the brain: Controversies, questions, and where to go from here. *Annual Review of Psychology, 66*, 601–629.

Eisenberger, N. I., Jarcho, J. M., Lieberman, M. D., & Naliboff, B. D. (2006). An experimental study of shared sensitivity to physical pain and social rejection. *Pain, 126*, 132–138.

Eisenberger, N. I., Lieberman, M. D., & Williams, K. D. (2003). Does rejection hurt? An fMRI study of social exclusion. *Science, 302*(5643), 290–292.

Ekman, P. (1972). Universal and cultural differences in facial expression of emotion. In J. R. Cole (Ed.), *Nebraska Symposium on Motivation*. Lincoln: Nebraska University Press.

Ekman, P. (1992). An argument for basic emotions. *Cognition and Emotion, 6*, 169–200.

Ekman, P., & Friesen, W. V. (1976). *Pictures of Facial Affect*. Palo Alto, CA: Consulting Psychologists Press.

Ekman, P., Friesen, W. V., & Ellsworth, P. (1972). *Emotion in the Human Face: Guidelines for Research and an Integration of Findings*. New York: Pergamon.

Elias, M. (1981). Serum cortisol, testosterone, and testosterone-binding globulin responses to competitive fighting in human males. *Aggressive Behavior, 7*, 215–224.

Ellis, H. D., & Lewis, M. B. (2001). Capgras delusion: A window on face recognition. *Trends in Cognitive Sciences, 5*, 149–156.

Ellis, H. D., & Young, A. W. (1990). Accounting for delusional misidentifications. *British Journal of Psychiatry, 157*, 239–248.

Ellis, H. D., Young, A. W., Quayle, A. H., & DePauw, K. W. (1997). Reduced autonomic responses to faces in Capgras delusion. *Proceedings of the Royal Society of London B, 264*, 1085–1092.

Emery, N. J., Seed, A. M., von Bayern, A. M. P., & Clayton, N. S. (2007). Cognitive adaptations of social bonding in birds. *Philosophical Transactions of the Royal Society B-Biological Sciences, 362*(1480), 489–505.

Engel, A. K., Konig, P., & Singer, W. (1991). Direct physiological evidence for scene segmentation by temporal encoding. *Proceedings of the National Academy of Science, USA, 88*, 9136–9140.

Engel, A. K., Moll, C. K. E., Fried, I., & Ojemann, G. A. (2005). Invasive recordings from the human brain: Clinical insights and beyond. *Nature Reviews Neuroscience, 6*, 35–47.

Enticott, P. G., Kennedy, H. A., Rinehart, N. J., Bradshaw, J. L., Tonge, B. J., Daskalakis, Z., J., . . . Fitzgerald, P. B. (2013). Interpersonal motor resonance in autism spectrum disorder: Evidence against a global "mirror system" deficit. *Frontier in Human Neuroscience, 7*, e218.

Epley, N., Akalis, S., Waytz, A., & Cacioppo, J. T. (2008). Creating social connection through inferential reproduction – Loneliness and perceived agency in gadgets, gods, and greyhounds. *Psychological Science, 19*(2), 114–120.

Erickson, K., Drevets, W., & Schulkin, J. (2003). Glucocorticoid regulation of diverse cognitive functions in normal and pathological emotional states. *Neuroscience and Biobehavioral Reviews, 27*(3), 233–246.

Ernst, M., Pine, D. S., & Hardin, M. (2006). Triadic model of the neurobiology of motivated behavior in adolescence. *Psychological Medicine, 36*(3), 299–312.

Ernst, M., & Spear, L. P. (2009). Reward systems. In M. D. Haan & M. R. Gunnar (Eds.), *Handbook of Developmental Social Neuroscience*. New York: The Guilford Press.

Eslinger, P. J., & Damasio, A. R. (1985). Severe disturbance of higher cognition after bilateral frontal ablation: Patient EVR. *Neurology, 35*, 1731–1741.

Ethofer, T., Bretscher, J., Gschwind, M., Kreifelts, B., Wildgruber, D., & Vuilleumier, P. (2012). Emotional voice areas: Anatomic location, functional properties, and structural connections revealed by combined fMRI/DTI. *Cerebral Cortex, 22*(1), 191–200

Everitt, B. J., & Robbins, T. W. (2005). Neural systems of reinforcement for drug addiction: From actions to habits to compulsion. *Nature Neuroscience, 8*(11), 1481–1489.

Fabes, R. A., Eisenberg, N., Karbon, M., Troyer, D., & Switzer, G. (1994). The relations of childrens emotion regulation to their vicarious emotional responses and comforting behaviors. *Child Development, 65*(6), 1678–1693.

Fadiga, L., Fogassi, L., Pavesi, G., & Rizzolatti, G. (1995). Motor facilitation during action observation – a magnetic stimulation study. *Journal of Neurophysiology, 73*(7), 2608–2611.

Falk, A., Fehr, E., & Fischbacher, U. (2005). Driving forces behind informal sanctions. *Econometrica, 73*(6), 2017–2030.

Fan, Y., Duncan, N. W., de Greck, M., & Northoff, G. (2011). Is there a core neural network in empathy? An fMRI based quantitative meta-analysis. *Neuroscience and Biobehavioral Review, 35*(3), 903–911.

Farah, M. J. (2005). Neuroethics: The practical and the philosophical. *Trends in Cognitive Sciences, 9*(1), 34–40.

Farah, M. J., Rabinowitz, C., Quinn, G. E., & Liu, G. T. (2000). Early commitment of neural substrates for face recognition. *Cognitive Neuropsychology, 17*(1–3), 117–123.

Farrer, C., & Frith, C. D. (2002). Experiencing oneself vs another person as being the cause of an action: The neural correlates of the experience of agency. *NeuroImage, 15*, 596–603.

Farroni, T., Csibra, G., Simion, G., & Johnson, M. H. (2002). Eye contact detection in humans from birth. *Proceedings of the National Academy of Science, USA, 99*, 9602–9605.

Farroni, T., Mansfield, E. M., Lai, C., & Johnson, M. H. (2003). Infants perceiving and acting on the eyes: Tests of an evolutionary hypothesis. *Journal of Experimental Child Psychology, 85*(3), 199–212.

Farroni, T., Massaccesi, S., Menon, E., & Johnson, M. H. (2007). Direct gaze modulates face recognition in young infants. *Cognition, 102*(3), 396–404.

Fazio, R. H., Jackson, J. R., Dunton, B. C., & Williams, C. J. (1995). Variability in automatic activation as an unobtrusive measure of racial-attitudes – a bona-fide pipeline. *Journal of Personality and Social Psychology, 69*(6), 1013–1027.

Fedorenko, E., Duncan, J., & Kanwisher, N. (2013). Broad domain generality in focal regions of frontal and parietal cortex. *Proceedings of the National Academy of Sciences of the United States of America, 110*(41), 16616–16621.

Fehr, E., & Fischbacher, U. (2004). Social norms and human cooperation. *Trends in Cognitive Sciences, 8*(4), 185–190.

Fehr, E., & Gachter, S. (2000). Cooperation and punishment in public goods experiments. *American Economic Review, 90*(4), 980–994.

Fehr, E., & Gachter, S. (2002). Altruistic punishment in humans. *Nature, 415*, 137–140.

Feldman Barrett, L. (2006). Are emotions natural kinds?. *Perspectives on Psychological Science, 1*, 28–58.

Fellows, L.K., & Farah, M.J. (2003). Ventromedial frontal cortex mediates affective shifting in humans: Evidence from a reversal learning paradigm. *Brain, 126*, 1830–1837.

Ferrari, P.F., Vanderwert, R.E., Paukner, A., Bower, S., Suomi, S.J., & Fox, N.A. (2012). Distinct EEG amplitude suppression to facial gestures as evidence for a mirror mechanism in newborn monkeys. *Journal of Cognitive Neuroscience, 24*(5), 1165–1172.

Ferrari, P. F., Visalberghi, E., Paukner, A., Fogassi, L., Ruggiero, A., & Suomi, S. J. (2006). Neonatal imitation in rhesus macaques. *PLoS Biology, 4*, 1501–1508.

Ferstl, E. C., & von Cramon, D. Y. (2002). What does the frontomedian cortex contribute to language processing: Coherence or theory of mind? *NeuroImage, 17*, 1599–1612.

Festinger, L., Schacter, S., & Back, K. (1950). *Social Pressures in Informal Groups: A Study of Human Factors in Housing*. New York: Harper.

Feys, J. (1991). Briefly induced belongingness to self and preference. *European Journal of Social Psychology, 21*(6), 547–552.

Fiedler, K., Messner, C., & Bluemke, M. (2006). Unresolved problems with the "I", the "A" and the "T": A logical and psychometric critique of the Implicit Association Test (IAT). *European Review of Social Psychology, 17*, 74–147.

Fine, C., Lumsden, J., & Blair, R. J. R. (2001). Dissociation between 'theory of mind' and executive functions in a patient with early left amygdala damage. *Brain, 124*, 287–298.

Finger, E. C., Marsh, A. A., Kamel, N., Mitchell, D. G. V., & Blair, J. R. (2006). Caught in the act: The impact of audience on the neural response to morally and socially inappropriate behavior. *Neuroimage, 33*(1), 414–421.

Fischbacher, U., Gachter, S., & Fehr, E. (2001). Are people conditionally cooperative? Evidence from a public goods experiment. *Economics Letters, 71*(3), 397–404.

Fiske, S. T., Cuddy, A. J. C., Glick, P., & Xu, J. (2002). A model of (often mixed) stereotype content: Competence and warmth respectively follow from perceived status and competition. *Journal of Personality and Social Psychology, 82*(6), 878–902.

Fitch, W. T., Hauser, M. D., & Chomsky, N. (2005). The evolution of the language faculty: Clarifications and implications. *Cognition, 97*, 179–210.

Fitch, W. T., & Reby, D. (2001). The descended larynx is not uniquely human. *Proceedings of the Royal Society of London B, 268*, 1669–1675.

Fletcher, P. C., Happe, F., Frith, U., Baker, S. C., Dolan, R. J., Frackowiak, R. S. J., & Frith, C. D. (1995). Other minds in the brain: A functional imaging study of 'theory of mind' in story comprehension. *Cognition, 57*, 109–128.

Flor, H., Birbaumer, N., Hermann, C., Ziegler, S., & Patrick, C. J. (2002). Aversive Pavlovian conditioning in psychopaths: Peripheral and central correlates. *Psychophysiology, 39*(4), 505–518.

Fodor, J. A. (1983). *The Modularity of Mind*. Cambridge, MA: MIT Press.

Fodor, J. A. (1992). A theory of the child's theory of mind. *Cognition, 44*, 283–296.

Fogassi, L., Ferrari, P. F., Gesierich, B., Rozzi, S., Chersi, F., & Rizzolatti, G. (2005). Parietal lobe: From action organization to intention understanding. *Science, 308*(5722), 662–667.

Forster, P. (2004). Ice ages and the mitochondrial DNA chronology of human dispersals: A review. *Philosophical Transactions of the Royal Society B, 359*, 255–264.

Fowler, J. H., & Christakis, N. A. (2010). Cooperative behavior cascades in human social networks. *Proceedings of the National Academy of Sciences of the United States of America, 107*(12), 5334–5338.

Fox, C. J., Moon, S. Y., Iaria, G., & Barton, J. S. J. (2009). The correlates of subjective perception of identity and expression in the face network: An fMRI adaptation study. *NeuroImage, 44*, 569–580.

Fraedrich, E. M., Lakatos, K., & Spangler, G. (2010). Brain activity during emotion perception: The role of attachment representation. *Attachment and Human Development, 12*, 231–248.

Fraley, R. C., Waller, N. G., & Brennan, K. A. (2000). An item-response theory analysis of self-report measures of adult attachment. *Journal of Personality and Social Psychology, 78*, 350–365.

Francis, D., Diorio, J., Liu, D., & Meaney, M. J. (1999). Nongenomic transmission across generations of maternal behavior and stress responses in the rat. *Science, 286*(5442), 1155–1158.

Fraser, O. N., & Bugnyar, T. (2010). Do ravens show consolation? Responses to distressed others. *PLoS One, 5*(5).

Fraser, O. N., & Bugnyar, T. (2011). Ravens reconcile after aggressive conflicts with valuable partners. *PLoS One, 6*(3).

Fredrikson, M., Hursti, T., Salmi, P., Bojeson, S., Furst, C., Peterson, C., & Steineck, G. (1993). Conditioned nausea after cancer chemotherapy and autonomic system conditionability. *Scandanavian Journal of Psychology, 34*, 318–327.

Frick, P. J., & White, S. F. (2008). Research review: The importance of callous-unemotional traits for developmental models of aggressive and antisocial behavior. *Journal of Child Psychology and Psychiatry, 49*(4), 359–375.

Freud, S. (1922). *Group psychology and the analysis of the ego*: International Psychoanalytic Press.

Fridlund, A. J., & Cacioppo, J. T. (1986). Guidelines for human electromyographic research. *Psychophysiology, 23*, 567–589.

Frischen, A., Bayliss, A. P., & Tipper, S. P. (2007). Gaze cueing of attention: Visual attention, social cognition, and individual differences. *Psychological Bulletin, 133*(4), 694–724.

Frischen, A., Eastwood, J. D., & Smilek, D. (2008). Visual search for faces with emotional expressions. *Psychological Bulletin, 134*(5), 662–676.

Friston, K. (2002). Beyond phrenology: What can neuroimaging tell us about distributed circuitry? *Annual Review of Neuroscience, 25*, 221–250.

Frith, C. D. (1992). *The Cognitive Neuropsychology of Schizophrenia*. Hove: Psychology Press.

Frith, C. D. (2007). The social brain? In N. Emery, N. Clayton & C. Frith (Eds.), *Social Intelligence: From Brain to Culture*. Oxford: Oxford University Press.

Frith, C. D., Blakemore, S. J., & Wolpert, D. M. (2000). Explaining the symptoms of schizophrenia: Abnormalities in the awareness of action. *Brain Research Reviews, 31*(2–3), 357–363.

Frith, C. D., & Done, D. J. (1989). Experiences of alien control in schizophrenia reflect a disorder in the central monitoring of action. *Pschological Medicine, 19*, 521–530.

Frith, C. D., & Frith, U. (1999). Interacting minds – a biological basis. *Science, 286*, 1692–1695.

Frith, C. D., & Frith, U. (2008). Implicit and explicit processes in social cognition. *Neuron, 60*(3), 503–510.

Frith, U. (1989). *Autism: Explaining the Enigma*. Oxford: Blackwell.

Frith, U. (2012). The 38th Sir Frederick Bartlett Lecture Why we need cognitive explanations of autism. *Quarterly Journal of Experimental Psychology, 65*(11), 2073–2092.

Frith, U., & Frith, C. D. (2003). Development and neurophysiology of mentalising. *Philosophical Transactions of the Royal Society of London B, 358*, 459–472.

Fritsch, G. T., & Hitzig, E. (1870). On the electrical excitability of the cerebrum. In G. V. Bonin (Ed.), *Some Papers on the Cerebral Cortex*. Springfield, IL: Charles C. Thomas.

Fultz, J., Batson, C. D., Fortenbach, V. A., McCarthy, P. M., & Varney, L. L. (1986). Social evaluation and the empathy altruism hypothesis. *Journal of Personality and Social Psychology, 50*(4), 761–769.

Fusar-Poli, P., & Broome, M. R. (2007). Love and brain: From mereological fallacy to "folk" neuroimaging. *Psychiatry Research, 154*, 285–286.

Fuster, J. M. (1989). *The Prefrontal Cortex: Anatomy, Physiology, and Neuropsychology of the Frontal Lobe (Second Edition)*. New York: Raven Press.

Gaffan, D. (1992). Amygdala and the memory of reward. In J. P. Aggleton (Ed.), *The Amygdala: Neurobiological Aspects of Emotion, Memory and Mental Dysfunction*. New York: Wiley-Liss.

Gaillard, W. D., Grandon, C. B., & Xu, B. (2001). Developmental aspects of pediatric fMRI: Considerations for image acquisition, analysis and interpretation. *NeuroImage, 13*, 239–249.

Galinsky, A.D., Magee, J.C., Inesi, M.E., & Gruenfeld, D.H. (2006). Power and perspectives not taken. *Psychological Science, 17*, 1068–1074.

Gallagher, H. L., Happe, F., Brunswick, N., Fletcher, P. C., Frith, U., & Frith, C. D. (2000). Reading the mind in cartoons and stories: An fMRI study of 'theory of mind' in verbal and nonverbal tasks. *Neuropsychologia, 38*, 11–21.

Gallagher, S. (2000). Philosophical conceptions of the self: Implications for cognitive science. *Trends in Cognitive Sciences, 4*(1), 14–21.

Gallagher, S. (2007). Simulation trouble. *Social Neuroscience, 2*(3–4), 353–365.

Gallardo-Pujol, D., Andres-Pueyo, A., & Maydeu-Olivares, A. (2013). MAOA genotype, social exclusion and aggression: An experimental test of a gene-environment interaction. *Genes Brain and Behavior, 12*(1), 140–145.

Gallegos, D. R., & Tranel, D. (2005). Positive facial affect facilitates the identification of famous faces. *Brain and Language, 93*, 338–348.

Gallese, V. (2001). The 'shared manifold' hypothesis: From mirror neurons to empathy. *Journal of Consciousness Studies, 8*, 33–50.

Gallese, V. (2003). The manifold nature of interpersonal relations: The quest for a common mechanism. *Philosophical Transactions of the Royal Society of London B, 358*, 517–528.

Gallese, V., & Goldman, A. (1998). Mirror neurons and the simulation theory of mind-reading. *Trends in Cognitive Sciences, 2*, 493–501.

Gallup, G. G. J. (1970). Chimpanzees: Self-recognition. *Science, 167*, 86–87.

Ganel, T., Valyear, K. F., Goshen-Gottstein, Y., & Goodale, M. A. (2005). The involvement of the "fusiform face area" in processing facial expression. *Neuropsychologia, 43*, 1645–1654.

Gardner, M., & Steinberg, L. (2005). Peer influence on risk taking, risk preference, and risky decision making in adolescence and adulthood: An experimental study. *Developmental Psychology, 41*(4), 625–635.

Garnica, O. (1977). Some prosodic and paralinguistic features of speech to young children. In C. E. Snow & C. A. Ferguson (Eds.), *Talking to Children: Language Input and Acquisition*. Cambridge: Cambridge University Press.

Gauthier, I., Skudlarski, P., Gore, J. C., & Anderson, A. W. (2000). Expertise for cars and birds recruits brain areas involved in face recognition. *Nature Neuroscience, 3*, 191–197.

Gauthier, I., & Tarr, M. J. (1997). Becoming a 'Greeble' expert: Exploring mechanisms for face recognition. *Vision Research, 37*, 1673–1682.

Gauthier, I., Tarr, M. J., Anderson, A. W., Skudlarski, P., & Gore, J. C. (1999). Activation of middle fusiform 'face area' increases with expertise in recognizing novel objects. *Nature Neuroscience, 2*, 568–573.

Gazzola, V., Aziz-Zadeh, L., & Keysers, C. (2006). Empathy and the somatotopic auditory mirror system in humans. *Current Biology, 16*, 1824–1829.

Gehring, W. J., Goss, B., Coles, M. G. H., Meyer, D. E., & Donchin, E. (1993). A neural system for error detection and compensation. *Psychological Science, 4*, 385–390.

Georgme M. S., & Aston-Jones, G. (2009). Noninvasive techniques for probing neurocircuitry and treating illness: Vagus nerve stimulation (VNS), transcranial magnetic stimulation (TMS) and transcranial direct current stimulation (tDCS). *Neuropsychopharmacology, 35*, 301–316.

Gergely, G., Bekkering, H., & Kiraly, I. (2002). Rational imitation in preverbal infants. *Nature, 415*, 755–755.

Gergely, G., & Csibra, G. (2003). Teleological reasoning in infancy: The naive theory of rational action. *Trends in Cognitive Sciences, 7*, 287–292.

Geschwind, D.H., & Levitt, P. (2007). Autism spectrum disorders: Developmental disconnection syndromes. *Current Opinion in Neurobiology, 17*(1), 103–111.

Gianaros, P. J., Horenstein, J. A., Cohen, S., Matthews, K. A., Brown, S. M., Flory, J. D., Critchley, H. D., Manuck, S. B., & Hariri, A. R. (2007). Perigenual anterior cingulate morphology covaries with perceived social standing. *Social Cognitive and Affective Neuroscience, 2*(3), 161–173.

Gibbons, A., S. V. I. P.-P.M. (2004). Tracking the evolutionary history of a "warrior" gene. *Science, 304*(5672), 819–819.

Giedd, J. N., Blumenthal, J., Jeffries, N. O., Castellanos, F. X., Liu, H., Zijdenbos, A., Paus, T., Evans, A. C., & Rapoport, J. L. (1999). Brain development during childhood and adolescence: A longitudinal MRI study. *Nature Neuroscience, 2*, 861–863.

Gilbert, S. J., Bird, G., Brindley, R., Frith, C. D., & Burgess, P. W. (2008). Atypical recruitment of medial prefrontal cortex in autism spectrum disorders: An fMRI study of two executive function tasks. *Neuropsychologia, 46*(9), 2281–2291.

Gilbert, S. J., Swencionis, J. K., & Amodio, D. M. (2012). Evaluative vs. trait representation in intergroup social judgments: Distinct roles of anterior temporal lobe and prefrontal cortex. *Neuropsychologia, 50*(14), 3600–3611.

Gillath, O., Bunge, S. A., Shaver, P. R., Wendelken, C., & Mikulincer, M. (2005). Attachment-style differences in the ability to suppress negative thoughts: Exploring the neural correlates. *NeuroImage, 28*, 835–847.

Gillihan, S. J., & Farah, M. J. (2005). Is self special? A critical review of evidence from experimental psychology and cognitive neuroscience. *Psychological Bulletin, 131*(1), 76–97.

Gleichgerrcht, E., Torralva, T., Rattazzi, A., Marenco, V., Roca, M., & Manes, F. (2013). Selective impairment of cognitive empathy for moral judgment in adults with high functioning autism. *Social Cognitive and Affective Neuroscience, 8*(7), 780–788.

Glocker, M. L., Langleben, D. D., Ruparel, K., Loughead, J. W., Valdez, J. N., Griffin, M. D., . . . Gur, R. C. (2009). Baby schema modulates the brain reward system in nulliparous women. *Proceedings of the National Academy of Sciences of the United States of America, 106*(22), 9115–9119.

Gobrogge, K. L., Liu, Y., Jia, X. X., & Wang, Z. X. (2007). Anterior hypothalamic neural activation and neurochemical associations with aggression in pair-bonded male prairie voles. *Journal of Comparative Neurology, 502*(6), 1109–1122.

Goddings, A. L., Heyes, S. B., Bird, G., Viner, R. M., & Blakemore, S. J. (2012). The relationship between puberty and social emotion processing. *Developmental Science, 15*(6), 801–811

Gogtay, N., Giedd, J. N., Lusk, L., Hayashi, K. M., Greenstein, D., Vaituzis, A. C., Nugent, T. F., Herman, D. H., Clasen, L. S., Toga, A. W., Rapoport, J. L., & Thompson, P. M. (2004). Dynamic mapping of human cortical development during childhood through early adulthood. *Proceedings of the National Academy of Sciences of the United States of America, 101*(21), 8174–8179.

Golarai, G., Ghahremani, D. G., Whitfield-Gabrieli, S., Reiss, A., Eberhardt, J. L., Gabrieli, J. D. E., & Grill-Spector, K. (2007). Differential development of high-level visual cortex correlates with category-specific recognition memory. *Nature Neuroscience, 10*(4), 512–522.

Golby, A. J., Gabrieli, J. D. E., Chiao, J. Y., & Eberhardt, J. L. (2001). Differential responses in the fusiform region to same-race and other-race faces. *Nature Neuroscience, 4*(8), 845–850.

Goldberg, E. (2001). *The Executive Brain: Frontal Lobes and the Civilised Mind.* Oxford: Oxford University Press.

Goldman, A. (2006). *Simulating Minds: The Philosophy, Psychology and Neuroscience of Mindreading.* Oxford: Oxford University Press.

Gonzalez, A., Atkinson, L., & Fleming, A. S. (2009). Attachment and comparative psychobiology of mothering. In M. D. Haan & M. R. Gunnar (Eds.), *Handbook of Developmental Social Neuroscience.* New York: The Guilford Press.

Goodenough, O. R. (2004). Responsibility and punishment: whose mind? A response. *Philosophical Transactions of the Royal Society of London Series B-Biological Sciences, 359*(1451), 1805–1809.

Gopnik, M., & Wellman, H. (1992). Why the child's theory of mind really is a theory. *Mind and Language, 7*, 145–171.

Gorenstein, E. E. (1982). Frontal-lobe functions in psychopaths. *Journal of Abnormal Psychology, 91*(5), 368–379.

Gosselin, N., Peretz, I., Johnsen, E., & Adolphs, R. (2007). Amygdala damage impairs emotion recognition from music. *Neuropsychologia, 45*, 236–244.

Gotlib, I. H., & Joormann, J. (2010). Cognition and depression: current status and future directions. In S. Nolen-Hoeksema, T. D. Cannon & T. Widiger (Eds.), *Annual Review of Clinical Psychology, Vol 6* (Vol. 6, pp. 285–312). Palo Alto: Annual Reviews.

Gould, S. J. (1991). Exaptation: A crucial tool for evolutionary psychology. *Journal of Social Issues, 47*, 43–65.

Gozzi, M., Raymont, V., Solomon, J., Koenigs, M., & Grafman, J. (2009). Dissociable effects of prefrontal and anterior temporal cortical lesions on stereotypical gender attitudes. *Neuropsychologia, 47*(10), 2125–2132. 002

Grafman, J., Schwab, K., Warden, D., Pridgen, A., Brown, H. R., & Salazar, A. M. (1996). Frontal lobe injuries, violence, and aggression: A report of the Vietnam head injury study. *Neurology, 46*, 1231–1238.

Graham, J., Haidt, J., & Nosek, B. A. (2009). Liberals and conservatives rely on different sets of moral foundations. *Journal of Personality and Social Psychology, 96*(5), 1029–1046.

Graham, M. D., Rees, S. L., Steiner, M., & Fleming, A. S. (2006). The effects of adrenalectomy and corticosterone replacement on maternal memory in postpartum rats. *Hormones and Behavior, 49*(3), 353–361.

Grandin, T. (1995). *Thinking in Pictures: And Other Reports from My Life with Autism.* New York: Double Day.

Granqvist, P., Fredrikson, M., Unge, P., Hagenfeldt, A., Valind, S., Larhammar, D., & Larsson, M. (2005). Sensed presence and mystical experiences are predicted by suggestibility, not by the application of transcranial weak complex magnetic fields. *Neuroscience Letters, 379*(1), 1–6.

Gray, J. A., & McNaughton, N. (2000). *The Neuropsychology of Anxiety: An Enquiry into the Functions of the Septo-Hippocampal System (2nd ed.).* New York, NY: Oxford University Press.

Graziano, M. S. A. (1999). Where is my arm? The relative role of vision and proprioception in the neuronal representation of limb position. *Proceedings of the National Academy of Sciences of the United States of America, 96*, 10418–10421.

Graziano, M. S. A., Cooke, D. F., & Taylor, C. S. R. (2000). Coding the location of the arm by sight. *Science, 290*, 1782–1786.

Grecucci, A., Giorgetta, C., van't Wout, M., Bonini, N., & Sanfey, A. G. (2013). Reappraising the Ultimatum: An fMRI study of emotion regulation and decision making. *Cerebral Cortex, 23*(2), 399–410.

Greene, J. D. (2008). *The Secret Joke of Kant's Soul.* Cambridge, MA: MIT Press.

Greene, J. D., Morelli, S. A., Lowenberg, K., Nvstrom, L. E., & Cohen, J. D. (2008). Cognitive load selectively interferes with utilitarian moral judgment. *Cognition, 107*(3), 1144–1154.

Greene, J. D., Nystrom, L. E., Engell, A. D., Darley, J. M., & Cohen, J. D. (2004). The neural bases of cognitive conflict and control in moral judgment. *Neuron, 44*(2), 389–400.

Greene, J. D., Sommerville, R. B., Nystrom, L. E., Darley, J. M., & Cohen, J. D. (2001). An fMRI investigation of emotional engagement in moral judgment. *Science, 293*(5537), 2105–2108.

Greenwald, A. G., Banaji, M. R., Rudman, L. A., Farnham, S. D., Nosek, B. A., & Mellott, D. S. (2002). A unified theory of implicit attitudes, stereotypes, self-esteem, and self-concept. *Psychological Review, 109*(1), 3–25.

Greenwald, A. G., McGhee, D. E., & Schwartz, J. L. K. (1998). Measuring individual differences in implicit cognition: The implicit association test. *Journal of Personality and Social Psychology, 74*(6), 1464–1480.

Greimel, E., Schulte-Ruther, M., Fink, G. R., Piefke, M., Herpertz-Dahlmann, B., & Konrad, K. (2010). Development of neural correlates of empathy from childhood to early adulthood: An fMRI study in boys and adult men. *Journal of Neural Transmission, 117*(6), 781–791.

Gross, C. G., Rocha-Miranda, C. E., & Bender, D. B. (1972). Visual properties of neurons in the inferotemporal cortex of the macaque. *Journal of Neurophysiology, 35*, 96–111.

Grossman, E., Donnelly, M., Price, R., Pickens, D., Morgan, V., Neighbor, G., & Blake, R. (2000). Brain areas involved in perception of biological motion. *Journal of Cognitive Neuroscience, 12*(5), 711–720.

Grossman, K. E. (1988). Longitudinal and systematic approaches to the study of biological high- and low-risk groups. In M. Rutter (Ed.), *Studies of Psychosocial Risk: The Power of Longitudinal Data.* Cambridge: Cambridge University Press.

Grossmann, T., Oberecker, R., Koch, S. P., & Friederici, A. D. (2010). The developmental origins of voice processing in the human brain. *Neuron, 65*(6), 852–858.

Grossmann, T., Striano, T., & Friederici, A. D. (2005). Infants' electric brain responses to emotional prosody. *Neuroreport, 16*(16), 1825–1828.

Gruenfeld, D.H., Inesi, M.E., Magee, J.C., & Galinsky, A. D. (2008). Power and the objectification of social targets. *Journal of Personality and Social Psychology, 95*(1), 111–127.

Gunnar, M. R., Morison, S. J., Chisholm, K., & Schuder, M. (2001). Salivary cortisol levels in children adopted from Romanian orphanages. *Development and Psychopathology, 13*(3), 611–628.

Gupta, U., & Singh, P. (1982). An exploratory study of love and liking and types of marriage. *Indian Journal of Applied Psychology, 19*, 92–97.

Guroglu, B., van den Bos, W., & Crone, E. A. (2009). Fairness considerations: Increasing understanding of intentionality during adolescence. *Journal of Experimental Child Psychology, 104*(4), 398–409.

Gutchess, A. H., Welsh, R. C., Boduroglu, A., & Park, D. C. (2006). Cultural differences in neural function associated with object processing. *Cognitive Affective & Behavioral Neuroscience, 6*(2), 102–109.

Guth, W., Schmittberger, R., & Schwarze, B. (1982). An experimental analysis of ultimatum bargaining. *Journal of Economics, Behavior, & Organizations, 3*, 367–388.

Guthrie, S. (1993). *Faces in the Clouds: A New Theory of Religion.* New York: Oxford University Press.

Gweon, H., Dodell-Feder, D., Bedny, M., & Saxe, R. (2012). Theory of mind performance in children correlates with functional specialization of a brain region for thinking about thoughts. *Child Development, 83*(6), 1853–1868.

Hadjikhani, N., & de Gelder, B. (2003). Seeing fearful body expressions activates the fusiform cortex and amygdala. *Current Biology, 13*(24), 2201–2205.

Hadjikhani, N., Joseph, R. M., Snyder, J., & Tager-Flusberg, H. (2006). Anatomical differences in the mirror neuron system and social cognition network in autism. *Cerebral Cortex, 16*(9), 1276–1282.

Hadjikhani, N., Zurcher, N.R., Rogier, O., Hippolyte, L., Lemonnier, E., Ruest, T., . . . Gillberg, C. (2014). Emotional contagion for pain is intact in autism spectrum disorders. *Translational Psychiatry, 5*, e343.

Haidt, J. (2001). The emotional dog and its rational tail: A social intuitionist approach to moral judgment. *Psychological Review, 108*(4), 814–834.

Haggard, P. (2008). Human volition: Towards a neuroscience of will. *Nature Reviews Neuroscience, 9*, 934–946.

Haidt, J. (2003). The moral emotions. In R. J. Davidson, K. R. Scherer & H. H. Goldsmith (Eds.), *Handbook of Affective Sciences.* Oxford: Oxford University Press.

Haidt, J. (2007). The new synthesis in moral psychology. *Science, 316*(5827), 998–1002.

Haidt, J. (2012). *The Righteous Mind: Why Good People are Divided by Politics and Religion.* New York: Pantheon Books.

Hajcak, G., MacNamara, A., & Olvet, D.M. (2010). Event-related potentials, emotion, and emotion regulation: An integrative review. *Developmental Neuropsychology, 35*(2), 129–155.

Hallmayer, J., Cleveland, S., Torres, A., Phillips, J., Cohen, B., Torigoe, T., . . . Risch, N. (2011). Genetic heritability and shared environmental factors among twin pairs with autism. *Archives of General Psychiatry, 68*(11), 1095–1102.

Hamann, S., & Canli, T. (2004). Individual differences in emotion processing. *Current Opinion in Neurobiology, 14*(2), 233–238.

Hamilton, A. F. D., Brindley, R. M., & Frith, U. (2007). Imitation and action understanding in autistic spectrum disorders: How valid is the hypothesis of a deficit in the mirror neuron system? *Neuropsychologia, 45*(8), 1859–1868.

Hamilton, W. D. (1964). The genetical evolution of social behaviour. Parts I, II. *Journal of Theoretical Biology, 7,* 1–52.

Han, S., Northoff, G., Vogeley, K., Wexler, B. E., Kitayama, S., & Varnum, M. E. W. (2013). A cultural neuroscience approach to the biosocial nature of the human brain. *Annual Review of Psychology, 64*, 335–359.

Han, S. H., & Northoff, G. (2008). Culture-sensitive neural substrates of human cognition: A transcultural neuroimaging approach. *Nature Reviews Neuroscience, 9*(8), 646–654.

Happe, F. (1999). Autism: Cognitive deficit or cognitive style? *Trends in Cognitive Sciences, 3*, 216–222.

Happe, F., Ehlers, S., Fletcher, P., Frith, U., Johansson, M., Gillberg, C., Dolan, R., Frackowiak, R., & Frith, C. (1996). 'Theory of mind' in the brain. Evidence from a PET scan study of Asperger syndrome. *Neuroreport, 8*(1), 197–201.

Happe, F. G. (1995). The role of age and verbal ability in the theory of mind task performance of subjects with autism. *Child Development, 66*, 843–855.

Happe, F. G. E. (1995). Understanding minds and metaphors: Insights from the study of figurative language in autism. *Metaphor and Symbolic Activity, 10*, 275–295.

Harbaugh, W. T., Mayr, U., & Burghart, D. R. (2007). Neural responses to taxation and voluntary giving reveal motives for charitable donations. *Science, 316*(5831), 1622–1625.

Hare, B., Brown, M., Williamson, C., & Tomasello, M. (2002). The domestication of social cognition in dogs. *Science, 298*(5598), 1634–1636.

Hare, R. D. (1980). A research scale for the assessment of psychopathy in criminal populations. *Personality and Individual Differences, 1*, 111–119.

Hare, T. A., O'Doherty, J., Camerer, C. F., Schultz, W., & Rangel, A. (2008). Dissociating the role of the orbitofrontal cortex and the striatum in the computation of goal values and prediction errors. *Journal of Neuroscience, 28*(22), 5623–5630.

Harden, K. P., & Tucker-Drob, E. M. (2011). Individual differences in the development of sensation seeking and impulsivity during adolescence: Further evidence for a dual systems model. *Developmental Psychology, 47*(3), 739–746.

Hari, R., Parkkonen, L., & Nangini, C. (2010). The brain in time: Insights from neuromagnetic recordings. In A. Kingstone & M. B. Miller (Eds.), *Year in Cognitive Neuroscience 2010* (Vol. 1191, pp. 89–109). Malden: Wiley-Blackwell.

Hariri, A. R., Bookheimer, S. Y., & Mazziotta, J. C. (2000). Modulating emotional responses: Effects of a neocortical network on the limbic system. *Neuroreport, 11*(1), 43–48.

Hariri, A. R., Mattay, V. S., Tessitore, A., Kolachana, B., Fera, F., Goldman, D., . . . Weinberger, D. R. (2002). Serotonin transporter genetic variation and the response of the human amygdala. *Science, 297*(5580), 400–403.

Harlow, H. F. (1958). The nature of love. *American Psychologist, 13*, 673–685.

Harlow, J. M. (1993). Recovery from the passage of an iron bar through the head. *History of Psychiatry, 4*, 271–281 (reprint of original published in 1848 in Publications of the Massachusetts Medical Society).

Harris, L. T., & Fiske, S. T. (2006). Dehumanizing the lowest of the low – Neuroimaging responses to extreme outgroups. *Psychological Science, 17*(10), 847–853.

Harris, L. T., & Fiske, S. T. (2007). Social groups that elicit disgust are differentially processed in mPFC. *Social Cognitive and Affective Neuroscience, 2*(1), 45–51.

Harris, L. T., & Fiske, S. T. (2011). Dehumanized perception a psychological means to facilitate atrocities, torture, and genocide? *Zeitschrift Fur Psychologie-Journal of Psychology, 219*(3), 175–181.

Harris, L. T., Lee, V. K., Capestany, B. H., & Cohen, A. O. (2014). Assigning economic value to people results in dehumanization brain response. *Journal of Neuroscience Psychology and Economics, 7*(3), 151–163.

Harris, S., Kaplan, J. T., Curiel, A., Bookheimer, S. Y., Iacoboni, M., & Cohen, M. S. (2009). The neural correlates of religious and nonreligious belief. *PLoS One, 4*(10).

Harrison, V., & Hole, G. J. (2009). Evidence for a contact-based explanation of the own-age bias in face recognition. *Psychonomic Bulletin & Review, 16*(2), 264–269.

Hart, A. J., Whalen, P. J., Shin, L. M., McInerney, S. C., Fischer, H., & Rauch, S. L. (2000). Differential response in the human amygdala to racial outgroup vs ingroup face stimuli. *Neuroreport, 11*(11), 2351–2355.

Hart, S. D., & Hare, R. D. (1996). Psychopathy and antisocial personality disorder. *Current Opinion in Psychiatry, 9*(2), 129–132.

Haselhuhn, M. P., & Wong, E. M. (2012). Bad to the bone: Facial structure predicts unethical behaviour. *Proceedings of the Royal Society B: Biological Sciences, 279*, 571–576.

Hassin, R., & Trope, Y. (2000). Facing faces: Studies on the cognitive aspects of physiognomy. *Journal of Personality and Social Psychology, 78*(5), 837–852.

Hatfield, T., Han, J. S., Conley, M., Gallagher, M., & Holland, P. (1996). Neurotoxic lesions of basolateral, but not central, amygdala interfere with Pavlovian second-order conditioning and reinforcer devaluation effects. *Journal of Neuroscience, 16*(16), 5256–5265.

Hauser, M. D. (2006). *Moral Minds.* London: Abacus.

Hauser, M. D. (2009). The possibility of impossible cultures. *Nature, 460*, 190–196.

Hauser, M. D., Chomsky, N., & Fitch, W. T. (2002). The faculty of language: What is it, who has it, and how did it evolve? *Science, 298*, 1569–1579.

Hawkley, L. C., Masi, C. M., Berry, J. D., & Cacioppo, J. T. (2006). Loneliness is a unique predictor of age-related differences in systolic blood pressure. *Psychology and Aging, 21*(1), 152–164.

Hawley, P. H. (1999). The ontogenesis of social dominance: A strategy-based evolutionary perspective. *Developmental Review, 19*(1), 97–132.

Hawley, P. H. (2002). Social dominance and prosocial and coercive strategies of resource control in preschoolers. *International Journal of Behavioral Development, 26*(2), 167–176.

Hawley, P. H. (2003). Prosocial and coercive configurations of resource control in early adolescence: A case for the well-adapted Machiavellian. *Merrill-Palmer Quarterly-Journal of Developmental Psychology, 49*(3), 279–309.

Hawley, P. H., Little, T. D., & Rodkin, P. C. (2007). *Aggression and Adaptation: The Bright Side to Bad Behavior*. Hillsdale, NJ: Lawrence Erlbaum Associates.

Haxby, J. V., Hoffman, E. A., & Gobbini, M. I. (2000). The distributed human neural system for face perception. *Trends in Cognitive Sciences, 4*(6), 223–233.

Hayes, J. P., Morey, R. A., Petty, C. M., Seth, S., Smoski, M. J., McCarthy, G., . . . LaBar, K. S. (2010). Staying cool when things get hot: Emotion regulation modulates neural mechanisms of memory encoding. *Frontiers in Human Neuroscience, 4*.

Haynes, J. D., & Rees, G. (2006). Decoding mental states from brain activity in humans. *Nature Reviews Neuroscience, 7*, 523–534.

Hazan, C., & Shaver, P. (1987). Romantic love conceptualized as an attachment process. *Journal of Personality and Social Psychology, 52*, 511–524.

Healy, S. D., & Rowe, C. (2007). A critique of comparative studies of brain size. *Proceedings of the Royal Society B-Biological Sciences, 274*(1609), 453–464.

Heatherton, T. F., & Wagner, D. D. (2011). Cognitive neuroscience of self-regulation failure. *Trends in Cognitive Sciences, 15*(3), 132–139.

Heberlein, A. S., & Adolphs, R. (2007). Neurobiology of emotion recognition: Current evidence for shared substrates. In E. Harmon-Jones & P. Winkielman (Eds.), *Social Neuroscience*. New York, NY: Guilford Press.

Heberlein, A. S., Padon, A. A., Gillihan, S. J., Farah, M. J., & Fellows, L. K. (2008). Ventromedial Frontal Lobe plays a critical role in facial emotion recognition. *Journal of Cognitive Neuroscience, 20*(4), 721–733.

Heider, F., & Simmel, M. (1944). An experimental study of apparent behavior. *American Journal of Psychology, 57*, 243–259.

Hein, G., Silani, G., Preuschoff, K., Batson, C. D., & Singer, T. (2010). Neural responses to ingroup and outgroup members' suffering predict individual differences in costly helping. *Neuron, 68*(1), 149–160.

Heinrichs, M., Baumgartner, T., Kirschbaum, C., & Ehlert, U. (2003). Social support and oxytocin interact to suppress cortisol and subjective responses to psychosocial stress. *Biological Psychiatry, 54*, 1389–1398.

Henderson, L. M., Yoder, P. J., Yale, M. E., & McDuffie, A. (2002). Getting to the point: Electrophysiological correlates of protodeclarative pointing. *International Journal of Developmental Neuroscience, 20*, 449–458.

Henrich, J., Boyd, R., Bowles, S., Camerer, C., Fehr, E., Gintis, H., McElreath, R., Alvard, M., Barr, A., Ensminger, J., Henrich, N. S., Hill, K., Gil-White, F., Gurven, M., Marlowe, F. W., Patton, J. Q., & Tracer, D. (2005). "Economic man" in cross-cultural perspective: Behavioral experiments in 15 small-scale societies. *Behavioral and Brain Sciences, 28*(6), 795-+.

Henrich, J., Ensminger, J., McElreath, R., Barr, A., Barrett, C., Bolyanatz, A., Cardenas, J. C., Gurven, M., Gwako, E., Henrich, N., Lesorogol, C., Marlowe, F. W., Tracer, D., & Ziker, J. (2010). Markets, religion, community size, and the evolution of fairness and punishment. *Science, 327*(5972), 1480–1484.

Henrich, J., McElreath, R., Barr, A., Ensminger, J., Barrett, C., Bolyanatz, A., Cardenas, J. C., Gurven, M., Gwako, E., Henrich, N., Lesorogol, C., Marlowe, F. W., Tracer, D., & Ziker, J. (2006). Costly punishment across human societies. *Science, 312*(5781), 1767–1770.

Herman, L. H. (2002). Vocal, social and self-imitation by bottlenosed dolphins. In K. Dautenhahn & C. L. Nehaniv (Eds.), *Imitation in Animals and Artifacts*. Cambridge, MA: MIT Press.

Hermans, E. J., Bos, P. A., Ossewaarde, L., Ramsey, N. F., Fernandez, G., & van Honk, J. (2010). Effects of exogenous testosterone on the ventral striatal BOLD response during reward anticipation in healthy women. *NeuroImage, 52*, 277–283.

Herrmann, E., Call, J., Hernandez-Lloreda, M. V., Hare, B., & Tomasello, M. (2007). Humans have evolved specialized skills of social cognition: The cultural intelligence hypothesis. *Science, 317*(5843), 1360–1366.

Hess, U. (2009). Facial EMG. In E. Harmon-Jones & J. S. Beer (Eds.), *Methods in Social Neuroscience*. New York: The Guilford Press.

Hess, U., & Blairy, S. (2001). Facial mimicry and emotional contagion to dynamic emotional facial expressions and their influence on decoding accuracy. *International Journal of Psychophysiology, 40*, 129–141.

Heyes, C. (2010). Where do mirror neurons come from? *Neuroscience and Biobehavioral Reviews, 34*(4), 575–583.

Heyes, C. M., & Galef, B. G. (1996). *Social Learning in Animals: The Roots of Culture*. San Diego: Academic Press.

Hickok, G. (2014). *The Myth of Mirror Neurons: The Real Science of Communication and Cognition*. New York: W. W. Norton and Company.

Hietanen, J. K. (1999). Does your gaze direction and head orientation shift my visual attention? *Neuroreport, 10*(16), 3443–3447.

Hihara, S., Notoya, T., Tanaka, M., Ichinose, S., Ojima, H., Obayashi, S., Fujii, N., & Iriki, A. (2006). Extension of corticocortical afferents into the anterior bank of the intraparietal sulcus by tool-use training in adult monkeys. *Neuropsychologia, 44*, 2636–2646.

Hihara, S., Obayashi, S., Tanaka, M., & Iriki, A. (2003). Rapid learning of sequential tool use by macaque monkeys. *Physiology and Behavior, 78*, 427–434.

Hill, E., Berthoz, S., & Frith, U. (2004). Cognitive processing of own emotions in individuals with autistic spectrum disorder and in their relatives. *Journal of Autism and Developmental Disorders, 34*(2), 229–235.

Hill, E. L., & Frith, U. (2003). Understanding Autism: Insights from mind and brain. *Philosophical Transactions of the Royal Society of London B, 358*, 281–289.

Hill, R. A., & Dunbar, R. I. M. (2003). Social network size in humans. *Human Nature: An Interdisciplinary Biological Perspective, 14*, 53–72.

Hoffman, E., & Haxby, J. (2000). Distinct representations of eye gaze and identity in the distributed human neural system for face perception. *Nature Neuroscience, 3*, 80–84.

Hofmann, W., Gawronski, B., Gschwendner, T., Le, H., & Schmitt, M. (2005). A meta-analysis on the correlation between the implicit association test and explicit self-report measures. *Personality and Social Psychology Bulletin, 31*(10), 1369–1385.

Hoge, E. A., Pollack, M. H., Kaufman, R. E., Zak, P. J., & Simon, N. M. (2008). Oxytocin levels in social anxiety disorder. *CNS Neuroscience & Therapeutics, 14*(3), 165–170.

Hoge, R. D., & Pike, G. B. (2001). Quantitative measurement using fMRI. In P. Jezzard, P. M. Matthews & S. M. Smith (Eds.), *Functional MRI*. Oxford: Oxford University Press.

Hogeveen, J., Inzlicht, M., & Obhi, S. S. (2014). Power changes how the brain responds to others. *Journal of Experimental Psychology. General, 143*(2), 755–762.

Hogg, M. A., & Vaughan, G. M. (2011). *Social Psychology (Sixth Edition)*. Harlow, UK: Pearson.

Holekamp, K. E., Sakai, S. T., & Lundrigan, B. L. (2007). Social intelligence in the spotted hyena (Crocuta crocuta). *Philosophical Transactions of the Royal Society B-Biological Sciences, 362*(1480), 523–538.

Holmes, N. P., Calvert, G. A., & Spence, C. (2007). Tool use changes multisensory interactions in seconds: Evidence from the crossmodal congruency task. *Experimental Brain Research, 183*, 465–476.

Holmes, W. G., & Sherman, P. W. (1982). The ontogeny of kin recognition in 2 species of ground-squirrels. *American Zoologist, 22*(3), 491–517.

Hooker, C. I., Paller, K. A., Gitelman, D. R., Parrish, T. B., Mesulam, M. M., & Reber, P. J. (2003). Brain networks for analyzing eye gaze. *Cognitive Brain Research, 17*(2), 406–418.

Hoorens, V., & Nuttin, J. M. (1993). Overvaluation of own attributes – mere ownership or subjective frequency. *Social Cognition, 11*(2), 177–200.

Hopkins, W. D., Keebaugh, A. C., Reamer, L. A., Schaeffer, J., Schapiro, S. J., & Young, L. J. (2014). Genetic influences on receptive joint attention in chimpanzees (Pan troglodytes). *Scientific Reports, 4*.

Hornak, J., Bramham, J., Rolls, E. T., Morris, R. G., O'Doherty, J., Bullock, P. R., & Polkey, C. E. (2003). Changes in emotion after circumscribed surgical lesions of the orbitofrontal and cingulate cortices. *Brain, 126*, 1691–1712.

Hornak, J., Rolls, E. T., & Wade, D. (1996). Face and voice expression identification inpatients with emotional and behavioural changes following ventral frontal lobe damage. *Neuropsychologia, 34*(4), 247–261.

Horner, V., & Whiten, A. (2005). Causal knowledge and imitation/emulation switching in chimpanzees (Pan troglodytes) and children. *Animal Behavior, 64*, 851–859.

Hosobuchi, Y., Adams, J. E., & Linchitz, R. (1977). Pain relief by electrical-stimulation of central gray-matter in humans and its reversal by naloxone. *Science, 197*, 183–186.

Huber, D., Veinante, P., & Stoop, R. (2005). Vasopressin and oxytocin excite distinct neuronal populations in the central amygdala. *Science, 308*(5719), 245–248.

Hughes, C., Russell, J., & Robbins, T. W. (1994). Evidence for executive dysfunction in autism. *Psychological Medicine, 27*, 209–220.

Humphrey, N. K. (1976). The social function of intellect. In P. Bateson & R. A. Hinde (Eds.), *Growing Points in Ethology*. Cambridge: Cambridge University Press.

Hunt, G. R., & Gray, R. D. (2003). Diversification and cumulative evolution in New Caledonian crow tool manufacture. *Proceedings of the National Academy of Science, USA, 270*, 867–874.

Hunt, P. S., Holloway, J. L., & Scordalakes, E. M. (2001). Social interaction with an intoxicated sibling can result in increased intake of ethanol by periadolescent rats. *Developmental Psychobiology, 38*, 101–109.

Hurley, S., Clark, A., & Kiverstein, J. (2008). The shared circuits model (SCM): How control, mirroring, and simulation can enable imitation, deliberation, and mindreading. *Behavioral and Brain Sciences, 31*(1), 1–+.

Huttenlocher, P. R., & Dabholkar, A. S. (1997). Regional differences in synaptogenesis in human cerebral cortex. *Journal of Comparative Neurology, 387*, 167–178.

Iacoboni, M. (2009). Imitation, empathy, and mirror neurons. *Annual Review of Psychology, 60*, 653–670.

Iacoboni, M., & Dapretto, M. (2006). The mirror neuron system and the consequences of its dysfunction. *Nature Reviews Neuroscience, 7*(12), 942–951.

Iacoboni, M., Molnar-Szakacs, I., Gallese, V., Buccino, G., Mazziotta, J. C., & Rizzolatti, G. (2005). Grasping the intentions of others with one's own mirror neuron system. *PLoS Biology, 3*, 529–535

Iacoboni, M., Woods, R., Brass, M., Bekkering, H., Mazziotta, J. C., & Rizzolatti, G. (1999). Cortical mechanisms of human imitation. *Science, 286*, 2526–2528.

Ickes, W. (1993). Empathic accuracy. *Journal of Personality, 61*(4), 587–610.

Ickes, W., Gesn, P. R., & Graham, T. (2000). Gender differences in empathic accuracy: Differential ability or differential motivation? *Personal Relationships, 7*(1), 95–110.

Inagaki, T. K., & Eisenberger, N. I. (2013). Shared neural mechanisms underlying social warmth and physical warmth. *Psychological Science, 24*(11), 2272–2280.

Insel, T. R., & Harbaugh, C. R. (1989). Lesions of the hypothalamic paraventricular nucleus disrupt the initiation of maternal-behavior. *Physiology & Behavior, 45*(5), 1033–1041.

Insel, T. R., & Shapiro, L. E. (1992). Oxytocin receptor distribution reflects social-organization in monogamous and polygamous voles. *Proceedings of the National Academy of Sciences of the United States of America, 89*(13), 5981–5985.

Iriki, A. (2006). The neural origins and implications of imitation, mirror neurons and tool use. *Current Opinion in Neurobiology, 16*, 660–667.

Iriki, A., & Sakura, O. (2008). The neuroscience of primate intellectual evolution: Natural selection and passive and intentional niche construction. *Philosophical Transactions of the Royal Society B, 363*, 2229–2241.

Iriki, A., Tanaka, M., & Iwamura, Y. (1996). Coding of modified body schema during tool use by macaque postcentral neurons. *NeuroReport, 7*, 2325–2330.

Isen, A. M., & Levin, P. F. (1972). Effect of feeling good on helping – cookies and kindness. *Journal of Personality and Social Psychology, 21*(3), 384–388.

Ishibashi, H., Hihara, S., Takahashi, M., Heike, T., Yokota, T., & Iriki, A. (2002). Tool-use learning selectively induces expression of brain-derived neurotrophic factor, its receptor trkB, and neurotrophin 3 in the intraparietal cortex of monkeys. *Cognitive Brain Research, 14*, 3–9.

Israel, S., Lerer, E., Shalev, I., Uzefovsky, F., Reibold, M., Bachner-Melman, R., . . . Ebstein, R. P. (2008). Molecular genetic studies of the arginine vasopressin 1a receptor (AVPR1a) and the oxytocin receptor (OXTR) in human behaviour: From autism to altruism with some notes in between. *Advances in Vasopressin and Oxytocin: From Genes to Behaviour to Disease, 170*, 435–449.

Ito, T. A., & Urland, G. R. (2003). Race and gender on the brain: Electrocortical measures of attention to the race and gender of multiply categorizable individuals. *Journal of Personality and Social Psychology, 85*(4), 616–626.

Ito, T. A., & Urland, G. R. (2005). The influence of processing objectives on the perception of faces: An ERP study of race and gender perception. *Cognitive Affective & Behavioral Neuroscience, 5*(1), 21–36.

Jabbi, M., Swart, M., & Keysers, C. (2007). Empathy for positive and negative emotions in gustatory cortex. *NeuroImage, 34*, 1744–1753.

Jack, R. E., Blais, C., Scheepers, C., Schyns, P. G., & Caldara, R. (2009). Cultural confusions show that facial expressions are not universal. *Current Biology, 19*(18), 1543–1548.

Jackson, P. L., Meltzoff, A. N., & Decety, J. (2005). How do we perceive the pain of others? A window into the neural processes involved in empathy. *Neuroimage, 24*(3), 771–779.

Jahoda, G. (1982). *Psychology and Anthropology: A Psychological Perspective*. London: Academic Press.

James, W. (1884). What is an emotion? *Mind, 9*, 188–205.

Jankoviak, W. R., & Fischer, E. F. (1992). A cross-cultural perspective on romantic love. *Ethology, 31*, 149–155.

Jellema, T., Baker, C. I., Wicker, B., & Perrett, D. I. (2000). Neural representation for the perception of the intentionality of actions. *Brain and Cognition, 44*(2), 280–302.

Jellema, T., Maassen, G., & Perrett, D. I. (2004). Single cell integration of animate form, motion and location in the superior temporal cortex of the macaque monkey. *Cerebral Cortex, 14*(7), 781–790.

Jellema, T. & Perrett, D. I. (2005). Neural basis for the perception of goal-directed actions. In A. Easton and N. J. Emery (Eds.), *The Cognitive Neuroscience of Social Behavior*. Hove, UK: Psychology Press.

Jenkins, J. M., & Astington, J. W. (1996). Cognitive factors and family structure associated with theory of mind development in young children. *Developmental Psychobiology, 32*, 70–78.

Joffe, T. H. (1997). Social pressures have selected for an extended juvenile period in primates. *Journal of Human Evolution, 32*, 593–605.

Johnson, D. W., & Johnson, F. P. (1987). *Joining Together: Group Theory and Group Skills*. Englewood Cliffs, NJ: Prentice Hall.

Johnson, J. G., Cohen, P., Smailes, E. M., Kasen, S., & Brook, J. S. (2002). Television viewing and aggressive behavior during adolescence and adulthood. *Science, 295*(5564), 2468–2471.

Johnson, M. H., Dziurawiec, S., Ellis, H. D., & Morton, J. (1991). Newborns' preferential tracking of face-like stimuli and its subsequent decline. *Cognition, 40*, 1–19.

Johnson, M. H., Griffin, R., Csibra, G., Halit, H., Farroni, T., de Haan, M., Tucker, L. A., Baron-Cohen, S., & Richards, J. (2005). The emergence of the social brain network: Evidence from typical and atypical development. *Developmental Psychopathology, 17*, 599–619.

Johnson-Laird, P. N., & Oatley, K. (1992). Basic emotions, rationality, and folk theory. *Cognition & Emotion, 6*(3–4), 201–223.

Johnston, L. (2002). Behavioral mimicry and stigmatization. *Social Cognition, 20*(1), 18–35.

Jones, A. P., Happe, F. G. E., Gilbert, F., Burnett, S., & Viding, E. (2010). Feeling, caring, knowing: Different types of empathy deficit in boys with psychopathic tendencies and autism spectrum disorder. *Journal of Child Psychology and Psychiatry, 51*(11), 1188–1197.

Jorgensen, B. W., & Cervone, J. C. (1978). Affect enhancement in the pseudorecognition task. *Personality and Social Psychology Bulletin, 4*, 285–288.

Josephs, O., & Henson, R. N. A. (1999). Event-related functional magnetic resonance imaging: Modelling, inference and optimization. *Philosophical Transactions of the Royal Society B, 354*, 1215–1228.

Kahana-Kalman, R., & Walker-Andrews, A. S. (2001). The role of person familiarity in young infants' perception of emotional expressions. *Child Development, 72*(2), 352–369.

Kalin, N. H., Shelton, S. E., & Barksdale, C. M. (1988). Opiate modulation of separation-induced distress in non-human primates. *Brain Research, 440*(2), 285–292.

Kanai, R., Bahrami, B., Roylance, R., & Rees, G. (2012). Online social network size is reflected in human brain structure. *Proceedings of the Royal Society B: Biological Sciences, 279*, 1327–1334.

Kanai, R., & Rees, G. (2011). The structural basis of inter-individual differences in human behaviour and cognition. *Nature Reviews Neuroscience, 12*(4), 231–242.

Kanner, L. (1943). Autistic disturbances of affective contact. *Nervous Child, 2*, 217–250.

Kanwisher, N. (2000). Domain specificity in face perception. *Nature Neuroscience, 3*, 759–763.

Kanwisher, N., McDermott, J., & Chun, M. M. (1997). The fusiform face area: A module in human extrastriate cortex specialised for face perception. *Journal of Neuroscience, 17*, 4302–4311.

Kanwisher, N., & Wojciulik, E. (2000). Visual attention: Insights from brain imaging. *Nature Reviews Neuroscience, 1*, 91–100.

Kanwisher, N., & Yovel, G. (2006). The fusiform face area: A cortical region specialized for the perception of faces. *Philosophical Transactions of the Royal Society B-Biological Sciences, 361*(1476), 2109–2128.

Kaplan, J. T., & Iacoboni, M. (2006). Getting a grip on other minds: Mirror neurons, intention understanding, and cognitive empathy. *Social Neuroscience, 1*(3–4), 175–183.

Kavanagh, L. C., Suhler, C. L., Churchland, P. S., & Winkielman, P. (2011). When it's an error to mirror: The surprising reputational costs of mimicry. *Psychological Science, 22*(10), 1274–1276.

Kawai, M. (1965). Newly-acquired pre-cultural behavior of the natural troop of Japanese monkeys on Koshima Islet. *Primates, 6*, 1–30.

Kawamura, S. (1959). The process of sub-culture propagation among Japanese macaques. *Primates, 2*, 43–60.

Kaye, K., & Fogel, A. (1980). The temporal structure of face-to-face communication between mothers and infants. *Developmental Psychology, 16*(5), 454–464.

Keizer, K., Lindenberg, S., & Steg, L. (2008). The spreading of disorder. *Science, 322*(5908), 1681–1685.

Kelley, W. M., Macrae, C. N., Wyland, C. N., Caglar, S., Inati, S., & Heatherton, T. F. (2002). Finding the self? An event related fMRI study. *Journal of Cognitive Neuroscience, 14*, 785–794.

Kellogg, C. K., Awatramani, G. B., & Piekut, D. T. (1998). Adolescent development alters stressor-induced Fos immunoreactivity in rat brain. *Neuroscience, 83*(3), 681–689.

Kensinger, E. A., Garoff-Eaton, R. J., & Schacter, D. L. (2007). How negative emotion enhances the visual specificity of a memory. *Journal of Cognitive Neuroscience, 19*(11), 1872–1887.

Kenward, B., Weir, A. A. S., Rutz, C., & Kacelnik, A. (2005). Tool manufacture by naive juvenile crows. *Nature, 433*, 121.

Kerns, J. G., Cohen, J. D., MacDonal, A. W., Cho, R. Y., Stenger, V. A., & Carter, C. S. (2004). Anterior cingulate conflict monitoring and adjustments in control. *Science, 303*, 1023–1026.

Kerth, G., Perony, N., & Schweitzer, F. (2011). Bats are able to maintain long-term social relationships despite the high fission-fusion dynamics of their groups. *Proceedings of the Royal Society B-Biological Sciences, 278*(1719), 2761–2767.

Kiehl, K. A. (2008). *Without Morals: The Cognitive Neuroscience of Criminal Psychopaths*. Cambridge, MA: MIT Press.

Kiehl, K. A., Smith, A. M., Hare, R. D., Mendrek, A., Forster, B. B., Brink, J., & Liddle, P. F. (2001). Limbic abnormalities in affective processing by criminal psychopaths as revealed by functional magnetic resonance imaging. *Biological Psychiatry, 50*, 677–684.

Killgore, W. D. S., & Yurgelun-Todd, D. A. (2010). Cerebral correlates of amygdala responses during non-conscious perception of facial affect in adolescent and pre-adolescent children. *Cognitive Neuroscience, 1*, 33–43.

Kilner, J. M. (2011). More than one pathway to action understanding. *Trends in Cognitive Sciences, 15*(8), 352–357.

Kilner, J. M., & Lemon, R. N. (2013). What we know currently about mirror neurons. *Current Biology, 23*(23), R1057–R1062.

Kim, H. S., & Sasaki, J. Y. (2014). Cultural neuroscience: Biology of the mind in cultural contexts. *Annual Review of Psychology, 65*, 487–514.

King, M., & Wilson, A. (1975). Evolution at two levels in humans and chimpanzees. *Science, 188*, 107–116.

King-Casas, B., Tomlin, D., Anen, C., Camerer, C. F., Quartz, S. R., & Montague, P. R. (2005). Getting to know you: Reputation and trust in a two-person economic exchange. *Science, 308*(5718), 78–83.

Kinzler, K. D., & Spelke, E. S. (2011). Do infants show social preferences for people differing in race? *Cognition, 119*(1), 1–9.

Kipps, C. M., Duggins, A. J., McCusker, E. A., & Calder, A. J. (2007). Disgust and happiness recognition correlate with anteroventral insula and amygdala volume respectively in preclinical Huntington's disease. *Journal of Cognitive Neuroscience, 19*, 1206–1217.

Kirsch, P., Esslinger, C., Chen, Q., Mier, D., Lis, S., Siddhanti, S., Gruppe, H., Mattay, V. S., Gallhofer, B., & Meyer-Lindenberg, A. (2005). Oxytocin modulates neural circuitry for social cognition and fear in humans. *Journal of Neuroscience, 25*(49), 11489–11493.

Klein, J. T., & Platt, M. L. (2013). Social information signaling by neurons in primate striatum. *Current Biology, 23*(8), 691–696.

Klein, S. B., & Lax, M. L. (2010). The unanticipated resilience of trait self-knowledge in the face of neural damage. *Memory, 18*(8), 918–948.

Klein, S. B., Loftus, J., & Kihlstrom, J. F. (1996). Self-knowledge of an amnesic patient: Toward a neuropsychology of personality and social psychology. *Journal of Experimental Psychology-General, 125*(3), 250–260.

Klein, S. B., Rozendal, K., & Cosmides, L. (2002). A social-cognitive neuroscience analysis of the self. *Social Cognition, 20*, 105–135.

Klinnert, M. D., Campos, J. J., & Source, J. (1983). Emotions as behavior regulators: Social referencing in infancy. In R. Plutchik & H. Kellerman (Eds.), *Emotions in Early Development*. New York: Academic Press.

Klobusicky, E., & Ross, L. A. (2013). Social cognition and the anterior temporal lobes: A review and theoretical framework. *Social Cognitive and Affective Neuroscience, 8*(2), 123–133.

Kluver, H., & Bucy, P. C. (1939). Preliminary analysis of functions of the temporal lobes in monkeys. *Archives of Neurology and Psychiatry, 42*, 979–1000.

Knoch, D., Pascual-Leone, A., Meyer, K., Treyer, V., & Fehr, E. (2006). Diminishing reciprocal fairness by disrupting the right prefrontal cortex. *Science, 314*(5800), 829–832.

Knoch, D., Schneider, F., Schunk, D., Hohmann, M., & Fehr, E. (2009). Disrupting the prefrontal cortex diminishes the human ability to build a good reputation. *Proceedings of the National Academy of Sciences of the United States of America, 106*(49), 20895–20899.

Knutson, B., Adams, C. M., Fong, G. W., & Hommer, D. (2001). Anticipation of increasing monetary reward selectively recruits nucleus accumbens. *Journal of Neuroscience, 21*(16), art. no.-RC159.

Knutson, K. M., Mah, L., Manly, C. F., & Grafman, J. (2007). Neural correlates of automatic beliefs about gender and race. *Human Brain Mapping, 28*(10), 915–930.

Knutson, K. M., Wood, J. N., Spampinato, M. V., & Grafman, J. (2006). Politics on the brain: An MRI investigation. *Social Neuroscience, 1*(1), 25–40.

Kobayashi, C., Glover, G. H., & Temple, E. (2006). Cultural and linguistic influence on neural bases of 'theory of mind': An fMRI study with Japanese bilinguals. *Brain and Language, 98*(2), 210–220.

Koenigs, M., & Tranel, D. (2007). Irrational economic decision-making after ventromedial prefrontal damage: Evidence from the ultimatum game. *Journal of Neuroscience, 27*, 951–956.

Koenigs, M., Young, L., Adolphs, R., Tranel, D., Cushman, F., Hauser, M., & Damasio, A. (2007). Damage to the prefrontal cortex increases utilitarian moral judgements. *Nature, 446*(7138), 908–911.

Kohlberg, L., Levine, C., & Hewer, A. (1983). Moral stages: A current formulation and response to critics. In J. A. Meacham (Ed.), *Contributions to Human Development*. Basel: Karger.

Kolb, B., & Whishaw, I. Q. (2002). *Fundamentals of human neuropsychology* (5th edition). New York: Worth/Freeman

Kontaris, I., Wiggett, A. J., & Downing, P. E. (2009). Dissociation of extrastriate body and biological-motion selective areas by manipulation of visual-motor congruency. *Neuropsychologia, 47*(14), 3118–3124.

Koob, G. F. (1992). Dopamine, addiction and reward. *Seminars in the Neurosciences, 4*, 139–148.

Koppensteiner, M. (2013). Motion cues that make an impression Predicting perceived personality by minimal motion information. *Journal of Experimental Social Psychology, 49*(6), 1137–1143.

Kosfeld, M., Heinrichs, M., Zak, P. J., Fischbacher, U., & Fehr, E. (2005). Oxytocin increases trust in humans. *Nature, 435*(7042), 673–676.

Kosslyn, S. M. (1999). If neuroimaging is the answer, what is the question? *Philosophical Transactions of the Royal Society of London B, 354*, 1283–1294.

Koster-Hale, J., Saxe, R., Dungan, J., & Young, L. L. (2013). Decoding moral judgments from neural representations of intentions. *Proceedings of the National Academy of Sciences of the United States of America, 110*(14), 5648–5653.

Kotelchuck, M., Zelazo, P. R., Akgan, J., & Spelke, E. (1975). Infant reaction to parental separations when left with familiar and unfamiliar adults. *Journal of Genetic Psychology, 126*(2), 255–262.

Kouneiher, F., Charron, S., & Koechlin, E. (2009). Motivation and cognitive control in the human prefrontal cortex. *Nature Neuroscience, 12*(7), 939-U167.

Kramer, R. S. S., Arend, I., & Ward, R. (2010). Perceived health from biological motion predicts voting behaviour. *Quarterly Journal of Experimental Psychology, 63*(4), 625–632.

Kraus, M. W., Cote, S., & Keltner, D. (2010). Social class, contextualism, and empathic accuracy. *Psychological Science, 21*(11), 1716–1723.

Krienen, F. M., Tu, P. C., & Buckner, R. L. (2010). Clan mentality: Evidence that the medial prefrontal cortex responds to close others. *Journal of Neuroscience, 30*(41), 13906–13915.

Kringelbach, M. L. (2005). The human orbitofrontal cortex: Linking reward to hedonic experience. *Nature Reviews Neuroscience, 6*, 691–702.

Kringelbach, M. L., & Rolls, E. T. (2003). Neural correlates of rapid = context-dependent reversal learning in a simple model of human social interaction. *NeuroImage, 20*, 1371–1383.

Krueger, F., Barbey, A. K., & Grafman, J. (2009). The medial prefrontal cortex mediates social event knowledge. *Trends in Cognitive Sciences, 13*(3), 103–109.

Krueger, F., McCabe, K., Moll, J., Kriegeskorte, N., Zahn, R., Strenziok, M., Heinecke, A., & Grafman, J. (2007). Neural correlates of trust. *Proceedings of the National Academy of Sciences of the United States of America, 104*(50), 20084–20089.

Kubota, J. T., Banaji, M. R., & Phelps, E. A. (2012). The neuroscience of race. *Nature Neuroscience, 15*(7), 940–948.

Kuhl, P. K. (2007). Is speech learning 'gated' by the social brain? *Developmental Science, 10*(1), 110–120.

Kuhl, P. K., Coffey-Corina, S., Padden, D., & Dawson, G. (2005). Links between social and linguistic processing of speech in preschool children with autism: Behavioral and electrophysiological measures. *Developmental Science, 8*(1), F1–F12.

Kuhl, P. K., Tsao, F. M., & Liu, H. M. (2003). Foreign-language experience in infancy: Effects of short-term exposure and social interaction on phonetic learning. *Proceedings of the National Academy of Sciences of the United States of America, 100*, 9096–9101.

Kumsta, R., Kreppner, J., Rutter, M., Beckett, C., Castle, J., Stevens, S., & Sonuga-Barke, E. J. (2010). III. Deprivation-specific psychological patterns. *Monographs of the Society for Research in Child Development, 75*(1), 48–78.

LaBar, K. S., Gatenby, J. C., Gore, J. C., LeDoux, J. E., & Phelps, E. A. (1998). Human amygdala activation during conditioned fear acquisition and extinction: A mixed-trial fMRI study. *Neuron, 20*(5), 937–945.

Lakin, J. L., & Chartrand, T. L. (2003). Using nonconscious behavioral mimicry to create affiliation and rapport. *Psychological Science, 14*, 334–339.

Lamm, C., Batson, C. D., & Decety, J. (2007). The neural substrate of human empathy: Effects of perspective-taking and cognitive appraisal. *Journal of Cognitive Neuroscience, 19*(1), 42–58.

Lamm, C., & Decety, J. (2008). Is the extrastriate body area (EBA) sensitive to the perception of pain in others? *Cerebral Cortex, 18*(10), 2369–2373.

Lang, P. J., Bradley, M. M., & Cuthbert, B. N. (1990). Emotion, attention and the startle reflex. *Psychological Review, 97*, 377–395.

Langlois, J. H., & Roggman, L. A. (1990). Attractive faces are only average. *Psychological Science, 1*(2), 115–121.

Langton, S. R. H., & Bruce, V. (1999). Reflexive visual orienting in response to the social attention of others. *Visual Cognition, 6*, 541–567.

Laurent, H. K., & Ablow, J. C. (2012). A cry in the dark: Depressed mothers show reduced neural activation to their own infant's cry. *Social Cognitive and Affective Neuroscience, 7*(2), 125–134.

Laurent, H. K., & Ablow, J. C. (2012). The missing link: Mothers' neural response to infant cry related to infant attachment behaviors. *Infant Behavior & Development, 35*(4), 761–772.

Lavie, N. (1995). Perceptual load as a necessary condition for selective attention. *Journal of Experimental Psychology: Human Perception and Performance, 21*, 451–468.

Lawrence, A. D., Calder, A. J., McGowan, S. V., & Grasby, P. M. (2002). Selective disruption of the recognition of facial expressions of anger. *Neuroreport, 13*(6), 881–884.

Lawrence, J. H., & DeLuca, C. J. (1983). Myoelectric signal versus force relationship in different human muscles. *Journal of Applied Physiology, 54*, 1653–1659.

Lea, S. E. G., & Webley, P. (2006). Money as tool, money as drug: The biological psychology of a strong incentive. *Behavioral and Brain Sciences, 29*, 161–209.

Leakey, R. E. (1994). *The Origin of Humankind*. London: Weidenfield & Nicolson.

Leavens, D. A., Hopkins, W. D., & Bard, K. A. (2005). Understanding the point of chimpanzee pointing – Epigenesis and ecological validity. Current Directions in Psychological Science, 14(4), 185–189.

Le Bihan, D., Mangin, J. F., Poupon, C., Clark, C. A., Pappata, S., Molko, N., . . . Chabriat, H. (2001). Diffusion tensor imaging: Concepts and applications. *Journal of Magnetic Resonance Imaging, 13*, 534–546.

LeDoux, J. E. (1996). *The Emotional Brain*. New York: Simon and Schuster.

LeDoux, J. E. (2002). *Synaptic Self: How Our Brains Become who we are*. New York: Viking.

LeDoux, J. E., Iwata, J., Cicchetti, P., & Reis, D. (1988). Differential projections of the central amygdaloid nucleus mediate autonomic and behavioral correlates of conditioned fear. *Journal of Neuroscience, 8*, 2517–2529.

LeDoux, J. E. (2000). Emotion circuits in the brain. *Annual Review of Neuroscience, 23*, 155–184.

Leekam, S. R., & Perner, J. (1991). Does the autistic-child have a metarepresentational deficit? *Cognition, 40*, 203–218.

Lefevre, C. E., Lewis, G. J., Bates, T. C., Dzhelyova, M., Coetzee, V., Deary, I. J., . . . Perrett, D. I. (2012). No evidence for sexual dimorphism of facial width-to-height ratio in four large adult samples. *Evolution and Human Behavior, 33*(6), 623–627.

Lefevre, C. E., Lewis, G. J., Perrett, D. I., & Penke, L. (2013). Telling facial metrics: Facial width is associated with testosterone levels in men. *Evolution and Human Behavior, 34*(4), 273–279.

Lefevre, C. E., Wilson, V. A. D., Morton, F. B., Brosnan, S. F., Paukner, A., & Bates, T. C. (2014). Facial width-to-height ratio relates to alpha status and assertive personality in capuchin monkeys. *PLoS One, 9*(4).

Lefebvre, L., & Bouchard, J. (2003). Social learning about food in birds. In D. M. Fragazsy & S. Perry (Eds.), *The Biology of Traditions: Models and Evidence.* Cambridge, UK: Cambridge University Press.

Leibenluft, E., Gobbini, M.I., Harrison, T., & Haxby, J. V. (2004). Mothers' neural activation in response to pictures of their children and other children. *Biological Psychiatry, 56*(4), 225–232.

Lemche, E., Giampietro, V. P., Surguladze, S. A., Amaro, E. J., Andrew, C. M., Williams, S. C. R., Brammer, M. J., Lawrence, N., Maier, M. A., Russell, T. A., Simmons, A., Ecker, C., Joraschky, P., & Phillips, M. L. (2006). Human attachment security is mediated by the amygdala: Evidence from combined fMRI and psychophysiological measures. *Human Brain Mapping, 27*, 623–635.

Lenggenhager, B., Tadi, T., Metzinger, T., & Blanke, O. (2007). Video ergo sum: Manipulating bodily self-consciousness. *Science, 317*, 1096–1099.

Leppanen, J. M., Moulson, M. C., Vogel-Farley, V. K., & Nelson, C. A. (2007). An ERP study of emotional face processing in the adult and infant brain. *Child Development, 78*(1), 232–245.

Leslie, A. M. (1987). Pretence and representation: The origins of "Theory of Mind". *Psychological Review, 94*, 412–426.

Leslie, A. M., Mallon, R., & Di Corcia, J. A. (2006). Transgressors, victims and cry babies: Is basic moral judgment spared in autism? *Social Neuroscience, 1*, 270–283.

Lesch, K. P., Bengel, D., Heils, A., Sabol, S. Z., Greenberg, B.D., Petri, S., . . . Murphy, D.L. (1996). Association of anxiety-related traits with a polymorphism in the serotonin transporter gene regulatory region. *Science, 274*(5292), 1527–1531.

Lewis, G., Lefevre, C., & Bates, T. (2012). Facial width-to-height ratio predicts achievement drive in US presidents. *Personality and Individual Differences, 52*, 855–857.

Lewis, M., & Brooks-Gunn, J. (1979). *Social Cognition and the Acquisition of Self.* New York: Plenum.

Lewis, M., & Carmody, D. P. (2008). Self-representation and brain development. *Developmental Psychology, 44*, 1329–1334.

Libet, B. (1985). Unconscious cerebral initiative and the role of conscious will in voluntary action. *Behavioral and Brain Sciences, 8*, 529–566.

Libet, B., Gleason, C. A., Wright, E. W., & Pearl, D. K. (1983). Time of conscious intention to act in relation to onset of cerebral activity (readiness potential): The unconscious initiation of a freely voluntary act. *Brain, 102*, 623–642.

Lieberman, M. D., & Cunningham, W. (2009). Type I and Type II error concerns in fMRI research: Re-balancing the scale. *Social Cognitive and Affective Neuroscience, 4*, 423–428.

Lieberman, M. D., Eisenberger, N. I., Crockett, M. J., Tom, S. M., Pfeifer, J. H., & Way, B. M. (2007). Putting feelings into words – Affect labeling disrupts amygdala activity in response to affective stimuli. *Psychological Science, 18*(5), 421–428.

Lieberman, M. D., Hariri, A., Jarcho, J. M., Eisenberger, N. I., & Bookheimer, S. Y. (2005). An fMRI investigation of race-related amygdala activity in African-American and Caucasian-American individuals. *Nature Neuroscience, 8*(6), 720–722.

Light, S. N., Coan, J. A., Zahn-Waxler, C., Frye, C., Goldsmith, H. H., & Davidson, R. J. (2009). Empathy is associated with dynamic change in prefrontal brain electrical activity during positive emotion in children. *Child Development, 80*(4), 1210–1231.

Lin, Z. C., & Han, S. H. (2009). Self-construal priming modulates the scope of visual attention. *Quarterly Journal of Experimental Psychology, 62*(4), 802–813.

Lin, Z. C., Lin, Y., & Han, S. H. (2008). Self-construal priming modulates visual activity: Underlying global/local perception. *Biological Psychology, 77*(1), 93–97.

Lindquist, K. A., & Barrett, L. F. (2012). A functional architecture of the human brain: Emerging insights from the science of emotion. *Trends in Cognitive Sciences, 16*(11), 533–540.

Lipps, T. (1903). Einfuhlung, inner nachahnubg, und organ-empfindungen. *Archiv für die Gesamte Psychologie, 1*, 185–204.

Little, A. C., Burt, D. M., & Perrett, D. I. (2006). What is good is beautiful: Face preference reflects desired personality. *Personality and Individual Differences, 41*(6), 1107–1118.

Liu, Y., & Wang, Z. X. (2003). Nucleus accumbens oxytocin and dopamine interact to regulate pair bond formation in female prairie voles. *Neuroscience, 121*(3), 537–544.

Livingston, R.W., & Pearce, N.A. (2009). The teddy-bear effect: Does having a baby face benefit black chief executive officers? *Psychological Science, 20*(10), 1229–1236.

Ljungberg, T., Apicella, P., & Schultz, W. (1992). Responses of monkey dopamine neurons during learning of behavioral reactions. *Journal of Neurophysiology, 67*(1), 145–163.

Lloyd-Fox, S., Blasi, A., & Elwell, C. E. (2010). Illuminating the developing brain: The past, present and future of functional near-infrared spectroscopy. *Neuroscience and Biobehavioral Review, 34*, 269–284.

Lohmann, H., & Tomasello, M. (2003). The role of language in the development of false belief understanding: A training study. *Child Development, 74*, 1130–1144.

Lorberbaum, J.P., Newman, J.D., Horwitz, A.R., Dubno, J.R., Lydiard, R.B., Hamner, M.B., . . . George, M.S. (2002). A potential role for thalamocingulate circuitry in human maternal behavior. *Biological Psychiatry, 51*(6), 431–445.

Lorenz, K. (1966). *On Aggression.* New York: Harcourt Brace.

Lutchmaya, S., Baron-Cohen, S., Raggatt, P., Knickmeyer, R., & Manning, J. T. (2004). 2nd to 4th digit ratios, fetal testosterone and estradiol. *Early Human Development, 77*(1–2), 23–28.

Ly, M., Haynes, M. R., Barter, J. W., Weinberger, D. R., & Zink, C. F. (2011). Subjective socioeconomic status predicts human ventral striatal responses to social status information. *Current Biology, 21*(9), 794–797.

Lykken, D. T. (1957). A study of anxiety in the sociopathic personality. *Journal of Abnormal and Social Psychology, 55*, 6–10.

Ma, D. Q., Salyakina, D., Jaworski, J. M., Konidari, I., Whitehead, P. L., Andersen, A. N., . . . Pericak-Vance, M. A. (2009). A genome-wide association study of autism reveals a common novel risk locus at 5p14.1. *Annals of Human Genetics, 73*, 263–273.

Macchi Cassia, V., Turati, C., & Simion, F. (2004). Can a non-specific bias toward top-heavy patterns explain new-borns face preference? *Psychological Science, 15*, 379–383.

MacDonald, G., & Leary, M. R. (2005). Why does social exclusion hurt? The relationship between social and physical pain. *Psychological Bulletin, 103*, 202–223.

MacLean, P. D. (1949). Psychosomatic disease and the 'visceral brain': Recent developments bearing on the Papez theory of emotion. *Psychosomatic Medicine, 11*, 338–353.

MacLoed, C. M., & MacDonald, P. A. (2000). Interdimensional interference in the Stroop effect: Uncovering the cognitive and neural anatomy of attention. *Trends in Cognitive Sciences, 4*, 383–391.

Macmillan, M. B. (1986). A wonderful journey through skull and brains: The travels of Mr. Gage's tamping iron. *Brain and Cognition, 5*, 67–107.

Macrae, C. N., & Bodenhausen, G. V. (2000). Social cognition: Thinking categorically about others. *Annual Review of Psychology, 51*, 93–120.

Macrae, C. N., Moran, J. M., Heatherton, T. F., Banfield, J. F., & Kelley, W. M. (2004). Medial prefrontal activity predicts memory for self. *Cerebral Cortex, 14*(6), 647–654.

Maia, T. V., & McClelland, J. L. (2004). A reexamination of the evidence for the somatic marker hypothesis: What participants really know in the Iowa gambling task. *Proceedings of the National Academy of Science, USA, 101*, 16075–16080.

Marazziti, D. (2009). Neurobiology and hormonal aspects of romantic relationships. In M. D. Haan & M. R. Gunnar (Eds.), *Handbook of Developmental Social Neuroscience.* New York: The Guilford Press.

Marazziti, D., Akiskal, H. S., Rossi, A., & Cassano, G. B. (1999). Alteration of the platelet serotonin transporter in romantic love. *Psychological Medicine, 29*(3), 741–745.

Marazziti, D., & Canale, D. (2004). Hormonal changes when falling in love. *Psychoneuroendocrinology, 29*(7), 931–936.

Marcar, V. L., Strassle, A. E., Loenneker, T., Schwarz, U., & Martin, E. (2004). The influence of cortical maturation on the BOLD response: An fMRI study of visual cortex in children. *Pediatric Research, 56*, 967–974.

Marcus, G. B. (1986). Stability and change in political attitudes: Observe, recall and 'explain'. *Political Behavior, 8*, 21–44.

Marcus-Newhall, A., Pedersen, W. C., Carlson, M., & Miller, N. (2000). Displaced aggression is alive and well: A meta-analytic review. *Journal of Personality and Social Psychology, 78*(4), 670–689.Markus, H. R., & Kitayama, S. (1991). Culture and the self – implications for cognition, emotion, and motivation. *Psychological Review, 98*(2), 224–253.

Markus, H. R., Uchida, Y., Omoregie, H., Townsend, S. S. M., & Kitayama, S. (2006). Going for the gold – Models of agency in Japanese and American contexts. *Psychological Science, 17*(2), 103–112.

Martinelli, P., Sperduti, M., & Piolino, P. (2013). Neural substrates of the self-memory system: New insights from a meta-analysis. *Human Brain Mapping, 34*(7), 1515–1529.

Mascaro, J. S., Hackett, P. D., & Rilling, J. K. (2013). Testicular volume is inversely correlated with nurturing-related brain activity in human fathers. *Proceedings of the National Academy of Sciences of the United States of America, 110*(39), 15746–15751.

Mascaro, J. S., Hackett, P. D., & Rilling, J. K. (2014). Differential neural responses to child and sexual stimuli in human fathers and non-fathers and their hormonal correlates. *Psychoneuroendocrinology, 46*, 153–163.

Maslow, A. H. (1943). A theory of human motivation. *Psychological Review, 50*, 370–396.

Master, S. L., Eisenberger, N. I., Taylor, S. E., Naliboff, B. D., Shirinyan, D., & Lieberman, M. D. (2009). A picture's worth: Partner photographs reduce experimentally induced pain. *Psychological Science, 20*(11), 1316–1318.

Masuda, T., & Nisbett, R. E. (2006). Culture and change blindness. *Cognitive Science, 30*(2), 381–399.

Materna, S., Dicke, P. W., & Thier, P. (2008). The posterior superior temporal sulcus is involved in social communication not specific for the eyes. *Neuropsychologia, 46*(11), 2759–2765.

Matsumoto, D., Yoo, S. H., Fontaine, J., Anguas-Wong, A. M., Arriola, M., Ataca, B., Bond, M. H., Boratav, H. B., Breugelmans, S. M., Cabecinhas, R., Chae, J., Chin, W. H., Comunian, A. L., Degere, D. N., Djunaidi, A., Fok, H. K., Friedlmeier, W., Ghosh, A., Glamcevski, M., Granskaya, J. V., Groenvynck, H., Harb, C., Haron, F., Joshi, R., Kakai, H., Kashima, E., Khan, W., Kurman, J., Kwantes, C. T., Mahmud, S. H., Mandaric, M., Nizharadze, G., Odusanya, J. O. T., Ostrosky-Solis, F., Palaniappan, A. K., Papastylianou, D., Safdar, S., Setiono, K., Shigemasu, E., Singelis, T. M., Iva, P. S., Spiess, E., Sterkowicz, S., Sunar, D., Szarota, P., Vishnivetz, B., Vohra, N., Ward, C., Wong, S., Wu,

R. X., Zebian, S., Zengeya, A., Altarriba, J., Bauer, L. M., Mogaji, A., Siddiqui, R. N., Fulop, M., Bley, L., Alexandre, J., Garcia, F. M., & Grossi, E. (2008). Mapping expressive differences around the world – The relationship between emotional display rules and individualism versus collectivism. *Journal of Cross-Cultural Psychology, 39*(1), 55–74.

Matthews, G., & Wells, A. (1999). The cognitive science of attention and emotion. In T. Dalgleish & M. J. Power (Eds.), *Handbook of Cognition and Emotion*. New York: Wiley.

Maynard Smith, J. (1982). *Evolution and the Theory of Games*. Cambridge: Cambridge University Press.

Mazur, A., & Booth, A. (1998). Testosterone and dominance in men. *Behavioral and Brain Sciences, 21*(3), 353–397.

Mazur, A., Booth, A., & Dabbs, J. (1992). Testosterone and chess competition. *Social Psychology Quarterly, 55*, 70–77.

McCabe, K., Houser, D., Ryan, L., Smith, V., & Trouard, T. (2001a). A functional imaging study of cooperation in two-person reciprocal exchange. *Proceedings of the National Academy of Sciences, USA, 98*, 11832–11835.

McCabe, K., Houser, D., Ryan, L., Smith, V., & Trouard, T. (2001b). A functional imaging study of cooperation in two-person reciprocal exchange. *Proceedings of the National Academy of Science, USA, 98*, 11832–11835.

McClure, S., Lee, J., Tomlin, D., Cypert, K., Montague, L., & Montague, P. R. (2004). Neural correlates of behavioural preferences for culturally familiar drinks. *Neuron, 44*, 379–387.

McComb, K., Moss, C., Durant, S. M., Baker, L., & Sayialel, S. (2001). Matriarchs as repositories of social knowledge in African elephants. *Science, 292*(5516), 491–494.

McConahay, J. B. (1986). Modern racism, ambivalence, and the Modern Racism Scale. In J. F. Dovidio & S. L. Gaertner (Eds.), *Prejudice, Discrimination and Racism*. New York: Academic Press.

McConnell, A. R., & Leibold, J. M. (2001). Relations among the implicit association test, discriminatory behavior, and explicit measures of racial attitudes. *Journal of Experimental Social Psychology, 37*(5), 435–442.

McGowan, P. O., Sasaki, A., D'Alessio, A. C., Dymov, S., Labonte, B., Szyf, M., . . . Meaney, M. J. (2009). Epigenetic regulation of the glucocorticoid receptor in human brain associates with childhood abuse. *Nature Neuroscience, 12*(3), 342–348.

McLoed, P., Dittrich, W., Driver, J., Perrett, D., & Zihl, J. (1996). Preserved and impaired detection of structure from motion by a "motion-blind" patient. *Visual Cognition, 3*, 363–391.

McNeil, J. E., & Warrington, E. K. (1993). Prosopagnosia: A face-specific disorder. *Quarterly Journal of Experimental Psychology, 46A*, 1–10.

McQuaid, R. J., McInnis, O. A., Matheson, K., & Anisman, H. (2015). Distress of ostracism: Oxytocin receptor gene polymorphism confers sensitivity to social exclusion. *Social Cognitive and Affective Neuroscience, 10*(8), 1153–1159.

McQuaid, R. J., McInnis, O. A., Stead, J. D., Matheson, K., & Anisman, H. (2013). A paradoxical association of an oxytocin receptor gene polymorphism: Early-life adversity and vulnerability to depression. *Frontiers in Neuroscience, 7*, 7.

Meaney, M. J. (2001). Maternal care, gene expression, and the transmission of individual differences in stress reactivity across generations. *Annual Review of Neuroscience, 24*, 1161–1192.

Meeren, H. K. M., van Heijnsbergen, C., & de Gelder, B. (2005). Rapid perceptual integration of facial expression and emotional body language. *Proceedings of the National Academy of Sciences of the United States of America, 102*(45), 16518–16523.

Meffert, H., Gazzola, V., den Boer, J. A., Bartels, A. A. J., & Keysers, C. (2013). Reduced spontaneous but relatively normal deliberate vicarious representations in psychopathy. *Brain, 136*, 2550–2562.

Mehler, J., Bertoncini, J., Barriere, M., & Jassikgerschenfeld, D. (1978). Infant recognition of mothers voice. *Perception, 7*(5), 491–497.

Meltzoff, A. N. (1995). Understanding the intentions of others: Re-enactment of intended acts by 18-month-old children. *Developmental Psychology, 31*, 838–850.

Meltzoff, A. N. (2007). 'Like me': A foundation for social cognition. *Developmental Science, 10*, 126–134.

Meltzoff, A. N., & Borton, R. W. (1979). Intermodal matching by human neonates. *Nature, 282*, 403–404.

Meltzoff, A. N., & Decety, J. (2003). What imitation tells us about social cognition: A rapprochement between developmental psychology and cognitive neuroscience. *Philosophical Transactions of the Royal Society of London Series B-Biological Sciences, 358*(1431), 491–500.

Meltzoff, A. N., & Moore, M. K. (1977). Imitation of facial and manual gestures by human neonates. *Science, 198*, 75–78.

Meltzoff, A. N., & Moore, M. K. (1983). Newborn infants imitate adult facial gestures. *Child Development, 54*, 702–709.

Menon, V., & Uddin, L. Q. (2010). Saliency, switching, attention and control: A network model of insula function. *Brain Structure & Function, 214*(5–6), 655–667.

Mertins, V., Schote, A. B., Hoffeld, W., Griessmair, M., & Meyer, J. (2011). Genetic susceptibility for individual cooperation preferences: The role of monoamine oxidase a gene (MAOA) in the voluntary provision of public goods. *PLoS One, 6*(6).

Meyer-Lindenberg, A., Buckholtz, J. W., Kolachana, B., Hariri, A. R., Pezawas, L., Blasi, G., . . . Weinberger, D. R. (2006). Neural mechanisms of genetic risk for impulsivity and violence in humans. *Proceedings of the National Academy of Sciences of the United States of America, 103*(16), 6269–6274.

Milgram, S. (1963). Behavioral study of obedience. *Journal of Abnormal and Social Psychology, 67*, 371–378.

Miller, E. K., & Cohen, J. D. (2001). An integrative theory of prefrontal cortex function. *Annual Review of Neuroscience, 24*, 167–202.

Miller, G. (2010). The seductive allure of behavioral epigenetics. *Science, 329*(5987), 24–27.

Mills, D., & Conboy, B. T. (2009). Early communicative development and the social brain. In M. D. Haan & M. R. Gunnar (Eds.), *Handbook of Developmental Social Neuroscience.* New York: The Guilford Press.

Mills, K. L., Lalonde, F., Clasen, L. S., Giedd, J. N., & Blakemore, S. J. (2014). Developmental changes in the structure of the social brain in late childhood and adolescence. *Social Cognitive and Affective Neuroscience, 9*(1), 123–131.

Mineka, S., & Cook, M. (1993). Mechanisms involved in the observational conditioning of fear. *Journal of Experimental Psychology: General, 122*, 23–38.

Mitchell, J. P. (2008). Activity in right temporo-parietal junction is not selective for theory-of-mind. *Cerebral Cortex, 18*(2), 262–271.

Mitchell, J. P. (2009). Social psychology as a natural kind. *Trends in Cognitive Sciences, 13*, 246–251.

Mitchell, J. P., Banaji, M. R., & Macrae, C. N. (2005a). General and specific contributions of the medial prefrontal cortex to knowledge about mental states. *Neuroimage, 28*(4), 757–762.

Mitchell, J. P., Banaji, M. R., & Macrae, C. N. (2005b). The link between social cognition and self-referential thought in the medial prefrontal cortex. *Journal of Cognitive Neuroscience, 17*, 1306–1315.

Mitchell, J. P., Heatherton, T. F., & Macrae, C. N. (2002). Distinct neural systems subserve person and object knowledge. *Proceedings of the National Academy of Sciences of the United States of America, 99*(23), 15238–15243.

Mitchell, J. P., Macrae, C. N., & Banaji, M. R. (2004). Encoding-specific effects of social cognition on the neural correlates of subsequent memory. *Journal of Neuroscience, 24*(21), 4912–4917.

Mitchell, J. P., Nosek, B. A., & Banaji, M. R. (2003). Contextual variations in implicit evaluation. *Journal of Experimental Psychology-General, 132*(3), 455–469.

Mitchell, P., & Ropar, D. (2004). Visuo-spatial abilities in autism: A review. *Infant and Child Development, 13*(3), 185–198.

Mitchell, R. W., & Anderson, J. R. (1993). Discrimination learning of scratching, but failure to obtain imitation and self-recognition in a long-tailed macaque. *Primates, 34*, 301–309.

Mithen, S. (2007). Did farming arise from a misapplication of social intelligence? *Philosophical Transactions of the Royal Society B, 362*, 705–718.

Mogenson, G. J., Jones, D. L., & Yim, C. Y. (1980). From motivation to action: Functional interface between the limbic system and the motor system. *Progress in Neurobiology, 14*, 69–97.

Molenberghs, P., & Morrison, S. (2014). The role of the medial prefrontal cortex in social categorization. *Social Cognitive and Affective Neuroscience, 9*(3), 292–296.

Moll, J., de Oliveira-Souza, R., Eslinger, P. J., Bramati, I. E., Mourao-Miranda, J., Andreiuolo, P. A., & Pessoa, L. (2002). The neural correlates of moral sensitivity: A functional magnetic resonance imaging investigation of basic and moral emotions. *Journal of Neuroscience, 22*(7), 2730–2736.

Moll, J., de Oliveira-Souza, R., Moll, F. T., Ignacio, F. A., Bramati, I. E., Caparelli-Daquer, E. M., & Eslinger, P. J. (2005). The moral affiliations of disgust – A functional MRI study. *Cognitive and Behavioral Neurology, 18*(1), 68–78.

Moll, J., de Oliveira-Souza, R., Zahn, R., & Grafman, J. (2008). *The Cognitive Neuroscience of Moral Emotions.* Cambridge, MA: MIT Press.

Moll, J., Krueger, F., Zahn, R., Pardini, M., de Oliveira-Souzat, R., & Grafman, J. (2006). Human fronto-mesolimbic networks guide decisions about charitable donation. *Proceedings of the National Academy of Sciences of the United States of America, 103*(42), 15623–15628.

Moll, J., & Schulkin, J. (2009). Social attachment and aversion in human moral cognition. *Neuroscience and Biobehavioral Reviews, 33*(3), 456–465.

Moll, J., Zahn, R., de Oliveira-Souza, R., Krueger, F., & Grafman, J. (2005). The neural basis of human moral cognition. *Nature Reviews Neuroscience, 6*(10), 799–809.

Montague, P. R., Berns, G. S., Cohen, J. D., McClure, S. M., Pagnoni, G., Dhamala, M., . . . Fisher, R. E. (2002). Hyperscanning: Simultaneous fMRI during linked social interactions. *NeuroImage, 16*(4), 1159–1164.

Moran, J. M., Macrae, C. N., Heatherton, T. F., Wyland, C. L., & Kelley, W. M. (2006). Neuroanatomical evidence for distinct cognitive and affective components of self. *Journal of Cognitive Neuroscience, 18*(9), 1586–1594.

Moran, J. M., Young, L. L., Saxe, R., Lee, S. M., O'Young, D., Mavros, P. L., & Gabrieli, J. D. (2011). Impaired theory of mind for moral judgment in high-functioning autism. *Proceedings Of The National Academy Of Sciences Of The United States Of America, 108*(7), 268–269.

Moreland, R. L., & Beach, S. R. (1992). Exposure effects in the classroom: The development of affinity amongst students. *Journal of Experimental Social Psychology, 28*, 255–276.

Moreno, J. D. (2003). Neuroethics: An agenda for neuroscience and society. *Nature Reviews Neuroscience, 4*(2), 149–153.

Moretti, L., & Di Pellegrino, G. (2010). Disgust selectively modulates reciprocal fairness in economic interactions. *Emotion, 10*, 169–180.

Morishima, Y., Schunk, D., Bruhin, A., Ruff, C. C., & Fehr, E. (2012). Linking brain structure and activation in temporoparietal junction to explain the neurobiology of human altruism. *Neuron, 75*(1), 73–79.

Moro, V., Urgesi, C., Pernigo, S., Lanteri, P., Pazzaglia, M., & Aglioti, S. M. (2008). The neural basis of body form and body action agnosia. *Neuron, 60*(2), 235–246.

Morris, J., Friston, K. J., Buechel, C., Frith, C. D., Young, A. W., Calder, A. J., & Dolan, R. J. (1998). A neuromodulatory role for the human amygdala in processing emotional facial expressions. *Brain, 121*, 47–57.

Morris, J., Frith, C. D., Perrett, D., Rowland, D., Young, A. W., Calder, A. J., & Dolan, R. J. (1996). A differential neural response in the human amygdala to fearful and happy facial expressions. *Nature, 383*, 812–815.

Morris, J. S., Ohmann, A., & Dolan, R. (1999). A sub-cortical pathway to the right amygdala mediating 'unseen' fear. *Proceedings of the National Academy of Science, USA, 96*, 1680–1685.

Morton, J., & Johnson, M. H. (1991). CONSPEC and CONLERN – A 2-process theory of infant face recognition. *Psychological Review, 98*(2), 164–181.

Moses, L. J., Baldwin, D. A., Rosicky, J. G., & Tidball, G. (2001). Evidence for referential understanding in the emotions domain at twelve and eighteen months. *Child Development, 72*(3), 718–735.

Moutsiana, C., Fearon, P., Murray, L., Cooper, P., Goodyer, I., Johnstone, T., . . . Halligan, S. (2014). Making an effort to feel positive: Insecure attachment in infancy predicts the neural underpinnings of emotion regulation in adulthood. *Journal of Child Psychology and Psychiatry, 55*(9), 999–1008.

Mukamel, R., Ekstrom, A. D., Kaplan, J. T., Iacoboni, M., & Fried, I. (2010). Single-neuron responses in humans during execution and observation of actions. *Current Biology, 8*, 750–756.

Mummery, C. J., Patterson, K., Price, C. J., Ashburner, J., Frackowiak, R. S. J., & Hodges, J. R. (2000). A voxel-based morphometry study of semantic dementia: Relationship between temporal lobe atrophy and semantic memory. *Annals of Neurology, 47*, 36–45.

Munafo, M. R., Yalcin, B., Willis-Owen, S. A., & Flint, J. (2008). Association of the dopamine D4 receptor (DRD4) gene and approach-related personality traits: Meta-analysis and new data. *Biological Psychiatry, 63*(2), 197–206.

Mundy, P., Card, J., & Fox, N. (2000). EEG correlates of the development of infant joint attention skills. *Developmental Psychobiology, 36*(4), 325–338.

Murnighan, J. K., & Saxon, M. S. (1998). Ultimatum bargaining by children and adults. *Journal of Economic Psychology, 19*(4), 415–445.

Murray, E. A., & Baxter, M. G. (2006). Cognitive neuroscience and nonhuman primates: Lesion studies. In C. Senior, T. Russell & M. S. Gazzaniga (Eds.), *Methods in Mind*. Cambridge, MA: MIT Press.

Neave, N. (2008). *Hormones and Behaviour: A Psychological Approach*. Cambridge, UK: Cambridge University Press.

Neisser, U. (1988). Five kinds of self-knowledge. *Philosophical Psychology, 1*, 35–59.

Nelson, C. A., & DeHaan, M. (1996). Neural correlates of infants' visual responsiveness to facial expressions of emotion. *Developmental Psychobiology, 29*(7), 577–595.

Neumann, I. D., Toschi, N., Ohl, F., Torner, L., & Kromer, S. A. (2001). Maternal defence as an emotional stressor in female rats: Correlation of neuroendocrine and behavioural parameters and involvement of brain oxytocin. *European Journal of Neuroscience, 13*(5), 1016–1024.

Newcomb, T. M. (1961). *The Acquaintance Process*. New York: Holt, Rinehart & Winston.

Newheiser, A.-K., & Olson, K. R. (2012). White and Black American children's implicit intergroup bias. *Journal of Experimental Social Psychology, 48*(1), 264–270.

Newman, J. P., & Lorenz, A. R. (2002). Response modulation and emotion processing: Implications for psychopathy and other dysregulatory psychopathology. In R. J. Davidson, J. Scherer & H. H. Goldsmith (Eds.), *Handbook of Affective Sciences*. Oxford: Oxford University Press.

Nicholls, M. E. R., Ellis, B. E., Clement, J. G., & Yoshino, M. (2004). Detecting hemifacial asymmetries in emotional expression with three-dimensional computerized image analysis. *Proceedings of the Royal Society of London Series B-Biological Sciences, 271*(1540), 663–668.

Nieder, A. (2005). Counting on neurons: The neurobiology of numerical competence. *Nature Reviews Neuroscience, 6*, 1–14.

Nisbett, R. E., & Cohen, D. (1996). *The Culture of Honor: The Psychology of Violence in the South*. Boulder, CO: Westview Press.

Nisbett, R. E., Peng, K. P., Choi, I., & Norenzayan, A. (2001). Culture and systems of thought: Holistic versus analytic cognition. *Psychological Review, 108*(2), 291–310.

Nisbett, R. E., & Wilson, T. D. (1977). Halo effect – Evidence for unconscious alteration of judgments. *Journal of Personality and Social Psychology, 35*(4), 250–256.

Nitsche, M. A., Cohen, L. G., Wassermann, E. M., Priori, A., Lang, N., Antal, A., . . . Pascual-Leone, A. (2008). Transcranial direct current stimulation: State of the art 2008. *Brain Stimulation, 1*, 206–223.

Nitsche, M. A., Liebetanz, D., Lang, N., Antal, A., Tergau, F., & Paulus, W. (2003). Safety criteria for transcranial direct current stimulation (tDCS) in humans. *Clinical Neurophysiology, 114*(11), 2220–2222.

Norenzayan, A., Gervais, W. M., & Trzesniewski, K. H. (2012). Mentalizing deficits constrain belief in a personal god. *PLoS One, 7*(5).

Norman, D. A., & Shallice, T. (1986). Attention to action. In R. J. Davidson, G. E. Schwartz & D. Shapiro (Eds.), *Consciousness and Self Regulation*. New York: Plenum Press.

Norman, L., Lawrence, N., Iles, A., Benattayallah, A., & Karl, A. (2015). Attachment-security priming attenuates amygdala activation to social and linguistic threat. *Social Cognitive and Affective Neuroscience, 10*(6), 832–839.

Norscia, I., & Palagi, E. (2011). Yawn contagion and empathy in homo sapiens. *PLoS One, 6*(12).

Nosek, B. A., Banaji, M. R., & Greenwald, A. G. (2002a). Harvesting implicit group attitudes and beliefs from a demonstration website. *Group Dynamics, 6*(1), 101–115.

Nosek, B. A., Banaji, M. R., & Greenwald, A. G. (2002b). Math = male, me = female, therefore math not equal me. *Journal of Personality and Social Psychology, 83*(1), 44–59.

Nowak, M., & Sigmund, K. (1993). A strategy of win stay, lose shift that outperforms tit-for-tat in the prisoners-dilemma game. *Nature, 364*(6432), 56–58.

Nowak, M. A. (2006). Five rules for the evolution of cooperation. *Science, 314*(5805), 1560–1563.

Nowak, M. A., & Sigmund, K. (2005). Evolution of indirect reciprocity by image scoring. *Nature, 437*, 1291–1298.

Nunez, P. L. (1981). *Electric Fields of the Brain: The Neurophysics of EEG*. London: Oxford University Press.

Nuttin, J. M. (1985). Narcissism beyond gestalt and awareness – The name letter effect. *European Journal of Social Psychology, 15*(3), 353–361.

O'Connell, R. G., Dockree, P. M., Bellgrove, M. A., Kelly, S. P., Hester, R., Garavan, H., Robertson, I. H., & Foxe, J. J. (2007). The role of cingulate cortex in the detection of errors with and without awareness: A high-density electrical mapping study. *European Journal of Neuroscience, 25*(8), 2571–2579.

O'Connor, M. F., Wellisch, D. K., Stanton, A. L., Eisenberger, N. I., Irwin, M. R., & Lieberman, M. D. (2008). Craving love? Enduring grief activates brain's reward center. *Neuroimage, 42*(2), 969–972.

O'Connor, T. G., Bredenkamp, D., & Rutter, M. (1999). Attachment disturbances and disorders in children exposed to early severe deprivation. *Infant Mental Health Journal, 20*(1), 10–29.

O'Connor, T. G., & Rutter, M. (2000). Attachment disorder behavior following early severe deprivation: Extension and longitudinal follow-up. *Journal of the American Academy of Child and Adolescent Psychiatry, 39*(6), 703–712.

O'Doherty, J., Kringelbach, M. L., Rolls, E. T., Hornak, J., & Andrews, C. (2001). Abstract reward and punishment representations in the human orbitofrontal cortex. *Nature Neuroscience, 4*, 95–102.

Oberman, L. M., Hubbard, E. M., McCleery, J. P., Altschuler, E. L., Ramachandran, V. S., & Pineda, J. A. (2005). EEG evidence for mirror neuron dysfunction in autism spectrum disorders. *Cognitive Brain Research, 24*, 190–198.

Oberman, L. M., & Ramachandran, V. S. (2007). The simulating social mind: The role of the mirror neuron system and simulation in the social and communicative deficits of autism spectrum disorders. *Psychological Bulletin, 133*(2), 310–327.

Oberman, L. M., Winkielman, P., & Ramachandran, V. S. (2007). Face to face: Blocking facial mimicry can selectively impair recognition of emotional expressions. *Social Neuroscience, 2*(3–4), 167–178.

Obhi, S. S., Swiderski, K., & Brubacher, S. (2012). Induced power changes the sense of agency. *Consciousness & Cognition, 21*(3), 1547–1550.

Ochsner, K. N., Knierim, K., Ludlow, D. H., Hanelin, J., Ramachandran, T., Glover, G., . . . Mackey, S. C. (2004). Reflecting upon feelings: An fMRI study of neural systems supporting the attribution of emotion to self and other. *Journal of Cognitive Neuroscience, 16*(10), 1746–1772.

Ochsner, K., & Lieberman, M. D. (2001). The emergence of social cognitive neuroscience. *American Psychologist, 56*, 717–734.

Ochsner, K. N., Bunge, S. A., Gross, J. J., & Gabrieli, J. D. E. (2002). Rethinking feelings: An fMRI study of the cognitive regulation of emotion. *Journal of Cognitive Neuroscience, 14*(8), 1215–1229.

Ochsner, K. N., Silvers, J. A., & Buhle, J. T. (2012). Functional imaging studies of emotion regulation: A synthetic review and evolving model of the cognitive control of emotion. *Year in Cognitive Neuroscience, 1251*, E1–E24.

Ochsner, K. N., Ray, R. D., Cooper, J. C., Robertson, E. R., Chopra, S., Gabrieli, J. D. E., & Gross, J. J. (2004). For better or for worse: Neural systems supporting the cognitive down- and up-regulation of negative emotion. *Neuroimage, 23*(2), 483–499.

Ogawa, S., Lee, T. M., Kay, A. R., & Tank, D. W. (1990). Brain magnetic resonance imaging with contrast dependent on blood oxygenation. *Proceedings of the National Academy of Science USA, 87*, 9862–9872.

Ohman, A., Flykt, A., & Esteves, F. (2001). Emotion drives attention: Detecting the snake in the grass. *Journal of Experimental Psychology: General, 130*, 466–478.

Ohman, A., & Soares, J. J. F. (1994). Unconscious anxiety: Phobic responses to masked stimuli. *Journal of Abnormal Psychology, 102*, 121–132.

Olds, J. (1956). Pleasure centers of the brain. *Scientific American, 195*, 105–116.

Olds, J., & Milner, P. (1954). Positive reinforcement produced by electrical stimulation of septal area and other regions of the rat brain. *Journal of Comparative and Physiological Psychology, 47*, 419–427.

Olson, I. R., McCoy, D., Klobusicky, E., & Ross, L. A. (2013). Social cognition and the anterior temporal lobes: a review and theoretical framework. *Social Cognitive and Affective Neuroscience, 8*(2), 123–133.

Olson, K. R., & Spelke, E. S. (2008). Foundations of cooperation in young children. *Cognition, 108*(1), 222–231.

Olsson, A., & Phelps, E. A. (2004). Learned fear of "unseen" faces after Pavlovian, observational, and instructed fear. *Psychological Science, 15*(12), 822–828.

Omura, K., Constable, R.T., & Canli, T. (2005). Amygdala gray matter concentration is associated with extraversion and neuroticism. *NeuroReport, 16*(17), 1905–1908.

Öngür, D., & Price, J. L. (2000). The organization of networks within the orbital and medial prefrontal cortex of rats, monkeys and humans. *Cerebral Cortex, 10*, 206–219.

O'Nions, E., Sebastian, C.L., McCrory, E., Chantiluke, K., Happe, F., & Viding, E. (2014). Neural bases of Theory of Mind in children with autism spectrum disorders and children with conduct problems and callous-unemotional traits *Developmental Science, 17*(5), 786–796

Onishi, K. H., & Baillargeon, R. (2005). Do 15-month-old infants understand false beliefs? *Science, 308*, 255–258.

Oosterhof, N. N., & Todorov, A. (2008). The functional basis of face evaluation. *Proceedings of the National Academy of Sciences of the United States of America, 105*(32), 11087–11092.

Ortony, A., Clore, G. L., & Collins, A. (1988). *The Cognitive Structure of Emotions*. New York: Cambridge University Press.

Ortony, A., & Turner, T. J. (1990). What's basic about basic emotions. *Psychological Review, 97*(3), 315–331.

Otten, L. J., & Rugg, M. D. (2005). Interpreting event-related brain potentials. In T. C. Handy (Ed.), *Event-related Potentials: A Methods Handbook*. Cambridge, MA: MIT Press.

Ozonoff, S., Pennington, B. F., & Rogers, S. J. (1991). Executive function deficits in high-functioning autistic individuals: Relationship to theory of mind. *Journal of Child Psychology and Psychiatry, 32*, 1081–1105.

Paik, H., & Comstock, G. (1994). The effects of television violence on antisocial-behavior – A meta-analysis. *Communication Research, 21*(4), 516–546.

Panksepp, J. (2005). Why does separation distress hurt? Comment on MacDonald and Leary (2005). *Psychological Bulletin, 131*(2), 224–230.

Panksepp, J. (2007). Neurologizing the psychology of affects how appraisal-based constructivism and basic emotion theory can coexist. *Perspectives on Psychological Science, 2*, 281–296.

Panksepp, J., Herman, B. H., Vilberg, T., Bishop, P., & DeEskinazi, F. G. (1980). Endogenous opioids and social behavior. *Neuroscience and Biobehavioral Reviews, 4*, 473–487.

Papez, J. W. (1937). A proposed mechanism of emotion. *Archives of Neurology and Psychiatry, 38*(4), 725–743.

Park, J., Baek, Y.M., & Cha, M. (2014). Cross-cultural comparison of nonverbal cues in emoticons on twitter: Evidence from big data analysis. *Journal of Communication, 64*(2), 333–354.

Park, K. A., & Waters, E. (1989). Security of attachment and preschool friendships. *Child Development, 60*, 1076–1081.

Parkman, J. M., & Groen, G. (1971). Temporal aspects of simple additions and comparison. *Journal of Experimental Psychology, 92*, 437–438.

Pascalis, O., de Haan, M., Nelson, C. A., & de Schonen, S. (1998). Long-term recognition memory for faces assessed by visual paired comparison in 3- and 6-month-old infants. *Journal of Experimental Psychology-Learning Memory and Cognition, 24*(1), 249–260.

Pascual-Leone, A., Bartres-Faz, D., & Keenan, J. P. (1999). Transcranial magnetic stimulation: Studying the brain-behaviour relationship by induction of 'virtual lesions'. *Philosophical Transactions of the Royal Society of London B, 354*, 1229–1238.

Patterson, K. (2007). The reign of typicality in semantic memory. *Philosophical Transactions of the Royal Society B-Biological Sciences, 362*(1481), 813–821.

Patterson, K., Nestor, P.J., & Rogers, T. T. (2007). Where do you know what you know? The representation of semantic knowledge in the human brain. *Nature Reviews Neuroscience, 8*(12), 976–987.

Paul, L.K., Corsello, C., Tranel, D., & Adolphs, R. (2010). Does bilateral damage to the human amygdala produce autistic symptoms? *Journal of Neurodevelopmental Disorders, 2*(3), 165–173.

Paulus, M. P., Rogalsky, C., Simmons, A., Feinstein, J. S., & Stein, M. B. (2003). Increased activation in the right insula during risk-taking decision making is related to harm avoidance and neuroticism. *Neuroimage, 19*(4), 1439–1448.

Payne, B. K. (2001). Prejudice and perception: The role of automatic and controlled processes in misperceiving a weapon. *Journal of Personality and Social Psychology, 81*(2), 181–192.

Pedersen, C. A., Ascher, J. A., Monroe, Y. L., & Prange, A. J. (1982). Oxytocin induces maternal-behavior in virgin female rats. *Science, 216*(4546), 648–650.

Peelen, M. V., & Downing, P. E. (2005). Selectivity for the human body in the fusiform gyrus. *Journal of Neurophysiology, 93*(1), 603–608.

Peelen, M. V., & Downing, P. E. (2007). The neural basis of visual body perception. *Nature Reviews Neuroscience, 8*(8), 636–648.

Pelham, B. W., Mirenberg, M. C., & Jones, J. T. (2002). Why Susie sells seashells by the seashore: Implicit egotism and major life decisions. *Journal of Personality and Social Psychology, 82*(4), 469–487.

Pelphrey, K. A., Morris, J.P., & McCarthy, G. (2005). Neural basis of eye gaze processing deficits in autism. *Brain, 128*, 1038–1048.

Pelphrey, K. A., Singerman, J. D., Allison, T., & McCarthy, G. (2003). Brain activation evoked by perception of gaze shifts: The influence of context. *Neuropsychologia, 41*(2), 156–170.

Pempek, T. A., Yermolayeva, Y. A., & Calvert, S. L. (2009). College students' social networking experiences on

Facebook. *Journal of Applied Developmental Psychology,* 30(3), 227–238.

Pena-Gomez, C., Vidal-Pineiro, D., Clemente, I.C., Pascual-Leone, A., & Bartres-Faz, D. (2011). Down-regulation of negative emotional processing by transcranial direct current stimulation: Effects of personality characteristics. *PLoS One, 6*(7).

Penn, D. C., & Povinelli, D. J. (2007). On the lack of evidence that non-human animals possess anything remotely resembling a 'theory of mind'. *Philosophical Transactions of the Royal Society B, 362,* 731–744.

Penton-Voak, I. S., Pound, N., Little, A. C., & Perrett, D. I. (2006). Personality judgments from natural and composite facial images: More evidence for a "kernel of truth" in social perception. *Social Cognition, 24*(5), 607–640.

Pepperberg, I. M. (2000). *The Alex Studies: Cognitive and Communicative Abilities of Grey Parrots.* Cambridge, MA: Harvard University Press.

Perner, J., Aichhorn, M., Kronbichler, M., Staffen, W., & Ladurner, G. (2006). Thinking of mental and other representations: The roles of left and right temporo-parietal junction. *Social Neuroscience, 1*(3–4), 245–258.

Perner, J., Frith, U., Leslie, A. M., & Leekam, S. R. (1989). Exploration of the autistic child's theory of mind: Knowledge, belief and communication. *Child Development, 60,* 689–700.

Perner, J., & Ruffman, T. (1995). Episodic memory and autonoetic consciousness – Developmental evidence and a theory of childhood amnesia. *Journal of Experimental Child Psychology, 59*(3), 516–548.

Perner, J., & Wimmer, H. (1985). 'John thinks that Mary thinks that . . .': attribution of second-order beliefs by 5- to 10-year-old children. *Journal of Experimental Child Psychology, 39,* 437–471.

Perrett, D., & Mistlin, A. (1990). Perception of facial characteristics by monkeys. In W. Stebbins & M. Berkley (Eds.), *Comparative Perception Volume 2: Complex Signals.* New York: John Wiley and Son.

Perrett, D., Smith, P., Potter, D., Mistlin, A., Head, A., Milner, A., & Jeeves, M. (1985). Visual cells in the temporal cortex sensitive to face view and gaze direction. *Proceedings of the Royal Society of London B, 223,* 293–317.

Perrett, D. I., Harries, M. H., Bevan, R., Thomas, S., Benson, P. J., Mistlin, A. J., Chitty, A. J., Hietanen, J. K., & Ortega, J. E. (1989). Frameworks of analysis for the neural representation of animate objects and actions. *Journal of Experimental Biology, 146,* 87–113.

Perrett, D. I., Hietanen, J. K., Oram, M. W., & Benson, P. J. (1992). Organisation and functions of cells responsive to faces in the temporal cortex. *Philosophical Transactions of the Royal Society London B, 335,* 23–30.

Perrett, D. I., May, K. A., & Yoshikawa, S. (1994). Facial shape and judgments of female attractiveness. *Nature, 368*(6468), 239–242.

Perrin, F., Maquet, P., Peigneux, P., Ruby, P., Degueldre, C., Balteau, E., . . . Laureys, S. (2005). Neural mechanisms involved in the detection of our first name: A combined ERPs and PET study. *Neuropsychologia, 43*(1), 12–19.

Perry, S., Baker, M., Fedigan, L., Gros-Louis, J., Jack, K., MacKinnon, K. C., Manson, J. H., Panger, M., & Rose, L. (2003). Social conventions in white-face capuchins monkeys: Evidence for behavioral traditions in a neotropical primate. *Current Anthropology, 44,* 241–268.

Persaud, N., McLeod, P., & Cowey, A. (2007). Post-decision wagering objectively measures awareness. *Nature Neuroscience, 10*(2), 257–261.

Persinger, M. A. (1983). Religious and mystical experiences as artifacts of temporal lobe function: A general hypothesis. *Perceptual and Motor Skills, 57,* 1255–1262.

Peterson, C. C., & Siegal, M. (1995). Deafness, conversation and theory of mind. *Journal of Child Psychology and Psychiatry and Allied Disciplines, 36*(3), 459–474.

Petrinovich, L., & O'Neill, P. (1996). Influence of wording and framing effects on moral intuitions. *Ethology and Sociobiology, 17,* 145–171.

Pfaus, J. G., Damsma, G., Nomikos, G. G., Wenkstern, D. G., Blaha, C. D., Phillips, A. G., & Fibiger, H. C. (1990). Sexual-behavior enhances central dopamine transmission in the male-rat. *Brain Research, 530*(2), 345–348.

Phelps, E. A. (2006). Emotion and cognition: Insights from studies of the human amygdala. *Annual Review of Psychology, 57,* 27–53.

Phelps, E. A., Cannistraci, C. J., & Cunningham, W. A. (2003). Intact performance on an indirect measure of race bias following amygdala damage. *Neuropsychologia, 41*(2), 203–208.

Phelps, E. A., O'Connor, K. J., Cunningham, W. A., Funayama, E. S., Gatenby, J. C., Gore, J. C., & Banaji, M. R. (2000). Performance on indirect measures of race evaluation predicts amygdala activation. *Journal of Cognitive Neuroscience, 12,* 729–738.

Phillips, M. L., Young, A. W., Senior, C., Brammer, M., Andrews, C., Calder, A. J., Bullmore, E. T., Perrett, D. I., Rowland, D., Williams, S. C. R., Gray, J. A., & David, A. S. (1997). A specific neural substrate for perceiving facial expressions of disgust. *Nature, 389,* 495–498.

Phillips, R. G., & Ledoux, J. E. (1992). Differential contribution of amygdala and hippocampus to cued and contextual fear conditioning. *Behavioral Neuroscience, 106*(2), 274–285.

Phillips, W. A., Zeki, S., & Barlow, H. B. (1984). Localisation of function in the cerebral cortex: Past, present and future. *Brain, 107,* 327–361.

Piaget, J. (1932). *The Moral Judgment of the Child.* London: Routledge & Kegan Paul.

Piazza, M., Izard, V., Pinel, P., Le Bihan, D., & Dehaene, S. (2004). Tuning curves for approximate numerosity in the human intraparietal sulcus. *Neuron, 44,* 547–555.

Pica, P., Lemer, C., Izard, V., & Dehaene, S. (2004). Exact and approximate arithmetic in an Amazonion indigene group with a reduced number lexicon. *Science, 306*, 499–503.

Piliavin, J. A., Dovidio, J. F., Gaertner, S. L., & Clark, R. D. (1981). *Emergency Intervention*. New York: Academic Press.

Pillemer, D. B., & White, S. H. (1989). Childhood events recalled by children and adults. In H. W. Reese (Ed.), *Advances in Child Development and Behavior*. New York: Academic Press.

Pine, D. S., Cohen, P., Gurley, D., Brook, J., & Ma, Y. J. (1998). The risk for early-adulthood anxiety and depressive disorders in adolescents with anxiety and depressive disorders. *Archives of General Psychiatry, 55*(1), 56–64.

Pineda, J. A. (2005). The functional significance of mu rhythms: Translating "seeing" and "hearing" into "doing". *Brain Research Reviews, 50*(1), 57–68.

Pinker, S., & Bloom, P. (1990). Natural-language and natural-selection. *Behavioral and Brain Sciences, 13*, 707–726.

Pinker, S., & Jackendoff, R. (2005). The faculty of language: What's special about it? *Cognition, 95*, 201–236.

Pitcher, D., Garrido, L., Walsh, V., & Duchaine, B. C. (2008). Transcranial magnetic stimulation disrupts the perception and embodiment of facial expressions. *Journal of Neuroscience, 28*(36), 8929–8933.

Plant, E. A., & Devine, P. G. (1998). Internal and external motivation to respond without prejudice. *Journal of Personality and Social Psychology, 75*(3), 811–832.

Plassmann, H., O'Doherty, J., Shiv, B., & Rangel, A. (2008). Marketing actions can modulate neural representations of experienced pleasantness. *Proceedings of the National Academy of Sciences of the United States of America, 105*(3), 1050–1054.

Platek, S. M., Critton, S. R., Myers, T. E., & Gallup, G. G. (2003). Contagious yawning: The role of self-awareness and mental state attribution. *Cognitive Brain Research, 17*(2), 223–227.

Plomin, R., DeFries, J. C., & Loehlin, J. C. (1977). Genotype-environment interaction and correlation in the analysis of human behavior. *Psychological Bulletin, 84*, 309–322.

Plutchik, R. (1980). *Emotion: A Psychoevolutionary Synthesis*. New York: Harper & Row.

Poldrack, R. A. (2006). Can cognitive processes be inferred from neuroimaging data? *Trends in Cognitive Sciences, 10*, 59–63.

Posner, M. I. (1978). *Chonometric Explorations of Mind*. Hillsdale, NJ: Lawrence Earlbaum Associates.

Pound, N., Penton-Voak, I. S., & Surridge, A. K. (2009). Testosterone responses to competition in men are related to facial masculinity. *Proceedings of the Royal Society B-Biological Sciences, 276*(1654), 153–159.

Pourtois, G., Schettino, A., & Vuilleumier, P. (2013). Brain mechanisms for emotional influences on perception and attention: What is magic and what is not. *Biological Psychology, 92*(3), 492–512.

Povinelli, D. J., & Simon, B. B. (1998). Young children's understanding of briefly versus extremely delayed images of the self: Emergence of the autobiographical stance. *Developmental Psychology, 34*(1), 188–194.

Powell, L. J., Macrae, C. N., Cloutier, J., Metcalfe, J., & Mitchell, J. P. (2010). Dissociable neural substrates for agentic versus conceptual representations of self. *Journal of Cognitive Neuroscience, 22*(10), 2186–2197.

Premack, D., & Woodruff, G. (1978). Does the chimpanzee have a theory of mind? *Behavioral and Brain Sciences, 1*, 515–526.

Preston, S. D., & de Waal, F. B. M. (2002). Empathy: Its ultimate and proximate bases. *Behavioral and Brain Sciences, 25*(1), 1–20.

Provine, R. R. (1996). Contagious yawning and laughter: Significance for sensory feature detection, motor pattern generation, imitation, and the evolution of social behavior. In C. M. Heyes & B. G. Galef (Eds.), *Social Learning in Animals: The Roots of Culture*. San Diego: Academic Press.

Purves, D. (1994). *Neural Activity and the Growth of the Brain*. Cambridge: Cambridge University Press.

Putman, P., Hermans, E., & van Honk, J. (2004). Emotional Stroop performance for masked angry faces: It's BAS, not BIS. *Emotion, 4*(3), 305–311.

Quinn, P. C., Yahr, J., Kuhn, A., Slater, A. M., & Pascalis, O. (2002). Representation of the gender of human faces by infants: A preference for female. *Perception, 31*(9), 1109–1121.

Quiroga, R. G., Reddy, L., Kreiman, G., Koch, C., & Fried, I. (2005). Invariant visual representation by single neurons in the human brain. *Nature, 435*, 1102–1107.

Raafat, R. M., Chater, N., & Frith, C. (2009). Herding in humans. *Trends in Cognitive Sciences, 13*(10), 420–428.

Rabbie, J. M., & Horwitz, M. (1969). Arousal of ingroup-outgroup bias by a chance win or loss. *Journal of Personality and Social Psychology, 13*(3), 269-&.

Raichle, M. E. (1987). Circulatory and metabolic correlates of brain function in normal humans. In F. Plum & V. Mountcastle (Eds.), *Handbook of Physiology: The Nervous System*. Baltimore: Williams and Wilkins.

Raichle, M. E., MacLoed, A. M., Snyder, A. Z., Powers, W. J., Gusnard, D. A., & Shulman, G. L. (2001). A default mode of brain function. *Proceedings of the National Academy of Science, USA, 98*, 676–682.

Ramachandran, V. S. (2000). *Mirror Neurons and Imitation Learning as the Driving Force Behind "the Freat Leap Forward" in Human Evolution*. Available: www.edge.org.

Ramachandran, V. S., & Oberman, L. M. (2006). Broken mirrors – A theory of autism. *Scientific American, 295*(5), 62–69.

Ramamurthi, B. (1988). Stereotactic operation in behaviour disorders: Amygdalotomy and hypothalamotomy. *Acta Neurochirurgica, Supplementum (Wein), 44*, 152–157.

Rameson, L. T., Morelli, S. A., & Lieberman, M. D. (2012). The neural correlates of empathy: Experience, automaticity, and prosocial behavior. *Journal of Cognitive Neuroscience, 24*(1), 235–245.

Rand, D. G., Dreber, A., Ellingsen, T., Fudenberg, D., & Nowak, M. A. (2009). Positive interactions promote public cooperation. *Science, 325*(5945), 1272–1275.

Ratner, K. G., & Amodio, D. M. (2013). Seeing "us vs. them": Minimal group effects on the neural encoding of faces. *Journal of Experimental Social Psychology, 49*(2), 298–301.

Ratner, K. G., Dotsch, R., Wigboldus, D. H. J., van Knippenberg, A., & Amodio, D. M. (2014). Visualizing minimal ingroup and outgroup faces: Implications for impressions, attitudes, and behavior. *Journal of Personality and Social Psychology, 106*(6), 897–911.

Rausch, J., Johnson, M. E., Li, J. Q., Hutcheson, J., Carr, B. M., Corley, K. M., . . . Smith, J. (2005). Serotonin transport kinetics correlated between human platelets and brain synaptosomes. *Psychopharmacology, 180*(3), 391–398.

Ray, E., & Heyes, C. (2011). Imitation in infancy: The wealth of the stimulus. *Developmental Science, 14*(1), 92–105.

Raznahan, A., Toro, R., Daly, E., Robertson, D., Murphy, C., Deeley, Q., . . . Murphy, D. G. M. (2010). Cortical anatomy in autism spectrum disorder: An in Vivo MRI study on the effect of age. *Cerebral Cortex, 20*(6), 1332–1340.

Read, S. J., Monroe, B. M., Brownstein, A. L., Yang, Y., Chopra, G., & Miller, L. C. (2010). A neural network model of the structure and dynamics of human personality. *Psychological Review, 117*(1), 61–92.

Reader, S. M., & Laland, K. N. (2002). Social intelligence, innovation, and enhanced brain size in primates. *Proceedings of the National Academy of Science, USA, 99*, 4436–4441.

Reby, D., McComb, K., Cargnelutti, B., Darwin, C., Fitch, W. T., & Clutton-Brock, T. (2005). Red deer stags use formants as assessment cues during intrasexual agonistic interactions. *Proceedings of the Royal Society of London B, 272*, 941–947.

Redcay, E., & Courchesne, E. (2005). When is the brain enlarged in autism? A meta-analysis of all brain size reports. *Biological Psychiatry, 58*(1), 1–9.

Redcay, E., Dodell-Feder, D., Pearrow, M. J., Mavros, P. L., Kleiner, M., Gabrieli, J. D. E., & Saxe, R. (2010). Live face-to-face interaction during fMRI: A new tool for social cognitive neuroscience. *NeuroImage, 50*(4), 1639–1647.

Redlich, R., Grotegerd, D., Opel, N., Kaufmann, C., Zwitserlood, P., Kugel, H., . . . Dannlowski, U. (2015). Are you gonna leave me? Separation anxiety is associated with increased amygdala responsiveness and volume. *Social Cognitive and Affective Neuroscience, 10*(2), 278–284.

Reeb-Sutherland, B. C., Levitt, P., & Fox, N. A. (2012). The predictive nature of individual differences in early associative learning and emerging social behavior. *PLoS One, 7*(1).

Reicher, S. D., & Haslam, S. A. (2006). Rethinking the psychology of tyranny: The BBC prison study. *British Journal of Social Psychology, 45*, 1–40.

Reiss, A. L., Abrams, M. T., Singer, H. S., Ross, J. L., & Denckla, M. B. (1996). Brain development, gender and IQ in children: A volumetric imaging study. *Brain, 119*, 1763–1774.

Reiss, D., & Marino, L. (2001). Mirror self-recognition in the bottlenose dolphin: A case of cognitive convergence. *Proceedings of the National Academy of Science, USA, 98*, 5937–5942.

Renfrew, C. (2007). *Prehistory: Making of the Human Mind.* London: Weidenfield & Nicolson.

Repa, J. C., Muller, J., Apergis, J., Desrochers, T. M., Zhou, Y., & LeDoux, J. E. (2001). Two different lateral amygdala cell populations contribute to the initiation and storage of memory. *Nature Neuroscience, 4*(7), 724–731.

Reti, I. M., Xu, J. Z., Yanofski, J., McKibben, J., Uhart, M., Cheng, Y. J., . . . Nestadt, G. (2011). Monoamine oxidase A regulates antisocial personality in whites with no history of physical abuse. *Comprehensive Psychiatry, 52*(2), 188–194.

Reuter, M., Frenzel, C., Walter, N. T., Markett, S., & Montag, C. (2011). Investigating the genetic basis of altruism: The role of the COMT Val158Met polymorphism. *Social Cognitive and Affective Neuroscience, 6*(5), 662–668.

Rhodes, G. (2006). The evolutionary psychology of facial beauty. *Annual Review of Psychology, 57*, 199–226.

Rhodes, G., Sumich, A., & Byatt, G. (1999). Are average facial configurations attractive only because of their symmetry? *Psychological Science, 10*(1), 52–58.

Richardson, M. P., Strange, B. A., & Dolan, R. J. (2004). Encoding of emotional memories depends on amygdala and hippocampus and their interactions. *Nature Neuroscience, 7*, 278–285.

Richeson, J. A., Todd, A. R., Trawalter, S., & Baird, A. A. (2008). Eye-gaze direction modulates race-related amygdala activity. *Group Processes & Intergroup Relations, 11*(2), 233–246.

Ridley, M. (2003). *Nature via Nurture.* London: Fourth Estate.

Riem, M. M. E., Bakermans-Kranenburg, M. J., van Ijzendoorn, M. H., Out, D., & Rombouts, S. (2012). Attachment in the brain: Adult attachment representations predict amygdala and behavioral responses to infant crying. *Attachment & Human Development, 14*(6), 533–551.

Rilling, J. K., Glenn, A. L., Jairam, M. R., Pagnoni, G., Goldsmith, D. R., Elfenbein, H. A., & Lilienfeld, S. O. (2007). Neural correlates of social cooperation and noncooperation as a function of psychopathy. *Biological Psychiatry, 61*(11), 1260–1271.

Rilling, J. K., Goldsmith, D. R., Glenn, A. L., Jairam, M. R., Elfenbein, H. A., Dagenais, J. E., Murdock, C. D., & Pagnoni, G. (2008). The neural correlates of the affective response to unreciprocated cooperation. *Neuropsychologia, 46*(5), 1256–1266.

Rilling, J. K., Gutman, D. A., Zeh, T. R., Pagnoni, G., Berns, G. S., & Kilts, C. D. (2002). A neural basis for social cooperation. *Neuron, 35*(2), 395–405.

Rilling, J. K., Sanfey, A. G., Aronson, J. A., Nystrom, L. E., & Cohen, J. D. (2004). The neural correlates of theory of mind within interpersonal interactions. *Neuroimage, 22*(4), 1694–1703.

Rilling, J. K., & Young, L. J. (2014). The biology of mammalian parenting and its effect on offspring social development. *Science, 345*(6198), 771–776.

Rizzolatti, G. (2005). The mirror neuron system and imitation. In S. Hurley & N. Chater (Eds.), *Perspectives on Imitation: From Neuroscience to Social Science* (Vol. 1). Cambridge, MA: MIT Press.

Rizzolatti, G., & Craighero, L. (2004). The mirror-neuron system. *Annual Review of Neuroscience, 27*, 169–192.

Rizzolatti, G., & Fabbri-Destro, M. (2010). Mirror neurons: From discovery to autism. *Experimental Brain Research, 200*(3–4), 223–237.

Rizzolatti, G., Fadiga, L., Fogassi, L., & Gallese, V. (1996). Premotor cortex and the recognition of motor actions. *Cognitive Brain Research, 3*, 131–141.

Rizzolatti, G., Fogassi, L., & Gallese, V. (2002). Motor and cognitive functions of the ventral premotor cortex. *Current Opinion in Neurobiology, 12*, 149–154.

Rizzolatti, G., Fogassi, L., & Gallese, V. (2006). Mirrors in the mind. *Scientific American, Nov.*, 30–37.

Robbins, T. W., Cador, M., Taylor, J. R., & Everitt, B. J. (1989). Limbic-striatal interactions in reward-related processes. *Neuroscience and Biobehavioral Reviews, 13*(2–3), 155–162.

Robertson, E. M., Theoret, H., & Pascual-Leone, A. (2003). Studies in cognition: The problems solved and created by transcranial magnetic stimulation. *Journal of Cognitive Neuroscience, 15*, 948–960.

Robins, L. N., Tipp, J., & Przybeck, T. (1991). Antisocial personality. In L. N. Robins & D. A. Reiger (Eds.), *Psychiatric Disorders in America*: New York: Free Press.

Roca, M., Torralva, T., Gleichgerrcht, E., Woolgar, A., Thompson, R., Duncan, J., & Manes, F. (2011). The role of Area 10 (BA10) in human multitasking and in social cognition: A lesion study. *Neuropsychologia, 49*(13), 3525–3531.

Rodrigues, S. M., Saslow, L. R., Garcia, N., John, O. P., & Keltner, D. (2009). Oxytocin receptor genetic variation relates to empathy and stress reactivity in humans. *Proceedings of the National Academy of Sciences of the United States of America, 106*, 21437–21441.

Roether, C. L., Omlor, L., Christensen, A., & Giese, M. A. (2009). Critical features for the perception of emotion from gait. *Journal of Vision, 9*(6).

Rolls, E. T. (1996). The orbitofrontal cortex. *Philosophical Transactions of the Royal Society of London B, 351*, 1433–1444.

Rolls, E. T. (2005). *Emotion Explained*. Oxford: Oxford University Press.

Rolls, E. T., Hornak, J., Wade, D., & McGrath, J. (1994). Emotion-related learning in patients with social and emotional changes associated with frontal damage. *Journal of Neurology, Neurosurgery and Psychiatry, 57*, 1518–1524.

Rolls, E. T., & Tovee, M. J. (1995). Sparseness of the neuronal representation of stimuli in the primate temporal visual cortex. *Journal of Neurophysiology, 73*, 713–726.

Rorden, C., & Karnath, H. O. (2004). Using human brain lesions to infer function: A relic from a past era in the fMRI age? *Nature Reviews Neuroscience, 5*, 813–819.

Ross, M., & Wilson, A. E. (2002). It feels like yesterday: Self-esteem, valence of personal past experiences, and judgments of subjective distance. *Journal of Personality and Social Psychology, 82*(5), 792–803.

Rossi, S., Hallett, M., Rossini, P. M., Pascual-Leone, A., & Safety TMS Consensus Group. (2009). Safety, ethical considerations, and application guidelines for the use of transcranial magnetic stimulation in clinical practice and research. *Clinical Neurophysiology, 120*(12), 2008–2039.

Rosvold, H. E., Mirsky, A. F., & Pribram, K. H. (1954). Influence of amygdalactomy on social behaviour in monkeys. *Journal of Comparative Physiological Psychology, 47*, 173–178.

Rotshtein, P., Henson, R. N. A., Treves, A., Driver, J., & Dolan, R. J. (2005). Morphing Marilyn into Maggie dissociates physical and identity face representations in the brain. *Nature Neuroscience, 8*, 107–113.

Rousselet, G. A., Mace, M. J.-M., & Thorpe, M. F. (2004). Animal and human faces in natural scenes: How specific to human faces is the N170 ERP component? *Journal of Vision, 4*, 13–21.

Roy, P., Rutter, M., & Pickles, A. (2004). Institutional care: associations between overactivity and lack of selectivity in social relationships. *Journal of Child Psychology and Psychiatry, 45*(4), 866–873.

Rozin, P., Haidt, J., & McCauley, C. R. (1993). Disgust. In M. Lewis & J. M. Haviland (Eds.), *Handbook of Emotions*. New York: Guilford Press.

Rubin, K. H., Bukowski, W., & Parker, J. G. (2006). Peer interactions, relationships, and groups. In W. Damon, R. M. Lerner & N. Eisenberg (Eds.), *Handbook of Child Psychology: Vol. 3, Social, Emotional, and Personality Development*. New York: Wiley.

Ruby, P., & Decety, J. (2004). How would you feel versus how do you think she would feel? A neuroimaging study of perspective-taking with social emotions. *Journal of Cognitive Neuroscience, 16*(6), 988–999.

Rudebeck, P. H., Buckley, M. J., Walton, M. E., & Rushworth, M. F. S. (2006). A role for the macaque anterior cingulate gyrus in social valuation. *Science, 313*(5791), 1310–1312.

Rudebeck, P. H., Walton, M. E., Smyth, A. N., Bannerman, D. M., & Rushworth, M. F. S. (2006). Separate neural pathways process different decision costs. *Nature Neuroscience, 9*(9), 1161–1168.

Ruff, C.C., & Fehr, E. (2014). The neurobiology of rewards and values in social decision making. *Nature Reviews Neuroscience, 15*(8), 549–562.

Rupniak, N. M. J., Carlson, E. C., Harrison, T., Oates, B., Seward, E., Owen, S., de Felipe, C., Hunt, S., & Wheeldon, A. (2000). Pharmacological blockade or genetic deletion of substance P (NK1) receptors attenuates neonatal vocalisation in guinea-pigs and mice. *Neuropharmacology, 39*, 1413–1421.

Rushworth, M. F. S., Behrens, T. E. J., Rudebeck, P. H., & Walton, M. E. (2007). Contrasting roles for cingulate and orbitofrontal cortex in decisions and social behaviour. *Trends in Cognitive Sciences, 11*(4), 168–176.

Russell, J. (1997). *Autism as an Executive Disorder*. Oxford: Oxford University Press.

Russell, J. A., & Barrett, L. F. (1999). Core affect, prototypical emotional episodes, and other things called emotion: Dissecting the elephant. *Journal of Personality and Social Psychology and Aging, 76*, 805–819.

Sabbagh, M. A., Bowman, L. C., Evraire, L. E., & Ito, J. M. B. (2009). Neurodevelopmental correlates of theory of mind in preschool children. *Child Development, 80*(4), 1147–1162.

Sagiv, N., & Bentin, S. (2001). Structural encoding of human and schematic faces: Holistic and part based processes. *Journal of Cognitive Neuroscience, 13*, 1–15.

Sahdra, B., & Ross, M. (2007). Group identification and historical memory. *Personality and Social Psychology Bulletin, 33*(3), 384–395.

Sai, F. Z. (2005). The role of the mother's voice in developing mother's face preference: Evidence for intermodal perception at birth. *Infant and Child Development, 14*(1), 29–50.

Said, C. P., Sebe, N., & Todorov, A. (2009). Structural resemblance to emotional expressions predicts evaluation of emotionally neutral faces. *Emotion, 9*(2), 260–264.

Sally, D., & Hill, E. (2006). The development of interpersonal strategy: Autism, theory-of-mind, cooperation and fairness. *Journal of Economic Psychology, 27*(1), 73–97.

Samson, D. (2009). Reading other people's mind: Insights from neuropsychology. *Journal of Neuropsychology, 3*, 3–16.

Samson, D., Apperly, I. A., Braithwaite, J. J., Andrews, B. J., & Scott, S. E. B. (2010). Seeing it their way: Evidence for rapid and involuntary computation of what other people see. *Journal of Experimental Psychology-Human Perception and Performance, 36*(5), 1255–1266.

Samson, D., Apperly, I. A., Chiavarino , C., & Humphreys, G. W. (2004). Left temporoparietal junction is necessary for representing someone else's belief. *Nature Neurosecience, 7*, 499–500.

Sanfey, A., Rilling, J., Aaronson, J., Nystron, L., & Cohen, J. (2003). Probing the neural basis of economic decision-making: An fMRI investigation of the ultimatum game. *Science, 300*, 1755–1758.

Santiesteban, I., Banissy, M. J., Catmur, C., & Bird, G. (2012). Enhancing social ability by stimulating right temporoparietal junction. *Current Biology, 22*(23), 2274–2277.

Sato, A., & Yasuda, A. (2005). Illusion of sense of self-agency: Discrepancy between the predicted and actual sensory consequences of actions modulates the sense of self-agency, but not the sense of self-ownership. *Cognition, 94*(3), 241–255.

Saver, J. L., & Damasio, A. R. (1991). Preserved access and processing of social knowledge in a patient with acquired sociopathy due to ventromedial frontal damage. *Neuropsychologia, 29*, 1241–1249.

Saxe, R. (2006). Uniquely human social cognition. *Current Opinion in Neurobiology, 16*(2), 235–239.

Saxe, R., & Kanwisher, N. (2003). People thinking about thinking people: The role of the temporo-parietal junction in 'theory of mind'. *NeuroImage, 19*, 1835–1842.

Saxe, R., & Powell, L. J. (2006). It's the thought that counts: Specific brain regions for one component of theory of mind. *Psychological Science, 17*, 692–699.

Saxe, R., & Wexler, A. (2005). Making sense of another mind: The role of the right temporo-parietal junction. *Neuropsychologia, 43*(10), 1391–1399.

Schaafsma, S. M., Pfaff, D. W., Spunt, R. P., & Adolphs, R. (2015). Deconstructing and reconstructing theory of mind. *Trends in Cognitive Sciences, 19*(2), 65–72.

Schacter, S., & Singer, J. E. (1962). Cognitive, social, and physiological determinants of emotional state. *Psychology Review, 69*, 379–399.

Schaffer, H. R. (1996). *Social Development*. Oxford: Blackwell.

Schaffer, H. R., & Emerson, P. E. (1964). The development of social attachments in infancy. *Monographs of the Society for Research in Child Development, 29*(3), 1–77.

Scherer, K. R., Banse, R., & Wallbott, H. G. (2001). Emotion inferences from vocal expression correlate across languages and cultures. *Journal of Cross-Cultural Psychology, 32*, 76–92.

Schilbach, L., Timmermans, B., Reddy, V., Costall, A., Bente, G., Schlicht, T., . . . Vogeley, K. (2013). Toward a second-person neuroscience. *Behavioral and Brain Sciences, 36*(4), 393–414.

Schilhab, T. S. S. (2004). What mirror self-recognition in nonhumans can tell us about aspects of self. *Biology and Philosophy, 19*, 111–126.

Schjødta, U., Stødkilde-Jørgensenb, H., Geertza, A. W., & Roepstorff, A. (2008). Rewarding prayers. *Neuroscience Letters, 443*, 165–168.

Schjoedt, U., Stdkilde-Jorgensen, H., Geertz, A. W., & Roepstorff, A. (2009). Highly religious participants recruit areas of social cognition in personal prayer. *Social Cognitive and Affective Neuroscience, 4*(2), 199–207.

Schneider, K., Pauly, K. D., Gossen, A., Mevissen, L., Michel, T. M., Gur, R. C., . . . Habel, U. (2013). Neural correlates of moral reasoning in autism spectrum disorder. *Social Cognitive and Affective Neuroscience, 8*(6), 702–710.

Schneider, W., & Shiffrin, R. M. (1977). Controlled and automatic human information processing: I. Detection, search and attention. *Psychological Review, 84*, 1–66.

Schneiderman, I., Zagoory-Sharon, O., Leckman, J.F., & Feldman, R. (2012). Oxytocin during the initial stages of romantic attachment: Relations to couples' interactive reciprocity. *Psychoneuroendocrinology, 37*(8), 1277–1285.

Schneiderman, I., Zilberstein-Kra, Y., Leckman, J.F., & Feldman, R. (2011). Love alters autonomic reactivity to emotions. *Emotion, 11*(6), 1314–1321.

Scholz, J., Triantafyllou, C., Whitfield-Gabrieli, S., Brown, E. N., & Saxe, R. (2009). Distinct regions of right temporo-parietal junction are selective for theory of mind and exogenous attention. *Plos One, 4*(3), 7.

Schultheiss, O.C., & Stanton, S.J. (2009). Assessment of salivary hormones. In E. Harmon-Jones & J. S. Beer (Eds.), *Methods in Social Neuroscience*. New York: The Guilford Press.

Schultheiss, O.C., Wirth, M.M., & Stanton, S.J. (2004). Effects of affiliation and power motivation arousal on salivary progesterone and testosterone. *Hormones and Behavior, 46*(5), 592–599. 005

Schultz, W., Apicella, P., Scarnati, E., & Ljungberg, T. (1992). Neuronal-activity in monkey ventral striatum related to the expectation of reward. *Journal of Neuroscience, 12*(12), 4595–4610.

Schultz, W., Dayan, P., & Montague, P. R. (1997). A neural substrate of prediction and reward. *Science, 275*(5306), 1593–1599.

Schurz, M., Radua, J., Aichhorn, M., Richlan, F., & Perner, J. (2014). Fractionating theory of mind: A meta-analysis of functional brain imaging studies. *Neuroscience and Biobehavioral Review, 42*, 9–34.

Scott, L. S., & Monesson, A. (2009). The origin of biases in face perception. *Psychological Science, 20*(6), 676–680.

Scott, S., Young, A. W., Calder, A. J., Hellawell, D. J., Aggleton, J. P., & Johnson, M. (1997). Auditory recognition of emotion after amygdalactomy: Impairment of fear and anger. *Nature, 385*, 254–227.

Scoville, W. B., & Milner, B. (1957). Loss of recent memory after bilateral hippocampal lesions. *Journal of Neurology, Neurosurgery and Psychiatry, 20*, 11–21.

Sebanz, N., & Shiffrar, M. (2009). Detecting deception in a bluffing body: The role of expertise. *Psychonomic Bulletin & Review, 16*(1), 170–175.

Sebastian, C., Viding, E., Williams, K. D., & Blakemore, S. J. (2010). Social brain development and the affective consequences of ostracism in adolescence. *Brain and Cognition, 72*(1), 134–145.

Seed, A., Emery, N., & Clayton, N. (2009). Intelligence in corvids and apes: A case of convergent evolution? *Ethology, 115*(5), 401–420.

Sellaro, R., Guroglu, B., Nitsche, M.A., van den Wildenberg, W.P.M., Massaro, V., Durieux, J., . . . Colzato, L.S. (2015). Increasing the role of belief information in moral judgments by stimulating the right temporoparietal junction. *Neuropsychologia, 77*, 400–408.

Senholzi, K.B., & Ito, T.A. (2013). Structural face encoding: How task affects the N170's sensitivity to race. *Social Cognitive and Affective Neuroscience, 8*(8), 937–942.

Senju, A., Southgate, V., White, S., & Frith, U. (2009). Mindblind eyes: An absence of spontaneous theory of mind in asperger syndrome. *Science, 325*(5942), 883–885.

Sergent, J., & Signoret, J.-L. (1992). Varieties of functional deficits in prosopagnosia. *Cerebral Cortex, 2*, 375–388.

Serino, A., Bassolino, M., Farne, A., & Ladavas, E. (2007). Extended multisensory space in blind cane users. *Psychological Science, 18*, 642–648.

Seyfarth, R. M., & Cheney, D. L. (2002). What are big brains for? *Proceedings of the National Academy of Science, USA, 99*, 4141–4142.

Seyfarth, R.M., & Cheney, D.L. (2012). The evolutionary origins of friendship. *Annual Review of Psychology, 63*, 153–177.

Shah, A., & Frith, U. (1983). Islet of ability in autistic-children: A research note. *Journal of Child Psychology and Psychiatry and Allied Disciplines, 24*, 613–620.

Shamay-Tsoory, S.G. (2011). The neural bases for empathy. *Neuroscientist, 17*(1), 18–24.

Shamay-Tsoory, S. G., Aharon-Peretz, J., & Perry, D. (2009). Two systems for empathy: A double dissociation between emotional and cognitive empathy in inferior frontal gyrus versus ventromedial prefrontal lesions. *Brain, 132*, 617–627.

Shamay-Tsoory, S. G., Tibi-Elhanany, Y., & Aharon-Peretz, J. (2006). The ventromedial prefrontal cortex is involved in understanding affective but not cognitive theory of mind stories. *Social Neuroscience, 1*(3–4), 149–166.

Shatz, M., Wellman, H. M., & Silber, S. (1983). The acquisition of mental verbs – a systematic investigation of the 1st reference to mental state. *Cognition, 14*, 301–321.

Shaver, P. R., Morgan, H. J., & Wu, S. (1996). Is love a 'basic' emotion?. *Personal Relationships, 3*, 81–96.

Shellock, F. G. (2014). *Reference Manual for Magnetic Resonance Safety, Implants and Devices*. Los Angeles, CA: Biomedical Research Publishing Company.

Shepher, J. (1971). Mate selection amongst second generation Kibbutz adolescents and adults: Incest avoidance and negative imprinting. *Archives of Sexual Behavior, 1*, 293–307.

Shepherd, S.V., Klein, J.T., Deaner, R.O., & Platt, M.L. (2009). Mirroring of attention by neurons in macaque parietal cortex. *Proceedings of the National Academy of Sciences of the United States of America, 106*(23), 9489–9494.

Sherman, P. W. (1977). Nepotism and evolution of alarm calls. *Science, 197*(4310), 1246–1253.

Shih, M., Pittinsky, T. L., & Ambady, N. (1999). Stereotype susceptibility: Identity salience and shifts in quantitative performance. *Psychological Science, 10*(1), 80–83.

Siegel, A., Roeling, T. A. P., Gregg, T. R., & Kruk, M. R. (1999). Neuropharmacology of brain-stimulation-evoked aggression. *Neuroscience and Biobehavioral Reviews, 23*(3), 359–389.

Sigman, M., Mundy, P., Ungerer, J., & Sherman, T. (1986). Social interactions of autistic, mentally retarded, and normal children and their caregivers. *Journal of Child Psychology and Psychiatry, 27*, 647–656.

Silani, G., Bird, G., Brindley, R., Singer, T., Frith, C., & Frith, U. (2008). Levels of emotional awareness and autism: An fMRI study. *Social Neuroscience, 3*(2), 97–112.

Simner, M. L. (1971). Newborns' response to the cry of another infant. *Developmental Psychology, 5*, 136–150.

Simons, D. J., & Chabris, C. F. (1999). Gorillas in our midst: Sustained inattentional blindness for dynamic events. *Perception, 28*, 1059–1074.

Simpson, J. A. (1990). Influence of attachment styles on romantic relationships. *Journal of Personality and Social Psychology, 59*, 971–980.

Singer, T., Critchley, H. D., & Preuschoff, K. (2009). A common role of insula in feelings, empathy and uncertainty. *Trends in Cognitive Sciences, 13*(8), 334–340.

Singer, T., Kiebel, S. J., Winston, J. S., Dolan, R. J., & Frith, C. D. (2004a). Brain responses to the acquired moral status of faces. *Neuron, 41*(4), 653–662.

Singer, T., & Klimecki, O. M. (2014). Empathy and compassion. *Current Biology, 24*(18), R875–R878.

Singer, T., Seymour, B., O'Doherty, J., Kaube, H., Dolan, R. J., & Frith, C. D. (2004b). Empathy for pain involves the affective but not the sensory components of pain. *Science, 303*, 1157–1162.

Singer, T., Seymour, B., O'Doherty, J. P., Stephan, K. E., Dolan, R. J., & Frith, C. D. (2006). Empathic neural responses are modulated by the perceived fairness of others. *Nature, 439*, 466–469.

Singh, L., Morgan, J. L., & Best, C. T. (2002). Infants' listening preferences: Baby talk or happy talk? *Infancy, 3*(3), 365–394.

Slater, M., Antley, A., Davison, A., Swapp, D., Guger, C., Barker, C., . . . Sanchez-Vives, M. V. (2006). A virtual reprise of the stanley milgram obedience experiments. *PLoS One, 1*(1).

Small, D. M., Gregory, M. D., Mak, Y. E., Gitelman, D., Mesulam, M. M., & Parrish, T. (2003). Dissociation of neural representation of intensity and affective valuation in human gustation. *Neuron, 39*(4), 701–711.

Small, D. M., Zatorre, R. J., Dagher, A., Evans, A. C., & Jones-Gotman, M. (2001). Changes in brain activity related to eating chocolate: From pleasure to aversion. *Brain, 124*, 1720–1733.

Smeltzer, M. D., Curtis, J. T., Aragona, B. J., & Wang, Z. X. (2006). Dopamine, oxytocin, and vasopressin receptor binding in the medial prefrontal cortex of monogamous and promiscuous voles. *Neuroscience Letters, 394*(2), 146–151.

Smetana, J. G. (1981). Preschool children's conceptions of moral and social rules. *Child Development, 52*(4), 1333–1336.

Smetana, J. G. (1985). Preschool childrens conceptions of transgressions – effects of varying moral and conventional domain-related attributes. *Developmental Psychology, 21*(1), 18–29.

Smith, C. A., & Lazarus, R. S. (1990). Emotion and adaptation. In L. A. Pervin (Ed.), *Handbook of Personality: Theory and Research*. New York: Guilford.

Sodian, B., & Frith, U. (1992). Deception and sabotage in autistic, retarded and normal children. *Journal of Child Psychology and Psychiatry, 33*, 591–605.

Soken, N. H., & Pick, A. D. (1999). Infants' perception of dynamic affective expressions: Do infants distinguish specific expressions? *Child Development, 70*(6), 1275–1282.

Southgate, V., & Hamilton, A. F. C. (2008). Unbroken mirrors: Challenging a theory of autism. *Trends in Cognitive Sciences, 12*, 225–229.

Southgate, V., Johnson, M. H., El Karoui, I., & Csibra, G. (2010). Motor system activation reveals infants' on-line prediction of others' goals. *Psychological Science, 21*, 355–359.

Sowell, E. R., Thompson, P. M., Holmes, C. J., Batth, R., Jernigan, T. L., & Toga, A. W. (1999). Localizing age-related changes in brain structure between childhood and adolescence using statistical parametric mapping. *Neuroimage, 9*(6), 587–597.

Spangler, G., & Grossmann, K. E. (1993). Biobehavioral organization in securely and insecurely attached infants. *Child Development, 64*(5), 1439–1450.

Sparrow, B., & Chatman, L. (2013). Social cognition in the internet age: Same as it ever was? *Psychological Inquiry, 24*(4), 273–292.

Spengler, S., von Cramon, D. Y., & Brass, M. (2009). Control of shared representations relies on key processes involved in mental state attribution. *Human Brain Mapping, 30*(11), 3704–3718.

Spielberger, C. D., Jacobs, G., Russell, S., & Crane, R. S. (1983). Assessment of anger: The state-trait anger scale. In J. N. Butcher & C. D. Spielberger (Eds.), *Advances in Personality Assessment, Volume 2*. Hillsdale, NJ: Lawrence Erlbaum Associates.

Spinella, M. (2005). Prefrontal substrates of empathy: Psychometric evidence in a community sample. *Biological Psychology, 70*(3), 175–181.

Sprengelmeyer, R., Young, A. W., Calder, A. J., Karnat, A., Lange, H., Homberg, V., Perrett, D., & Rowland, D. (1996). Loss of disgust: Perception of faces and emotions in Huntington's disease. *Brain, 119*, 1647–1665.

Sprengelmeyer, R., Young, A. W., Sprengelmeyer, A., Calder, A. J., Rowland, D., Perrett, D., Homberg, V., & Lange, H. (1997). Recognition of facial expression: Selective impairment of specific emotions in Huntington's disease. *Cognitive Neuropsychology, 14*, 839–879.

Squire, L. R. (1992). Memory and the hippocampus: A synthesis from findings with rats, monkeys and humans. *Psychological Review, 99*, 195–231.

Stagg, C. J., & Nitsche, M. A. (2011). Physiological basis of transcranial direct current stimulation. *Neuroscientist, 17*(1), 37–53.

Stanley, D. A., & Adolphs, R. (2013). Toward a neural basis for social behavior. *Neuron, 80*(3), 816–826.

Stanton, S. J., Beehner, J. C., Saini, E. K., Kuhn, C. M., & LaBar, K. S. (2009). Dominance, politics, and physiology: Voters' testosterone changes on the night of the 2008 united states presidential election. *PLoS One, 4*, e7543.

Stanton, S. J., Wirth, M. M., Waugh, C. E., & Schultheiss, O. C. (2009). Endogenous testosterone levels are associated with amygdala and ventromedial prefrontal cortex responses to anger faces in men but not women. *Biological Psychology, 81*, 118–122.

Sternberg, R. J. (1986). A triangular theory of love. *Psychological Review, 93*(2), 119–135.

Sternberg, R. J. (1988). *The Triangle of Love*. New York: Basic Books.

Stewart, L., Battelli, L., Walsh, V., & Cowey, A. (1999). Motion perception and perceptual learning studied by magnetic stimulation. *Electroencephalography and Clinical Neurophysiology, 3*, 334–350.

Stillwell, A. M., Baumeister, R. F., & Del Priore, R. E. (2008). We're all victims here: Toward a psychology of revenge. *Basic and Applied Social Psychology, 30*(3), 253–263.

Stirrat, M., & Perrett, D. (2012). Face structure predicts cooperation men with wider faces are more generous to their in-group when out-group competition is salient. *Psychological Science, 23*, 718–722.

Stirrat, M., & Perrett, D. I. (2010). Valid facial cues to cooperation and trust male facial width and trustworthiness. *Psychological Science, 21*, 349–354.

Stolzenberg, D. S., & Numan, M. (2011). Hypothalamic interaction with the mesolimbic DA system in the control of the maternal and sexual behaviors in rats. *Neuroscience and Biobehavioral Review, 35*, 826–847.

Strafella, A. P., & Paus, T. (2000). Modulation of cortical excitability during action observation: A transcranial magnetic stimulation study. *Experimental Brain Research, 11*, 2289–2292.

Strathearn, L., Fonagy, P., Amico, J., & Montague, P. R. (2009). Adult attachment predicts maternal brain and oxytocin response to infant cues. *Neuropsychopharmacology, 34*(13), 2655–2666.

Striano, T., Henning, A., & Stahl, D. (2005). Sensitivity to social contingencies between 1 and 3 months of age. *Developmental Science, 8*, 509–519.

Strobel, A., Zimmermann, J., Schmitz, A., Reuter, M., Lis, S., Windmann, S., & Kirsch, P. (2011). Beyond revenge: Neural and genetic bases of altruistic punishment. *NeuroImage, 54*(1), 671–680.

Stroop, J. R. (1935). Studies of interference in serial verbal reactions. *Journal of Experimental Psychology: General, 106*, 404–426.

Stuss, D. T., & Alexander, M. P. (2007). Is there a dysexecutive syndrome? *Philosophical Transactions of the Royal Society B-Biological Sciences, 362*(1481), 901–915.

Stuss, D. T., & Benson, D. F. (1986). *The Frontal Lobes*. New York: Raven Press.

Stuss, D. T., Floden, D., Alexander, M. P., Levine, B., & Katz, D. (2001). Stroop performance in focal lesion patients: Dissociation of processes and frontal lobe lesion location. *Neuropsychologia, 39*, 771–786.

Suchan, B., Busch, M., Schulte, D., Groenermeyer, D., Herpertz, S., & Vocks, S. (2010). Reduction of gray matter density in the extrastriate body area in women with anorexia nervosa. *Behavioural Brain Research, 206*(1), 63–67.

Suhler, C. L., & Churchland, P. S. (2009). Control: conscious and otherwise. *Trends in Cognitive Sciences, 13*(8), 341–347.

Sui, J., Rotshtein, P., & Humphreys, G. W. (2013). Coupling social attention to the self forms a network for personal significance. *Proceedings of the National Academy of Sciences of the United States of America, 110*(19), 7607–7612.

Summerfield, J. J., Hassabis, D., & Maguire, E. A. (2009). Cortical midline involvement in autobiographical memory. *NeuroImage, 44*(3), 1188–1200.

Susskind, J. M., Lee, D. H., Cusi, A., Feiman, R., Grabski, W., & Anderson, A. K. (2008). Expressing fear enhances sensory acquisition. *Nature Neuroscience, 11*(7), 843–850.

Sutherland, C. A. M., Oldmeadow, J. A., Santos, I. M., Towler, J., Burt, D. M., & Young, A. W. (2013). Social inferences from faces: Ambient images generate a three-dimensional model. *Cognition, 127*(1), 105–118.

Sutton, J., Smith, P. K., & Swettenham, J. (1999). Social cognition and bullying: Social inadequacy or skilled manipulation? *British Journal of Developmental Psychology, 17*, 435–450.

Swain, J. E., Kim, P., Spicer, J., Ho, S. S., Dayton, C. J., Elmadih, A., & Abel, K. M. (2014). Approaching the biology of human parental attachment: Brain imaging, oxytocin and coordinated assessments of mothers and fathers. *Brain Research, 1580*, 78–101.

Swami, V., Furnham, A., Amin, R., Chaudhri, J., Joshi, K., Jundi, S., Miller, R., Mirza-Begum, J., Begum, F. N., Sheth, P., & Tovee, M. J. (2008). Lonelier, lazier, and teased: The stigmatizing effect of body size. *Journal of Social Psychology, 148*(5), 577–593.

Swingler, M. M., Sweet, M. A., & Carver, L. J. (2010). Brain-behavior correlations: Relationships between mother-stranger face processing and infants' behavioral responses to a separation from mother. *Developmental Psychology, 46*(3), 669–680.

Tager-Flusberg, H. (1992). Autistic childrens talk about psychological states: Deficits in the early acquisition of a theory of mind. *Child Development, 63*, 161–172.

Tager-Flusberg, H. (2003). Developmental disorders of genetic origin. In M. De Haan & M.H. Johnson (Eds.), *The Cognitive Neuroscience of Development*. New York: Psychology Press.

Tajfel, H., Billig, M.G., Bundy, R.P., & Flament, C. (1971). Social categorization and intergroup behavior. *European Journal of Social Psychology, 1*(2), 149–177.

Tajfel, H., & Turner, J. C. (1986). The social identity theory of intergroup behavior. In S. Worchel & W. Austin (Eds.), *Psychology of Intergroup Relations*. Chicago: Nelson Hall.

Takahashi, H., Yahata, N., Koeda, M., Matsuda, T., Asai, K., & Okubo, Y. (2004). Brain activation associated with evaluative processes of guilt and embarrassment: An fMRI study. *Neuroimage, 23*(3), 967–974.

Talairach, J., & Tournoux, P. (1988). *A Co-planar Stereotactic Atlas of the Human Brain*. Stuttgart: Thieme Verlag.

Tallis, F. (2005). *Love Sick: Love as a Mental Illness*. London: de Capo Press.

Tamietto, M., Castelli, L., Vighetti, S., Perozzo, P., Geminiani, G., Weiskrantz, L., . . . de Gelder, B. (2009). Unseen facial and bodily expressions trigger fast emotional reactions. *Proceedings of the National Academy of Sciences of the United States of America, 106*(42), 17661–17666.

Tamietto, M., & De Gelder, B. (2010). Neural bases of the non-conscious perception of emotional signals. *Nature Reviews Neuroscience, 11*, 697–709.

Tang, Y., Zhang, W., Chen, K., Feng, S., Ji, Y., Shen, J., Reiman, E. M., & Liu, Y. (2006). Arithmetic processing in the brain shaped by cultures. *Proceedings of the National Academy of Science, USA, 103*, 10775–10780.

Tankersley, D., Stowe, C. J., & Huettel, S. A. (2007). Altruism is associated with an increased neural response to agency. *Nature Neuroscience, 10*(2), 150–151.

Taylor, S. E., Saphire-Bernstein, S., & Seeman, T. E. (2010). Are plasma oxytocin in women and plasma vasopressin in men biomarkers of distressed pair-bond relationships? *Psychological Science, 21*, 3–7.

Taylor, J. C., Wiggett, A. J., & Downing, P. E. (2007). Functional MRI analysis of body and body part representations in the extrastriate and fusiform body areas. *Journal of Neurophysiology, 98*(3), 1626–1633.

Telzer, E.H., Flannery, J., Shapiro, M., Humphreys, K.L., Goff, B., Gabard-Durman, L., . . . Tottenham, N. (2013). Early experience shapes amygdala sensitivity to race: An international adoption design. *Journal of Neuroscience, 33*(33), 13484–1388.

Telzer, E.H., Humphreys, K.L., Shapiro, M., & Tottenham, N. (2013). Amygdala sensitivity to race is not present in childhood but emerges over adolescence. *Journal of Cognitive Neuroscience, 25*(2), 234–244.

Theoret, H., Halligan, E., Kobayashi, M., Fregni, F., Tager-Flusberg, H., & Pascual-Leone, A. (2005). Impaired motor facilitation during action observation in individuals with autism spectrum disorder. *Current Biology, 15*(3), R84–R85.

Thomas, K. M., & Casey, B. J. (2003). Methods for imaging the developing brain. In M. De Haan & M. H. Johnson (Eds.), *The Cognitive Neuroscience of Development*. New York: Psychology Press.

Thomas, K. M., & Nelson, C. A. (1996). Age related changes in the electrophysiological response to visual stimulus novelty: A topographical approach. *Electroencephalography and Clinical Neurophysiology, 98*, 294–308.

Thompson, P. M., Schwartz, C., Lin, R. T., Khan, A. A., & Toga, A. W. (1996). Three-dimensional statistical analysis of sulcal variability in the human brain. *Journal of Neuroscience, 16*, 4261–4274.

Thomson, J. J. (1976). Killing, letting die, and the trolley problem. *Monist, 59*, 204–207.

Thomson, J. J. (1986). *Rights, Restitution and Risk: Essays in Moral Theory*. Cambridge, MA: Harvard University Press.

Tiihonen, J., Rossi, R., Laakso, M. P., Hodgins, S., Testa, C., Perez, J., Repo-Tiihonen, E., Vaurio, O., Soininen, H., Aronen, H. J., Kononen, M., Thompson, P. A., & Frisoni, G. B. (2008). Brain anatomy of persistent violent offenders: More rather than less. *Psychiatry Research-Neuroimaging, 163*(3), 201–212.

Tinbergen, N. (1951). *The Study of Instinct*. London: Oxford University Press.

Titchener, E. B. (1909). *Lectures on the Experimental Psychology of the thought Processes*. New York: Macmillan.

Tizard, B., & Rees, J. (1975). Effect of early institutional rearing on behavior problems and affectional relationships of 4-year-old children. *Journal of Child Psychology and Psychiatry and Allied Disciplines, 16*(1), 61–73.

Todorov, A., Dotsch, R., Porter, J.M., Oosterhof, N.N., & Falvello, V.B. (2013). Validation of data-driven computational models of social perception of faces. *Emotion, 13*(4), 724–738.

Todorov, A., & Duchaine, B. (2008). Reading trustworthiness in faces without recognizing faces. *Cognitive Neuropsychology, 25*(3), 395–410.

Todorov, A., & Engell, A. D. (2008). The role of the amygdala in implicit evaluation of emotionally neutral faces. *Social Cognitive and Affective Neuroscience, 3*(4), 303–312.

Todorov, A., Mandisodza, A. N., Goren, A., & Hall, C. C. (2005). Inferences of competence from faces predict election outcomes. *Science, 308*(5728), 1623–1626.

Todorov, A., Olivola, C. Y., Dotsch, R., & Mende-Siedlecki, P. (2015). Social attributions from faces: Determinants, consequences, accuracy, and functional significance. *Annual Review of Psychology, Vol 66, 66*, 519–545.

Todorov, A., Said, C. P., Engell, A. D., & Oosterhof, N. N. (2008). Understanding evaluation of faces on social dimensions. *Trends in Cognitive Sciences, 12*(12), 455–460.

Tomasello, M. (1999). *The Cultural Origins of Human Cognition*. Boston, MA: Harvard University Press.

Tomasello, M. (2009). *Why we Cooperate*. Cambridge, MA: Bradford Books.

Tomasello, M., Hare, B., Lehmann, H., & Call, J. (2007). Reliance on head versus eyes in the gaze following of great apes and human infants: The cooperative eye hypothesis. *Journal of Human Evolution, 52*(3), 314–320.

Tomasello, M., & Todd, J. (1983). Joint attention and lexical acquisition style. *First Language, 4*, 197–212.

Toth, N. (1985). Archeoligical evidence for preferential right-handedness in the Lower Pleistocene, and its possible implications. *Journal of Human Evolution, 14*, 607–614.

Tranel, D., Bechara, A., & Denburg, N. L. (2002). Asymmetric functional roles of right and left ventromedial prefrontal cortices in social conduct, decision-making, and emotional processing. *Cortex, 38*(4), 589–612.

Tranel, D., Damasio, A. R., & Damasio, H. (1988). Intact recognition of facial expression, gender and age in patients with impaired recognition of face identity. *Neurology, 38*, 690–696.

Tranel, D., & Damasio, H. (1995). Neuroanatomical correlates of electrodermal skin conductance responses. *Psychophysiology, 31*, 427–438.

Tranel, D., Damasio, H., & Damasio, A. R. (1995). Double dissociation between overt and covert face recognition. *Journal of Cognitive Neuroscience, 7*, 425–432.

Tranel, D., Fowles, D. C., & Damasio, A. R. (1985). Electrodermal discrimination of familiar and unfamiliar faces – A methodology. *Psychophysiology, 22*(4), 403–408.

Trivers, R. L. (1971). Evolution of reciprocal altruism. *Quarterly Review of Biology, 46*(1), 35–57.

Trivers, R. L., & Hare, H. (1975). Haplodiploidy and evolution of social insects. *Science, 191*(4224), 249–263.

Tseng, K. Y., & O'Donnell, P. (2005). Post-pubertal emergence of prefrontal cortical up states induced by D-1 NMDA co-activation. *Cerebral Cortex, 15*(1), 49–57.

Tsukiura, T., & Cabeza, R. (2008). Orbitofrontal and hippocampal contributions to memory for face-name associations: The rewarding power of a smile. *Neuropsychologia, 46*(9), 2310–2319.

Tsukiura, T., & Cabeza, R. (2011). Shared brain activity for aesthetic and moral judgments: Implications for the Beauty-is-Good stereotype. *Social Cognitive and Affective Neuroscience, 6*(1), 138–148.

Tulving, E. (1983). *Elements of Episodic Memory*. Oxford: Oxford University Press.

Turati, C., Cassia, V. M., Simion, F., & Leo, I. (2006). Newborns' face recognition: Role of inner and outer facial features. *Child Development, 77*(2), 297–311.

Turiel, E. (1983). *The Development of Social Knowledge: Morality and Convention*. Cambridge, UK: Cambridge University Press.

Tybur, J. M., Lieberman, D., & Griskevicius, V. (2009). Microbes, mating, and morality: Individual differences in three functional domains of disgust. *Journal of Personality and Social Psychology, 97*(1), 103–122.

Tzourio-Mazoyer, N., De Schonen, S., Crivello, F., Reutter, B., Aujard, Y., & Mazoyer, B. (2002). Neural correlates of woman face processing by 2-month-old infants. *Neuroimage, 15*(2), 454–461.

Uchino, B. N., Cacioppo, J. T., & KiecoltGlaser, J. K. (1996). The relationship between social support and physiological processes: A review with emphasis on underlying mechanisms and implications for health. *Psychological Bulletin, 119*(3), 488–531.

Uddin, L. Q., Iacoboni, M., Lange, C., & Keenan, J. P. (2007). The self and social cognition: The role of cortical midline structures and mirror neurons. *Trends in Cognitive Sciences, 11*, 153–157.

Ulrich, R. E., & Azrin, N. H. (1962). Reflexive fighting in response to aversive stimulation. *Journal of the Experimental Analysis of Behavior, 5*(4), 511–520.

Umilta, M. A., Escola, L., Intskirveli, I., Grammont, F., Rochat, M., Caruana, F., Jezzini, A., Gallese, V., & Rizzolatti, G. (2008). When pliers become fingers in the monkey motor system. *Proceedings of the National Academy of Sciences of the United States of America, 105*(6), 2209–2213.

Umilta, M. A., Kohler, E., Gallese, V., Fogassi, L., Fadiga, L., Keysers, C., & Rizzolatti, G. (2001). I know what you are doing: A neurophysiological study. *Neuron, 25*, 287–295.

Ungerleider, L. G., & Mishkin, M. (1982). Two cortical systems. In D. J. Ingle, M. A. Goodale & R. J. W. Mansfield (Eds.), *Analysis of Visual Behaviour*. Cambridge, MA: MIT Press.

Urgesi, C., Berlucchi, G., & Aglioti, S. M. (2004). Magnetic stimulation of extrastriate body area impairs visual processing of nonfacial body parts. *Current Biology, 14*(23), 2130–2134.

Urgesi, C., Candidi, M., Ionta, S., & Aglioti, S.M. (2007). Representation of body identity and body actions in extrastriate body area and ventral premotor cortex. *Nature Neuroscience, 10*(1), 30–31.

Uttal, W.R. (2001). *The New Phrenology: The Limits of Localising Cognitive Processes in the Brain*. Cambridge, MA: MIT Press.

Vaish, A., Carpenter, M., & Tomasello, M. (2009). Sympathy through affective perspective taking and its relation to prosocial behavior in toddlers. *Developmental Psychology, 45*(2), 534–543.

Valentine, T., Darling, S., & Donnelly, M. (2004). Why are average faces attractive? The effect of view and averageness on the attractiveness of female faces. *Psychonomic Bulletin & Review, 11*(3), 482–487.

van Baaren, R., Holland, R. W., Kawakami, K., & van Knippenberg, A. (2004). Mimicry and prosocial behavior. *Psychological Science, 15*(1), 71–74.

van Baaren, R., Janssen, L., Chartrand, T. L., & Dijkster-huis, A. (2009). Where is the love? The social aspects of mimicry. *Philosophical Transactions of the Royal Society B-Biological Sciences, 364*(1528), 2381–2389.

Van Bavel, J. J., Packer, D. J., & Cunningham, W. A. (2011). Modulation of the fusiform face area following minimal exposure to motivationally relevant faces: Evidence of in-group enhancement (not out-group disregard). *Journal of Cognitive Neuroscience, 23*(11), 3343–3354.

Vanderwert, R. E., Fox, N. A., & Ferrari, P. F. (2013). The mirror mechanism and mu rhythm in social development. *Neuroscience Letters, 540*, 15–20.

van Elst, L. T., Woermann, F. G., Lemieux, L., Thompson, P. J., & Trimble, M. R. (2000). Affective aggression in patients with temporal lobe epilepsy – A quantitative MRI study of the amygdala. *Brain, 123*, 234–243.

van Erp, A. M. M., & Miczek, K. A. (2000). Aggressive behavior, increased accumbal dopamine, and decreased cortical serotonin in rats. *Journal of Neuroscience, 20*(24), 9320–9325.

Van Hoesen, G. W., Morecraft, R. J., & Vogt, B. A. (1993). Connections of the monkey cingulate cortex. In B. A. Vogt & M. Gabriel (Eds.), *The Neurobiology of the Cingulate Cortex and Limbic Thalamus: A Comprehensive Handbook*. Boston, MA: Birkhauser.

van Honk, J., & Schutter, D. L. G. (2007). Vigilant and avoidant responses to angry facial expressions: Dominance and submission motives. In E. Harmon-Jones & P. Winkielman (Eds.), *Social Neuroscience: Integrating Biological and Psychological Explanations of Social Behavior*. New York: The Guilford Press.

van Honk, J., Tuiten, A., Hermans, E., Putman, P., Koppeschaar, H., Thijssen, J., Verbaten, R., & van Doornen, L. (2001a). A single administration of testosterone induces cardiac accelerative responses to angry faces in healthy young women. *Behavioral Neuroscience, 115*(1), 238–242.

van Honk, J., Tuiten, A., van den Hout, M., Putman, P., de Haan, E., & Stam, H. (2001b). Selective attention to unmasked and masked threatening words: Relationships to trait anger and anxiety. *Personality and Individual Differences, 30*(4), 711–720.

Van Ijzendoorn, M. H., & Kroonenberg, P. M. (1988). Cross-cultural patterns of attachment – A meta-analysis of the strange situation. *Child Development, 59*(1), 147–156.

Vanman, E. J., Paul, B. Y., Ito, T. A., & Miller, N. (1997). The modern face of prejudice and structural features that moderate the effect of cooperation on affect. *Journal of Personality and Social Psychology, 73*(5), 941–959.

van Schaik, C. P., Isler, K., & Burkart, J. M. (2012). Explaining brain size variation: From social to cultural brain. *Trends in Cognitive Sciences, 16*(5), 277–284.

van't Wout, M., Kahn, R. S., Sanfey, A. G., & Aleman, A. (2006). Affective state and decision-making in the Ultimatum Game. *Experimental Brain Research, 169*, 564–568.

Varlinskaya, E. I., & Spear, L. P. (2002). Acute effects of ethanol on social behavior of adolescent and adult rats: Role of familiarity of the test situation. *Alcoholism-Clinical and Experimental Research, 26*(10), 1502–1511.

Veit, R., Flor, H., Erb, M., Hermann, C., Lotze, M., Grodd, W., & Birbaumer, N. (2002). Brain circuits involved in emotional learning in antisocial behavior and social phobia in humans. *Neuroscience Letters, 328*(3), 233–236.

Volkow, N. D., Fowler, J. S., Wang, G.-J., Telang, F., Logan, J., Jayne, M., . . . Swanson, J. M. (2010). Cognitive control of drug craving inhibits brain reward regions in cocaine abusers. *NeuroImage, 49*(3), 2536–2543.

Voorn, P., Vanderschuren, L., Groenewegen, H. J., Robbins, T. W., & Pennartz, C. M. A. (2004). Putting a spin on the dorsal-ventral divide of the striatum. *Trends in Neurosciences, 27*(8), 468–474.

Vritcka, P., Anderson, F., Grandjean, D., Sander, D., & Vuilleumier, P. (2008). Individual differences in attachment style modulates human amygdala and striatum activity during social appraisal. *PLoS One, 3*, e2868.

Vrticka, P., Bondolfi, G., Sander, D., & Vuilleumier, P. (2012). The neural substrates of social emotion perception and regulation are modulated by adult attachment style. *Social Neuroscience, 7*(5), 473–493.

Vrticka, P., & Vuilleumier, P. (2012). Neuroscience of human social interactions and adult attachment style. *Frontiers in Human Neuroscience, 6*.

Vuilleumier, P. (2005). How brains beware: Neural mechanisms of emotional attention. *Trends in Cognitive Sciences, 9*, 585–594.

Vuilleumier, P., Armony, J. L., Driver, J., & Dolan, R. J. (2001). Effects of attention and emotion on face processing in the human brain: An event-related fMRI study. *Neuron, 30*(3), 829–841.

Vul, E., Harris, C., Winkielman, P., & Pashler, H. (2009). Puzzlingly high correlations in fMRI studies of emotion, personality, and social cognition. *Perspectives on Psychological Science, 4*(3), 274–290.

Wager, T. D., Atlas, L. Y., Lindquist, M. A., Roy, M., Woo, C. W., & Kross, E. (2013). An fMRI-based neurologic signature of physical pain. *New England Journal of Medicine, 368*(15), 1388–1397.

Wagner, A. D., Schacter, D. L., Rotte, M., Koutstaal, W., Maril, A., Dale, A. M., Rosen, B. R., & Buckner, R. I. (1998). Building memories: Remembering and forgetting of verbal experiences as predicted by brain activity. *Science, 281*, 1188–1191.

Wahba, A., & Bridgewell, L. (1976). Maslow reconsidered: A review of research on the need hierarchy theory. *Organizational Behavior and Human Performance, 15*, 212–240.

Walker, A. E. (1940). A cytoarchitectural study of the prefrontal area of the macaque monkey. *Journal of Comparative Neurology, 73*, 59–86.

Wallace, B., Cesarini, D., Lichtenstein, P., & Johannesson, M. (2007). Heritability of ultimatum game responder behavior. *Proceedings of the National Academy of Sciences of the United States of America, 104*(40), 15631–15634.

Walsh, V., & Cowey, A. (1998). Magnetic stimulation studies of visual cognition. *Trends in Cognitive Sciences, 2*, 103–110.

Walster, E., Aronson, V., Abrahams, D., & Rottman, L. (1966). Importance of physical attractiveness in dating behavior. *Journal of Personality and Social Psychology, 4*, 508–516.

Walton, M. E., Bannerman, D. M., Alterescu, K., & Rushworth, M. F. S. (2003). Functional specialization within medial frontal cortex of the anterior cingulate for evaluating effort-related decisions. *Journal of Neuroscience, 23*(16), 6475–6479.

Wang, A. T., Lee, S. S., Sigman, M., & Dapretto, M. (2006). Neural basis of irony comprehension in children with autism: The role of prosody and context. *Brain, 129*, 932–943.

Wang, G., Mao, L. H., Ma, Y. N., Yang, X. D., Cao, J. Q., Liu, X., . . . Han, S. H. (2012). Neural representations of close others in collectivistic brains. *Social Cognitive and Affective Neuroscience, 7*(2), 222–229.

Ward, J. (2015). The Student's Guide to Cognitive Neuroscience Third Edition. New York: Psychology Press.

Warneken, F., Hare, B., Melis, A. P., Hanus, D., & Tomasello, M. (2007). Spontaneous altruism by chimpanzees and young children. *Plos Biology, 5*(7), 1414–1420.

Warneken, F., & Tomasello, M. (2006). Altruistic helping in human infants and young chimpanzees. *Science, 311*(5765), 1301–1303.

Warneken, F., & Tomasello, M. (2008). Extrinsic rewards undermine altruistic tendencies in 20-month-olds. *Developmental Psychology, 44*(6), 1785–1788.

Wassermann, E. M. (1996). Risk and safety of transcranial magnetic stimulation: Report and suggested guidelines from the international workshop on the safety of repetitive transcranial magnetic stimulation, June 5–7. *Electroencephalogy and Clinical Neurophysiology, 108*, 1–16.

Wassermann, E. M., Cohen, L. G., Flitman, S. S., Chen, R., & Hallett, M. (1996). Seizures in healthy people with repeated "safe" trains of transcranial magnetic stimulation. *Lancet, 347*, 825–826

Waters, E., Merrick, S., Treboux, D., Crowell, J., & Albersheim, L. (2000). Attachment security in infancy and early adulthood: A twenty-year longitudinal study. *Child Development, 71*(3), 684–689.

Watson, K. K., & Platt, M. L. (2012). Social signals in primate orbitofrontal cortex. *Current Biology, 22*(23), 2268–2273.

Way, B. M., & Lieberman, M. D. (2010). Is there a genetic contribution to cultural differences? Collectivism, individualism and genetic markers of social sensitivity. *Social Cognitive and Affective Neuroscience, 5*(2–3), 203–211.

Way, B. M., & Taylor, S. E. (2010). Social influences on health: Is serotonin a critical mediator? *Psychosomatic Medicine, 72*(2), 107–112.

Way, B. M., Taylor, S. E., & Eisenberger, N. I. (2009). Variation in the mu-opioid receptor gene (OPRM1) is associated with dispositional and neural sensitivity to social rejection. *Proceedings of the National Academy of Science, USA, 106*, 15079–15084.

Weaver, I. C. G., Cervoni, N., Champagne, F. A., D'Alessio, A. C., Charma, S., Seckl, J., . . . Meaney, M. J. (2004). Epigenetic programming by maternal behavior. *Nature Neuroscience, 7*, 847–854.

Wegner, D. M. (2002). *The Illusion of Conscious Will*. Cambridge, MA: MIT Press.

Weiskrantz, L. (1956). Behavioral changes associated with ablations of the amygdaloid complex in monkeys. *Journal of Comparative Physiological Psychology, 49*, 381–391.

Wellman, H. M. (2002). Understanding the psychological world: Developing a theory of mind. In U. Goswami (Ed.), *Childhood Cognitive Development*. Oxford: Blackwell.

Wellman, H. M., & Lagattuta, K. H. (2000). Developing understandings of mind. In S. Baron-Cohen, H. Tager-Flusberg & D. Cohen (Eds.), *Understanding Other Minds: Perspectives from Developmental Cognitive Neuroscience*. Oxford: Oxford University Press.

Westermarck, E. (1891). *The History of Human Marriage*. London: Macmillan.

Wheeler, M. E., & Fiske, S. T. (2005). Controlling racial prejudice – Social-cognitive goals affect amygdala and stereotype activation. *Psychological Science, 16*(1), 56–63.

White, S., Hill, E., Winston, J., & Frith, U. (2006). An islet of social ability in Asperger Syndrome: Judging social attributes from faces. *Brain and Cognition, 61*(1), 69–77.

White, S., O'Reilly, H., & Frith, U. (2009). Big heads, small details and autism. *Neuropsychologia, 47*(5), 1274–1281.

White, S. J., Burgess, P. W., & Hill, E. L. (2009). Impairments on "open-ended" executive function tests in Autism. *Autism Research, 2*(3), 138–147.

Whitehead, H., & Rendell, L. (2014). *The Cultural Lives of Whales and Dolphins*. Chicago, IL: Chicago University Press.

Whiten, A., & Byrne, R. W. (1988). The Machiavellian intelligence hypothesis. In R. W. Byrne & A. Whiten (Eds.), *Maciavellian Intelligence: Social Complexity and the Evolution of Intellect in Monkeys, Apes and Humans*. Oxford: Oxford University Press.

Whiten, A., Horner, V., & de Waal, F. B. M. (2005). Conformity to cultural norms of tool use in chimpanzee. *Nature, 437*, 737–740.

Whiten, A., & van Schaik, C. P. (2007). The evolution of animal 'cultures' and social intelligence. *Philosophical Transactions of the Royal Society B, 362*, 603–620.

Wicker, B., Keysers, C., Plailly, J., Royet, J. P., Gallese, V., & Rizzolatti, G. (2003). Both of us disgusted in my insula:

The common neural basis of seeing and feeling disgust. *Neuron, 40*, 655–664.

Wiese, H., Wolff, N., Steffens, M.C., & Schweinberger, S.R. (2013). How experience shapes memory for faces: An event-related potential study on the own-age bias. *Biological Psychology, 94*(2), 369–379.

Wilkinson, G. S. (1988). Reciprocal altruism in bats and other mammals. *Ethology and Sociobiology, 9*(2–4), 85–100.

Wilkinson, R., & Pickett, K. (2009). *The Spirit Level: Why Equality is Better for Everyone.* London: Penguin.

Williams, G. C. (1966). *Adaptation and Natural Selection: A Critique of Some Current Evolutionary Thought.* Princeton: Princeton University Press.

Williams, J.H.G., Waiter, G.D., Perra, O., Perrett, D.I., & Whiten, A. (2005). An fMRI study of joint attention experience. *NeuroImage, 25*(1), 133–140.

Williams, J. H. G., Whiten, A., & Singh, T. (2004). A systematic review of action imitation in autistic spectrum disorder. *Journal of Autism and Developmental Disorders, 34*(3), 285–299.

Williams, J. R., Catania, K. C., & Carter, C. S. (1992). Development of partner preferences in female prairie voles (microtus-ochrogaster) – the role of social and sexual experience. *Hormones and Behavior, 26*(3), 339–349.

Wilms, M., Schilbach, L., Pfeiffer, U., Bente, G., Fink, G.R., & Vogeley, K. (2010). It's in your eyes-using gaze-contingent stimuli to create truly interactive paradigms for social cognitive and affective neuroscience. *Social Cognitive and Affective Neuroscience, 5*(1), 98–107.

Willingham, D. T., & Dunn, E. W. (2003). What neuroimaging and brain localization can do, cannot do, and should not do for social psychology. *Journal of Personality and Social Psychology, 85*(4), 662–671.

Wilson, R. S., Krueger, K. R., Arnold, S. E., Schneider, J. A., Kelly, J. F., Barnes, L. L., Tang, Y. X., . . . Bennett, D. A. (2007). Loneliness and risk of Alzheimer disease. *Archives of General Psychiatry, 64*(2), 234–240.

Wimmer, H., & Perner, J. (1983). Beliefs about beliefs: Representation and the constraining function of wrong beliefs in young children's understanding of deception. *Cognition, 13*, 103–128.

Winston, J. S., Gottfried, J. A., Kilner, J. M., & Dolan, R. J. (2005). Integrated neural representations of odor intensity and affective valence in human amygdala. *Journal of Neuroscience, 25*(39), 8903–8907.

Winston, J. S., O'Doherty, J., & Dolan, R. J. (2003). Common and distinct neural responses during direct and incidental processing of multiple facial emotions. *NeuroImage, 20*, 84–97.

Winston, J. S., Strange, B. A., O'Doherty, J., & Dolan, R. J. (2002). Automatic and intentional brain responses during evaluation of trustworthiness of faces. *Nature Neuroscience, 5*(3), 277–283.

Wohlschlager, A., Gattis, M., & Bekkering, H. (2003). Action generation and action perception in imitation: An instance of the ideomotor principle. *Philosophical Transactions of the Royal Society of London – B, 358*, 501–515.

Wolf, A. P. (1995). *Sexual Attraction and Childhood Associations.* Stanford, CA: Stanford University Press.

Wolpert, D. M., Ghahramani, Z., & Jordan, M. I. (1995). An internal model for sensorimotor integration. *Science, 269*, 1880–1882.

Wommack, J. C., Liu, Y., & Zuoxin, W. (2009). Animal models of romantic relationships. In M. D. Haan & M. R. Gunnar (Eds.), *Handbook of Developmental Social Neuroscience.* New York: The Guilford Press.

Wu, S. P., Jia, M. X., Ruan, Y., Liu, J., Guo, Y. Q., Shuang, M., . . . Zhang, D. (2005). Positive association of the oxytocin receptor gene (OXTR) with autism in the Chinese Han population. *Biological Psychiatry, 58*(1), 74–77.

Yik, M. S. M., Russell, J. A., & Barrett, L. F. (1999). Structure of self-reported current affect: Integration and beyond. *Journal of Personality and Social Psychology, 77*(3), 600–619.

Young, A. W., Hellawell, D. J., Van de Wal, C., & Johnson, M. (1996). Facial expression processing after amygdalactomy. *Neuropsychologia, 34*, 31–39.

Young, L., Bechara, A., Tranel, D., Damasio, H., Hauser, M., & Damasio, A. (2010). Damage to ventromedial prefrontal cortex impairs judgment of harmful intent. *Neuron, 65*(6), 845–851.

Younger, J., Aron, A., Parke, S., Chatterjee, N., & Mackey, S. (2010). Viewing pictures of a romantic partner reduces experimental pain: Involvement of neural reward systems. *Plos One, 5*(10), 7.

Yovel, G., & Belin, P. (2013). A unified coding strategy for processing faces and voices. *Trends in Cognitive Sciences, 17*(6), 263–271.

Yovel, G., & Kanwisher, N. (2005). The neural basis of the behavioral face-inversion effect. *Current Biology, 15*(24), 2256–2262.

Yovel, G., Tambini, A., & Brandman, T. (2008). The asymmetry of the fusiform face area is a stable individual characteristic that underlies the left-visual-field superiority for faces. *Neuropsychologia, 46*(13), 3061–3068.

Zahavi, A. (1975). Mate selection: A selection for a handicap. *Journal of Theoretical Biology, 53*, 205–214.

Zahavi, A. (1995). Altruism as a handicap – the limitations of kin selection and reciprocity. *Journal of Avian Biology, 26*(1), 1–3.

Zahn, R., Moll, J., Krueger, F., Huey, E. D., Garrido, G., & Grafman, J. (2007). Social concepts are represented in the superior anterior temporal cortex. *Proceedings of the National Academy of Sciences of the USA, 104*, 6430–6435.

Zajonc, R. B. (1980). Feeling and thinking – preferences need no inferences. *American Psychologist, 35*(2), 151–175.

Zak, P. J., Kurzban, R., Ahmadi, S., Swerdloff, R. S., Park, J., Efremidze, L., Redwine, K., Morgan, K., & Matzner, W. (2009). Testosterone administration decreases generosity in the ultimatum game. *PLoS One, 4,* e8330.

Zak, P.J., Kurzban, R., & Matzner, W.T. (2005). Oxytocin is associated with human trustworthiness. *Hormones and Behavior, 48*(5), 522–527.

Zaki, J., & Ochsner, K. (2012). The neuroscience of empathy: Progress, pitfalls and promise. *Nature Neuroscience, 15*(5), 675–680.

Zaki, J., Weber, J., Bolger, N., & Ochsner, K. (2009). The neural bases of empathic accuracy. *Proceedings of the National Academy of Sciences of the United States of America, 106*(27), 11382–11387.

Zangl, R., & Mills, D. L. (2007). Increased brain activity to infant-directed speech in 6-and 13-month-old infants. *Infancy, 11*(1), 31–62.

Zebrowitz, L. A., Andreoletti, C., Collins, M. A., Lee, S. Y., & Blumenthal, J. (1998). Bright, bad, babyfaced boys: Appearance stereotypes do not always yield self-fulfilling prophecy effects. *Journal of Personality and Social Psychology, 75*(5), 1300–1320.

Zebrowitz, L.A., Luevano, V.X., Bronstad, P. M., & Aharon, I. (2009). Neural activation to babyfaced men matches activation to babies. *Social Neuroscience, 4*(1), 1–10.

Zebrowitz, L. A., & McDonald, S. M. (1991). The impact of litigants baby-facedness and attractiveness on adjudications in small claims courts. *Law and Human Behavior, 15*(6), 603–623.

Zelkowitz, P., Gold, I., Feeley, N., Hayton, B., Carter, C. S., Tulandi, T., . . . Levin, P. (2014). Psychosocial stress moderates the relationships between oxytocin, perinatal depression, and maternal behavior. *Hormones and Behavior, 66*(2), 351–360.

Zhong, S. F., Israel, S., Shalev, I., Xue, H., Ebstein, R. P., & Chew, S. H. (2010). Dopamine D4 receptor gene associated with fairness preference in ultimatum game. *PLoS One, 5*(11).

Zhu, Y., Zhang, L., Fan, J., & Han, S. H. (2007). Neural basis of cultural influence on self-representation. *Neuroimage, 34*(3), 1310–1316.

Zihl, J., Von Cramon, D., & Mai, N. (1983). Selective disturbance of movement vision after bilateral brain damage. *Brain, 106,* 313–340.

Zimbardo, P. G. (1972). *The Stanford Prison Experiment: A Simulation Study of the Psychology of Imprisonment.* Palo Alto, CA: Stanford University.

Zimbardo, P. G., Maslach, C., & Haney, C. (1999). Reflections on the Stanford prison experiment: Genesis, transformations, consequences. In T. Blass (Ed.), *Obedience to Authority: Current Perspectives on the Milgram Paradigm.* Mahwah, NJ: Earlbaum.

Author index

Subject index

Note: Page numbers in **bold** indicate where a term is defined.